IUTAM Symposium on Computational Mechanics of Solid Materials at Large Strains

SOLID MECHANICS AND ITS APPLICATIONS
Volume 108

Series Editor: G.M.L. GLADWELL
Department of Civil Engineering
University of Waterloo
Waterloo, Ontario, Canada N2L 3GI

Aims and Scope of the Series

The fundamental questions arising in mechanics are: *Why?*, *How?*, and *How much?* The aim of this series is to provide lucid accounts written by authoritative researchers giving vision and insight in answering these questions on the subject of mechanics as it relates to solids.

The scope of the series covers the entire spectrum of solid mechanics. Thus it includes the foundation of mechanics; variational formulations; computational mechanics; statics, kinematics and dynamics of rigid and elastic bodies: vibrations of solids and structures; dynamical systems and chaos; the theories of elasticity, plasticity and viscoelasticity; composite materials; rods, beams, shells and membranes; structural control and stability; soils, rocks and geomechanics; fracture; tribology; experimental mechanics; biomechanics and machine design.

The median level of presentation is the first year graduate student. Some texts are monographs defining the current state of the field; others are accessible to final year undergraduates; but essentially the emphasis is on readability and clarity.

IUTAM Symposium on Computational Mechanics of Solid Materials at Large Strains

Proceedings of the IUTAM Symposium held in Stuttgart, Germany, 20–24 August 2001

Edited by

CHRISTIAN MIEHE

University of Stuttgart, Germany

KLUWER ACADEMIC PUBLISHERS

DORDRECHT / BOSTON / LONDON

A C.I.P. Catalogue record for this book is available from the Library of Congress.

ISBN 1-4020-1170-9

Published by Kluwer Academic Publishers,
P.O. Box 17, 3300 AA Dordrecht, The Netherlands.

Sold and distributed in North, Central and South America
by Kluwer Academic Publishers,
101 Philip Drive, Norwell, MA 02061, U.S.A.

In all other countries, sold and distributed
by Kluwer Academic Publishers,
P.O. Box 322, 3300 AH Dordrecht, The Netherlands.

Printed on acid-free paper

Printed in the Netherlands.

Contents

Multiscale Analyses and Homogenization Methods

Advanced Numerical Methods

Advanced Computational Applications

Preface

The steady increase in computational power induces an equally steady increase in the complexity of the engineering models and associated computer codes. This particularly affects the modeling of the mechanical response of materials. Material behavior is nowadays modeled in the strongly nonlinear range by taking into account finite strains, complex hysteresis effects, fracture phenomena and multiscale features. Progress in this field is of fundamental importance for many engineering disciplines, especially those concerned with material testing, safety, reliability and serviceability analyses of engineering structures.

In recent years many important achievements have been made in the field of the theoretical formulation, the mathematical analysis and the numerical implementation of deformation processes in solids. Computational methods and simulation techniques today play a central role in advancing the understanding of complex material behavior. Research in the field of "Computational Mechanics of Materials" is concerned with the development of mathematical models and numerical solution techniques for the simulation of material response. It is a very broad interdisciplinary field of science with inputs from traditional fields such as Applied Mechanics, Applied Mathematics, Materials Science, Solid State Physics and Information Technology.

The intention of the IUTAM Symposium "Computational Mechanics of Solid Materials at Large Strains", held at the University of Stuttgart, Germany, from August 20–24, 2001, was to give a state of the art and a survey about recent developments in this field and to create perspectives for future research trends. The first key concern of the Symposium was a discussion of *models for solid materials undergoing deformations large strains*. Typical applications are simulations of forming processes and failure mechanisms of structures. These types of models are mostly governed by the geometrically nonlinear theory of continuum mechanics. The setting up of material models in this context and their mathematical and computational analysis is highly challenging due to the nonlinear geometry. The second key aspect of the Symposium dealt with the *multiscale characteristics of materials*. These multiscale properties are often a priori given by the structure of the material on different length scales, but may also be generated during the large strain deformation process as a consequence of phase transitions and the development of microstructures. The discussion of mathematical and computational analyses of such phenomena on the basis of *homogenization approaches* and *variational methods* was an important intention of the conference.

The IUTAM Symposium set a state of the art and a survey about

- theoretical and computational approaches to the modeling of large-strain elastic and inelastic deformations of solids applied to metals, polymers and geomaterials,
- mathematical analysis of finite inelastic deformations of solids, incremental variational formulations, treatment of non-convex problems and microstructures,
- alternative continuous and discrete micromechanical approaches to deformation mechanisms in solid materials on different phenomenological scales,
- homogenization methods and adaptive computational tools for the determination of effective overall properties of heterogeneous materials such as polycrystals, composites and granular materials.

The scientific programme consisted of 45 lectures including eight keynote presentations given by leading international scientists who were invited by the Scientific Committee. About 100 scientists from 14 nations attended the Symposium and actively took part in the discussions. The Symposium provided a forum for the interaction among young and established researchers from solid mechanics, applied mathematics and materials science.

This book contains the printed contributions of the IUTAM Symposium "Computational Mechanics of Solid Materials at Large Strains". The time and effort spent by the authors for preparing their manuscripts is greatly appreciated. The scientific and social success of the Symposium was critically based on the financial support of sponsors who also provided funds for international participants. It is my pleasant duty to give my thanks to the organizations and companies which made this conference possible by supporting it in a very generous way: the IUTAM Bureau, the German Research Foundation (DFG), the University of Stuttgart and the diverse companies listed below. The organization of the Symposium was in the hand of the local organizing committee. The dedicated work of the conference secretaries Dr.-Ing. Matthias Lambrecht and Dipl.-Ing. Jan Schotte is deeply acknowledged. Also, special thanks go to Kluwer Academic Publishers, Dordrecht, for their efforts to produce this attractive Symposium volume.

May 2002 *Christian Miehe*

International Scientific Committee

Sponsors of the IUTAM Symposium

DFG Deutsche Forschungsgemeinschaft, Bonn, Germany
IUTAM International Union of Theoretical and Applied Mechanics
Kanzlei Bardehle & Partner, München, Germany
Hilti AG, Schaan, Principality of Liechtenstein
Robert Bosch GmbH, Stuttgart, Germany
Landesbank Baden-Württemberg, Stuttgart, Germany
INA Wälzlager Schaeffler oHG, Herzogenaurach, Germany
Audi AG, Ingolstadt, Germany
Siemens AG, Stuttgart, Germany
DaimlerChrysler AG, Stuttgart, Germany
CAD-FEM GmbH, Grafing bei München, Germany
Dr. Koch Computertechnik AG, Tübingen, Germany
Leonhardt Andrä und Partner GmbH, Stuttgart, Germany
Andreas Stihl AG & Co., Waiblingen, Germany
Festo AG & Co., Esslingen, Germany
Kluwer Academic Publishers, Dordrecht, The Netherlands
Konrad Wittwer GmbH & Co. KG, Stuttgart, Germany
Ed. Züblin AG, Stuttgart, Germany
Debeka, Stuttgart, Germany

List of Participants

Prof. Dr. Hans-Dieter Alber, *TU Darmstadt, Fachbereich Mathematik, Schlossgartenstraße 7, 64289 Darmstadt, Germany,* alber@mathematik.tu-darmstadt.de

Prof. Lallit Anand, *Massachusetts Institute of Technology, Mechanical Engineering Department, 77 Massachusetts Avenue, Cambridge, MA 02139, U.S.A.,* anand@mit.edu

Markus Andre, *Continental AG, Strategische Technologie - 51120, Jädekamp 30, 30419 Hannover, Germany,* markus.andre@conti.de

Dipl.-Ing. Nikolas Apel, *Universität Stuttgart, Institut für Mechanik (Bauwesen), Pfaffenwaldring 7, 70550 Stuttgart, Germany,* apel@mechbau.uni-stuttgart.de

Dr. Francois Auslender, *Ecole Polytechnique / CNRS, Laboratoire de Mécanique des Solides, 91128 Palaiseau Cedex, France,* auslende@lms.polytechnique.fr

Dr.-Ing. Herbert Baaser, *Technische Universität Darmstadt, Institut für Mechanik, Hochschulstraße 1, 64289 Darmstadt, Germany,* baaser@mechanik.tu-darmstadt.de

Dr.-Ing. Franz-Joseph Barthold, *TU Braunschweig, Course Director of M.Sc. Programme CSE, Bültenweg 17, 38106 Braunschweig, Germany,* FJ.Barthold@tu-bs.de

Dipl.-Ing. Claus G. Bayreuther, *Universität Stuttgart, Institut für Mechanik (Bauwesen), Pfaffenwaldring 7, 70550 Stuttgart, Germany,* bayr@mechbau.uni-stuttgart.de

Prof. Ted Belytschko, *Northwestern University, Department of Mechanical Engineering, 2145 Sheridan Road, Evanston, IL 60208-3111, U.S.A.,* tedbelytschko@nwu.edu

Prof. Dr.-Ing. Albrecht Bertram, *Otto-von-Guericke Universität Magdeburg, Institut für Mechanik, Universitätsplatz 2, 39016 Magdeburg, Germany,* bertram@mb.uni-magdeburg.de

Prof. Dr.-Ing. Dieter Besdo, *Universität Hannover, Institut für Mechanik, Appelstraße 11, 30167 Hannover, Germany,* besdo@ifm.uni-hannover

Dr.-Ing. Thomas Böhlke, *Otto-v.-Guericke Universität Magdeburg, Institut für Mechanik, Universitätsplatz 2, 39016 Magdeburg, Germany,* thomas.boehlke@mb.uni-magdeburg.de

Prof. Mary C. Boyce, *Massachusetts Institute of Technology, Mechanical Engineering Department, 77 Massachusetts Avenue, Cambridge, MA 02139, U.S.A.,* mboyce@mit.edu

Priv.-Doz. Dr.-Ing. Michael Brünig, *Universität Dortmund, Lehrstuhl für Baumechanik-Statik, August-Schmidt-Straße 8, 44221 Dortmund, Germany,* michael.bruenig@uni-dortmund.de

Prof. Dr.-Ing. Otto T. Bruhns, *Ruhr-Universität Bochum, Lehrstuhl für Technische Mechanik, Universitätsstraße 150, 44780 Bochum, Germany,* bruhns@tm.bi.ruhr-uni-bochum.de

Prof. Dr. Carsten Carstensen, *Institut für Angewandte und Numerische Mathematik, Technische Universität Wien, Wiedner Hauptstraße 8–10, 1040 Vienna, Austria,* Carsten.Carstensen@tuwien.ac.at

Dr. Werner Daves, *Montanuniversität Leoben, Institut für Mechanik, Franz-Josef-Straße 18, 8700 Leoben, Austria,* daves@hal.unileoben.ac.at

Prof. René de Borst, *Delft University of Technology, Aerospace Engineering / Koiter Institute, PO Box 5058, 2600 GA Delft, The Netherlands,* R.deborst@lr.tudelft.nl

Dr. Arnaud Delaplace, *Laboratoire de Mécanique et Technologie, Ecole normale supérieure de Cachan, 61 Avenue du Président Wilson, 94235 Cachan Cedex, France,* Arnaud.Delaplace@lmt.ens-cachan.fr

Dr. Antonio DeSimone, *Max-Planck-Institute for Mathematics in the Sciences, Inselstraße 22-24, 04103 Leipzig, Germany,* desimone@mis.mpg.de

Dipl.-Ing. Joachim Dettmar, *Universität Stuttgart, Institut für Mechanik (Bauwesen), Pfaffenwaldring 7, 70550 Stuttgart, Germany,* dettmar@mechbau.uni-stuttgart.de

Prof. Dr. techn. Josef Eberhardsteiner, *Technische Universität Wien, Institut für Festigkeitslehre, Karlsplatz 13/202, 1040 Wien, Austria,* Josef.Eberhardsteiner@tuwien.ac.at

Prof. Dr.-Ing. Wolfgang Ehlers, *Universität Stuttgart, Institut für Mechanik (Bauwesen), Pfaffenwaldring 7, 70550 Stuttgart, Germany,* ehlers@mechbau.uni-stuttgart.de

Dipl.-Ing. Bernhard Eidel, *TU Darmstadt, Institut für Statik, Alexanderstraße 7, 64283 Darmstadt, Germany,* eidel@statik.tu-darmstadt.de

Prof. Jüri Engelbrecht, *Tallinn Technical University, Institute of Cybernetics, Akadeemia tee 21, 12618 Tallinn, Estonia,* je@ioc.ee

Prof. Dr. techn. Franz Dieter Fischer, *Montanuniversität Leoben, Institut für Mechanik, Franz-Josef-Straße 18, 8700 Leoben, Austria,* fischer@unileoben.ac.at

Dr. Hans-Peter Gänser, *Hilti AG, Feldkircherstraße 100, 9494 Schaan, Principality of Liechtenstein,* gaenser@hilti.com

Prof. Dr. Huajian Gao, *Max-Planck-Institut für Metallforschung, Theorie mesoskopischer Phänomene, Seestraße 92, 70174 Stuttgart, Germany,* hjgao@mf.mpg.de

Prof. Sanjay Govindjee, *University of California at Berkeley, Structural Engineering, Mechanics and Materials, 721 Davis Hall, Berkeley, CA 94720, U.S.A.,* sanjay@ce.berkeley.edu

Prof. Dr.-Ing. Friedrich Gruttmann, *Technische Universität Darmstadt, Institut für Statik, Alexanderstraße 7, 64283 Darmstadt, Germany,* gruttmann@statik.tu-darmstadt.de

Dipl.-Math. Sonja Gutmann, *Universität Stuttgart, Institut für Mechanik (Bauwesen), Pfaffenwaldring 7, 70550 Stuttgart, Germany,* gutmann@mechbau.uni-stuttgart.de

Prof. Dr. rer. nat. Klaus Hackl, *Ruhr-Universität Bochum, Lehrstuhl für Allgemeine Mechanik, Gebäude IA 3/126, 44780 Bochum, Germany,* hackl@am.bi.ruhr-uni-bochum.de

Dr. Halim Haddadi, *Université Paris-Nord, Institut Galilée, Avenue J.-B. Clèment, 93430 Villetaneuse, France,* halim.haddadi@lpmtm.univ-paris13.fr

Dr.-Ing. Stefan Hartmann, *Universität Kassel, Institut für Mechanik, Mönchebergstraße 7, 34109 Kassel, Germany,* hart@ifm.maschinenbau.uni-kassel.de

Prof. Dr.-Ing. Peter Haupt, *Universität Kassel, Institut für Mechanik, Mönchebergstraße 7, 34109 Kassel, Germany,* haupt@ifm.maschinenbau.uni-kassel.de

Andrew Heath, *ESI GmbH, Frankfurter Straße 13–15, 65760 Eschborn, Germany,* ah@esigmbh.de

Dr. Andreas Hönig, *Universität Bonn, Institut für angewandte Mathematik, Wegelerstraße 10, 53115 Bonn, Germany,* hoenig@iam.uni-bonn.de

Prof. Gerhard Holzapfel, *Technische Universität Graz, Institut für Baustatik, Schiesstattgasse 14b, 8010 Graz, Austria,* Gerhard.Holzapfel@biomech.tu-graz.ac.at

Dr.-Ing. Kurt Hornberger, *Robert Bosch GmbH, FV/PLK3, Postfach 1131, 71301 Waiblingen, Germany,* kurt.hornberger@de.bosch.com

Prof. Adnan Ibrahimbegovic, *Laboratoire de Mécanique et Technologie, Ecole normale supérieure de Cachan, 61 Avenue du Président Wilson, 94235 Cachan Cedex, France,* ai@lmt.ens-cachan.fr

Prof. Tatsuo Inoue, *Kyoto University, Department of Energy Conversion Science, Yoshida-honmachi, Sakyo-ku, Kyoto 606-8501, Japan,* inoue@energy.kyoto-u.ac.jp

Dr. Michael Kaliske, *Continental AG, Strategische Technologie - 51120, Jädekamp 30, 30419 Hannover, Germany,* michael.kaliske@conti.de

Prof. Dr.-Ing. Rolf Katzenbach, *Technische Universität Darmstadt, Institut und Versuchsanstalt für Geotechnik, Petersenstraße 13, 64287 Darmstadt, Germany,* katzenbach@geotechnik.tu-darmstadt.de

Dipl.-Ing. Thomas Kersten, *Universität Kassel, Institut für Mechanik, Mönchebergstraße 7, 34109 Kassel, Germany,* kersten@ifm.maschinenbau.uni-kassel.de

Prof. Michał Kleiber, *Polish Academy of Sciences, Institute of Fundamental Technological Research, Swietokrzyska 21, 00-049 Warsaw, Poland,* mkleiber@ippt.gov.pl

Dipl.-Ing. Andreas Koch, *Universität Stuttgart, Institut für Mechanik (Bauwesen), Pfaffenwaldring 7, 70550 Stuttgart, Germany,* koch@mechbau.uni-stuttgart.de

Prof. em. Dr.-Ing. Dr.-Ing. E.h. Franz Gustav Kollmann, *Mauerkircherstraße 16, 81679 München, Germany,* fg.kollmann@t-online.de

Dr.-Ing. Matthias Lambrecht, *Universität Stuttgart, Institut für Mechanik (Bauwesen), Pfaffenwaldring 7, 70550 Stuttgart, Germany,* ml@mechbau.uni-stuttgart.de

Dipl.-Ing. Frank Lulei, *Universität Stuttgart, Institut für Mechanik (Bauwesen), Pfaffenwaldring 7, 70550 Stuttgart, Germany,* lulei@mechbau.uni-stuttgart.de

Prof. Dr. techn. Dr. h.c. Herbert A. Mang, *Technische Universität Wien, Institut für Festigkeitslehre, Karlsplatz 13/202, 1040 Wien, Austria,* Herbert.Mang@tuwien.ac.at

Prof. Dr.-Ing. Anton Matzenmiller, *Universität Kassel, Institut für Mechanik, Mönchebergstraße 7, 34109 Kassel, Germany,* amat@ifm.maschinenbau.uni-kassel.de

Prof. Gérard A. Maugin, *Université Pierre et Marie Curie, Laboratoire de Modélisation en Mécanique, Case 162, 8 rue du Capitaine Scott, 75015 Paris, France,* gam@ccr.jussieu.fr

Dipl.-Ing. Andreas Menzel, *Universität Kaiserslautern, Lehrstuhl für Technische Mechanik, Postfach 3049, 67653 Kaiserslautern, Germany,* amenzel@rhrk.uni-kl.de

Prof. Dr.-Ing. Christian Miehe, *Universität Stuttgart, Institut für Mechanik (Bauwesen), Pfaffenwaldring 7, 70550 Stuttgart, Germany,* cm@mechbau.uni-stuttgart.de

Prof. Dr. rer. nat. Alexander Mielke, *Universität Stuttgart, Mathematisches Institut A, Pfaffenwaldring 57, 70569 Stuttgart, Germany,* mielke@mathematik.uni-stuttgart.de

Prof. Eiji Nakamachi, *Osaka Institute of Technology, Department of Mechanical Engineering, 5-16-1 Omiya, Asahi, Osaka 535-8585, Japan,* nakamati@med.oit.ac.jp

Prof. Alan Needleman, *Brown University, Division of Engineering, Barus & Holley Building, Box M, Providence, RI 02912, U.S.A.,* needle@engin.brown.edu

Dr. rer. nat. Patrizio Neff, *Technische Universität Darmstadt, Fachbereich Mathematik, AG6, Schloßgartenstraße 7, 64289 Darmstadt, Germany,* neff@mathematik.tu-darmstadt.de

Prof. Sia Nemat-Nasser, *University of California at San Diego, Center of Excellence for Advanced Materials, 9500 Gilman Drive, La Jolla, CA 92093-0416, U.S.A.,* sia@halebopp.ucsd.edu

Dr.-Ing. Claus Oberste-Brandenburg, *Ruhr-Universität Bochum, Lehrstuhl f. Technische Mechanik, Universitätsstraße 150, 44780 Bochum, Germany,* cob@tm.bi.ruhr-uni-bochum.de

Prof. Michael Ortiz, *California Institute of Technology, Graduate Aeronautical Laboratories, 1200 East California Boulevard, Pasadena, CA 91125, U.S.A.,* ortiz@aero.caltech.edu

Prof. D. Roger J. Owen, *University of Wales Swansea, Department of Civil Engineering, Singleton Park, Swansea SA2 8PP, United Kingdom,* D.R.J.Owen@swansea.ac.uk

Prof. Djordje Perić, *University of Wales Swansea, Department of Civil Engineering, Singleton Park, Swansea SA2 8PP, United Kingdom,* D.Peric@swansea.ac.uk

Prof. Pedro Ponte-Castañeda, *University of Pennsylvania, Mechanical Engineering and Applied Mechanics, 220 S. 33rd Street, Philadelphia, PA 19104, U.S.A.*, ponte@seas.upenn.edu

Prof. Dr.-Ing. Ekkehard Ramm, *Universität Stuttgart, Institut für Baustatik, Pfaffenwaldring 7, 70569 Stuttgart, Germany,* eramm@statik.uni-stuttgart.de

Prof. Mark M. Rashid, *University of California at Davis, Department of Civil and Environmental Engineering, 1 Shields Avenue, Davis, CA 95616, U.S.A.*, mmrashid@ucdavis.edu

Prof. Daya Reddy, *University of Cape Town, Faculty of Science, PD Hahn Building, 7701 Rondebosch, South Africa,* bdr@maths.uct.ac.za

Prof. Dr.-Ing. Stefanie Reese, *Ruhr-Universität Bochum, Numerische Mechanik und Simulationstechnik, Gebäude IA 01/128, 44780 Bochum, Germany,* reese@nm.ruhr-uni-bochum.de

Priv.-Doz. Dr. Volker Reitmann, *Max-Planck-Institut für Physik komplexer Systeme, Nöthnitzer Straße 38, 01187 Dresden, Germany,* reitmann@rcs.urz.tu-dresden.de

Dipl.-Ing. Andreas Rieger, *Universität Stuttgart, Institut für Mechanik (Bauwesen), Pfaffenwaldring 7, 70550 Stuttgart, Germany,* rieger@mechbau.uni-stuttgart.de

Prof. Dr. Karl Roll, *DaimlerChrysler AG, Produktionsplanung Mercedes-Benz Pkw, HPC: B510, Abt.: PWT/VEO, 71059 Sindelfingen, Germany,* karl.roll@daimlerchrysler.com

Dr. Franz Roters, *Max-Planck-Institut für Eisenforschung, Abteilung Mikrostrukturphysik und Umformtechnik, Max-Planck-Straße 1, 40237 Düsseldorf, Germany,* roters@mpie.de

Prof. Kenneth Runesson, *Chalmers University of Technology, Department of Solid Mechanics, Hörsalsvägen 7, 41296 Göteborg, Sweden,* keru@solid.chalmers.se

Dr.-Ing. habil. Carlo Sansour, *Universität Karlsruhe, Institut für Baustatik, Kaiserstraße 12, 76131 Karlsruhe, Germany,* carlo.sansour@bs.uni-karlsruhe.de

Dipl.-Ing. Gerald Scheday, *Universität Stuttgart, Institut für Mechanik (Bauwesen), Pfaffenwaldring 7, 70550 Stuttgart, Germany,* scheday@mechbau.uni-stuttgart.de

Prof. Dr.-Ing. Dr. h.c. Werner Schiehlen, *Universität Stuttgart, Institut B für Mechanik, Pfaffenwaldring 9, 70569 Stuttgart, Germany,* wos@mechb.uni-stuttgart.de

Prof. Dr. rer. nat. Siegfried Schmauder, *Universität Stuttgart, Staatliche Materialprüfanstalt, Pfaffenwaldring 32, 70569 Stuttgart, Germany,* Siegfried.Schmauder@mpa.uni-stuttgart.de

Dr.-Ing. Ingo Schmidt, *Technische Universität Darmstadt, Institut für Mechanik (IV), Hochschulstraße 1, 64289 Darmstadt, Germany,* i.schmidt@mechanik.tu-darmstadt.de

Dipl.-Ing. Jan Schotte, *Universität Stuttgart, Institut für Mechanik (Bauwesen), Pfaffenwaldring 7, 70550 Stuttgart, Germany,* schotte@mechbau.uni-stuttgart.de

Prof. Dr.-Ing. Jörg Schröder, *Universität Essen, Institut für Mechanik, Universitätsstraße 15, 45117 Essen, Germany,* j.schroeder@uni-essen.de

Prof. Miroslav Šilhavý, *Academy of Sciences of the Czech Republic, Mathematical Institute, Žitná 25, 11567 Prague 1, Czech Republic,* silhavy@math.cas.cz

Prof. em. Dr.-Ing. Dr.-Ing. E.h. Dr. h.c. mult. Erwin Stein, *Universität Hannover, Institut für Baumechanik und Numerische Mechanik, Appelstraße 9A, 30167 Hannover, Germany,* stein@ibnm.uni-hannover.de

Prof. Dr.-Ing. Paul Steinmann, *Universität Kaiserslautern, Lehrstuhl für Technische Mechanik, Postfach 3049, 67653 Kaiserslautern, Germany,* ps@rhrk.uni-kl.de

Prof. Pierre M. Suquet, *Laboratoire de Mécanique et d'Acoustique / C.N.R.S., 31 chemin Joseph Aiguier, 13402 Marseille Cedex 20, France,* suquet@lma.cnrs-mrs.fr

Prof. Dr. rer. nat. Bob Svendsen, *Universität Dortmund, Fakultät Maschinenbau, Lehrstuhl für Mechanik, Leonhard-Euler-Straße 5, 44221 Dortmund, Germany,* bob.svendsen@mech.mb.uni-dortmund.de

Prof. Cristian Teodosiu, *Université Paris-Nord, Institut Galilée, Avenue J.-B. Clèment, 93430 Villetaneuse, France,* teodosiu@galilee.univ-paris13.fr

Dr. Kathrin Thiele, *Continental AG, Strategische Technologie - 51120, Jädekamp 30, 30419 Hannover, Germany,* kathrin.thiele@conti.de

Prof. Yoshihiro Tomita, *Kobe University, Graduate School of Science and Technology, 1-1 Rokkodai-cho Nada-ku Kobe, Hyogo 657-8501, Japan,* tomita@mech.kobe-u.ac.jp

Prof. Viggo Tvergaard, *Technical University of Denmark, Department of Mechanical Engineering, Solid Mechanics, Nils Koppels Allé, Building 404, 2800 Kgs. Lyngby, Denmark,* viggo@fam.dtu.dk

Dr.-Ing. Wolfgang A. Wall, *Universität Stuttgart, Institut für Baustatik, Pfaffenwaldring 7, 70569 Stuttgart, Germany,* wall@statik.uni-stuttgart.de

Prof. André Zaoui, *Ecole Polytechnique / CNRS, Laboratoire de Mécanique des Solides, 91128 Palaiseau Cedex, France,* zaoui@lms.polytechnique.fr

Dipl.-Ing. Holger Zecha, *Universität Stuttgart, Institut für Mechanik (Bauwesen), Pfaffenwaldring 7, 70550 Stuttgart, Germany,* zecha@mechbau.uni-stuttgart.de

Dr. Zisu Zhao, *Max-Planck-Institut für Eisenforschung, Abteilung Mikrostrukturphysik und Umformtechnik, Max-Planck-Straße 1, 40237 Düsseldorf, Germany,* zhao@mpie.de

Prof. Dr. techn. Dr. h.c. Franz Ziegler, *Technische Universität Wien, Institut für Allgemeine Mechanik, Wiedner-Hauptstraße 8–10, 1040 Wien, Austria,* franz.ziegler@tuwien.ac.at

Variational Principles and Non-Convex Problems

Nonconvex Energy Minimization and Relaxation in Computational Material Science

C. Carstensen
Institute for Applied Mathematics and Numerical Analysis, Vienna University of Technology
Wiedner Hauptstraße 8–10, 1040 Vienna, Austria
Carsten.Carstensen@tuwien.ac.at

Abstract: The mathematical modeling of microstructures has important applications in material science, e.g. advanced materials or phase transitions, micromagnetism, homogenization, and optimization. It is typical in those models that the continuous minimization problem (M) is nonconvex and lacks classical solutions. There exist infimising sequences in (M) that have a weak limit, but non-(quasi-)convexity implies, in general, that the weak limit u is *not* a solution of the problem (M). In physical and in numerical experiments, we observe oscillations of strains which form a macroscopic or averaged quantity Du. The efficient numerical simulation on the macroscopic level aims to compute the weak limit u as a solution of a related *Relaxed Problem*. This is in contrast to microscopic mechanisms which are directly approached by a finite element minimization of (M). The two relaxations discussed are based on Young measures (G) and on quasiconvexification (Q). The presentation discusses adapted finite element strategies for relaxed problems such as the 2-well problem in one-dimension, in higher dimensions, in linearized elasticity (with hysteresis) or in related topics in micromagnetism and homogenization problems.

Keywords: Computational material science, microstructures, adaptive finite element schemes, relaxation, non-convex minimization, quasiconvexification, phase transitions.

1. Example of microstructures in 1D

This section is devoted to an introduction to induced microstructures enforced by minimizing nonconvex energy densities in the simplest 1D setting. The subsequent well-known examples are essentially due to Bolza from the beginning of the last century and were advertised very much in the context of optimal control by Young long before measure-valued solution concepts entered the mathematical analysis of fluid and solid mechanics.

Throughout this section, let $\Omega = (0, 1)$ be a one-dimensional body and W an energy density,

$$W(F) = (F^2 - 1)^2 \quad \text{for} \quad F \in \mathbb{R} \tag{1}$$

(in terms of a strain F). The nonconvex energy density W displayed in Figure 1 is a very smooth non-negative function with exactly two global minimizers at $F = \pm 1$. Hence, the material energetically favors the two states or phases $F = \pm 1$.

C. Miehe (ed.), IUTAM Symposium on Computational Mechanics of Solid Materials at Large Strains, 3-20.

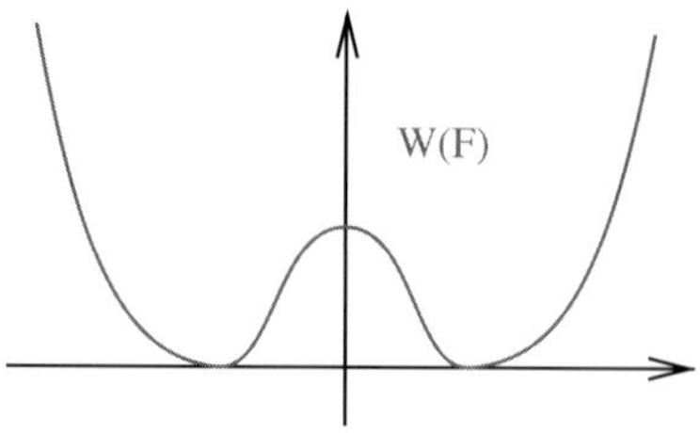

Figure 1. Nonconvex 1D energy density $W(F)$ of Section 1.

Because of the fourth order growth of the energy density W, the correct energy space is the Sobolev space $W^{1,4}(\Omega)$ defined as the set of all absolutely continuous functions $v : \Omega \to \mathbb{R}$ such that the pointwise derivative v' (which, then, exists almost everywhere as a limit of difference quotients) is summable in its fourth power,

$$W^{1,4}(\Omega) = \left\{ v : \Omega \to \mathbb{R} : \ v \text{ abs. continuous and } \int_\Omega |v'|^4 \, dx < \infty \right\}. \quad (2)$$

1.1. Example 1

Given a certain subset $\mathcal{A} \subset W^{1,4}(\Omega)$, Example 1 reads:

$$(M) \text{ Minimize } \quad E(u) := \int_0^1 W(u') \, dx \quad \text{over} \quad u \in \mathcal{A}. \quad (3)$$

Theorem. *Example 1 allows infinitely many minimizers u in*

$$\mathcal{A} := W_0^{1,4}(0,1) := \left\{ u \in W^{1,4}(0,1) : u(0) = 0 = u(1) \right\} \quad (4)$$

characterized by $u \in \mathcal{A}$ and $u' = \pm 1$.

A proof of this theorem is indicated in Figure 2: Lipschitz continuous functions are absolutely continuous and so the zig-zag functions of Figure 2 are admissible functions in $\mathcal{A}$. Their Lebesgue integral simply ignores the countable set of jumps and sees only the energy density $W(+1)$ or $W(-1)$, i.e. the energy density zero. Consequently, $E(u) = 0$ for any function u depicted in Figure 2 which, then, is a minimizer. Notice that there is no scale in the problem (e.g. caused by higher order terms which model surface energy etc.); hence the sea-saw functions may have an amplitude and period of any size. The situation can be completely different for other boundary conditions.

Theorem. *Example 1 allows a unique minimizer $u(x) = 2x$ of (M) in*

$$\mathcal{A} := \left\{ u \in W^{1,4}(0,1) : u(0) = 0, u(1) = 2 \right\}. \quad (5)$$

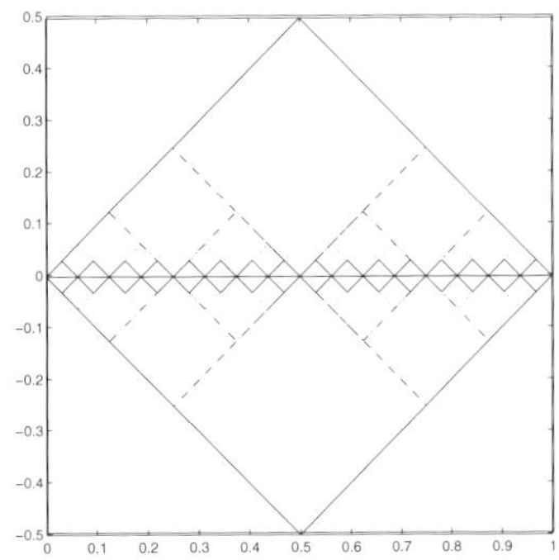

Figure 2. Various minimizers in Example 1 characterized by slopes ± 1.

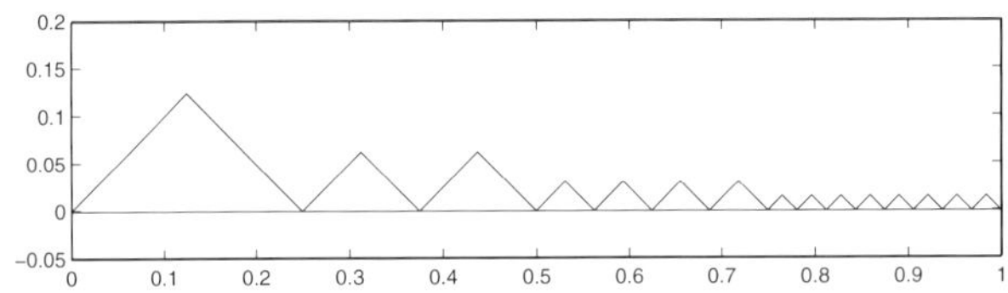

Figure 3. Typical zig-zag function u_ε in Example 2 characterized by slopes ± 1 and amplitudes $|u| < \varepsilon$.

This theorem is based on the strict convexity of W near $F = 2$: The tangential line $L(F)$ through $(F, W(F))$ at $F = 2$ leaves the rest of the entire graph over L, namely,

$$L(F) = 44\,F - 76 \leq W(F) \quad \text{for all} \quad F \in \mathbb{R}. \tag{6}$$

Since $v \in \mathcal{A}$ has a derivative v' with integral mean

$$\int_0^1 v'(x)\,dx = v(1) - v(0) = 2, \tag{7}$$

the function $u(x) = 2x$ yields the estimate

$$\begin{aligned} E(u) = 12 = \int_0^1 (44\,v'(x) - 76)\,dx &= \int_0^1 L(v'(x))\,dx \\ \leq \int_0^1 W(v'(x))\,dx = E(v) \end{aligned} \tag{8}$$

in which equality implies $u = v$.

1.2. Example 2

The situation is again completely different if low order terms arise. The mechanical interpretation of the low order term $|u(x)|^2$ is that of an elastic support of the 1D body Ω. Notice also that terms like $|u-f|^2$ typically steam from a time discretization of the parabolic problem with the stress contribution $DW(u')$. Then Example 2 reads:

$$(M) \text{ Minimize } \quad E(u) := \int_0^1 W(u')\,dx + \int_0^1 u(x)^2\,dx \quad \text{over} \quad u \in \mathcal{A}. \tag{9}$$

Theorem. *There exists no minimizer of* (M) *in* $\mathcal{A} := W_0^{1,4}(0,1)$ *for Example 2.*

The proof is of some interest for the interpretation of microstructures and hence discussed in some detail. In the first step, we observe

$$\inf E(\mathcal{A}) \geq 0 \tag{10}$$

because the two energy contributions are non-negative. In the next step we claim

$$\inf E(\mathcal{A}) = 0. \tag{11}$$

Indeed, given any small $\varepsilon > 0$ we can find a function $u_\varepsilon \in \mathcal{A}$ which, first, is a zig-zag curve with slopes ± 1, and, second, satisfies

$$|u_\varepsilon'(x)| < \varepsilon. \tag{12}$$

The second condition can be satisfied by finer and finer oscillations in Figure 2. A typical example is depicted in Figure 3. Form the two properties, we immediately derive

$$0 \leq \inf_{v \in \mathcal{A}} E(v) \leq E(u_\varepsilon) < \varepsilon^2. \tag{13}$$

Since ε may be arbitrary small, we have $\inf E(\mathcal{A}) = 0$, as announced.

The infimal energy is zero; we should not use the phrase "minimal energy" as the minimum is *not* attained. We prove this indirectly, by contradiction. Suppose, for a moment, that $u \in W^{1,4}(0,1)$ exists with $E(u) = 0$. The energy is a sum of two non-negative contributions and so each of the two integrals vanishes. In particular, $\int_0^1 u(x)^2\,dx = 0$ implies that $u \equiv 0$. But then $u' \equiv 0$ and so

$$E(u) = E(0) = \int_0^1 W(0)\,dx = W(0) = 1. \tag{14}$$

This contradicts our assumption $E(u) = 0$. Notice that nonconvexity enters here: If W was convex, $W(0) \leq \frac{1}{2}W(-1) + \frac{1}{2}W(1) = 0$ and the energy was minimized by $u \equiv 0$. Notice that the derivatives of u_ε do not tend to zero; they simply oscillate on finer and finer scales. The candidate $u \equiv 0$ for minimization does not see those oscillations in the limit and this is why there is no (classical) solution in Example 2.

1.3. GENERAL OBSERVATIONS FROM THE 1D EXAMPLES

The preceding examples allow some conclusions which are typical in the applications described in this paper.

- Existence and uniqueness of minimizers depend on boundary conditions in $\mathcal{A}$ and on low-order terms in E. This is not typical for PDEs where the leading derivatives dominate the existence theory. Here, the nonconvex energy densities are of changing type and globally degenerated: $W''(F)$ is positive for $|F|$ large and negative for $|F|$ small.
- Infimizing sequences do exist in $\mathcal{A}$, namely u_ε for smaller and smaller ε. They are bounded and weakly convergent

$$u_\varepsilon \rightharpoonup u \quad \text{in } W^{1,p}(\Omega) \tag{15}$$

(e.g. with $p = 4$), converge in weaker norms

$$u_\varepsilon \to u \quad \text{in } L^q(\Omega) \tag{16}$$

(e.g. with $q = 2$), but are *not* strongly convergent in $W^{1,p}(\Omega)$,

$$u_\varepsilon \not\to u \quad \text{in } W^{1,p}(\Omega). \tag{17}$$

- The weak limit u describes the macroscopic displacement, its derivative Du models averaged strain.
- Microscopic oscillations of Du_ε are *not* passed to the average Du; their limit is a measure rather than a Sobolev function. The interpretation in Example 2 is that the oscillations of $u'_\varepsilon = \pm 1$ in the neighborhood of any material point are to be statistically described: There are the slopes ± 1 and they relatively occur in 50% for both phases. This distribution is made rigorous by the expression

$$\nu_x = \frac{1}{2}\,\delta_{-1} + \frac{1}{2}\,\delta_{+1}. \tag{18}$$

Here δ_F is the Dirac measure with respect to the atom F. The measure ν_x is the same at any material point $x \in \Omega$. In Example 2, the energy $W(u'_\varepsilon)$ over a small volume element ω converges to

$$\lim_{\varepsilon \to 0} \int_\omega W(u'_\varepsilon)\, dx = \int_\omega \langle \nu_x, W\rangle\, dx; \tag{19}$$

recall that the action of a measure ν_x to W reads

$$\langle \nu_x, W\rangle = \left\langle \frac{1}{2}\,\delta_{-1} + \frac{1}{2}\,\delta_{+1}, W \right\rangle = \frac{1}{2}\,W(-1) + \frac{1}{2}\,W(1) = 0. \tag{20}$$

2. Fundamental theorem for Young measures

The examples of Section 1 serve as motivating illustrations of a rather abstract setting of Young measures (YM). For the purpose of an introduction it makes no harm to think of a Young measure ν as a convex combination of Dirac measures. The mathematical framework, however, is more delicate.

Let $L^\infty_w(\Omega;\mathcal{M}(\mathbb{R}^n))$ denote the Banach space of families of measures $(\nu_x : x\in\Omega)$ with $\nu_x\in\mathcal{M}(\mathbb{R}^n)$ for almost all $x\in\Omega$ and such that

$$\langle\nu;g\rangle:\Omega\to\mathbb{R},\quad x\mapsto\langle\nu_x;g\rangle:=\int_{\mathbb{R}^n} g(A)\,d\nu_x(A) \tag{21}$$

is measurable for each $g\in C_0(\mathbb{R}^n):=\{f\in C(\mathbb{R}^n):\lim_{|x|\to\infty} f(x)=0\}$; $\mathcal{M}(\mathbb{R}^n)$ is the dual of $C_0(\mathbb{R}^n)$. The norm in $L^\infty_w(\Omega;\mathcal{M}(\mathbb{R}^n))$ is $\operatorname{ess\,sup}_{x\in\Omega}\|\nu_x\|_{\mathcal{M}}$.

Let $\mathrm{YM}(\Omega;\mathbb{R}^n)$ be the set of all $\nu\in L^\infty_w(\Omega;\mathcal{M}(\mathbb{R}^n))$ which are probability measures (i.e. ν_x is non-negative with mass 1 for almost all $x\in\Omega$). Then, a sequence (F_k) in $L^p(\Omega;\mathbb{R}^n)$ is said to *generate a Young measure* (YM) ν if $\nu\in\mathrm{YM}(\Omega;\mathbb{R}^n)$ and

$$(g(F_k))\rightharpoonup^*\langle\nu;g\rangle\ \text{(weakly star) in } L^1(\Omega)\quad\text{for all } g\in C_0(\mathbb{R}^n); \tag{22}$$

that is, for all samples $\omega\subset\Omega$ and all $g\in C_0(\mathbb{R}^n)$ there holds

$$\lim_{k\to\infty}\int_\omega g(F_k)\,dx=\int_\omega\langle\nu_x,g\rangle\,dx. \tag{23}$$

Fundamental theorem for young measures. *For each sequence (F_j) bounded in $L^p(\Omega)$, $1\le p\le\infty$, there exists a Young measure $\nu=(\nu_x)_{x\in\Omega}$ and a subsequence (F_k) that generates ν.*

The proof can be found in Ball [1].

Remarks.

- The measure ν_x depends measurably on x; the integrals of $\langle g,\nu_x\rangle$ exist as Lebesgue integrals (Ball [1]).
- A YM $\nu=(\nu_x)_{x\in\Omega}$ which equals a Dirac measure $\delta_{Du(x)}$ for $x\in\omega\subset\Omega$ is called *trivial* on ω. In this way, the classical solutions are naturally embedded into the measure-valued solution concept.
- Pointwise convergence yields trivial Young measures: If (p_j) is bounded in $L^1(\Omega)$ and $(p_j(x))\to p(x)$ for almost all $x\in\omega$, then $\nu_x=\delta_{p(x)}$ for almost all $x\in\omega$.

- For almost all $x \in \Omega$, we have from Ball [1],

$$\nu_x = \lim_{\varepsilon \to 0} \lim_{j \to \infty} \frac{\int_{B(x,\varepsilon)} \delta_{p_j(y)}\, dy}{|B(x,\varepsilon)|}. \tag{24}$$

- The radiation condition $\lim_{|x|\to\infty} g(x) = 0$ on $g \in C_0(\mathbb{R}^n)$ can be weakened (Ball [1]).
- Let $K \subseteq \mathbb{R}^n$ be closed and suppose $(p_j) \to K$ in measure (i.e. for each open $V \supseteq K$, $|\{x \in \Omega : p_j(x) \notin V\}| \to 0$ as $j \to \infty$) then $\operatorname{supp} \nu_x \subseteq K$ for almost all $x \in \Omega$.
- The functional analytical setting of the fundamental theorem for YMs asserts

$$L^1(\Omega; C_0(\mathbb{R}^n))^* = L^\infty_\omega(\Omega; \mathcal{M}(\mathbb{R}^n)). \tag{25}$$

- We refer to Roubíček [28] for other classes of Young measures related to other growth conditions (instead of the radiation condition in$C_0(\mathbb{R})$ at infinity) and mention that other measures are certainly important such as DiPerna-Mayda or H-measures.
- In the preceding Example 2, *some* infimizing sequence, namely (u_ε), generates the YM $\nu_x = 1/2\,\delta_{-1} + 1/2\,\delta_{+1}$. One can prove that *any* infimizing sequence generates this YM.
- A sequence of gradients $(D\,u_j)$ generates a gradient Young Measure (GYM) $(\nu_x)_{x\in\Omega}$. The fact that ν_x is a GYM (and not just a YM) is highly relevant in vector-valued problems with deformations$u : \Omega \subset \mathbb{R}^n \to \mathbb{R}^m$ and $m, n \geq 2$ (Bethuel et al. [7]).

3. Relaxation theory

Throughout this section let $u : \Omega \subset \mathbb{R}^n \to \mathbb{R}^m$ and the energy E be

$$E(u) := \int_\Omega W(x, Du(x))\, dx + L(u) \tag{26}$$

with some smooth energy density $W(x, F)$ with p-growth in the variable $F \in \mathbb{R}^{m\times n}$ and some convex lower-order contribution $L(u)$ (e.g. from applied loads). With the set $\mathcal{A} \subset W^{1,p}(\Omega; \mathbb{R}^m)$ of admissible functions the minimization problem reads:

$$(M) \text{ Minimize } \int_\Omega W(x, Du(x))\, dx + L(u) \quad \text{over} \quad u \in \mathcal{A}. \tag{27}$$

Microstructures are linked with highly oscillatory infimizing sequences of gradients Du_j which are bounded and weakly convergent to some Du and which generate a YM $(\nu_x)_{x\in\Omega}$. Typically, for the weak limit there holds

$$\int_\Omega W(Du)\,dx \not\leq \lim_{j\to\infty}\int_\Omega W(Du_j)\,dx \tag{28}$$

because (quasi-)convexity of W is related to lower semicontinuity of E.

3.1. YOUNG MEASURE RELAXATION

The limit of the energy can be characterized with the fundamental theorem for Young measures. For the integrand W we obtain formally

$$\lim_{j\to\infty}\int_\Omega W(x, Du_j)\,dx = \int_\Omega \langle \nu_x, W(x,\cdot)\rangle\,dx \tag{29}$$

with the YM $(\nu_x)_{x\in\Omega}$ generated by Du_j. (The behavior of W at infinity is not as expected for test functions; however, this step can be made rigorous.) The limit of the energy densities is hence described by the statistics of oscillations rather than the single matrix of the averaged limit $Du(x)$. Replacing in the limit $W(x, Du(x))$ by $\langle \nu_x, W(x,\cdot)\rangle$ we obtain the Young measure relaxation:

$$(G)\ \text{Minimize}\ \int_\Omega \langle \nu_x, W(x,\cdot)\rangle\,dx + L(u)\quad\text{over}\quad (u,\nu)\in\mathcal{A}\times\text{GYM} \tag{30}$$

with two coupled unknowns (u,ν): $u\in\mathcal{A}$ models directly the macroscopic deformation and develops no oscillations at all. The information on statistics of oscillations in (M) is contained in the GYM $(\nu_x)_{x\in\Omega}$. The two variables are coupled by the restriction

$$\langle \nu_x, \text{Id}\rangle = Du(x)\quad\text{for a.e.}\quad x\in\Omega. \tag{31}$$

This condition states that the average of the oscillating deformations (i.e. the expected value of ν_x) equals the macroscopic quantity $Du(x)$. We refer to Bethuel et al. [7], Chipot [20], Dacorogna [23], Pedregal [27] and Roubíček [28] for more details and proofs.

3.2. QUASICONVEXIFICATION

It is of relevance in Problem (G) that, given a material point $x\in\Omega$ and a macroscopic gradient $Du(x)\in\mathbb{R}^{m\times n}$, the problem in the minimization for ν_x is local. The Young measure is simply determined as the minimizer of

$$\langle \nu_x, W(x,\cdot)\rangle\quad\text{amongst GYMs}\ \mu\ \text{with}\ \langle\mu,\text{Id}\rangle = Du(x) \tag{32}$$

and there is no interaction with low-order terms $L(u)$ or with other material points. If this minimum is known (or computed in sufficient accuracy) then it is a function of x and $Du(x)$, written

$$QW(x, Du(x)) := \min\{\langle\mu, W(x,\cdot)\rangle : \mu\ \text{GYM with}\ \langle\mu,\text{Id}\rangle = Du(x)\}. \tag{33}$$

It turns out that $QW(x,\cdot)$ is the quasiconvex envelope also defined by

$$QW(x,F) := \inf_{\phi\in W_0^{1,\infty}((0,1)^n;\mathbb{R}^m)} \int_{(0,1)^n} W(x,F+D\phi(x))\,dx. \quad (34)$$

If a formula for $QW(x,F)$ is known, $\langle \nu_x, W(x,\cdot)\rangle$ can be replaced in (G) by $QW(x,Du(x))$. This motivates the quasiconvexified problem:

$$(Q)\text{ Minimize } \int_\Omega QW(x,Du(x))\,dx + L(u) \quad \text{over} \quad u\in\mathcal{A}. \quad (35)$$

We refer to Bethuel et al. [7], Chipot [20], Dacorogna [23], Pedregal [27] and Roubíček [28] for more details and proofs.

3.3. Equivalence of (M), (G), (Q)

It is known (cf., e.g. Bethuel et al. [7], Chipot [20], Dacorogna [23], Pedregal [27] and Roubíček [28] for details and proofs) that (M), (G), and (Q) are pairwise equivalent. The examples of Section 1 show that (M) may have no (classical) solution while (G) and (Q) do have solutions: There exists minimizers in Problem (G) and (Q). The three Problems (M), (G), (Q) are essentially equivalent in the following sense:

- If u is the weak limit of an infimizing sequence which generates a GYM ν, then the pair (u,ν) solves (G) and u solves (Q).
- If (u,ν) is a minimizer in (G), then there exists an infimizing sequence in (M) with weak limit u and generated GYM ν, and, in particular, u solves (Q).
- If u minimizes (Q) then there exists an infimizing sequence in (M) with weak limit u which generates a GYM ν and (u,ν) solves (G).

Moreover, infimal energies, macroscopic deformations, and generated stress fields in (M), (G), (Q) coincide.

4. Numerical examples

Throughout this section, the 1D numerical experiments (taken from Carstensen [8]) concern (a shifted) Tartar's broken extremal example (Nicolaides and Walkington [25]).

4.1. EXAMPLE 3

The framework is as in Section 1. With the function $f(x) = -3/128(x - 0.5)^5 - 1/3(x-1/2)^3$, Example 3 reads:

$$(M) \text{ Minimize } E(u) := \int_0^1 W(u')\,dx + \int_0^1 |u-f|^2\,dx \text{ over } u \in \mathcal{A} \quad (36)$$

with $\mathcal{A} = \{v \in W^{1,4}(0,1) : v(0) = u(0) \text{ and } v(1) = u(1)\}$ and the generalized exact solution

$$\begin{aligned} u(x) &= f(x) && \text{for } \; 0 \le x \le 1/2 \\ u(x) &= 1/24\,(x-1/2)^3 + x - 1/2 && \text{for } \; 1/2 \le x \le 1. \end{aligned} \quad (37)$$

In the numerical examples below we shifted the aforementioned functions by $\pi/100$ to the left to avoid that the break point at $x_{\rm b} = 1/2 - \pi/100$ is aligned with any mesh. It is known that Problem (M) has no (classical) solution and that microstructures in form of high oscillations with zig-zag curves are expected in the region $(0, x_{\rm b})$ while the Young measures are trivial (hence no oscillations) on the right half $(x_{\rm b}, 1)$.

4.2. DISCRETIZATION (M_h)

Let $h := 1/N$ for a natural number N and let $\mathcal{T}$ denote the uniform mesh with size h on $\Omega = (0,1)$, so that $\mathcal{A}_h \subset \mathcal{S}^1(\mathcal{T})$ consists of continuous piecewise affine functions in $\mathcal{A}$. Then the discrete problem reads:

$$(M_h) \text{ Minimize } E(u_h) := \int_0^1 W(u_h')\,dx + \int_0^1 |u_h - f|^2\,dx \text{ over } u_h \in \mathcal{A}_h. \quad (38)$$

Figure 4 shows the finite element approximation u_h (black dotted line) for $N = 6$. Although the mesh is very coarse, one observes that u_h is oscillatory in the left half of the domain and seemingly smooth in the right half. The comparison with u (red solid line) shows that the behavior of u_h is very good.

Remarks.

- Discrete minimizers exist owing to local compactness caused by the finite dimension of $\mathcal{A}_h$.
- Weak convergence of u_h and the GYM of u_h' was established (in simple examples) by Luskin, Collins, Kinderlehrer, Chipot and others in the last decade [24, 20, 21, 6, 3].
- The discrete minimizer u_h cannot converge strongly in $W^{1,p}(\Omega)$ as this limit minimized (M).

- The discrete minimizer u_h is difficult and expensive to compute. The initial approximation of a Newton-Raphson scheme in Figure 4 (blue dashed line) is very close; otherwise the method may fail to work [8].

Descent methods stuck at local minimizers: The following model example (Carstensen [8]) illustrates that local minimizers of discrete energy the$E|_{\mathcal{A}_h}$ in (M_h) cluster around u_h and might cause numerical analysts' nightmares!

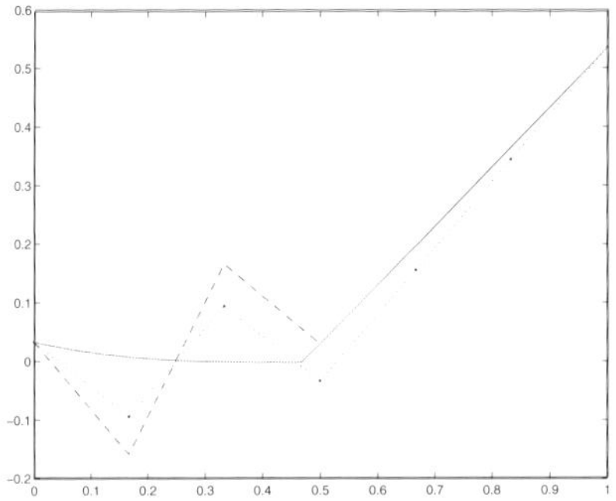

Figure 4. Approximation u_h from (M_h) in Example 3 (dotted black line) and exact generalized solution u (solid red line). The dashed blue line indicates the initial approximations for the Newton-Raphson solver.

Theorem. *For Example 2 and a uniform partition of* $\Omega = (0,1)$ *into* N *elements of size* $h = 1/N$ *and* P_1 *finite element space* $\mathcal{A}_h$ *(without boundary conditions,* $\mathcal{A} = W^{1,4}(0,1)$*) there holds*

$$E_h := h^2(1 - h^2/48)/12 = \min_{u_h \in \mathcal{A}_h} E = E(u_h) \tag{39}$$

for two global discrete minimizers u_h *with* $u_h' = \pm\sqrt{1 - h^2/24}$ *and alternating signs. Moreover, given any number* m *with* $9 \leq m^2 < N$*, there are at least* m^m *local minimizers* v_h *of* $E : \mathcal{A}_h \to \mathbb{R}$ *with*

$$\|u_h - v_h\|_{L^2(\Omega)} < h \quad and \quad E_h \leq E(v_h) \leq (1 + 24\,h)\,E_h. \tag{40}$$

Two distinct such local minimizers are separated by an energy threshold $\geq 12\,N\,E_h$.

4.3. DISCRETIZATION (G_h)

The finite element space $\mathcal{A}_h := \mathcal{S}^1(\mathcal{T})$ is as before, but the GYM variable requires a discretization $\mathrm{GYM}_h \subset \mathrm{GYM}$: On each finite element $T \in \mathcal{T}$, let

$$\nu_h|_T = \sum_{j=1}^{j_T} \mu_{T,j}\, \delta_{F_{T,j}} \tag{41}$$

for unknown convex coefficients $\mu_{T,j}$ and given atoms $F_{T,j}$. The resulting discrete problem reads:

$$(G_h)\ \text{Minimize} \int_\Omega \langle \nu_h, W(x,\cdot)\rangle\, dx + L(u_h) \text{ over } u_h \in \mathcal{A}_h,\ \nu_h \in \text{GYM}_h \qquad (42)$$
$$\text{subject to } \langle \nu_h, \text{Id}\rangle = Du_h.$$

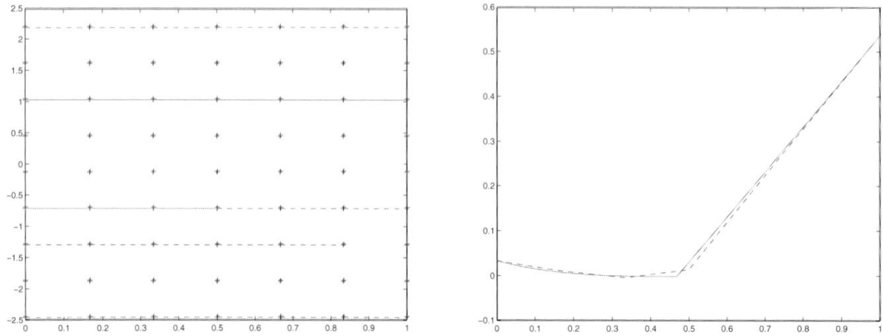

Figure 5. Approximation (u_h, ν_h) from (G_h) in Example 3 and exact generalized solution u (solid red line). Support of Young measure (left) and displacement u_h (right, dashed blue line) for approximate and continuous solution.

The discrete solution u_h is shown in Figure 5 (right, dashed blue line) and very close to the exact (generalized) solution u (solid red line). The discrete solution u_h approximates the averaged deformation directly and hence shows no oscillations at all. Information about oscillations in the original problem (M) are contained in the discrete YM ν_h. Its atoms are displayed in Figure 5 (left) where the small crosses indicate the possible 9 atoms on each of the 6 finite elements. The lines indicate the active atoms, namely those atoms whose coefficients were different from zero throughout the calculations. The final result shown in solid red is that two atoms near -0.6 and 1.1 are active for the first three elements and on the right half of the domain the atom near 1.1 remains.

The interpretation is that of approximated microstructures on the first three finite elements and a classical solution (with Dirac measures) on the remaining part of the domain.

Remarks.

- The discrete problem (G_h) leads to a quadratic programming problem (qp) with a huge number of degrees of freedom.
- Abstract (weak) convergence of u_h and ν_h has been established in Roubíček [28].

- The huge number $\geq h^{-2D}$ of unknowns causes difficulties. The remedy with adaptive multilevel active set strategy (Carstensen and Roubíček [19] and Bartels [4]) shows linear complexity!
- In the scalar situation at hand, YM=GYM. This is no longer the case when $QW \neq W^{**}$ for the vector case. Then the Young measure relaxation is problematic as the characterization of the GYM in the set of YM is difficult to handle. Then, it appears more promising to study related convexity notions and their hulls (Roubíček [28]).

4.4. Discretization (Q_h)

In situations where the quasiconvexified energy density QW is known, the discrete problem (Q_h) for a given finite element space $\mathcal{A}_h$ reads:

$$(Q_h) \quad \text{Minimize} \quad \int_\Omega QW(x, Du_h(x))\, dx + L(u_h) \quad \text{over} \quad u \in \mathcal{A}_h. \tag{43}$$

The convexification of the one-dimensional example W is simple, cf. Figure 6.

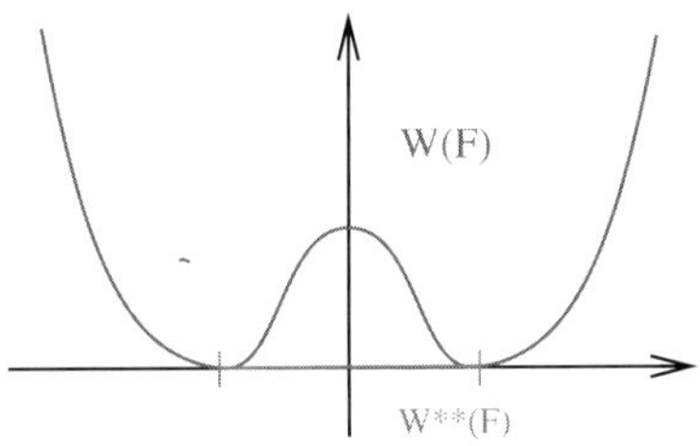

Figure 6. Convexified energy density $QW = W^{**}$ of W for (Q_h) in Example 3; W and W^{**} differ only between the wells for arguments F with $|F| < 1$.

The resulting finite element approximation u_h is displayed in Figure 7 (black dotted line) in good agreement with the exact (generalized) solution u (red solid line).

Remarks.

- In the scalar case with deformations $u : \Omega \subset \mathbb{R}^n \to \mathbb{R}^m$ and $m = 1$ or $n = 1$, the quasiconvexification QW of W equals the lower convex envelope $CW = W^{**}$ of W.
- The computation of QW leads to a local minimization problem (M) and can be extremely complicated (Chipot [20]).
- In contrast to the problem (M_h), the approximations u_h in (Q_h) can convergence strongly in $W^{1,p}$; an affirmative 1D example is given in Nicolaides and Walkington [25].

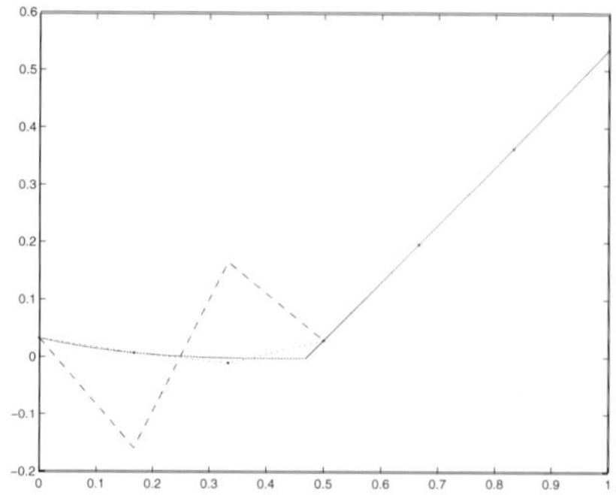

Figure 7. Approximation u_h from (Q_h) in Example 3 (dotted black line) and exact generalized solution u (solid red line). The dashed blue line indicates the initial approximations for the Newton-Raphson solver.

- The minimization of the convex (but degenerated) functional E^{**} is possible with standard software. Each local extremum is a global discrete minimizer (by convexity) and so clusters of local minimizers are not present.
- The finite element approximation u_h shows no oscillations; nevertheless, the YM can be recovered from u_h (Carstensen and Plecháč [14]).
- The approach of (Q_h) is limited by the request of QW and needs to be modified for related hulls which are accessible.

5. Applications

This section reviews a few more appealing applications where a relaxation via quasiconvexification was possible and then proved to be very effective.

5.1. Scalar 2-well problem

An anti-plane strain model of the Erickson-James energy density yields

$$W(F) = |F - F_1|^2 \cdot |F - F_2|^2 \quad \text{for} \quad F \in \mathbb{R}^n \tag{44}$$

with the two distinct phases F_1 and F_2. The proposed benchmark example on a rectangle Ω in 2D reads:

$$(M) \qquad \begin{aligned} \text{Minimize} \quad E(u) &:= \int_\Omega (W(\nabla u) + |u - f|^2)\, dx \\ \text{over} \qquad \mathcal{A} &:= u_D + W_0^{1,4}(\Omega). \end{aligned} \tag{45}$$

This is the best analyzed 2D example. A priori and a posteriori error estimates for (Q_h) were established in Carstensen and Plecháč [14]. The striking result is quasi-optimal convergence in the stress field variable on the expense of an unknown regularity of the exact solution u.

The a posteriori error control has recently been improved in Carstensen and Jochimsen [12] with the striking result that averaging techniques for a posteriori error control are highly accurate even for this degenerated problem from nonconvex minimization.

Strong convergence which implies strong convergence of the YMs will be shown in Bartels et al. [5] for stabilization techniques.

We refer to Bartels [3], Bartels et al. [5], Carstensen and Jochimsen [12] and Carstensen and Plecháč [14] for numerical examples.

5.2. Nonconvex problems in computational mechanics

The variational model of an elastic body which occupies a bounded Lipschitz-domain $\Omega \subset \mathbb{R}^n$ with two distinct zero-stress strain phases

$$F_1 \text{ and } F_2 \text{ in } \mathbb{R}^{n\times n}_{sym} := \{F \in \mathbb{R}^{n\times n} : F = F^T\} \tag{46}$$

defines a vectorial 2-well problem

$$W(F) = \min\{W_1(F), W_2(F)\} \tag{47}$$

where

$$W_j(F) := 1/2\,(F - F_j) : \mathbb{C}(F - F_j) + W_j^0 \tag{48}$$

for each well $j = 1, 2$, with minimal energy $W_j^0 \in \mathbb{R}$ and the fourth-order elasticity tensor $\mathbb{C}$.

The numerical simulation of a low-temperature super-conductor is performed in Carstensen and Plecháč [15] with the help of a formula for the quasiconvexification due to Kohn. Quasi-uniform error estimates are included for the stress variables and an adaptive mesh-refining algorithm is suggested as well. Figure 8 displays the modulus $|\sigma_h|$ of the discrete stress field in the deformed configuration computed with an adapted mesh indicated. The numerical analysis is based on a rigorous estimates (Carstensen and Plecháč [15]). An important feature of multi-well energy densities is that they model phase-transitions allowing hysteresis loops from one state into another. A numerical simulation of this time-depending problem is performed in Carstensen and Plecháč [16] for the aforementioned material. There is raising evidence that a typical time discretization in models for finite plasticity result in non-(quasi-)convex energy densities and so involve microstructures (Carstensen et al. [11], Ortiz and Repetto [26]).

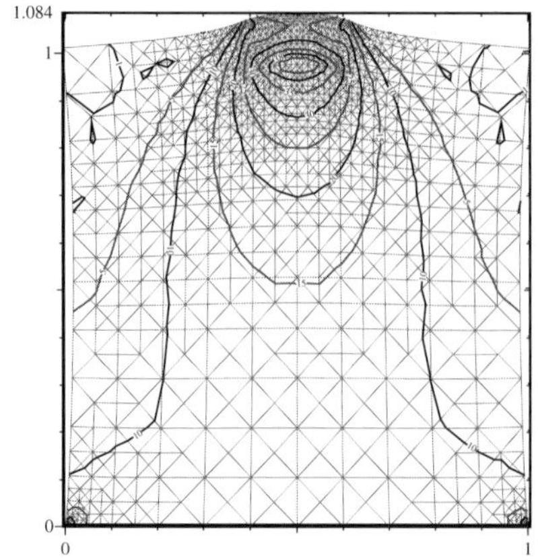

Figure 8. Stress approximation $|\sigma_h|$ from (Q_h) for loaded super-conductor from [15].

5.3. Time depending problems

Hysteresis involves time as a process time and many details are open in their numerical analysis (Carstensen and Plecháč [16]). Even more difficult is the simulation in finite plasticity (Carstensen et al. [11]) where convergence is widely an open question. Other models under consideration involve nonlinear wave equations which are either damped (Carstensen and Dolzmann [10]) or require a Young measure approximation (Carstensen and Rieger [18]).

5.4. Micromagnetism

The Landau-Lifshitz-Weiss model in micromagnetism leads to a nonconvex essentially scalar, minimization problem. The effective numerical treatment is suggested in Carstensen and Plecháč [17] in the spirit of (Q_h).

6. Conclusions and outlook

The paper concludes with a few observations and conclusions on the state of the art in computational material science which encounters many challenging phenomena (Ball [1] and Carstensen [8]).

Relaxation theory provides a set of essentially equivalent problems and completely different numerical counterparts (M_h), (G_h), and (Q_h). Depending on the target of a numerical simulation, the proper choice of one amongst the three schemes may be different. Provided that the target involves not more than the approximation of the averaged deformation, the stress field, and the Young measure, we suggest to relax via the quasiconvexification QW. Then, adaptive multilevel schemes work and stabilization improves the efficient computation: In our experience, state of the art tools such as hp-versions of the finite element method, nested iteration and multilevel methods, or adaptive mesh-refining algorithms are successfully applicable for (Q_h) and fail for (M_h). In the case of degenerated energy densities in (Q_h), difficulties arise from the lack of uniform or even strict convexity in relaxed problems. However, QW needs to be known or has to be computed. The latter leads to (M_h) and this can be a very challenging task. The regularity of solution (Carstensen and Müller [13, 22]) is an open problem and higher regularity cannot be expected (Carstensen and Jochimsen [12]). So far, rigorous (norm) error control is restricted to convex situations; the only error estimate for nonconvex minimization problem known to the author is for uniformly polyconvex material in Carstensen and Dolzmann [9].

Acknowledgments

The author acknowledges the support of his collaborators J.M. Ball, S. Bartels, G. Dolzmann, K. Jochimsen, K. Hackl, A. Mielke, S. Müller, P. Plecháč, A. Prohl, and T. Roubíček over the years as well as the support by the DFG Priority Program (Schwerpunktprogramm 1095) "Analysis, Modeling and Simulation of Multiscale Problem".

References

[1] Ball, J.M. (1989). A version of the fundamental theorem for young measures. In: *Partial differential equations and continuum models of phase transitions, Lecture Notes in Physics*, M. Slemrod, M. Rascle, D. Serre (eds.), Berlin, Springer, **344**, 207–215.

[2] Ball, J.M. (1999). Singularities and computation of minimizers for variational problems. In: *Proceedings of Foundations of Computational Mathematics Conference*, Oxford, Cambridge University Press.

[3] Bartels, S. (2001). *Numerical analysis of some non-convex variational problems*, PhD Thesis, Christian-Albrechts University Kiel, Germany.

[4] Bartels, S. (2002). Adaptive approximation of Young measure solutions in scalar nonconvex variational problems. In preparation.

[5] Bartels, S., Carstensen, C., Prohl, A. and Plechac, P. (2002). Work in progress on strong convergence by stabilization. In preparation.

[6] Bartels, S. and Prohl, A. (2002). Multiscale resolution in the computation of crystalline microstructure. Preprint available at "http://www.ima.umn.edu/preprints/feb02/feb02.html".

[7] Bethuel, F., Huisken, G., Müller, S. and Steffen, K. (1999). Variational models for microstructure and phase transitions. In: *Calculus of variations and geometric evolution problems*, S. Hildebrandt and M. Struwe (eds.), Springer, **1713**, 85–210.

[8] Carstensen, C. (2000). Numerical Analysis of Microstructure. In: *Theory and Numerics of Differential Equations, Durham 2000*, J.F. Blowey, J.P. Coleman and A.W. Craig (eds.), Springer Berlin Heidelberg (Univeritext), 59–126,

[9] Carstensen, C. and Dolzmann, G. (1999). An a priori error estimate for finite element discretizations in nonlinear elasticity for polyconvex materials under small loads. *Berichtsreihe*

des Mathematischen Seminars Kiel, **99-12**, preprint available at "http://www.numerik.uni-kiel.de/reports/1999/99-12.ps.gz".

[10] Carstensen, C. and Dolzmann, G. (2001). Time-space discretization of the nonlinear hyperbolic system $u_{tt} = div(\sigma(Du) + Du_t)$. *Publications of the Max Planck Institute for Mathematics in the Sciences*, Leipzig, Germany, **60**, preprint available at "http://www.mis.mpg.de/preprints/2001/prepr6001-abstr.html".

[11] Carstensen, C., Hackl, K. and Mielke, A. (2000). Nonconvex potentials and microstructures in finite-strain plasticity. *Proc. Roy. Soc. A.*, in press, preprint Nr. 72/2000 available at "http://www.mis.mpg.de/preprints/2000/".

[12] Carstensen, C. and Jochimsen, K. (2002). Adaptive finite element error control for nonconvex minimization problems: Numerical two-well model example allowing microstructures. Preprint.

[13] Carstensen, C. and Müller, S. (2001). Local stress regularity in scalar nonconvex variational problems. Preprint.

[14] Carstensen, C. and Plecháč, P. (1997). Numerical solution of the scalar double-well problem allowing microstructure. *Math. Comp.*, **66**, 997–1026.

[15] Carstensen, C. and Plecháč, P. (2000). Numerical analysis of compatible phase transitions in elastic solids. *SIAM J. Numer. Anal.*, **37(6)**, 2061–2081.

[16] Carstensen, C. and Plecháč, P. (2001). Numerical analysis of a relaxed variational model of hysteresis in two-phase solids. *M2AN*, in press, preprint available at "http://www.mis.mpg.de/ preprints/2000/prepr8000-abstr.html".

[17] Carstensen. C., Prohl, A. (2001). Numerical analysis of relaxed micromagnetics by penalized finite elements. *Numer. Math.*, published online May 4, 2001, DOI 10.1007/s002110100268.

[18] Carstensen, C. and Rieger, M.O. (2002). Work on Young measure approximation of nonlinear wave equation. In progress.

[19] Carstensen, C. and Roubíček, T. (2000). Numerical approximation of Young measures in nonconvex variational problems. *Numer. Math.*, **84**, 395–415.

[20] Chipot, M. (2001). *Elements of nonlinear analysis*, Birkhäuser.

[21] Chipot, M. and Müller, S. (1997). Sharp energy estimates for finite element approximation of nonconvex problems. Max-Plank-Institut Leipzig, preprint 1997-8.

[22] Chipot, M. and Evans, L.C. (1986). Linearisation at infinity and Lipschitz estimates for certain problems in the calculus of variations. *Proc. R. Soc. Edinburgh*, **102 A**, 291–303.

[23] Dacorogna, B. (1989). Direct methods in the calculus of variations. *Applied Mathematical Sciences*, Springer, **78**.

[24] Luskin, M. (1996). On the computation of crystalline microstructure. *Acta Numerica*, **5**, 191–257.

[25] Nicolaides, R.A. and Walkington, N.J. Strong Convergence of Numerical Solutions to Degenerate Variational Problems. *Math. Comp.*, **64**, 117–127.

[26] Ortiz, M. and Repetto, E.A. (1999). Nonconvex energy minimisation and dislocation in ductile single crystals. *Journal of the Mechanics and Physics of Solids*, **47**, 397–462.

[27] Pedregal, P. (1997). *Parametrised measures and variational principles*, Birkhäuser.

[28] Roubíček, T. (1997). Relaxation in Optimization Theory and Variational Calculus. *Series in Nonlinear Analysis and Applications*, Walter De Gruyter, New York, **4**.

Rank 1 Convex Hulls of SO(n)-Invariant Functions

M. Šilhavý

Mathematical Institute, Academy of Sciences of the Czech Republic

Žitná 25, 11567 Prague 1, Czech Republic

silhavy@math.cas.cz

Abstract: A rotationally invariant function f defined on the set $\mathbb{M}^{n\times n}$ of all real square matrices of order n has a representation $\tilde{f}$ on $\mathbb{R}^n$ through the signed singular values of the matrix argument $A \in \mathbb{M}^{n\times n}$. Using necessary and sufficient conditions for the rank 1 convexity of f in terms of $\tilde{f}$, three iterative procedures are derived to construct the rank 1 convex hull of f.

Keywords: Rank 1 convex functions, rotational invariance.

1. Introduction

Consider the energy functional

$$I(u) = \int_\Omega f(Du)\,dx$$

where $\Omega \subset \mathbb{R}^n$ is open and bounded, $u \in W^{1,p}(\Omega,\mathbb{R}^n)$ is a deformation with the gradient Du, $f : \mathbb{M}^{n\times n} \to \mathbb{R}\cup\{\infty\}$ is the energy defined on the set $\mathbb{M}^{n\times n}$ of all real square matrices of order n and $1 < p \le \infty$. If f is nonelliptic, i.e., not quasiconvex, the minimum problem

$$M = \inf\{I(u) : u \in \mathcal{A}\}$$

on the Dirichlet class $\mathcal{A} = \{u : u - v \in W_0^{1,p}(\Omega,\mathbb{R}^n)\}$ where $v \in W^{1,p}(\Omega,\mathbb{R}^n)$ is fixed, generally does not have a solution, i.e., there is no $u \in \mathcal{A}$ such that $M = I(u)$. Nonelliptic energies arise in the theory of phase transitions in solids, where the nonexistence of the solution corresponds to the formation of the microstructure: To reduce its energy, the material exhibits complicated patterns of spatially oscillating phases (Ball [2, 3], Chipot and Kinderlehrer [5], Dacorogna [6], Kohn and Strang [8] and Müller [12]). Recall that f is said to be quasiconvex (e.g., Morrey [11], Ball [1], Dacorogna [6], Roubíček [15], Šilhavý [16] and Pedragal [13]), if

$$|E|f(A) \le \int_E f(A + Dv(x))\,dx \tag{1}$$

for each $A \in \mathbb{M}^{n\times n}$, each bounded open $E \subset \mathbb{R}^n$ with $|\partial E| = 0$, and each $v \in W_0^{1,\infty}(E)$ such that the right-hand side in (1) makes sense as the Lebesgue

C. Miehe (ed.), IUTAM Symposium on Computational Mechanics of Solid Materials at Large Strains, 21-32.

integral. Closely related is the rank 1 convexity:

$$f((1-t)A+tB) \le (1-t)f(A)+tf(B)$$

for every $t \in [0,1]$ and every $A, B \in \mathbb{M}^{n\times n}$ with $\operatorname{rank}(A-B) \le 1$. The quasiconvexity implies rank 1 convexity for finite-valued functions. The relaxation theorem asserts that if f is continuous and satisfies suitable coercivity and growth conditions,

$$M = \min\{\bar{I}(u) : u \in \mathcal{A}\}$$

where $\bar{I}$ is the relaxed functional given by

$$\bar{I}(u) = \int_\Omega \mathsf{Q}f(Du)\,dx,$$

where $\mathsf{Q}f$ is the quasiconvex hull of f, i.e., the largest quasiconvex function not exceeding f. One defines the rank 1 convex hull $\mathsf{R}f$ similarly. For finite-valued functions, $\mathsf{Q}f \le \mathsf{R}f$, and it often happens that $\mathsf{Q}f = \mathsf{R}f$.

This paper describes some procedures for finding $\mathsf{R}f$ when f is rotationally invariant, i.e., when $f(A) = f(QAR)$ for all $A \in \mathbb{M}^{n\times n}$ and all Q, R proper orthogonal. If we define $\tilde{f}(\tau) = f(\operatorname{diag}(\tau))$ for any $\tau \in \mathbb{R}^n$ then $\tilde{f}$, called the representation of f, is symmetric and even, i.e., $\tilde{f}(P\tau) = \tilde{f}(\tau) = \tilde{f}(\epsilon\tau)$ for every $\tau \in \mathbb{R}^n$, every permutation matrix P and every diagonal proper orthogonal matrix ϵ. One finds that

$$f(A) = \tilde{f}(\tau)$$

$\tau = (\tau_1, \ldots, \tau_n)$ are the signed singular values of A, defined (see Rosakis [14] and Šilhavý [17]) as the unique $n$tuple such that $\tau_1, \ldots, \tau_{n-1}, |\tau_n|$ are the singular values of A, arranged in a nonincreasing way, and $\operatorname{sgn}\tau_n = \operatorname{sgn}\det A$.

In Šilhavý [17, 18, 19, 22] conditions equivalent to the rank 1 convexity have been given for invariant functions. This paper presents another condition, Theorem 2.1. Based on these conditions, three iterative procedures, with the steps denoted by $\mathsf{T}, \mathsf{U}, \mathsf{V}$, are described for evaluating the rank 1 convex hull of a rotationally invariant function. These are similar to the well-known procedure by Kohn and Strang [8] in their iterative nature, but differ from it in that they are formulated in terms of the representation $\tilde{f}$, which reduces the dimension of the representation space from n^2 to n. The step T consists of two parts, M, S, where M is the evaluation of the "Baker-Ericksen hull", i.e., the greatest function below $\tilde{f}$ that satisfies the Baker-Ericksen inequalities, while S is the evaluation of the infimum of convex combinations of the values of $\tilde{f}$ at pairs corresponding to specific rank 1 perturbations. The procedure U applies to invariant functions defined on $\mathbb{M}^{2\times 2}$ only, and integrates M, S into a single procedure. It is similar to the procedure defined in Šilhavý [20]. Finally, V evaluates the maximum of a family of specific rank 1 convex functions below $\tilde{f}$ on a certain region in the space of signed singular values. This may be also formulated in terms resembling a double dual (Fenchel) transformation for the convex hull known in the theory of convexity.

2. Invariant rank 1 convex functions

This section presents three conditions equivalent to the rank 1 convexity of invariant functions. The procedures T, U, V of Section 3 are in one-to-one correspondence with them.

Let $\mathbb{G}^n = \{\tau \in \mathbb{R}^n : \tau_1 \geq \tau_2 \geq \ldots \geq \tau_{n-1} \geq |\tau_n|\}$ and note that a $\tau \in \mathbb{R}^n$ is an n-tuple of signed singular values of some $A \in \mathbb{M}^{n\times n}$ if and only if $\tau \in \mathbb{G}^n$. For any $\alpha, \beta \in \mathbb{R}^n$ let $\alpha\beta = (\alpha_1\beta_1, \ldots, \alpha_n\beta_n), \alpha^2 = (\alpha_1^2, \ldots, \alpha_n^2)$, and write $\alpha \geq 0$ if $\alpha_1 \geq 0, \ldots, \alpha_n \geq 0$. For each $k \in \{1, \ldots, n\}$ denote by $S^k : \mathbb{R}^n \to \mathbb{R}$ the k-th elementary symmetric function of n variables,

$$S^k(\alpha_1, \ldots, \alpha_n) = \sum_{1\leq i_1 < \cdots < i_k \leq n} \alpha_{i_1} \ldots \alpha_{i_n}$$

and let $S(\alpha) = (S^1(\alpha), \ldots, S^n(\alpha))$. The pair $\alpha, \beta \in \mathbb{G}^n \cap \mathbb{R}^n_+$ is said to satisfy the *bilateral interlacing inequalities* (BIL) if

$$\beta_1 \geq \alpha_2, \quad \alpha_1 \geq \beta_2 \geq \alpha_3, \quad \ldots, \quad \alpha_{n-1} \geq \beta_n.$$

The pair $\alpha, \beta \in \mathbb{G}^n$ is said to satisfy the BIL if $(\alpha_1, \ldots, \alpha_{n-1}, |\alpha_n|)$, $(\beta_1, \ldots, \beta_{n-1}, |\beta_n|)$ satisfy the BIL. If $A = \operatorname{diag}(\alpha), \alpha \in \mathbb{G}^n$ then $\beta \in \mathbb{G}^n$ are the signed singular values of some rank 1 perturbation of $A \Leftrightarrow \alpha, \beta$ satisfy the BIL (Šilhavý [19], Proposition 3.1).

Theorem 2.1 *An invariant $f : \mathbb{M}^{n\times n} \to \mathbb{R}$ is rank 1 convex if and only if it satisfies the following two conditions:*

(i) *if $\alpha, \beta \in \mathbb{G}^n$ satisfy*

$$\prod_{i=1}^{k} \alpha_i \leq \prod_{i=1}^{k} \beta_i, \quad k = 1, \ldots, n-1, \quad \textit{and} \quad \prod_{i=1}^{n} \alpha_i = \prod_{i=1}^{n} \beta_i; \qquad (2)$$

then

$$\tilde{f}(\alpha) \leq \tilde{f}(\beta);$$

(ii) *if $\alpha, \beta \in \mathbb{G}^n$ satisfy the BIL and $\gamma \in \mathbb{G}^n, \epsilon \in \{-1, 1\}^n, t \in [0, 1]$ satisfy $\epsilon(\beta - \alpha) \geq 0$ and*

$$S(\epsilon\gamma) = (1-t)S(\epsilon\alpha) + tS(\epsilon\beta)$$

then

$$\tilde{f}(\gamma) \leq (1-t)\tilde{f}(\alpha) + t\tilde{f}(\beta).$$

The author does not know if the assertion holds for functions with values in $\mathbb{R}\cup\{\infty\}$. As explained in Šilhavý [21], Condition (i) is equivalent to the Baker-Ericksen inequalities if f is of class C^2. In Condition (ii), the requirement $\epsilon(\beta-\alpha)\geq 0$ determines $\epsilon_j\in\{-1,1\}$ uniquely as $\epsilon_j=\operatorname{sgn}(\beta_j-\alpha_j)$ if $\beta_j\neq\alpha_j$ while if $\beta_j=\alpha_j$ then both choices $\epsilon_j=\pm 1$ are possible. Item (ii) shows that the rank 1 line segments in $\mathbb{M}^{n\times n}$ translate into the representation space $\mathbb{G}^n$ as segments of curves with endpoints satisfying the BIL on which the elementary symmetric functions (composed with the sign matrix ϵ) behave like affine functions. For f of class C^1 Theorem 2.1 can be deduced from Šilhavý [18], Theorem 7.2, but the proof in the general case will be given elsewhere.

Theorem 2.2 *(Šilhavý [22], Theorem 5.3)*
Let $f:\mathbb{M}^{2\times 2}\to\mathbb{R}\cup\{\infty\}$ be invariant. Then f is rank 1 convex if and only if

$$\tilde f(\gamma)\leq(1-t)\tilde f(\alpha)+t\tilde f(\beta)$$

for every $\alpha,\beta,\gamma\in\mathbb{G}^2$ and $t\in[0,1]$ such that α,β satisfy the BIL and

$$\left.\begin{aligned}\gamma_1\gamma_2&=(1-t)\alpha_1\alpha_2+t\beta_1\beta_2,\\ \gamma_1+\epsilon\gamma_2&\leq(1-t)(\alpha_1+\epsilon\alpha_2)+t(\beta_1+\epsilon\beta_2),\end{aligned}\right\}\qquad(3)$$

where

$$\epsilon=\begin{cases}+1 & \text{if } (\alpha_1-\beta_1)(\alpha_2-\beta_2)\geq 0,\\ -1 & \text{if } (\alpha_1-\beta_1)(\alpha_2-\beta_2)<0.\end{cases}$$

This result joins Conditions (i), (ii) of Theorem 2.1 into one; the result holds for functions with values in $\mathbb{R}\cup\{\infty\}$.

Theorem 2.3 *(Šilhavý [19], Theorem 2.3)*
Let $f:\mathbb{M}^{n\times n}\to\mathbb{R}$ be invariant. Then f is rank 1 convex if and only if for every $\alpha\in\mathbb{G}^n$ there exists an $s\in\mathbb{R}^n$ satisfying

$$\left.\begin{aligned}&(s_i\alpha_i-s_j\alpha_j)(\alpha_i-\alpha_j)\geq 0,\\ &\alpha_i=\alpha_j\quad\Rightarrow\quad s_i=s_j,\\ &\alpha_i=-\alpha_j\quad\Rightarrow\quad s_i=-s_j,\end{aligned}\right\}\qquad(4)$$

$i\neq j$, such that for every $\beta\in\mathbb{G}^n,\epsilon\in\{-1,1\}^n$ satisfying the BIL and $\epsilon(\beta-\alpha)\geq 0$ we have

$$\tilde f(\beta)-\tilde f(\alpha)\geq\sum_{j=1}^n s_j(\beta_j-\alpha_j)\prod_i\frac{\epsilon_i\beta_i-\epsilon_j\alpha_j}{\epsilon_i\alpha_i-\epsilon_j\alpha_j}$$

with the product taken over all i for which the denominator is nonzero.

This provides a condition for the rank 1 convexity in a "subgradient form", with $s \in \mathbb{R}^n$ playing the role of the (nonlinear) subgradient. Applies only to finite-valued functions.

3. Rank 1 convexifications of invariant functions

Based on the conditions of the preceding section, three limiting procedures $\mathsf{T}, \mathsf{U}, \mathsf{V}$ are given for determining $\mathsf{R}f$. Recall that by the definition in Section 1, the rank 1 convex functions have their values in $\mathbb{R} \cup \{\infty\}$; thus to avoid the case when $\mathsf{R}f$ takes the value(s) $-\infty$, we have to impose the condition that f be bounded from below by a rank 1 convex function. Then if $f : \mathbb{M}^{n\times n} \to \mathbb{R} \cup \{\infty\}$ is invariant, also $\mathsf{R}f$ is invariant, and we denote by $\mathsf{R}\tilde{f}$ its representation.

To describe the first procedure, for each symmetric even function $\tilde{f} : \mathbb{R}^n \to \mathbb{R}$ denote by $\mathsf{M}\tilde{f} : \mathbb{R}^n \to \mathbb{R}$ the symmetric even function defined by

$$\mathsf{M}\tilde{f}(\alpha) = \inf\{\tilde{f}(\beta) : \beta \in \mathcal{R}(\alpha)\}, \quad \alpha \in \mathbb{G}^n,$$

where the infimum is taken over the set $\mathcal{R}(\alpha)$ of all $\beta \in \mathbb{G}^n$ satisfying (2). This is the largest function below $\tilde{f}$ that satisfies the general form of the Baker-Ericksen inequalities. It is noted that it suffices to define $\mathsf{M}\tilde{f}$ only for $\alpha \in \mathbb{G}^2$ since the function $\mathsf{M}\tilde{f}$ has a unique symmetric and even extension to $\mathbb{R}^n$. For each $\tilde{f} : \mathbb{R}^n \to \mathbb{R}$ define a symmetric even function $\mathsf{S}\tilde{f} : \mathbb{R}^n \to \mathbb{R}$ by

$$\mathsf{S}\tilde{f}(\alpha) = \inf\{(1-t)f(\beta) + tf(\gamma) : (t,\beta,\gamma) \in \mathcal{A}(\alpha)\}$$

$\alpha \in \mathbb{G}^n$, where the infimum is taken over the set $\mathcal{A}(\alpha)$ of all (t,β,γ) such that β, γ satisfy the BIL, $0 \le t \le 1$, and

$$S(\epsilon\alpha) = (1-t)S(\epsilon\beta) + tS(\epsilon\gamma)$$

where

$$\epsilon_j = \begin{cases} 1 & \text{if } \beta_j - \alpha_j \ge 0, \\ -1 & \text{if } \beta_j - \alpha_j < 0. \end{cases} \tag{5}$$

We finally set $\mathsf{T}\tilde{f} = \mathsf{SM}\tilde{f}$.

Proposition 3.1 *If $f : \mathbb{M}^{n\times n} \to \mathbb{R}$ is invariant and bounded from below by a rank 1 convex function then*

$$\mathsf{R}\tilde{f} = \lim_{k\to\infty} \mathsf{T}^k \tilde{f}. \tag{6}$$

Here and in what follows, the limits as in (6) are understood in the pointwise sense on $\mathbb{G}^n$.

Proof. Using Theorem 2.1 we find that $\mathsf{R}\tilde{f} \le \mathsf{M}\tilde{f}$ and $\mathsf{R}\tilde{f} \le \mathsf{SM}\tilde{f} = \mathsf{T}\tilde{f}$ and hence, by induction, $\mathsf{R}\tilde{f} \le \mathsf{T}^k\tilde{f} \le \tilde{f}$. Since the sequence $\mathsf{T}^k\tilde{f}$ is nonincreasing and bounded from below the limit $\tilde{l}$ of the right-hand side of (6) exists. The proof will be complete if we show that $\tilde{l}$ satisfies Conditions (i), (ii) of Theorem 2.1. From $\mathsf{T}^{k+1}\tilde{f} = \mathsf{T}\mathsf{T}^k\tilde{f}$ we find that $\tilde{l} = \mathsf{T}\tilde{l} \equiv \mathsf{SM}\tilde{l}$. Hence $\tilde{l} \le \mathsf{M}\tilde{l}$ and $\tilde{l} \le \mathsf{S}\tilde{l}$. These inequalities are easily seen to imply that $\tilde{l}$ satisfies Conditions (i), (ii) of Theorem 2.1. □

For $n = 2$ we define a procedure U, where for any symmetric even $\tilde{f} : \mathbb{R}^2 \to \mathbb{R} \cup \{\infty\}$ $\mathsf{U}\tilde{f}$ is a symmetric even function on $\mathbb{R}^2$ defined by

$$\mathsf{U}\tilde{f}(\gamma) = \inf\{(1-t)\tilde{f}(\alpha) + t\tilde{f}(\beta) : (t, \alpha, \beta) \in \mathcal{C}(\gamma)\}$$

$\gamma \in \mathbb{G}^2$. Here the infimum is taken over the set $\mathcal{C}(\gamma)$ of all (t, α, β) such that $\alpha, \beta \in \mathbb{G}^2$ satisfy the BIL, $t \in [0, 1]$ and (3) holds.

Proposition 3.2 *If* $f : \mathbb{M}^{2\times 2} \to \mathbb{R} \cup \{\infty\}$ *is invariant and bounded from below by a rank 1 convex function then*

$$\mathsf{R}\tilde{f} = \lim_{k\to\infty} \mathsf{U}^k\tilde{f}.$$

In contrast to Proposition 3.1, the above result holds for functions with values in $\mathbb{R} \cup \{\infty\}$.

Proof. This is based on Theorem 2.2 but otherwise is similar to the proof of Proposition 3.1. □

To describe the third procedure, for each $\alpha \in \mathbb{G}^n$ denote by $\mathcal{D}(\alpha)$ the set of all $\beta \in \mathbb{G}^n$ such that β, α satisfy the bilateral interlacing inequalities and by $\mathcal{E}(\alpha)$ the set of all $s \in \mathbb{R}^n$ satisfying (4). For each $\alpha \in \mathbb{G}^n$, and $s \in \mathcal{E}(\alpha)$ let $\psi_{\alpha,s} : \mathcal{D}(\alpha) \to \mathbb{R}$ be defined by

$$\psi_{\alpha,s}(\beta) = \sum_{j=1}^{n} s_j(\beta_j - \alpha_j) \prod_i \frac{\epsilon_i\beta_i - \epsilon_j\alpha_j}{\epsilon_i\alpha_i - \epsilon_j\alpha_j}, \quad \beta \in \mathcal{D}(\alpha),$$

where ϵ is given by (5) and the product is taken over all i for which the denominator is nonzero. For each symmetric even $\tilde{f} : \mathbb{R}^n \to \mathbb{R}$ let $\mathsf{V}\tilde{f} : \mathbb{R}^n \to \mathbb{R}$ be the symmetric even function defined by

$$\mathsf{V}\tilde{f}(\alpha) = \sup\{a \in \mathbb{R} : a + \psi_{\alpha,s} \le \tilde{f} \text{ on } \mathcal{D}(\alpha) \text{ for some } s \in \mathcal{E}(\alpha)\} \qquad (7)$$

if $\alpha \in \mathbb{G}^n$.

Proposition 3.3 *Let $f : \mathbb{M}^{n\times n} \to \mathbb{R}$ be invariant and bounded from below by a rank 1 convex function. Then*

(i) *the supremum in* (7) *is attained;*

(ii) *we have*

$$\mathsf{R}\tilde{f} = \lim_{k\to\infty} \mathsf{V}^k \tilde{f}. \tag{8}$$

Proof. (i): Let $\alpha \in \mathbb{G}^n$. Directly from the definition, $\tilde{f}(\alpha) \geq \mathsf{V}\tilde{f}(\alpha)$. Next prove $\mathsf{V}\tilde{f}(\alpha) \geq \mathsf{R}\tilde{f}(\alpha)$. We combine the rank 1 convexity of $\mathsf{R}f$ with Theorem 2.3 to note that there exists an $s \in \mathcal{E}(\alpha)$ such that

$$\tilde{f} \geq \mathsf{R}\tilde{f} \geq \mathsf{R}\tilde{f}(\alpha) + \psi_{\alpha,s} \text{ on } \mathcal{D}(\alpha).$$

Thus $\mathsf{V}\tilde{f}(\alpha) \geq \mathsf{R}\tilde{f}(\alpha)$. The definition of $\mathsf{V}\tilde{f}$ furthermore gives that there exist sequences $s^k \in \mathbb{R}^n$, $a^k \in \mathbb{R}$, $k = 1, \ldots,$ such that

$$\tilde{f} \geq a^k + \psi_{\alpha,s^k} \text{ on } \mathcal{D}(\alpha) \tag{9}$$

and $a^k \to \mathsf{V}\tilde{f}(\alpha) \in \mathbb{R}$. Prove that the sequence s^k is bounded. Assume, for simplicity, that α is in the interior of $\mathbb{G}^n$ and $\alpha_{n-1} > 0$. Then there exists a $\delta > 0$ such that $\beta^+_{(\gamma)} := (\alpha_1, \alpha_2, \ldots, \alpha_\gamma + \delta, \ldots, \alpha_n)$ and $\beta^-_{(\gamma)} := (\alpha_1, \alpha_2, \ldots, \alpha_\gamma - \delta, \ldots, \alpha_n)$ are in $\mathcal{D}(\alpha)$. Inequality (9) yields

$$\tilde{f}(\beta^+_{(\gamma)}) \geq s^k_\gamma \delta + a^k, \quad \tilde{f}(\beta^-_{(\gamma)}) \geq -s^k_\gamma \delta + a^k,$$

and the boundedness of a^k yields the boundedness of $s^k \in \mathbb{R}^n$. Thus $s^k \to s_0 \in \mathbb{R}^n$ for a subsequence. We have $s_0 \in \mathcal{E}(\alpha)$ and it is also easy to see that $a^k + \psi_{\alpha,s^k} \to a_0 + \psi_{\alpha,s_0}$ on $\mathcal{D}(\alpha)$ where $a_0 = \mathsf{V}\tilde{f}(\alpha)$. Thus the supremum in (7) is attained on $a = a_0$ and the corresponding s is s_0. (ii): By iterating the inequality $\mathsf{V}\tilde{f} \geq \mathsf{R}\tilde{f}$ we obtain $\mathsf{V}^k\tilde{f} \geq \mathsf{R}\tilde{f}$. Since the sequence $\mathsf{V}^k\tilde{f}$ is nonincreasing and bounded from below the limit $\tilde{l}$ of the right-hand side of (8) exists and satisfies $\tilde{l} = \mathsf{V}\tilde{l}$. Applying the attainment assertion of (i) to $\tilde{l}$, we see that with $a = \mathsf{V}\tilde{l}(\alpha) \equiv \tilde{l}(\alpha)$ we have $\tilde{l} \geq a + \psi_{\alpha,s}$ on $\mathcal{D}(\alpha)$ for some $s \in \mathcal{E}(\alpha)$. By Theorem 2.3 this guarantees that $\tilde{l}$ is rank 1 convex. □

The step V can be calculated as a double maximization in a way similar to the double Fenchel transformation. For each $\tilde{f} : \mathbb{R}^n \to \mathbb{R}$ and $\alpha \in \mathbb{G}^n$ define a function $\tilde{f}^*_\alpha : \mathcal{E}(\alpha) \to \mathbb{R}$ by

$$\tilde{f}^*_\alpha(s) := \sup\{\psi_{\alpha,s}(\beta) - \tilde{f}(\beta) : \beta \in \mathcal{D}(\alpha)\}, \quad s \in \mathcal{E}(\alpha),$$

and a symmetric even function $\tilde{f}^{**} : \mathbb{R}^n \to \mathbb{R}$ by

$$\tilde{f}^{**}(\alpha) := \sup\{-\tilde{f}^*_\alpha(s) : s \in \mathcal{E}(\alpha)\}, \quad \alpha \in \mathbb{G}^n.$$

Proposition 3.4 *Let f be an invariant function, bounded from below by a rank 1 convex function. Then*

$$\tilde{f}^{**}(\alpha) = \mathsf{V}\tilde{f}(\alpha), \quad \alpha \in \mathbb{G}^n.$$

Proof. Let $\alpha \in \mathbb{G}^n$. By Proposition 3.3(i), there exist $a_0 \in \mathbb{R}$, $s_0 \in \mathcal{E}(\alpha)$ such that

$$\tilde{f}(\beta) \geq a_0 + \psi_{\alpha,s_0}(\beta), \quad \beta \in \mathcal{D}(\alpha) \tag{10}$$

and $\mathsf{V}\tilde{f}(\alpha) = a_0$. By (10) then $-a_0 \geq \tilde{f}^*_\alpha(s_0)$ and from that $a_0 \leq \tilde{f}^{**}(\alpha)$. To prove the converse inequality, we note that $\tilde{f}(\beta) \geq \psi_{\alpha,s}(\beta) - \tilde{f}^*_\alpha(s)$ for each $s \in \mathcal{E}(\alpha)$ and $\beta \in \mathcal{D}(\alpha)$ and thus the definition (7) of $a_0 = \mathsf{V}\tilde{f}(\alpha)$ gives $a_0 \geq -\tilde{f}^*_\alpha(s)$ for each $s \in \mathcal{E}(\alpha)$ which implies $a_0 \geq \tilde{f}^{**}(\alpha)$. □

4. Outlook

The iterative nature of the procedures $\mathsf{T}, \mathsf{U}, \mathsf{V}$ presents an obvious difficulty in analytical computations. This drawback should not be exaggerated: also the calculations of the convex hull leads to a closed-form formula only exceptionally. Yet the procedure is well understood, and, if needed, can be implemented numerically. In the case of the rank 1 convexification procedures, the numerical implementation remains an open problem at the moment. Therefore, to explain the role and meaning of $\mathsf{T}, \mathsf{U}, \mathsf{V}$, this section discusses some of the known examples of explicit relaxations of invariant functions. In all of them, the procedures terminate in the first or the second step, say $\mathsf{T}f$ or $\mathsf{T}^2 f$, and the proof that this already gives $\mathsf{R}f$ proceeds via showing that $\mathsf{T}f$ or $\mathsf{T}^2 f$ are polyconvex functions. *Thus in all the examples below we have* $\mathsf{R}f = \mathsf{Q}f = \mathsf{P}f$. No proofs are given in this section.

The first three examples require two steps.

Example 4.1 (Optimal design, Kohn and Strang [8])
Let $f : \mathbb{M}^{2\times 2} \to \mathbb{R}$ be given by

$$f(A) = \begin{cases} 0 & \text{if } A = 0, \\ 1 + |A|^2 & \text{if } A \neq 0. \end{cases} \tag{11}$$

Then

$$\mathsf{R}f(A) = \begin{cases} 2\sqrt{|A|^2 + 2|\det A|} - 2|\det A| & \text{if } |A|^2 + 2|\det A| \leq 1, \\ 1 + |A|^2 & \text{if } |A|^2 + 2|\det A| \geq 1. \end{cases}$$

The reference (Kohn and Strang [8]) deals with f as in (11) on $\mathbb{M}^{m\times 2}$, $m \geq 2$, but this can be derived from the 2 by 2 case. It turns out that in this example,

$$\mathsf{R}f = \mathsf{S}^2 f = \mathsf{T}^2 f = \mathsf{U}^2 f, \tag{12}$$

so the procedures terminate after the second step. Actually, it is not even necessary to apply the procedure S in its full strength: it suffices to apply the convexification in the direction of the coordinate axes in the representation space $\mathbb{G}^2$.

Example 4.2 (St. Venant-Kirchhoff material, Le Dret and Raoult [9, 10])
Let $f : \mathbb{M}^{2\times 2} \to \mathbb{R}$ be represented by

$$\tilde{f}(\tau) = (\tau_1^2 - 1)^2 + (\tau_2^2 - 1)^2 + \frac{\nu}{1-2\nu}(\tau_1^2 + \tau_2^2 - 2)^2,$$

$\tau \in \mathbb{G}^2$, where $\nu \in [0, \frac{1}{2})$. Then

$$\mathsf{R}\tilde{f}(\tau) = \begin{cases} 0 & \text{if } \tau_1 \le 1, \\ \dfrac{1}{1-\nu}(\tau_1 - 1)^2 & \text{if } \tau_1 \ge 1,\ \nu\tau_1^2 + (1-\nu)\tau_2^2 \le 1, \\ \tilde{f}(\tau) & \text{otherwise,} \end{cases}$$

$\tau \in \mathbb{G}^2$. Here $\mathsf{C}f$ if the convex hull of f. We have again (12), and in each step, the convexification along the coordinate axes in $\mathbb{G}^2$ suffices.

Example 4.3 (A "double well potential", Šilhavý [20])
Let $f : \mathbb{M}^{2\times 2} \to \mathbb{R}$ be given by

$$f(A) = \min\{f_1(A), f_2(A)\}, \quad A \in \mathbb{M}^{2\times 2},$$

where $f_1, f_2 : \mathbb{M}^{2\times 2} \to \mathbb{R}$ are represented by

$$\tilde{f}_1(\tau) = (\tau_1 - a)^2 + (\tau_2 - a)^2, \quad \tilde{f}_2(\tau) = (\tau_1 - b)^2 + (\tau_2 - b)^2,$$

$\tau \in \mathbb{G}^2$, where $0 < a < b$. Then

$$\mathsf{R}f(A) = \begin{cases} a^2 - 2\det A & \text{if } s(A) \le a, \\ f_1(A) & \text{if } a \le s(A) \le \frac{1}{2}(3a+b), \\ (a+b)s(A) + c - 2\det A & \text{if } \frac{1}{2}(3a+b) \le s(A) \le \frac{1}{2}(a+3b), \\ f_2(A) & \text{if } s(A) \ge \frac{1}{2}(a+3b) \end{cases}$$

where $c = a^2 - \frac{1}{4}(a+b)(5a+b)$, and $s(A) = \tau_1 + \tau_2$, where τ are the signed singular values of A. Here,

$$\mathsf{R}f = \mathsf{T}^2 f = \mathsf{U}^2 f.$$

but in contrast to the preceding two examples, also M must be applied.

In the remaining three examples one application of M results in $\mathsf{R}f$, i.e.,

$$\mathsf{R}f = \mathsf{M}f. \tag{13}$$

Example 4.4 (The square of the distance from $SO(2)$, Šilhavý [20]) Let $f : \mathbb{M}^{2\times 2} \to \mathbb{R}$ be represented by

$$\tilde{f}(\tau) = (\tau_1 - 1)^2 + (\tau_2 - 1)^2, \quad \tau \in \mathbb{G}^2.$$

Then

$$\mathsf{R}f(A) = \begin{cases} 1 - 2\det A & \text{if } \sqrt{|A|^2 + 2\det A} \leq 1, \\ f(A) & \text{if } \sqrt{|A|^2 + 2\det A} \geq 1. \end{cases}$$

An easy application of the trace inequality shows that

$$f(A) = \operatorname{dist}^2(A, SO(2)).$$

In the following two examples, let $\gamma \in \mathbb{G}^n \cap \mathbb{R}^n_{++}$, $n = 2, 3$, be a fixed element with $\gamma_1 \ldots \gamma_n = 1$, and for any $\tau \in \mathbb{R}^n$ we abbreviate $\bar{\tau} := (\bar{\tau}_1, \ldots, \bar{\tau}_n)$ where $\bar{\tau}_i := \tau_i/\gamma_i$; moreover, we set $\det \tau := \tau_1 \cdots \tau_n$.

Example 4.5 (Nematic elastomers, Šilhavý [23], Theorem 3.2 and Section 4) Let $\tilde{f} : \mathbb{M}^{3\times 3}_+ \to \mathbb{R} \cup \{\infty\}$ be represented by

$$\tilde{f}(\tau) = h(\det \tau) + T(\bar{\tau}_1^3, \bar{\tau}_2^3, \bar{\tau}_3^3) \tag{14}$$

if $\tau \in \mathbb{G}^3$, $\det \tau > 0$ and $\tilde{f}(\tau) = \infty$ otherwise, where $h : \mathbb{R}_{++} \to \mathbb{R} \cup \{\infty\}$ is convex, T is a convex, symmetric, and strictly increasing function of class C^1 on $\mathbb{R}^n_{++}$. Then

$$\mathsf{R}\tilde{f}(\tau) = \begin{cases} h(\delta) + T(\delta, \delta, \delta) & \text{if } \bar{\tau}_1^3 \leq \delta \leq \bar{\tau}_3^3, \\ h(\delta) + T((\bar{\tau}_1\bar{\tau}_2)^{3/2}, (\bar{\tau}_1\bar{\tau}_2)^{3/2}, \bar{\tau}_3^3) & \text{if } \delta \geq \bar{\tau}_3^3, \bar{\tau}_2 \geq \bar{\tau}_1, \\ h(\delta) + T(\bar{\tau}_1^3, (\bar{\tau}_2\bar{\tau}_3)^{3/2}, (\bar{\tau}_2\bar{\tau}_3)^{3/2}) & \text{if } \bar{\tau}_1^3 \geq \delta, \bar{\tau}_2 \leq \bar{\tau}_3, \\ h(\delta) + T(\bar{\tau}_1^3, \bar{\tau}_2^3, \bar{\tau}_3^3) & \text{if } \bar{\tau}_1 \geq \bar{\tau}_2 \geq \bar{\tau}_3, \end{cases} \tag{15}$$

if $\det \tau > 0$ and $\mathsf{R}\tilde{f}(\tau) = \infty$ otherwise, where $\delta = \det \tau$.
If the repulsive potential h satisfies $h'(1) = -3T_1(1,1,1)$, the function f has a minimum at the isotropic well

$$SO(3)\operatorname{diag}(\gamma)SO(3),$$

which makes it different from the energies of rubbers, where the minimum is on $SO(3)$. As noted in DeSimone and Dolzmann [7], with $\gamma_1 = \gamma_2$, the scaled energies of the type (14) can model the nematic phase in some elastomers (Bladon et al. [4]). The relaxed energy (15) displays four regimes. In $(15)_1$ the microstructure leads to a full liquid-like behavior; this occurs if the singular values of the effective deformation are close to each other. In $(15)_{2,3}$ the effective behavior is mixed solid/liquid-like; if $\gamma_1 = \gamma_2$ then the region $(15)_2$ vanishes. Finally, in $(15)_4$ the phase displays a solid behavior, occurring if the singular values are sufficiently apart. The hypotheses cover

$$\tilde{f}(\tau) = \frac{\tau_1^p}{\gamma_1^p} + \frac{\tau_2^p}{\gamma_2^p} + \frac{\tau_3^p}{\gamma_3^p} + h(\det \tau),$$

$p \geq 3$. There is also an analog of the result for the incompressible case (Šilhavý [23], Theorem 3.2); in that case the hypotheses cover

$$\tilde{f}(\tau) = \begin{cases} \dfrac{\tau_1^p}{\gamma_1^p} + \dfrac{\tau_2^p}{\gamma_2^p} + \dfrac{\tau_3^p}{\gamma_3^p} & \text{if } \tau_1\tau_2\tau_3 = 1, \\ \infty & \text{otherwise,} \end{cases}$$

$p \geq 2$; this case is considered in DeSimone and Dolzmann [7].

Example 4.6 (Šilhavý [23])
For f represented by

$$\tilde{f}(\tau) = (\bar{\tau}_1 - 1)^2 + (\bar{\tau}_2 - 1)^2,$$

if $\det \tau > 0$ and $\tilde{f}(\tau) = \infty$ if $\det \tau \leq 0$, we have

$$\mathsf{R}\tilde{f}(\tau) = \begin{cases} \tilde{f}(\tau) & \text{if } \bar{\tau}_1 \geq \bar{\tau}_2,\ \bar{\tau}_1 + \bar{\tau}_2 \geq 1, \\ 2(\sqrt{\tau_1\tau_2} - 1)^2 & \text{if } \bar{\tau}_1 \leq \bar{\tau}_2,\ \tau_1\tau_2 \geq \dfrac{1}{4}, \\ 1 - 2\tau_1\tau_2 & \text{otherwise,} \end{cases}$$

if $\det \tau > 0$ and $\mathsf{R}\tilde{f}(\tau) = \infty$ if $\det \tau \leq 0$. We have (13) again. The function f can be interpreted as the square of a modified distance from $SO(2)\,\mathrm{diag}(\gamma)SO(2)$.

Acknowledgments

This research was supported by Grant 201/00/1516 of the Czech Republic.

References

[1] Ball, J.M. (1977). Convexity conditions and existence theorems in nonlinear elasticity. *Arch. Rational Mech. Anal.*, **63**, 337–403.

[2] Ball, J.M. and James, R.D. (1987). Fine phase mixtures as minimizers of energy. *Arch. Rational Mech. Anal.*, **100**, 13–52.

[3] Ball, J.M. and James, R.D. (1993). Proposed experimental tests of a theory of fine microstructure and the two-well problem. *Phil. Trans. Royal Soc. Lond.* **338**, 389–450.

[4] Bladon, P., Terentjev, E.M. and Warner, M. (1993). Transitions and instabilities in liquid-crystal elastomers. *Phys. Rev.* **E 47**, R3838–R3840.

[5] Chipot, M. and Kinderlehrer, D. (1988). Equilibrium configurations of crystals. *Arch. Rational Mech. Anal.* **103**, 237–277.

[6] Dacorogna, B. (1990). *Direct methods in the calculus of variations*, Springer, Berlin.

[7] DeSimone, A. and Dolzmann, G. (2001). Macroscopic response of nematic elastomers via relaxation of a class of $SO(3)$–invariant energies. Preprint No. 41 Max-Planck Institute, Leipzig.

[8] Kohn, R.V. and Strang, G. (1986). Optimal design and relaxation of variational problems, I, II, III. *Comm. Pure Appl. Math.*, *39*, 113–137, 139–182, 353–377.

[9] Le Dret, H. and Raoult, A. (1994). Envelope quasi-convexe de la densité d'énergie de Saint Venant-Kirchhoff. *C. R. Acad. Sci. Paris*, **318**, 93–98.

[10] Le Dret, H. and Raoult, A. (1995). The quasiconvex envelope of the Saint Venant-Kirchhoff stored energy function. *Proc. Royal Soc. Edinburgh*, **125A**, 1179–1192.

[11] Morrey Jr., C.B. (1966). *Multiple integrals in the calculus of variations*, Springer, New York.

[12] Müller, S. (1999). Variational models for microstructure and phase transitions. In *Calculus of variations and geometric evolution problems (Cetraro, 1996) Lecture notes in Math. 1713*, Berlin, Springer, 85–210.

[13] Pedregal, P. (1997). *Parameterized measures and variational principles*, Birkhäuser, Basel.

[14] Rosakis, P. (1997). Characterization of convex isotropic functions. *J. Elasticity*, **49**, 257–267.

[15] Roubíček, T. (1997). *Relaxation in optimization theory and variational calculus*, W. de Gruyter, Berlin.

[16] Šilhavý, M. (1997). *The mechanics and thermodynamics of continuous media*, Springer, Berlin.

[17] Šilhavý, M. (1999). Convexity conditions for rotationally invariant functions in two dimensions. In *Applied Nonlinear Analysis*, A. Sequeira et al. (ed.), Kluwer Academic, New York, 513–530.

[18] Šilhavý, M. (1999). On isotropic rank 1 convex functions. *Roy Soc. Edinburgh*, **129A**, 1081–1105.

[19] Šilhavý, M. (2001). Rotationally invariant rank 1 convex functions. *Applied Mathematics & Optimization*, **44**, 1–15.

[20] Šilhavý, M. (2001). Rank 1 Convex hulls of isotropic functions in dimension 2 by 2. *Mathematica Bohemica*, **126**, 521–529.

[21] Šilhavý, M. (2001). Monotonicity of rotationally invariant convex and rank 1 convex functions. *Proc. Royal Soc. Edinburgh.* To appear.

[22] Šilhavý, M. (2001). On the semiconvexity properties of rotationally invariant functions in two dimensions. To be published.

[23] Šilhavý, M. (2001). Relaxation in a class of $SO(n)$–invariant energies related to nematic elastomers. To be published.

Evolution of Rate-Independent Inelasticity with Microstructure using Relaxation and Young Measures

A. Mielke
Mathematisches Institut A, Universität Stuttgart
Pfaffenwaldring 57, 70569 Stuttgart, Germany
mielke@mathematik.uni-stuttgart.de

Abstract: We use the energetic formulation for rate-independent inelasticity which is based in the dissipation distance on the internal state. From this the incremental problems inherit a variational form which can be used to derive suitable relaxations. As a result we obtain a similar energetic formulation on the set of Young measures which then describes the evolution of microstructure.

Keywords: Rate-independent inelasticity, finite-strain elastoplasticity, dissipation distance, quasi-convexity, relaxation, Young measures.

1. Introduction

The theory of elastoplasticity at finite strain has undergone a rapid development in the last few decades, see for instance Simo and Ortiz [25], Simo [24], Miehe and Stein [9], Miehe [7], Ortiz and Stainier [20]. It can be understood as a special case of general inelastic material behavior (cf. Ziegler and Wehrli [27], Miehe [8]). For this general setting of inelasticity a new energetic formulation was introduced in Mielke et al. [16, 17], Carstensen et al. [3], Mielke [11]. This framework has the major advantage that it does not make any assumption on the smoothness of the processes to be described; hence, it is particularly suited to describe systems where formation of microstructure takes place. We note that this phenomenon can be seen as a phase transformation where the deformation gradients as well as the internal variables have jumps in space and time. To minimize the macroscopic energy in such situations the systems develop spatial oscillations on microscopic scales which need to be described by macroscopic quantities. In general, simple averages are not enough and therefore we use Young measures to describe these oscillations. The measures describe for each material point (or representative volume element) the joint distribution of the deformation gradient and the internal variables. The major task is then to find an evolution equation for these Young measures. Here we derive such equations via relaxation of the original model. Thus we provide a rather abstract evolution law which has all mathematically desirable properties but which is very complicated

C. Miehe (ed.), IUTAM Symposium on Computational Mechanics of Solid Materials at Large Strains, 33-44.

to use in practice. For more details, in particular to the specific application to elastoplasticity at finite strain, we refer to Mielke [13, 11, 12].

2. Rate-independent inelasticity

We consider a body $\Omega \subset \mathbb{R}^d$ which undergoes a deformation $\varphi : \Omega \mapsto \mathbb{R}^d$ such that the deformation gradient $\mathbf{F}(x) = \mathrm{D}\varphi(x) \in \mathbb{R}^{d\times d}$ lies in $\mathrm{GL}_+(\mathbb{R}^d) = \{\, \mathbf{F} \in \mathbb{R}^{d\times d} \mid \det \mathbf{F} > 0 \,\}$. Additionally there is a vector-valued internal variable $z = (z^1, \ldots, z^n) \in Z \subset \mathbb{R}^n$ which describes the properties of the material. The elastic properties are given by the dependence of the elastic potential ψ on $(x, \mathbf{F}, z)$ via $\psi = \widehat{\psi}(x, \mathbf{F}, z)$. For fixed (x, z) we assume that the function $\widehat{\psi}(x, \cdot, z)$: $\mathrm{GL}_+(\mathbb{R}^d) \mapsto [0, \infty)$ is coercive (i.e., $\widehat{\psi}(x, \mathbf{F}, z) \to \infty$ if $(\det \mathbf{F})^{-1} + |\mathbf{F}| \to \infty$) and quasicconvex. In addition, we describe boundary conditions and external loadings such that the total energy of a given state $(\varphi, z) : \Omega \mapsto \mathbb{R}^d \times \mathbb{R}^n$ at time t is

$$\mathcal{E}(t, \varphi, z) = \int_\Omega \widehat{\psi}(x, \mathrm{D}\varphi(x), z(x))\,\mathrm{d}x - \langle \ell(t), \varphi \rangle,$$

where $\ell(t)$ denotes the external loading and φ lies in the set $\mathcal{F}$ of admissible deformations

$$\mathcal{F} = \{\, \varphi : \Omega \mapsto \mathbb{R}^d \mid \varphi|_{\Gamma_{\mathrm{Dir}}} = \varphi_{\mathrm{Dir}},\ \mathrm{D}\varphi(x) \in \mathrm{GL}_+(\mathbb{R}^d) \text{ on } \Omega \,\}.$$

A basic assumption of our rate-independent model will be that a solution process has to be in a stable elastic equilibrium for all $t \in [0, T]$, i.e. $\varphi(t, \cdot) : \Omega \mapsto \mathbb{R}^d$ is a (local) minimizer of $\mathcal{E}(t, \cdot, z(t, \cdot))$ on $\mathcal{F}$. Changes of the internal variables during a slow loading or unloading process will give rise to internal friction which dissipates energy via the dissipation rate

$$\Delta = \widehat{\Delta}(x, z, \dot{z}) \geq 0 \text{ where } \dot{z}(t, x) = \tfrac{\partial}{\partial t} z(t, x).$$

It is assumed that there are no other dissipation mechanisms in the model. We call $\Delta : \Omega \times \mathrm{T}Z \mapsto [0, \infty]$ the dissipation potential (Ziegler and Wehrli [27]). Rate-independency is obtained by assuming homogeneity in $\dot{z}$ of degree 1, namely $\widehat{\Delta}(x, z, \alpha\dot{z}) = \alpha\widehat{\Delta}(x, z, \dot{z})$ for $\alpha \geq 0$. Furthermore, we assume that $\widehat{\Delta}(x, z, \cdot) : \mathrm{T}_z Z \mapsto [0, \infty]$ is convex and that $\widehat{\Delta}$ satisfies $\widehat{\Delta}(x, z, v) \geq c|v|$ for some $c > 0$. Considering a process $z : [0, T] \times \Omega \mapsto Z$ the dissipation on an interval $[t_0, t_1]$ is then

$$\mathrm{Diss}(z; [t_0, t_1]) = \int_{t_0}^{t_1} \int_\Omega \Delta(x, z(t, x), \dot{z}(t, x))\,\mathrm{d}x\,\mathrm{d}t.$$

For each $x \in \Omega$ the dissipation potential $\widehat{\Delta}(x, \cdot, \cdot)$ defines a distance metric on Z via

$$\widehat{D}(x; z_0, z_1) = \inf\{\, \int_0^1 \widehat{\Delta}(x, z(s), \dot{z}(s))\,\mathrm{d}s \mid z \in \mathrm{C}^1([0, 1], Z), z(0) = z_0, z(1) = z_1 \,\}.$$

This defines also a metric on the set of internal states $\mathcal{Z} = \{ z : \Omega \mapsto \mathbb{R}^n \mid z \text{ measurable} \}$ by setting $\mathcal{D}(z_0, z_1) = \int_\Omega \widehat{D}(x; z_0(x), z_1(x))\, \mathrm{d}x$ for $z_0, z_1 \in \mathcal{Z}$.

3. Energetic formulation of the model

Our model is completely described by the two constitutive relations $\psi = \widehat{\psi}(x, \mathbf{F}, z)$ and $\Delta = \widehat{\Delta}(x, z, \dot{z})$ and by the loadings $\ell(t)$.

Definition 3.1 *A process $(\varphi, z) : [0, T] \mapsto \mathcal{F}\times\mathcal{Z}$ is called a solution of the above rate-independent model, if (S) and (E) hold:*

(S) [Stability] *For all $t \in [0, T]$ we have*

$$\mathcal{E}(t, \varphi(t), z(t)) \leq \mathcal{E}(t, \widehat{\varphi}, \widehat{z}) + \mathcal{D}(z(t), \widehat{z}) \quad \textit{for all} \quad (\widehat{\varphi}, \widehat{z}) \in \mathcal{F}\times\mathcal{Z}$$

(E) [Energy inequality] *For all t_0, t_1 with $0 \leq t_0 < t_1 \leq T$ we have*

$$\mathcal{E}(t_1, \varphi(t_1), z(t_1)) + \mathrm{Diss}(z, [t_0, t_1]) \leq \mathcal{E}(t_0, \varphi(t_0), z(t_0)) - \int_{t_0}^{t_1} \langle \dot{\ell}(t), \varphi(t) \rangle \mathrm{d}t$$

The stability condition (S) has a clear mechanical interpretation. Letting $\widehat{z} = z(t)$ we have $\mathcal{D}(z(t), \widehat{z}) = 0$ and the condition implies that $\varphi(t)$ is the global minimizer of $\mathcal{E}(t, \cdot, z(t))$ on $\mathcal{F}$, which gives the elastic equilibrium. Moreover, changing the internal variable from $z(t)$ to $\widehat{z}$ (and adjusting $\widehat{\varphi}$ optimally) the dissipation must be at least as large as the elastic energy release. The internal variable z will change (and dissipate energy) as soon as the elastic energy release is large enough to compensate for the dissipation. Thus, (S) is equivalent to a principle of maximal dissipation. The energy inequality (E) has an obvious interpretation, since the work of the external forces is given by the last term. This abstract setting has applications in many rate-independent continuum models. We refer to Mielke et al. [16, 17], Kuczma et al. [6] and Govindjee et al. [4] for applications to phase transformations in shape memory alloys. Finally we note that the energetic formulation using (S) & (E) is equivalent to the classical local flow rules, when the solution processes are sufficiently smooth, see Mielke [13, 11].

One of the major advantages of the energetic formulation (S) & (E) is that it immediately gives rise to a natural incremental algorithm where each step is realized as a variational minimization problem.

We discretize the time interval $[0, T]$ via $0 = t_0 < t_1 < \ldots < t_N = T$ and give a stable initial condition $(\varphi_0, z_0) \in \mathcal{F}\times\mathcal{Z}$:

(IP) [Incremental Problem]
For $k = 1, \ldots, N$ find iteratively $(\varphi_k, z_k) \in \mathcal{F}\times\mathcal{Z}$ such that

$$(\varphi_k, z_k) \in \underset{(\varphi,z)\in\mathcal{F}\times\mathcal{Z}}{\operatorname{argmin}} \; \mathcal{E}(t_k, \varphi, z) + \mathcal{D}(z_{k-1}, z).$$

The fact that (IP) is very useful is manifested through the following result which states that incremental solutions are always stable and satisfy a discretized version of the energy inequality; for the simple proof see Mielke and Theil [15], Mielke et al. [17].

Theorem 3.1 *If $(\varphi_k, z_k)_{k=1,\ldots,N}$ is a solution of (IP), then for $k = 1, \ldots, N$ we have (i) stability of (φ_k, z_k) at time t_k, i.e.,*

$$\mathcal{E}(t_k, \varphi_k, z_k) \leq \mathcal{E}(t_k, \widehat{\varphi}, \widehat{z}) + \mathcal{D}(z_k, \widehat{z}) \quad \textit{for all} \quad (\widehat{\varphi}, \widehat{z}),$$

and (ii) the discretized energy estimate

$$\begin{aligned}&\mathcal{E}(t_k, \varphi_k, z_k) - \mathcal{E}(t_{k-1}, \varphi_{k-1}, z_{k-1}) + \mathcal{D}(z_{k-1}, z_k) \leq \\ &\int_{t_{k-1}}^{t_k} \tfrac{\partial}{\partial t}\mathcal{E}(s, \varphi_{k-1}, z_{k-1})\,\mathrm{d}s = \mathcal{E}(t_k, \varphi_{k-1}, z_{k-1}) - \mathcal{E}(t_{k-1}, \varphi_{k-1}, z_{k-1}).\end{aligned}$$

Another important feature of (IP) is the local occurrence of z (i.e. no gradients appear in the integrand defining $\mathcal{E}+\mathcal{D}$). This can be used to work out the minimization in z pointwise. We define the reduced potential Ψ^{red} via

$$\Psi^{\mathrm{red}}(z_{\mathrm{old}}; x, \mathbf{F}) = \min_{z\in Z} \Big[\widehat{\psi}(x, \mathbf{F}, z) + \widehat{D}(x, z_{\mathrm{old}}, z)\Big]$$

and choose $z = \widehat{Z}_{\mathrm{new}}(x, \mathbf{F}, z_{\mathrm{old}})$ such that this is a minimizer in the definition of Ψ^{red}. The new constitutive function Ψ^{red} is uniquely defined by $\widehat{\psi}$ and $\widehat{\Delta}$ and it contains the most important information on the combined effect of the elastic and plastic behavior of the material. Now we have

$$\begin{aligned}\mathcal{E}^{\mathrm{red}}_{z_{k-1}}(t_k, \varphi) \;&:=\; \min_{z\in\mathcal{Z}} \Big[\mathcal{E}(t_k, \varphi, z) + \mathcal{D}(z_{k-1}, z)\Big] \\ &=\; \int_\Omega \Psi^{\mathrm{red}}(z_{k-1}(x); x, \mathrm{D}\varphi(x))\,\mathrm{d}x - \langle \ell(t_k), \varphi\rangle.\end{aligned}$$

Thus, the kth step of (IP) is solved if we find a minimizer of $\mathcal{E}^{\mathrm{red}}_{z_{k-1}}(t_k, \cdot)$ on $\mathcal{F}$. Then, (φ_k, z_k) with $z_k(x) = \widehat{Z}_{\mathrm{new}}(x, \mathrm{D}\varphi_k(x), z_{k-1}(x))$ is the desired minimizer in (IP).

The minimization problem for $\mathcal{E}^{\mathrm{red}}_{z_{k-1}}(t_k, \cdot)$ has the standard form of a problem of nonlinear elasticity, where $\Psi^{\mathrm{red}}(z_{k-1}(x); x, \cdot)$ plays the role of the elastic potential. By construction we have frame indifference, i.e. $\Psi^{\mathrm{red}}(z; x, \mathbf{F}) = \Psi^{\mathrm{red}}(z; x, \mathbf{RF})$ for all $\mathbf{R} \in \mathrm{SO}(\mathbb{R}^d)$. However, other properties like the coercivity with respect to $\mathbf{F}$ are not nontrivial and depend on hardening properties.

If coercivity fails, the solvability of (IP) is not guaranteed, and we have to expect fracture or localization (shear bands), cf. Mielke [11].

Similarly, quasi- and rank-one convexity may no longer hold for $\Psi^{\mathrm{red}}(z;x,\cdot)$ and we have to expect the formation of microstructure in infimizing sequences, see Ortiz and Repetto [19], Carstensen et al. [3]. As a result, (IP) and consequently (S) & (E) may not have a solution. It is this case for which our theory below is developed.

4. Relaxation

Formation of microstructure occurs, if the minimization problem (IP) has no solution, cf. Ball and James [2], Carstensen et al. [3]. If the incremental problem (IP) has no solution, we may consider the following approximate incremental problem.

(AIP)$_\varepsilon$ *Given $z_0 \in \mathcal{Z}$ find $(\varphi_k^\varepsilon, z_k^\varepsilon) \in \mathcal{F}\times\mathcal{Z}$ with*

$$\mathcal{E}(t_k,\varphi_k^\varepsilon,z_k^\varepsilon)+\mathcal{D}(z_{k-1}^\varepsilon,z_k^\varepsilon) \le \varepsilon + \mathcal{E}(t_k,\varphi,z)+\mathcal{D}(z_{k-1}^\varepsilon,z)$$

for all $(\varphi,z)\in\mathcal{F}\times\mathcal{Z}$.

Obviously, this problem has, for all $\varepsilon > 0$, a solution and the question is how the solutions $(\varphi_k^\varepsilon, z_k^\varepsilon)$ behave for $\varepsilon \to 0$. We cannot expect pointwise convergence but certain macroscopic quantities should have limits for $\varepsilon \to 0$. To define an abstract notion of relaxation we introduce a generalized convergence "$\overset{\mathbb{W}}{\longrightarrow}$" on an enlarged space $\mathbb{W}$. This space is connected to $\mathcal{F}\times\mathcal{Z}$ via a continuous embedding $\mathcal{J}:\mathcal{F}\times\mathcal{Z}\mapsto\mathbb{W}$. Moreover, generalized functionals $\mathbb{E}:[0,T]\times\mathbb{W}\to\mathbb{R}$ and $\mathbb{D}:\mathbb{W}\times\mathbb{W}\to[0,\infty]$ replace the elastic functional $\mathcal{E}$ and the dissipation distance $\mathcal{D}$. We define the associated incremental problem for the initial datum $w_0\in\mathbb{W}$ and the time discretization $0=t_0<t_1<\ldots<t_N=T$ as follows.

($\mathbb{RIP}$) [Relaxed incremental problem]
For $k=1,\ldots,N$ find iteratively $w_k\in\mathbb{W}$ such that

$$w_k \in \operatorname*{argmin}_{w\in\mathbb{W}}\ \mathbb{E}(t_k,w)+\mathbb{D}(w_{k-1},w).$$

We do not ask for the conditions $\mathbb{D}(\mathcal{J}(0,z_0),\mathcal{J}(0,z_1)) = \mathcal{D}(z_0,z_1)$ and $\mathbb{E}(t,\mathcal{J}(\varphi,z))=\mathcal{E}(t,\varphi,z)$. Hence, the relaxation will not be an extension.

Definition 4.1 *A 4-tuple $(\mathbb{W},\mathcal{J},\mathbb{E},\mathbb{D})$ as defined above is called a* ***lower (or upper) incremental relaxation*** *of $(\mathcal{F}\times\mathcal{Z},\mathcal{E},\mathcal{D})$ if the following four conditions hold:*

(R1) *[Solvability] For each $w_0 \in \mathbb{W}$ the relaxed incremental problem ($\mathbb{RIP}$) has a solution.*

(R2) *[Minimality] $\mathcal{J}(\mathcal{F}\times\mathcal{Z})$ is dense in $\mathbb{W}$.*

(R3) *[Incremental consistency] If $(\varphi_k, z_k)_{k=1,\dots,N}$ is a solution of (IP), then $\mathcal{J}(\varphi_k, z_k)_{k=1,\dots,N}$ solves ($\mathbb{RIP}$); and vice versa, if $(w_k)_{k=1,\dots,N}$ satisfies $w_k = \mathcal{J}(\varphi_k, z_k)$ and solves ($\mathbb{RIP}$), then $(\varphi_k, z_k)_{k=1,\dots,N}$ solves (IP).*

(R4)$_{\text{low}}$ *[Lower increm. relaxation] For each solution $(w_k)_{k=1,\dots,N}$ of ($\mathbb{RIP}$), there exist solutions $(\varphi_k^\varepsilon, z_k^\varepsilon)_{k=1,\dots,N}$ of (AIP)$_\varepsilon$ with $\mathcal{J}(\varphi_k^\varepsilon, z_k^\varepsilon) \stackrel{\mathbb{W}}{\longrightarrow} w_k$ for $\varepsilon \to 0$.*

(R4)$_{\text{upp}}$ *[Upper incremental relaxation] If $(\varphi_k^\varepsilon, z_k^\varepsilon)_{k=1,\dots,N}$ solves (AIP)$_\varepsilon$ and $\mathcal{J}(\varphi_k^\varepsilon, z_k^\varepsilon) \stackrel{\mathbb{W}}{\longrightarrow} w_k$, then $(w_k)_{k=1,\dots,N}$ solves ($\mathbb{RIP}$).*

Our definition implies that the relaxed problem has to be of the same energetic kind as the original one; we just give up the clear distinction between $\varphi \in \mathcal{F}$ and $z \in \mathcal{Z}$. Condition (R1) forces us to consider only useful relaxations, namely those which have solutions. If the original problem is already solvable, then we can choose $\mathbb{W} = \mathcal{F}\times\mathcal{Z}$, $\mathbb{E} = \mathcal{E}$ and $\mathbb{D} = \mathcal{D}$, since no relaxation is necessary. Condition (R2) says that the new state space $\mathbb{W}$ should not be unnecessarily big. Condition (R3) is very important as it says that the relaxation must maintain classical solutions, if they exist for (IP) or if they are found by solving ($\mathbb{RIP}$). Conditions (R4)$_{\text{low}}$ and (R4)$_{\text{upp}}$ link the evolution of $(\mathcal{F}\times\mathcal{Z}, \mathcal{E}, \mathcal{D})$ to that of $(\mathbb{W}, \mathbb{E}, \mathbb{D})$ via the approximate incremental problem (AIP)$_\varepsilon$.

Moreover the relaxed incremental problem ($\mathbb{RIP}$) can be interpreted as the incremental problem associated to the following relaxed energetic formulation: The function $w : [0,T] \mapsto \mathbb{W}$ is called a solution process of the relaxed problem $(\mathbb{W}, \mathbb{E}, \mathbb{D})$ if ($\mathbb{S}$) and ($\mathbb{E}$) are satisfied:

($\mathbb{S}$) ***Stability*** *For all $t \in [0,T]$ and all $\widetilde{w} \in \mathbb{W}$ we have*

$$\mathbb{E}(t, w(t)) \le \mathbb{E}(t, \widetilde{w}) + \mathbb{D}(w(t), \widetilde{w}).$$

($\mathbb{E}$) ***Energy inequality*** *For all $0 \le t_1 < t_2 \le T$ we have*

$$\mathbb{E}(t_1, w(t_2)) + \text{Diss}^{\text{rel}}(w; [t_1, t_2]) \le \mathbb{E}(t_1, w(t_1)) - \int_{t_1}^{t_2} \langle \dot{\ell}(s), \mathbf{\Phi}(w(s)) \rangle \, \mathrm{d}s.$$

Here $\varphi = \mathbf{\Phi}(w)$ is the macroscopic deformation φ associated to the limit $w \in \mathbb{W}$. The relaxed dissipation is given by

$$\text{Diss}^{\text{rel}}(w; [t_1, t_2]) = \sup \sum_{j=1}^{M} \mathbb{D}(w(\tau_{j-1}), w(\tau_j)),$$

where the supremum is taken over all $M \in \mathbb{N}$ and all discretizations $t_1 \leq \tau_0 < \tau_1 < \ldots < \tau_{M-1} < \tau_M \leq t_2$.
A further desirable property for relaxations is the consistency for the time continuous problem:

(R5) [Time consistency] If $(\varphi, z) : [0,T] \mapsto \mathcal{F}\times\mathcal{Z}$ solves (S) & (E), then $\mathcal{J}(\phi, z) : [0,T] \mapsto \mathbb{W}$ solves ($\mathbb{S}$) & ($\mathbb{E}$); and vice versa, if $w : [0,T] \mapsto \mathbb{W}$ satisfies $w(t) = \mathcal{J}(\varphi(t), z(t))$ and solves ($\mathbb{S}$) & ($\mathbb{E}$), then $(\varphi, z) : [0,T] \mapsto \mathcal{F}\times\mathcal{Z}$ solves (S) & (E).

Another way to define relaxations for rate independent problems of the type (S) & (E) is proposed in Theil [26]. This definition avoids totally the usage of incremental problems and works directly with (S) & (E) and ($\mathbb{S}$) & ($\mathbb{E}$). An application of this theory to phase transformation is given in Mielke et al. [17] and Theil [26].

5. Young measures

A special relaxation can be given by using Young measures which were introduced into the field of continuum mechanics in Ball and James [2] and were further developed in the last 15 years, see Roubíček [23] and Müller [18]. One particular instance of this relaxation was studied in Mielke and Theil [14], Govindjee et al. [4] and Mielke et al. [17], where phase transformations in shape memory alloys were studied.
We refer to Ball [1], Pedregal [21], Roubíček [23] and Mielke [10] for the basic notions of Young measures and repeat here only the definitions. Denote by $\mathrm{Prob}(Z)$ the set of all probability measures on Z. The **Young measures** on Ω with values in Z are given by

$$\mathrm{YM}(\Omega, Z) := \{\, \mu : \Omega \mapsto \mathrm{Prob}(Z) \mid \mu \text{ is weakly measurable} \,\}.$$

We say that the sequence $(z^j)_{j\in\mathbb{N}}$ generates the Young measure μ (written $z^j \xrightarrow{\mathrm{YM}} \mu$) if for each $x_0 \in \Omega$ and sufficiently small ε and every function $g \in \mathrm{C}_0(Z, \mathbb{R})$ we have

$$\int_{|x_0-x|<\varepsilon} g(z^j(x))\,\mathrm{d}x \to \int_{|x_0-x|<\varepsilon} \int_Z g(z)\mu(x, \mathrm{d}z)\,\mathrm{d}x \text{ for } j \to \infty.$$

A function $x \mapsto z(x)$ can be considered as a Young measure by letting $\mu(x) = \delta_{z(x)}$ where δ_a denotes the point mass in a, i.e. $\delta_a(A) = 1$ if $a \in A$ and 0 else.
Given a (dissipation) metric $\widehat{D} : Z\times Z \mapsto [0, \infty]$, we have to generalize it to a metric $\mathbb{D}$ on the space of Young measures which satisfies $\mathbb{D}(\delta_{z_0}, \delta_{z_1}) = \mathcal{D}(z_0, z_1)$ for all $z_0, z_1 \in \mathcal{Z}$. Moreover, it should fit to the microscopic observation that microscopic changes from z_0 to z_1 can be arranged in the mesoscopic

region in an optimal way. We start with the Wasserstein metric D^{Wass} on $\text{Prob}(Z)$ which associates to $\widehat{D}$. For given $\nu_0, \nu_1 \in \text{Prob}(Z)$ we want to transport the mass, which at the beginning is distributed according to ν_0, in such a way that the final distribution is ν_1 and that the dissipation cost is minimal. Denoting by $\sigma \in \text{Prob}(Z\times Z)$ a probability measure such that $\sigma(\mathrm{d}z_0, \mathrm{d}z_1)$ gives the amount of mass which from a neighborhood of z_0 is transported into a neighborhood of z_1, we have to infimize with respect to σ under the given initial and final distributions:

$$D^{\text{Wass}}(\nu_0,\nu_1) = \inf\Big\{ \int_{Z\times Z} \widehat{D}(z_0,z_1)\sigma(\mathrm{d}z_0,\mathrm{d}z_1) \,\Big|\, \sigma \in \text{Prob}(Z\times Z),\\ \int_Z \sigma(\cdot,\mathrm{d}z_1) = \nu_0,\ \int_Z \sigma(\mathrm{d}z_0,\cdot) = \nu_1 \Big\}.$$

The Wasserstein metric has a dual representation (which is well-known in probability theory) using Lipschitz continuous functions, cf. Rachev [22] and Mielke [10]:

$$D^{\text{Wass}}(\nu_0,\nu_1) = \sup\{ \int_Z g(z)\,\nu_1(\mathrm{d}z) - \int_Z g(z)\,\nu_0(\mathrm{d}z)\\ \mid g \in \mathrm{C}_0(Z,\mathbb{R}),\ \text{Lip}_{\widehat{D}}(g) \le 1\}$$

where $\text{Lip}_{\widehat{D}}(g) = \sup_{z_1,z_2\in Z} |g(z_1)-g(z_2)|/\widehat{D}(z_1,z_2)$. Hence, D^{Wass} can be understood as the restriction of a Banach-space norm to the convex set $\text{Prob}(Z)$. For two Young measures $\mu_0, \mu_1 \in \text{YM}(\Omega, Z)$ we set

$$\mathbb{D}(\mu_0,\mu_1) = \int_{x\in\Omega} D^{\text{Wass}}(x,\mu_0(x),\mu_1(x))\,\mathrm{d}x,$$

which is an integral over a local function in the macroscopic variable $x \in \Omega$.

There is a special case of Young measures, namely those which are generated by sequences which are gradients of functions, like the deformation gradients in continuum mechanics $\mathbf{F}^j = \mathrm{D}\boldsymbol{\varphi}^j \in \mathbb{R}^{d\times d}$. The Young measures obtained as limits from gradients form a subset of all Young measures and are called **gradient Young measures** (GYM). They are characterized as follows (cf. Kinderlehrer and Pedregal [5], Pedregal [21] and Roubíček [23]):

$$\gamma \in \text{GYM}(\Omega,\mathbb{R}^{d\times d}) \iff\\ \begin{cases} \text{There exists } \boldsymbol{\varphi}:\Omega\mapsto\mathbb{R}^d \text{ such that for a.e. } x\in\Omega \text{ we have}\\ \text{(a) } \mathrm{D}\boldsymbol{\varphi}(x) = \int_{\mathbb{R}^{d\times d}} \mathbf{G}\,\gamma(x,\mathrm{d}G) \text{ and (b) } \gamma(x) \in \text{Prob}^{\text{grad}}(\mathbb{R}^{d\times d}), \end{cases}$$

where $\text{Prob}^{\text{grad}}(\mathbb{R}^{d\times d}) =$

$$\Big\{\nu \in \text{Prob}(\mathbb{R}^{d\times d}) \,\Big|\, \text{for all quasiconvex } Q:\mathbb{R}^{d\times d}\mapsto\mathbb{R} \text{ we have}\\ Q(\int_{\mathbb{R}^{d\times d}} \mathbf{G}\,\nu(\mathrm{d}\mathbf{G})) \le \int_{\mathbb{R}^{d\times d}} Q(\mathbf{G})\,\nu(\mathrm{d}\mathbf{G})\Big\}$$

We shortly write $\gamma \in \mathrm{GYM}_{\boldsymbol{\varphi}}(\Omega, \mathbb{R}^{d\times d})$ if (a) and (b) hold. The above condition (b) is essential as it reflects the information of compatibility of different gradients. For instance, $\nu = \theta\delta_A + (1-\theta)\delta_B$ with $\theta \in (0,1)$ and $A, B \in \mathbb{R}^{d\times d}$ lies in $\mathrm{Prob}^{\mathrm{grad}}(\mathbb{R}^{d\times d})$ if and only if $\mathrm{rank}(A-B) \leq 1$.

6. Separate relaxation

We propose a relaxation which uses the weak convergence in $\mathcal{F} \subset \mathrm{W}^{1,p}(\Omega, \mathbb{R}^d)$, and the Young measures $\mu \in \mathrm{YM}(\Omega, Z)$ to replace z. Hence, in the abstract relaxation setting we let

$$
\begin{aligned}
&w = (\boldsymbol{\varphi}, \mu) \in \mathcal{F}\times \mathrm{YM}(\Omega, Z) =: \mathbb{W}, \qquad \mathcal{J}(\boldsymbol{\varphi}, z) = (\boldsymbol{\varphi}, \delta_z),\\
&(\boldsymbol{\varphi}^j, \mu^j) \xrightarrow{\mathbb{W}} (\boldsymbol{\varphi}^\infty, \mu^\infty) \iff \\
&\begin{cases} \boldsymbol{\varphi}^j \rightharpoonup \boldsymbol{\varphi}^\infty \text{ in } \mathrm{W}^{1,p}(\Omega, \mathbb{R}^d), \\ \forall g \in \mathrm{C}_0(Z) : \int_Z g(z)\, \mu^j(\mathrm{d}z) \overset{*}{\rightharpoonup} \int_Z g(z)\, \mu^\infty(\mathrm{d}z) \text{ in } \mathrm{L}^1_{\mathrm{loc}}(\Omega). \end{cases}
\end{aligned}
$$

The relaxed functionals are abstractly defined as Young-measure relaxation separately for $\mathcal{E}$ and $\mathcal{D}$:

$$
\begin{aligned}
\mathbb{E}(t, \boldsymbol{\varphi}, \mu) &= \inf\{ \liminf_{m\to\infty} \mathcal{E}(t, \boldsymbol{\varphi}_m, z_m) \mid \mathcal{J}(\boldsymbol{\varphi}_m, z_m) \xrightarrow{\mathbb{W}} (\boldsymbol{\varphi}, \mu) \},\\
\mathbb{D}(\mu_0, \mu_1) &= \inf\{ \liminf_{m\to\infty} \mathcal{D}(z_{0,m}, z_{1,m}) \mid \mathcal{J}(\boldsymbol{\varphi}_{j,m}, z_{j,m}) \xrightarrow{\mathbb{W}} (\boldsymbol{\varphi}_j, \mu_j) \}.
\end{aligned}
$$

These functionals are given as the smallest possible limit of all sequences of classical functions producing the correct limit in $\mathbb{W}$. This abstract setting is not very useful for practical purposes unless the form of $\mathbb{E}$ and $\mathbb{D}$ is made more specific. However, we obtain a general result in terms of relaxation.

Theorem 6.1 *Under suitable technical assumptions the relaxation defined via $\mathbb{W}$, $\mathcal{J}$, $\mathbb{E}$ and $\mathbb{D}$ satisfies the properties (R1)–(R3) in Definition 4.1 as well as (R5).*

To make the above relaxation useful we need integral representations of the relaxed functionals. These follow from abstract results as given in Roubíček [23] and Mielke et al. [17].

Theorem 6.2 *Under suitable technical assumptions the relaxed functionals have the form*

$$
\begin{aligned}
\mathbb{E}(t, \boldsymbol{\varphi}, \mu) &= \int_\Omega \Psi_{\mathrm{rel}}(x, \mathrm{D}\boldsymbol{\varphi}(x), \mu(x))\, \mathrm{d}x - \langle \ell(t), \boldsymbol{\varphi} \rangle,\\
\mathbb{D}(\mu_0, \mu_1) &= \int_\Omega D^{\mathrm{Wass}}(x, \mu_0(x), \mu_1(x))\, \mathrm{d}x,
\end{aligned}
$$

where the relaxed potential is given by $\Psi_{\rm rel}(x,\mathbf{F},\nu) =$

$$\inf\Big\{ \int_{(0,1)^d} \widehat{\psi}(x,\mathbf{F}{+}\mathrm{D}\phi(y),z(y))\,\mathrm{d}y \;\Big|\; \phi\in \mathrm{W}_0^{1,\infty}((0,1)^d), \\ \nu(A)=\mathrm{vol}(\{\,y\,|\,z(y)\in A\,\})\ \textit{for all}\ A\subset Z \Big\}.$$

Unfortunately the definition of $\Psi_{\rm rel}$ is rather complicated. Note that it is not obtained by integrating $\psi(x,\mathbf{F},z)$ with respect to $\nu(\mathrm{d}z)$, since $\mathbf{F}$ is the macroscopic strain while ν relates to microscopic fluctuations in $z\in Z$ which have a counterpart in microscopic fluctuations of the strain. The definition of $\Psi_{\rm rel}$ shows that minimization with respect to the combined microscopic arrangements in (the representative volume element) $(0,1)^d$ of the internal variable z and the fluctuation strain $\mathrm{D}\phi$ is necessary.

Example 6.1 To illustrate the above concept we consider the case $d=1$, $Z=\mathbb{R}$, and $\widehat{\psi}(\mathbf{F},z)=(\mathbf{F}{-}z)^2$. We obtain $\Psi_{\rm rel}(\mathbf{F},\nu)=[\mathbf{F}{-}\int_{\mathbb{R}} z\,\nu(\mathrm{d}z)\,]^2$, since in the definition of the relaxed potential we may choose any $z:(0,1)\mapsto\mathbb{R}$ generating ν and then define ϕ via $\phi(0)=0$ and $\phi'(y)=z(y){-}\int_0^1 z(s)\,\mathrm{d}s$. Using $\int_0^1 z(s)\,\mathrm{d}s=\int_{\mathbb{R}} z\,\nu(\mathrm{d}z)$ the result follows. Moreover, we have

$$\Psi_{\rm rel}(\mathbf{F},\nu)=\int_{\mathbb{R}}\widehat{\psi}(\mathbf{F},z)\,\nu(\mathrm{d}z)-\Big(\int_{\mathbb{R}} z^2\nu(\mathrm{d}z)-\big[\int_{\mathbb{R}} z\,\nu(\mathrm{d}z)\big]^2\Big)$$

which shows that $\Psi_{\rm rel}$ is strictly less than the "averaged" energy density, if ν is not a Dirac mass.

Remark 6.1 Up to now the relaxed elastic potential can be computed only in a few cases, e.g., for applications in phase transformations in shape memory alloys, where $Z=\{1,2,\ldots,n\}$ is the discrete set of possible phases, cf. Mielke et al. [17] and Govindjee et al. [4]. There $\mathrm{Prob}(Z)$ can be identified with the polytope $P^n=\{\,(\theta_1,\ldots,\theta_n)\in\mathbb{R}^n\mid\theta_j\geq 0,\ \sum_1^n\theta_j=1\,\}$ and $\mathrm{YM}(\Omega,Z)=\mathrm{L}^\infty(\Omega,P^n)$. The dissipation distance $\mathbb{D}$ takes the form $\mathbb{D}(\mu_{\rm old},\mu_{\rm new})=\int_\Omega\|\theta_{\rm new}(x){-}\theta_{\rm old}(x)\|\,\mathrm{d}x$ for a suitable polyhedral vector norm $\|\cdot\|$ on $\mathbb{R}^n$.

Acknowledgments

The research was partially supported by DFG through SFB 404 "Multifield Problems". The author is grateful for stimulating discussions with Carsten Carstensen, Klaus Hackl and Christian Miehe.

References

[1] Ball, J. (1989). A version of the fundamental theorem for Young measures. In *PDEs and continuum models of phase transitions (Nice, 1988)*, Springer, Berlin, 207–215.

[2] Ball, J.M. and James, R.D. (1987). Fine phase mixtures as minimizers of energy. *Arch. Rational Mech. Anal.*, **100(1)**, 13–52.

[3] Carstensen, C., Hackl, K. and Mielke, A. (2001). Nonconvex potentials and microstructure in finite-strain plasticity. *Proc. Royal Soc. London Ser. A*, In press.

[4] Govindjee, S., Mielke, A. and Hall, G. (2001). The free-energy of mixing for n-variant martensitic phase transformations using quasi-convex analysis. *J. Mech. Physics Solids*. Submitted.

[5] Kinderlehrer, D. and Pedregal, P. (1994). Gradient Young measures generated by sequences in Sobolev spaces. *J. Geom. Anal.*, **4**, 59–90.

[6] Kuczma, M., Mielke, A. and Stein, E. (1999). Modelling of hysteresis in two-phase systems. *Archive of Mechanics (Warzawa)*, **51**, 693–715.

[7] Miehe, C. (1996). Multisurface thermoplasticity for single crystals at large strains in terms of Eulerian vector updates. *Int. J. Solids Struct.*, **33**, 1981–2004.

[8] Miehe, C. (2000). Elements of computational inelasticity at finite strains. In preparation.

[9] Miehe, C. and Stein, E. (1992). A canonical model of multiplicative elasto-plasticity. Formulation and aspects of numerical implementation. *Europ. J. Mech. A/Solids*, **11**, 25–43.

[10] Mielke, A. (1999). Flow properties for Young-measure solutions of semilinear hyperbolic problems. *Proc. Roy. Soc. Edinburgh Sect. A*, **129(1)**, 85–123.

[11] Mielke, A. (2001). Energetic formulation of multiplicative elastoplasticity using the dissipation metric. In preparation.

[12] Mielke, A. (2001). Evolution of microstructure in elastoplasticity using relaxation and Young measures. In preparation.

[13] Mielke, A. (2002). Finite plasticity, Lie groups and geodesics on SL(d). In Newton, P., Weinstein, A. and Holmes, P. editors, *Geometry, Dynamics, and Mechanics*, Springer–Verlag, 2002. SFB 404 Preprint 2000/14.

[14] Mielke, A. and Theil, F. (1999). A mathematical model for rate-independent phase transformations with hysteresis. In Alber, H.-D., Balean, R. and Farwig, R. editors, *Proceedings of the Workshop on* Models of Continuum Mechanics in Analysis and Engineering, Shaker-Verlag, 117–129.

[15] Mielke, A. and Theil, F. (2001). On rate-independent hysteresis models. *Nonl. Diff. Eqns. Appl. (NoDEA)*. To appear.

[16] Mielke, A., Theil, F. and Levitas, V. (1998). Mathematical formulation of quasistatic phase transformations with friction using an extremum principle. Preprint A8, Hannover.

[17] Mielke, A., Theil, F. and Levitas, V. (2000). A variational formulation of rate-independent phase transformations using an extremum principle. SFB 404 Preprint 2000/15, Stuttgart.

[18] Müller, S. (1999). Variational models for microstructure and phase transitions. In *Calculus of variations and geometric evolution problems (Cetraro, 1996)*, Springer, Berlin, 85–210.

[19] Ortiz, M. and Repetto, E. (1999). Nonconvex energy minimization and dislocation structures in ductile single crystals. *J. Mech. Phys. Solids*, **47(2)**, 397–462.

[20] Ortiz, M. and Stainier, L. (1999). The variational formulation of viscoplastic constitutive updates. *Comput. Methods Appl. Mech. Engrg.*, **171(3-4)**, 419–444.

[21] Pedregal, P. (1997). *Parameterized measures and variational principles*, Birkhäuser Verlag, Basel.

[22] Rachev, S. (1991). *Probability metrics and the stability of stochastic models*, John Wiley & Sons Ltd., Chichester.

[23] Roubíček, T. (1997). *Relaxation in optimization theory and variational calculus*, Walter de Gruyter & Co., Berlin.

[24] Simo, J. (1988). A framework for finite strain elastoplasticity based on maximum plastic dissipation and the multiplicative decomposition. I. Continuum formulation. *Comput. Methods Appl. Mech. Engrg.*, **66(2)**, 199–219.

[25] Simo, J. and Ortiz, M. (1985). A unified approach to finite deformation elastoplastic analysis based on the use of hyperelastic constitutive relations. *Comput. Methods Appl. Mech. Engrg.*, **49**, 221–245.

[26] Theil, F. (2001). Relaxation of rate independent evolution problems. *Proc. Roy. Soc. Edinb. A*. To appear.

[27] Ziegler, H. and Wehrli, C. (1987). The derivation of constitutive relations from the free energy and the dissipation function. In *Advances in applied mechanics, Vol. 25*, Academic Press, Orlando, FL, 183–237.

Mathematical Analysis of Constitutive Equations: Existence and Collapse of Solutions

H.D. Alber

Department of Mathematics, Darmstadt University of Technology

Schlossgartenstr. 7, 64289 Darmstadt, Germany

alber@mathematik.tu-darmstadt.de

Abstract: The article presents a recent existence result for quasistatic mathematical models for the inelastic behavior of metals at small strain. A very general existence theory is available for models with constitutive equations which contain terms modeling linear kinematic hardening. However, the majority of constitutive models do not contain such terms; for such models the existence theory is much less complete. The existence result in this article is valid for such constitutive models without terms of linear kinematic hardening and contributes to this theory. In the last section of the article the relation between collapse of the solution and a coercivity condition needed in the existence theorem is discussed. The article does not contain proofs. These are published elsewhere.

Keywords: Quasistatic initial-boundary value problem, inelastic behavior at small strains, constitutive equations of monotone type with positive semi definite free energy, evolution equations to monotone operators, existence of solutions.

1. Introduction

Let $\Omega \subseteq \mathbb{R}^3$ be the set of material points of a solid body. For many materials the history dependent deformation behavior of this body can be modeled by the equations

$$-\operatorname{div}_x T(x,t) = b(x,t) \tag{1}$$

$$T(x,t) = \mathcal{D}(\varepsilon(\nabla_x u(x,t) - \varepsilon_p(x,t))) \tag{2}$$

$$\frac{\partial}{\partial t}\varepsilon_p(x,t) = \tilde{g}_1(T(x,t), -\tilde{z}(x,t)) \tag{3}$$

$$\frac{\partial}{\partial t}\tilde{z}(x,t) = \tilde{g}_2(T(x,t), -\tilde{z}(x,t)) \tag{4}$$

$$\varepsilon_p(x,0) = \varepsilon_p^{(0)}(x), \quad \tilde{z}(x,0) = \tilde{z}^{(0)}(x), \tag{5}$$

with the Dirichlet boundary condition

$$u(x,t) = \gamma_D(x,t), \quad (x,t) \in \partial\Omega \times (0,\infty) \tag{6}$$

or the Neumann boundary condition

$$T(x,t)n(x) = \gamma_N(x,t), \quad (x,t) \in \partial\Omega \times (0,\infty). \tag{7}$$

C. Miehe (ed.), IUTAM Symposium on Computational Mechanics of Solid Materials at Large Strains, 45-53.

Here $u(x,t) : \Omega \times (0,\infty) \to \mathbb{R}^3$ denotes the displacement of the material point labeled x at time t. With the 3×3-matrix $\nabla_x u(x,t)$ of first order derivatives of u with respect to the components x_1, x_2, x_3 of x and with the transposed matrix $(\nabla_x u(x,t))^T$ the strain tensor is defined by

$$\varepsilon(\nabla_x u(x,t)) = \frac{1}{2}\Big(\nabla_x u(x,t) + (\nabla_x u(x,t))^T\Big) \in \mathcal{S}^3.$$

$\mathcal{S}^3$ is the set of symmetric 3×3-matrices. $T : \Omega \times (0,\infty) \to \mathcal{S}^3$ is the Cauchy stress tensor, $\varepsilon_p(x,t) \in \mathcal{S}^3$ is the plastic strain tensor, $\tilde{z}(x,t) \in \mathbb{R}^N$ is a vector of internal variables. Moreover, $\mathcal{D} : \mathcal{S}^3 \to \mathcal{S}^3$ is a linear, symmetric, positive definite mapping, the elasticity tensor, $b : \Omega \times (0,\infty) \to \mathbb{R}^3$ is a given volume force, $\gamma_D : \partial\Omega \times (0,\infty) \to \mathbb{R}^3$ is a given boundary displacement, $\gamma_N : \partial\Omega \times (0,\infty) \to \mathbb{R}^3$ is a given traction at the boundary, and $\varepsilon_p^{(0)}, \tilde{z}^{(0)}$ are given initial data. Finally, $n(x)$ in the Neumann boundary condition denotes the exterior unit normal to $\partial\Omega$ at x. The constitutive equations (3), (4) with given functions $\tilde{g}_1 : \mathcal{S}^3 \times \mathbb{R}^N \to \mathcal{S}^3$, $\tilde{g}_2 : \mathcal{S}^3 \times \mathbb{R}^N \to \mathbb{R}^N$ determine the inelastic behavior of the body.

A more general class of constitutive models consists of constitutive relations of monotone type. These relations have the form

$$z_t(x,t) \in g\Big(-\rho \nabla_z \psi(\varepsilon(\nabla_x u(x,t), z(x,t))\Big), \tag{8}$$

where $z = (\varepsilon_p, \tilde{z})$, and where $\rho > 0$ is the constant mass density. $g : \mathbb{R}^N \to 2^{\mathbb{R}^N}$ is a given monotone mapping satisfying $0 \in g(0)$, and ψ is the free energy, which is assumed to be a positive definite or positive semi-definite quadratic form

$$\rho\psi(\varepsilon, z) = \frac{1}{2}[D(\varepsilon - Bz)] \cdot (\varepsilon - Bz) + (Lz) \cdot z.$$

L is a symmetric, positive definite or positive semi-definite matrix.

The class of constitutive equations of monotone type was introduced in Alber [1]. It generalizes the class of constitutive equations of generalized standard materials introduced in Halphen and Nguyen [6]. For a generalized standard material, g is the gradient of a convex function.

It turns out that for most materials the free energy is only positive semi-definite, but not positive definite. In this case, if we assume that the constitutive relation is rate dependent and under a minor additional assumption the constitutive equation (8) of monotone type can be transformed to the equations (3) and (4) with a monotone vector field

$$(\varepsilon_p, \tilde{z}) \to (\tilde{g}_1(\varepsilon_p, \tilde{z}), \tilde{g}_2(\varepsilon_p, \tilde{z})) : \mathcal{S}^3 \times \mathbb{R}^N \to \mathcal{S}^3 \times \mathbb{R}^N.$$

For initial-boundary value problem to constitutive equations of monotone type with positive definite free energy a satisfactory and general existence theory is available, cf. Han and Reddy [7], Alber [2], Alber and Chełmiński [3]. Moreover, the dynamic initial-boundary value problem to positive semi-definite free energy has been treated in a number of articles, cf. Chełmiński [5] for example, to mention just one. The aim of this note is to present existence results for the quasi-static initial-boundary value problems (1)–(7) with positive semi-definite free energy. For such initial-boundary value problems the existence theory is much less complete.

Proofs are not contained in this note. These are to be published in Alber and Chełmiński [3].

2. Existence of solutions for initial-boundary value problems with history functionals

To prove existence of solutions to the initial-boundary value problems (1)–(7) it is first shown that a more general initial-boundary value problem with constitutive relations given by a history functional can be solved. Subsequentially it is proved that the constitutive equations (3)–(5) define such a history functional satisfying the conditions needed in the existence proof. To present this approach, we need a few notations:

Let $\Omega \subseteq \mathbb{R}^3$ be a bounded open set with C^1-boundary. T_e denotes a positive number (time of existence), and for $0 \leq t \leq T_e$ we set

$$Z_t = \Omega \times [0, t], \quad Z = Z_{T_e}.$$

By $1 < p, q < \infty$ we denote numbers with $\frac{1}{p} + \frac{1}{q} = 1$. The norms on $L^p(\Omega, \mathcal{S}^3)$ and $L^p(\Omega, \mathbb{R}^N)$ are denoted by

$$\|u\|_{p,\Omega} = \Big[\int_\Omega |u(x)|^p \, dx \Big]^{1/p}.$$

The same notation is used for Z instead of Ω. For the scalar product of two symmetric matrices $\sigma, \tau \in \mathcal{S}^3$ we write

$$\sigma \cdot \tau = \sum_{i,j=1}^{3} \sigma_{ij} \tau_{ij} \,.$$

With this notation the canonical bilinear forms on the product spaces $L^p(\Omega, \mathcal{S}^3) \times L^q(\Omega, \mathcal{S}^3)$ and $L^p(Z, \mathcal{S}^3) \times L^q(Z, \mathcal{S}^3)$ are given by

$$(\sigma, \tau)_\Omega = \int_\Omega \sigma(x) \cdot \tau(x) \, dx \,, \quad (\sigma, \tau)_Z = \int_Z \sigma(x, t) \cdot \tau(x, t) \, d(x, t) \,.$$

A history functional is defined as follows:

Definition 2.1 *Let $F(Z, \mathcal{S}^3)$ be the set of all functions from Z to $\mathcal{S}^3$. A mapping $\mathcal{H} : \Delta(\mathcal{H}) \to F(Z, \mathcal{S}^3)$ with $\Delta(\mathcal{H}) \subseteq F(Z, \mathcal{S}^3)$ is called a history functional on $F(Z, \mathcal{S}^3)$, if it has the following property: For all $0 \le t \le T_e$ and all $\tau_1, \tau_2 \in \Delta(\mathcal{H})$, which satisfy $\tau_1|_{Z_t} = \tau_2|_{Z_t}$, it follows that $\mathcal{H}[\tau_1]|_{Z_t} = \mathcal{H}[\tau_2]|_{Z_t}$.*

With a suitable history functional $\mathcal{H}$ on $F(Z, \mathcal{S}^3)$ the equations (1)–(5) can now be written in the form

$$-\mathrm{div}_x T = b \tag{9}$$

$$T = \mathcal{D}(\varepsilon(\nabla_x u) - \varepsilon_p) \tag{10}$$

$$\frac{\partial}{\partial t}\varepsilon_p = \mathcal{H}[T] \tag{11}$$

$$\varepsilon_p(x, 0) = \varepsilon_p^{(0)}(x). \tag{12}$$

To see this, let $\tilde{z}^{(0)} : \Omega \to \mathbb{R}^N$, $T : Z \to \mathcal{S}^3$ be given and let $(\tilde{z}, h) : Z \mapsto \mathbb{R}^N \times \mathcal{S}^3$ be a solution of the initial value problem

$$h(x, t) = \tilde{g}_1(T(x, t), -\tilde{z}(x, t)) \tag{13}$$

$$\frac{\partial}{\partial t}\tilde{z}(x, t) = \tilde{g}_2(T(x, t), -\tilde{z}(x, t)) \tag{14}$$

$$\tilde{z}(x, 0) = \tilde{z}^{(0)}(x), \tag{15}$$

for $x \in \Omega$ and $t \in [0, T_e]$. Then a history functional $\mathcal{H}_{\tilde{z}^{(0)}}$ on $F(Z, \mathcal{S}^3)$ is defined by

$$\mathcal{H}_{\tilde{z}^{(0)}}[T] = h. \tag{16}$$

Insertion of $\mathcal{H}_{\tilde{z}^{(0)}}$ for $\mathcal{H}$ into (11) reduces (9)–(12) to the equations (1)–(5).

Next we state the existence result for the Dirichlet or Neumann initial-boundary value problems to the equations (9)–(12) containing an abstract history functional. We need the following assumptions:

For numbers $2 \le p < \infty$ and $1 < q \le 2$ with $\frac{1}{p} + \frac{1}{q} = 1$ let the history functional be a mapping $\mathcal{H} : L^p(Z, \mathcal{S}^3) \to L^q(Z, \mathcal{S}^3)$, which satisfies four conditions:

(H1) There is a constant C such that for all $T \in L^p(Z, \mathcal{S}^3)$

$$\|\mathcal{H}(T)\|_{q,Z} \le C\Big(\|T\|_{p,Z}^{\frac{p}{q}} + 1\Big).$$

(H2) $\mathcal{H}$ is hemicontinuous and monotone with respect to the bilinear form $(\sigma, T)_Z$.

(H3) $\mathcal{H}$ satisfies a first coercivity condition:

$$\frac{\Big(T, \mathcal{H}[T]\Big)_Z}{\|\mathcal{H}[T]\|_{q,Z}} \to \infty \quad \text{for} \quad \|T\|_{p,Z} \to \infty\,.$$

(H4) $\mathcal{H}$ satisfies a second coercivity condition:

$$\frac{\Big(T, \mathcal{H}[T]\Big)_Z}{\|T\|_{p,Z}} \to \infty \quad \text{for} \quad \|T\|_{p,Z} \to \infty\,.$$

Theorem 2.1 *Let $\mathcal{H}$ satisfy the conditions* (H1)–(H4), *and let $b \subseteq L^p(Z, \mathbb{R}^3)$, $\varepsilon_p^{(0)} \in L^2(\Omega, \mathcal{S}^3)$ be given functions. Then the following assertions hold:*

(i) *For $\gamma_D \in L^p(0, T_e; H_1^p(\Omega, \mathbb{R}^3))$ there is a unique solution*

$$(u, T, \varepsilon_p) \in L^q(0, T_e; H_1^q(\Omega, \mathbb{R}^3)) \times L^p(Z, \mathcal{S}^3) \times H_1^q([0, T_e], L^q(\Omega, \mathcal{S}^3))$$

of the Dirichlet problem (9)–(12), (6).

(ii) *Assume that $\gamma_N \in L^p(\partial\Omega \times [0, T_e], \mathbb{R}^3)$ and that for all infinitesimal rigid motions $a + \omega \times x$ and for almost every $t \in [0, T_e]$ the equation*

$$\int_\Omega b(x,t) \cdot (a + \omega \times x)dx + \int_{\partial\Omega} \gamma_N(x,t) \cdot (a + \omega \times x)dS_x = 0 \qquad (17)$$

holds. Then there exists a solution

$$(u_0, T, \varepsilon_p) \in L^q(0, T_e; H_1^q(\Omega, \mathbb{R}^3)) \times L^p(Z, \mathcal{S}^3) \times H_1^q([0, T_e], L^q(\Omega, \mathcal{S}^3))$$

of the Neumann problem (9)–(12), (7). All solutions of this problem are obtained in the form $(u, T, \varepsilon_p) = (u_0, T, \varepsilon_p) + (w, 0, 0)$, where $w(x,t) = a(t) + \omega(t) \times x$ with $a, \omega \in L^q([0, T_e], \mathbb{R}^3)$ is an infinitesimal rigid motion.

The proof of this theorem is based on the reduction of the Dirichlet initial-boundary value problem (9)–(12), (6) to an initial value problem for an evolution equation in the space

$$H^q_{\mathrm{sol},D} = \big\{\sigma \in L^q(\Omega, \mathcal{S}^3) \mid \mathrm{div}(\mathcal{D}\sigma) = 0\big\},$$

the Neumann initial-boundary value problem is reduced to an initial value problem in the space

$$H^q_{\mathrm{sol},N} = \big\{\sigma \in L^q(\Omega, \mathcal{S}^3) \mid \mathrm{div}(\mathcal{D}\sigma) = 0\,,\ (\mathcal{D}\sigma)_{|\partial\Omega} n = 0\big\}\,.$$

This initial value problem is

$$\frac{\partial}{\partial t}\tau(t) = Q\mathcal{H}[-\mathcal{D}\tau + \hat{\sigma}](t), \qquad 0 \le t \le T_e \tag{18}$$

$$\tau(0) = Q\varepsilon_p^{(0)}, \tag{19}$$

where $Q : L^q(\Omega, \mathcal{S}^3) \to L^q(\Omega, \mathcal{S}^3)$ is the projection onto the closed subspace $H^q_{\mathrm{sol},D}$ with $\ker Q = \{\varepsilon(\nabla u) \mid u \in \overset{\circ}{H}{}^p_1(\Omega, \mathbb{R}^3)\}$ in the Dirichlet case, and onto the closed subspace $H^q_{\mathrm{sol},N}$ with $\ker Q = \{\varepsilon(\nabla u) \mid u \in H^p_1(\Omega, \mathbb{R}^3)\}$ in the Neumann case. The solution τ satisfies $\tau = Q\varepsilon_p$, and for almost every t the function $x \mapsto \hat{\sigma}(x,t)$ is the Cauchy stress in the solution of the linear boundary value problem, which consists of the equations (9), (10) with$\varepsilon_p = 0$, and the Dirichlet or Neumann boundary condition. Using arguments from the theory of evolution equations to monotone operators, cf. Barbu [4], it can be shown that the initial value problem (18), (19) has a unique solution.

3. Existence of solutions for two examples of constitutive models

We apply the preceding result to study the existence of solutions to the initial-boundary value problems (1)–(7) for two examples of constitutive models. The first example is the Norton-Hoff law:

$$-\mathrm{div}_x T = b \tag{20}$$

$$T = \mathcal{D}(\varepsilon(\nabla_x u) - \varepsilon_p) \tag{21}$$

$$\frac{\partial}{\partial t}\varepsilon_p = c|T|^r \frac{T}{|T|} \tag{22}$$

$$\varepsilon_p(0) = \varepsilon_p^{(0)}, \tag{23}$$

with constants $c > 0, r > 1$. The second example incorporates kinematic hardening:

$$-\mathrm{div}_x T = b \tag{24}$$

$$T = \mathcal{D}(\varepsilon(\nabla_x u) - \varepsilon_p) \tag{25}$$

$$\frac{\partial}{\partial t}\varepsilon_p = c_1|T - k(\varepsilon_p - \varepsilon_n)|^r \frac{T - k(\varepsilon_p - \varepsilon_n)}{|T - k(\varepsilon_p - \varepsilon_n)|} \tag{26}$$

$$\frac{\partial}{\partial t}\varepsilon_n = c_2(k|\varepsilon_p - \varepsilon_n|)^\gamma \frac{\varepsilon_p - \varepsilon_n}{|\varepsilon_p - \varepsilon_n|} \tag{27}$$

$$\varepsilon_p(0) = \varepsilon_p^{(0)}, \qquad \varepsilon_n(0) = \varepsilon_n^{(0)}, \tag{28}$$

where $c_1, c_2, k > 0$ and $r, \gamma > 1$ are constants, and where the internal variable $\varepsilon_n(x,t) \in \mathcal{S}^3$ is of the type of a strain tensor.

For both models it can be shown that the history functional defined by the constitutive equations satisfies the conditions (H1)–(H4). This yields the following results:

Theorem 3.1 (Norton-Hoff law) *Let $c > 0$ and $r > 1$ be constants, and let $p = 1+r$, $q = 1+\frac{1}{r}$. Then under the regularity assumptions for b, $\varepsilon_p^{(0)}$, γ_D and γ_N from Theorem 2.1 the assertions of that theorem also hold for the Dirichlet and Neumann initial-boundary value problems to the equations (20)–(23).*

Theorem 3.2 (Kinematic hardening) *Let c_1, c_2, k be positive constants and let the constants r and γ satisfy $\gamma > r > 1$. Set $p = 1+r$, $q = 1+\frac{1}{r}$, $\hat{p} = 1+\gamma$, $\hat{q} = 1+\frac{1}{\gamma}$. Suppose that $b \in L^p(Z, \mathbb{R}^3)$ and $\varepsilon_p^{(0)}, \varepsilon_n^{(0)} \in L^2(\Omega, \mathcal{S}^3)$. Then the following assertions hold:*
(i) *For $\gamma_D \in L^p\big(0, T_e; H_1^p(\Omega, \mathbb{R}^3)\big)$ there is a unique solution*

$$\begin{aligned} (u, T, \varepsilon_p, \varepsilon_n) &\in L^q\big(0, T_e; H_1^q(\Omega, \mathbb{R}^3)\big) \times L^p(Z, \mathcal{S}^3) \\ &\times H_1^q\big([0, T_e], L^q(\Omega, \mathcal{S}^3)\big) \times H_1^{\hat{q}}\big([0, T_e], L^{\hat{q}}(\Omega, \mathcal{S}^3)\big) \end{aligned}$$

of the Dirichlet problem to the equations (24)–(28).
(ii) *Assume that $\gamma_N \in L^p(\partial\Omega \times [0, T_e], \mathbb{R}^3)$ and that b and γ_N satisfy (17). Then there exists a solution*

$$\begin{aligned} (u_0, T, \varepsilon_p, \varepsilon_n) &\in L^q\big(0, T_e; H_1^q(\Omega, \mathbb{R}^3)\big) \times L^p(Z, \mathcal{S}^3) \\ &\times H_1^q\big([0, T_e], L^q(\Omega, \mathcal{S}^3)\big) \times H_1^{\hat{q}}\big([0, T_e]; L^{\hat{q}}(\Omega, \mathcal{S}^3)\big) \end{aligned}$$

of the Neumann problem to the equations (24)–(28). All solutions of this problem are obtained in the form $(u, T, \varepsilon_p, \varepsilon_n) = (u_0, T, \varepsilon_p, \varepsilon_n) + (w, 0, 0)$, where $w(x, t) = a(t) + \omega(t) \times x$ with $a, \omega \in L^q([0, T_e], \mathbb{R}^3)$.

4. The coercivity conditions

Here we discuss the meaning of the coercivity conditions (H3) and (H4). For simplicity we only consider the Neumann problem with homogeneous boundary data. Assume that (u, T, ε_p) is a solution of the Neumann initial-boundary value problem to the equations (9)–(12). For the positive semi-definite free energy

$$\rho\psi(\varepsilon\nabla_x u), \varepsilon_p) = \frac{1}{2}[\mathcal{D}(\varepsilon(\nabla_x u) - \varepsilon_p)] \cdot (\varepsilon(\nabla_x u) - \varepsilon_p)$$

we then obtain by formal differentiation and partial integration that

$$\begin{aligned} &\frac{d}{dt}\int_\Omega \rho\psi(\varepsilon(\nabla_x u), \varepsilon_p)\, dx = \int_\Omega T \cdot (\varepsilon(\nabla_x u_t) - \varepsilon_{pt})\, dx \\ &= \int_\Omega (-\operatorname{div}_x T) \cdot u_t - T \cdot \mathcal{H}[T]\, dx = \int_\Omega b \cdot u_t dx - (T, \mathcal{H}[T])_\Omega \,. \end{aligned}$$

It follows that $(T, \mathcal{H}[T])_Z$ is the energy dissipated during the time interval $[0, T_e]$ due to plastic deformation.

For the Norton-Hoff law the definition (13)–(16) of the history functional reduces to the equation

$$\mathcal{H}[T] = f(T)$$

with $f(T) = c|T|^r \frac{T}{|T|}$. Assume now that there is a volume force $b : \Omega \to \mathbb{R}^3$ with the following property: To the stress field $\hat{\sigma}$ in the solution $(\hat{u}, \hat{\sigma})$ of the (time independent) Neumann boundary value problem to the equations (20), (21) with data b and with $\varepsilon_p = 0$ there is a stress function $(x \to \sigma(x)) \in \mathcal{D}H^p_{\text{sol},N}$ such that $f(\sigma + \hat{\sigma})$ is a gradient field, i.e.

$$f(\sigma + \hat{\sigma}) \in \{\varepsilon(\nabla_x v) \mid v \in H^q_1(\Omega, \mathbb{R}^3)\}. \tag{29}$$

Thus, to these $\hat{\sigma}$ and σ there is a function v with $\varepsilon(\nabla_x v) = f(\sigma + \hat{\sigma})$. We set

$$\varepsilon_p^{(0)} = \varepsilon(\nabla \hat{u}) - \mathcal{D}^{-1}\sigma.$$

Then (u, T, ε_p) with

$$\begin{aligned} u(x,t) &= \hat{u}(x) + tv(x) \\ T(x,t) &= \sigma(x) + \hat{\sigma}(x) \\ \varepsilon_p(x,t) &= tf(\sigma(x) + \hat{\sigma}(x)) + \varepsilon_p^{(0)}(x) \end{aligned}$$

is a solution of the homogeneous Neumann initial-boundary value problem to the equations (20)–(23) with volume force $b(x,t) = b(x)$. Though the volume force b is constant in time, this solution shows indefinite plastic deformation. Condition (H3) excludes that a sequence $\sigma_n \in H^p_{\text{sol},N}$ exists with $\|\sigma_n\|_{p,\Omega} \to \infty$ for $n \to \infty$, such that $f(\sigma_n + \hat{\sigma}) \in \{\varepsilon(\nabla_x v) \mid v \in H^q_1(\Omega, \mathbb{R}^3)\}$. For, this would imply

$$(\sigma_n + \hat{\sigma}, \mathcal{H}[\sigma_n + \hat{\sigma}])_Z = (\sigma_n + \hat{\sigma}, f(\sigma_n + \hat{\sigma})_Z$$
$$= (\sigma_n + \hat{\sigma}, \varepsilon(\nabla_x v))_Z = (\hat{\sigma}, \varepsilon(\nabla_x v))_Z = (\hat{\sigma}, \mathcal{H}[\sigma_n + \hat{\sigma}])_Z,$$

which immediately shows that (H3) cannot be satisfied. Hence, (H3) excludes that solutions of the type just constructed exist for large stress fields $\sigma_n + \hat{\sigma}$.

These observations are connected to the collapse of solutions for elasto-plastic constitutive relations: To see this, replace (22) by a rate independent constitutive relation, for example by the Prandtl-Reuss law,

$$\frac{\partial}{\partial t}\varepsilon_p \in \partial\chi(T), \tag{30}$$

with the subdifferential $\partial\chi$ of the characteristic function $\chi : \mathcal{S}^3 \to [0, \infty]$ of a closed convex set $K \in \mathcal{S}^3$,

$$\chi(T) = \begin{cases} 0 & , \quad T \in K \\ \infty & , \quad T \in \mathcal{S}^3 \backslash K, \end{cases}$$

cf. Alber [1][p. 31]. Let $b, \hat{u}, \hat{\sigma}, \sigma, \varepsilon_p^{(0)}$ be as above, and instead of (29) suppose that there is a nonvanishing function $v \in H_1^q(\Omega, \mathbb{R}^3)$ such that

$$\varepsilon(\nabla_x v(x)) \in \partial\chi(\sigma(x) + \hat{\sigma}(x)) \tag{31}$$

for almost all $x \in \Omega$. Then for every function $\kappa : [0, T_e] \to [0, \infty)$ with $\kappa(0) = 0$ the function (u, T, ε_p) defined by

$$\begin{aligned} u(x,t) &= \hat{u}(x) + \kappa(t) v(x) \\ T(x,t) &= \sigma(x) + \hat{\sigma}(x) \\ \varepsilon_p(x,t) &= \kappa(t)\varepsilon(\nabla_x v(x)) + \varepsilon_p^{(0)}(x) \end{aligned}$$

solves the homogeneous Neumann initial-boundary value problem to the equations (20), (21), (30), (23). In particular, for every $T_e > 0$ we can choose κ such that $\kappa(t) \to \infty$ for $t \to T_e$. Thus, if (31) holds we can construct solutions which blow up (or collapse, in another terminology) in an arbitrarily short time. Therefore in any existence theory for the Prandtl-Reuss law the possibility that (31) can hold must be excluded by a save-load condition slightly stronger than the coercivity condition (H3). It is known that such a safe-load condition restricts the choice of the volume force and the boundary data, cf. Johnson [8].

References

[1] Alber, H.-D. (1998). Materials with memory. *Springer Lecture Notes in Mathematics*, **1682**, Berlin, Springer.

[2] Alber, H.-D. (2001). Existence of solutions to a class of quasi-static problems in viscoplasticity theory. In: *Optimal Control and Partial Differential Equations*, J.L. Menaldi, E. Rofman, A. Sulem (eds.), Amsterdam, IOS Press, 95–104.

[3] Alber, H.-D. and Chełmiński, K. (2001). *Quasi-static problems in viscoplasticity theory.* Manuscript to be published.

[4] Barbu, V. (1976). *Nonlinear semigroups and differential equations in Banach spaces.* Bucuresti, Editura Academiei and Leyden, Noordhoff.

[5] Chełmiński, K. (2001). Coercive approximation of viscoplasticity and plasticity. *Asymptotic Analysis*, **26**, 105–133.

[6] Halphen, B. and Nguyen, Q.S. (1975). Sur les matériaux standards généralisés. *J. Méc.*, **14**, 508–520.

[7] Han, W. and Reddy, B. (1999). *Plasticity. Mathematical theory and numerical analysis.* New York, Springer.

[8] Johnson, C. (1976). Existence theorems for plasticity problems. *J. Math. Pures Appl.*, **55**, 431–444.

Variational Methods in Non-Convex Plasticity

S. Aubry, M. Ortiz
Graduate Aeronautical Laboratories, California Institute of Technology
1200 East California Boulevard, Pasadena, CA 91125, U.S.A
ortiz@aero.caltech.edu

Abstract: We show how the theory of crystals with microstructure developed by Ortiz et al. can be applied to predict salient aspects of the body of experimental data compiled by Hughes et al. regarding lamellar dislocation structures in crystals deformed to large strains. The theory correctly predicts the statistics of misorientation angles and lamellar boundary spacings; and the scaling of the average misorientation and spacing with increasing macroscopic strain.

Keywords: Crystals, microstructure, laminates, lamellar boundaries, misorientation angle, lamellar-boundary spacings, plasticity, large deformations, latent hardening, relaxation.

1. Introduction

The dislocation structures which form in metallic crystals at large strains have been extensively investigated by Hansen and Hughes [9, 3, 10, 11, 12, 7, 13]. The dislocation structures observed at large strains in a wide variety of metals are regular lamellar structures containing arrays of so-called GNBs (geometrically necessary boundaries), i. e., dislocation walls generated as a result of the plastic deformation of the crystal.

Hughes et al. [14, 8] have studied the scaling of the misorientation angles across subgrain dislocation boundaries following varying amounts of deformation in cold-rolled aluminum and nickel, and in compressed stainless steel. Remarkably, when scaled by the average angle, the probability density functions for the small and large deformation regimes are nearly identical in all the materials studied, which led the authors to suggest that distribution may be, in some sense, universal. The scaling of deformation induced dislocation boundaries has been investigated by Godfrey and Hughes [2]. They have found that the probability density distribution for spacings exhibits a scaling behavior of nearly identical form to that observed for the probability density distribution of misorientation angles across boundaries.

The mechanistic understanding of these phenomena requires a theory of large plastic deformation of metallic crystals *which directly accounts for the formation of complex lamellar dislocation structures*. In particular, the theory must allow the structures to form under applied stress, and it must predict its geometry, layout, the orientation of the dislocation boundaries and of the intervening

C. Miehe (ed.), IUTAM Symposium on Computational Mechanics of Solid Materials at Large Strains, 55-66.

cells, the spacing between boundaries, and the attendant macroscopic behavior of the crystal.

Ortiz et al. [19, 20] have proposed a theory of single crystal plasticity which meets these requirements. In this theory, the formation of dislocation structures is the result of strong latent hardening, namely, the fact that crystals exhibit substantially higher hardening rates when deforming in multiple slip than when deforming in single slip. In this paper, a limiting case is studied. The crystals are assumed to exhibit infinitely strong latent hardening i. e. the crystal must necessarily deform in single slip at all material points. This requirement introduces a nonconvex constraint which renders the incremental problem nonconvex. Materials can attain plastic deformations other than single slip – and thus beat the single-slip constraint – by developing deformation microstructures. These microstructures may be expected to activate different slip systems over different regions of the crystal, possibly arranged in complex spatial patterns.

In this paper, we briefly recall the theory of single slip plasticity and show how it suffices to give mathematical description to the microstructures observed by Hughes and Hansen [9, 3, 10, 11, 12, 7, 13].

This paper has been condensed from a more extensive article by the authors (The Mechanics of Deformation-Induced Subgrain Dislocation Structures in Metallic Crystals at Large Strains, Proceedings of the Royal Society of London, Series A, submitted 2002), which may be consulted for further details.

2. Constitutive basis of the constrained theory

We begin by investigating the limiting case of crystals exhibiting infinite latent hardening. In this limit, the local kinematics of plastic deformation is restricted to processes of single slip. This results in a *constrained* theory, in the sense that the possible directions of the plastic rate-of-deformation tensor $\dot{\boldsymbol{F}}^p \boldsymbol{F}^{p-1}$ are restricted to a small set of values determined by crystallography.

2.1. Local constitutive framework

We assume a conventional multiplicative decomposition of the deformation gradient $\boldsymbol{F}$ of the form (Lee [17]):

$$\boldsymbol{F} = \boldsymbol{F}^e \boldsymbol{F}^p \tag{1}$$

where $\boldsymbol{F}^e$ is the elastic component of the deformation gradient and $\boldsymbol{F}^p$ is the plastic part.

The free energy of the crystal is of the form (Rice [21]):

$$A = W^e(\boldsymbol{F}^e) + W^p \tag{2}$$

where W^e is the material frame indifferent elastic strain-energy density,W^p is the stored energy of cold work. The first Piola-Kirchhoff stress tensor follows from the free energy as

$$\boldsymbol{P} = A,_{\boldsymbol{F}} = W^e,_{\boldsymbol{F}^e}(\boldsymbol{F}^e)\boldsymbol{F}^{p-T} \tag{3}$$

We shall additionally assume that the elastic strain-energy density is such that the only stress-free states are pure rotations.

In the limiting regime of infinite latent hardening investigated here, processes of multiple slip are barred altogether and the crystal is constrained to deform locally in single slip. Thus, locally at every material point$\boldsymbol{x}$, we must necessarily have

$$\dot{\boldsymbol{F}}^p(\boldsymbol{x})\boldsymbol{F}^{p-1}(\boldsymbol{x}) = \dot{\gamma}(\boldsymbol{x})\boldsymbol{s}(\boldsymbol{x}) \otimes \boldsymbol{m}(\boldsymbol{x}) \tag{4}$$

where $(\boldsymbol{s}(\boldsymbol{x}), \boldsymbol{m}(\boldsymbol{x})) \in \mathcal{S}$ is the sole slip system which is active at material point $\boldsymbol{x}$ and $\dot{\gamma}(\boldsymbol{x})$ is the corresponding slip strain. We have denoted$\mathcal{S}$ the set of all slip systems. It should be carefully noted that$\mathcal{S}$ is determined by crystallography and that the vectors $(\boldsymbol{s}, \boldsymbol{m})$ are constant and independent of deformation.

Slip as irreversible and we enforce the plastic irreversibility condition

$$\dot{\gamma} \geq 0 \tag{5}$$

We idealize processes of single-slip as being ideally plastic, i. e., we suppose the critical shear stresses τ_c to be constant. The local value of the stored energy is takes the simple form:

$$W^p(\boldsymbol{x}) = \tau_c \gamma(\boldsymbol{x}) \tag{6}$$

For simplicity, we also assume that the loading is slow enough that the rate sensitivity of the material may be neglected.

2.2. VARIATIONAL FORM OF INCREMENTAL CONSTITUTIVE RELATIONS

We assume that the local deformation is known at discrete times$t_0 = 0, \ldots,$ $t_{n+1} = t_n + \triangle t, \ldots$. The problem is to determine the corresponding evolution of the local state of the crystal. Suppose that the state of the crystal is known at time t_n, and that the deformation $\boldsymbol{F}_{n+1}$ is given at time t_{n+1}. The problem now is to determine the full state of the crystal at timet_{n+1}.

Under conditions of local incremental single slip, the plastic deformation$\boldsymbol{F}^p_{n+1}$ of the crystal at time t_{n+1} follows from the plastic deformation $\boldsymbol{F}^p_n$ of the crystal at time t_n as

$$\boldsymbol{F}^p_{n+1} = (\boldsymbol{I} + \triangle\gamma \boldsymbol{s} \otimes \boldsymbol{m})\boldsymbol{F}^p_n, \quad (\boldsymbol{s}, \boldsymbol{m}) \in \mathcal{S} \tag{7}$$

where

$$\triangle\gamma = \gamma_{n+1} - \gamma_n \tag{8}$$

is the incremental slip strain, and $(\boldsymbol{s}, \boldsymbol{m}) \in \mathcal{S}$ is the active slip system operating during the interval (t_n, t_{n+1}) at the material point under consideration.

For given $\boldsymbol{F}_{n+1}$, the optimal value of γ_{n+1} and the active slip system $(\boldsymbol{s}, \boldsymbol{m})$ are the solutions of the constrained minimization problem

$$W_n(\boldsymbol{F}_{n+1}) \equiv \min_{\substack{\gamma_{n+1},\ (\boldsymbol{s},\boldsymbol{m})\in\mathcal{S} \\ \text{subject to } \triangle\gamma\geq 0}} W^e(\boldsymbol{F}_{n+1}\boldsymbol{F}^{p-1}_{n+1}) + \tau_c \triangle\gamma \tag{9}$$

where $\boldsymbol{F}^p_{n+1}$ is given by (7) in terms of γ_{n+1} and $(\boldsymbol{s}, \boldsymbol{m})$. The Euler-Lagrange equation corresponding to γ_{n+1} is

$$-\tau_{n+1} + \tau_c = 0 \tag{10}$$

and thus, minimization in equation (9) with respect to γ_{n+1} is equivalent to enforcing the yield criterion.

The minimization problem governing the incremental state update may be solved simply as follows. We proceed by evaluating each possible slip system $(\boldsymbol{s}, \boldsymbol{m}) \in \mathcal{S}$ in turn. For each candidate slip system, we begin by assuming $\triangle\gamma = 0$, resulting in a purely elastic step. If the resulting resolved shear stress satisfies $\tau_{n+1} \leq \tau_c$, then the assumption is correct and the update is complete. Otherwise, $\triangle\gamma > 0$ and the step is plastic. Under these conditions, the value of γ_{n+1} follows from the yield condition (9), which may be solved, e. g., by a Newton-Raphson iteration. The active slip system then follows as that system for which $W_n(\boldsymbol{F}_{n+1})$ is minimized. In cases in which several systems deliver the same minimum value of $W_n(\boldsymbol{F}_{n+1})$, the local incremental update is set-valued.

3. Microstructure development and dislocation structures

Ortiz and Repetto [19] pointed out that the presence of strong latent hardening renders the incremental work-density function $W_n(\boldsymbol{F}_{n+1})$ nonconvex, with single-slip deformations strongly preferred over multiple-slip deformations. In order to accommodate an arbitrary deformation $\boldsymbol{F}_{n+1}$ by means of local deformations entailing single slip only, the crystal must necessarily develop microstructure. Within the present framework, microstructures may be allowed for by means of a suitable re-interpretation of (9). To this end, we consider a representative volume Ω of the crystal. Within this volume, the crystal is permitted to deform in accordance with a general deformation field $\boldsymbol{y}_{n+1} : \Omega \to \mathbb{R}^3$. This deformation field is required to match a macroscopic deformation $\bar{\boldsymbol{F}}_{n+1}$ on the boundary $\partial\Omega$ of the representative volume, i. e.,

$$\boldsymbol{y}_{n+1}(\boldsymbol{x}) = \bar{\boldsymbol{F}}_{n+1}\boldsymbol{x}, \quad \boldsymbol{x} \in \partial\Omega \tag{11}$$

In the presence of microstructure, the minimum principle (9) becomes

$$\bar{W}_n(\bar{\boldsymbol{F}}_{n+1}) = \inf_{\substack{\boldsymbol{y}_{n+1} \\ \text{subject to: } \boldsymbol{y}_{n+1|\partial\Omega}(\boldsymbol{x})=\bar{\boldsymbol{F}}_{n+1}\boldsymbol{x}}} \frac{1}{|\Omega|}\int_\Omega W_n(D\boldsymbol{y}_{n+1})dx \qquad (12)$$

where $\bar{W}_n$ is the *effective* or *macroscopic* work-of-deformation density, $D\boldsymbol{y}_{n+1}(\boldsymbol{x}) = \boldsymbol{F}_{n+1}(\boldsymbol{x})$ is the local deformation gradient, and $\gamma_{n+1}(\boldsymbol{x})$, $\boldsymbol{s}(\boldsymbol{x})$ and $\boldsymbol{m}(\boldsymbol{x})$ are now to be regarded as fields over the representative volume Ω.

4. Special microstructures: sequential laminates

Unfortunately, it is not known at present how to characterize the *optimal microstructures* $\boldsymbol{y}_{n+1}$ leading to the least possible incremental work of deformation $\bar{W}_n(\bar{\boldsymbol{F}}_{n+1})$. Instead, we content ourselves with investigating special microstructures which lend themselves to analytical treatment. Specifically, we consider microstructures which may be described as sequential laminates (Kohn and Strang [16], Kohn [15], Ortiz and Repetto [19], Ortiz et al. [20]).

Sequential laminates define a special microstructures which is composed by layers or lamellae and layers within layers. In each layer, a uniform deformation gradient can be defined. Since the macroscopic deformation has to be continuous in this microstructure, compatibility requirements are imposed to these deformation gradients. It can be shown (Ortiz and Repetto [19], Ortiz et al. [20]) that only a small number of parameters are necessary to determine explicitly a sequential laminate and ensure all compatibilities across boundary layers. Namely, the local deformation gradient $\boldsymbol{F}$, proportions of laminations $\lambda_i^{\pm}$, normals at the interface $\boldsymbol{N}_i^{\pm}$ and measures of the jump at the interface $\boldsymbol{a}_i^{\pm}$.

The macroscopic work-of-deformation function (12) is replaced by the explicitly known work-of-deformation of the sequential laminate $\bar{W}_n(\bar{\boldsymbol{F}}_{n+1})$ with is given by a minimization problem over variables: $\{\boldsymbol{a}_i, \lambda_i^{\pm}, \boldsymbol{N}_i\}$.

The Euler-Lagrange equations or equations of equilibrium corresponding to this minimization problem expresses the requirement of continuity of tractions across dislocation walls.

5. Nonlocal extension and microstructural size evolution

The model described in the foregoing is local, i. e., it contains no intrinsic length scale. In order to estimate the microstructural size and its evolution, e. g., the evolution of lamellar width with deformation, we take into account non local energy terms. As noted by Ortiz and Repetto [19] and Ortiz et al. [20], within this extension the size of the microstructure results from the competition of two energy terms: the dislocation wall energy; and the energy of the intervening boundary layers between laminates.

5.1. Dislocation-wall energy

Consider a lamella of thickness l, Ortiz et al. [19, 20] have shown that an approximation of the energy of the dislocation wall is given by the simple relation

$$W^p_{n+1} = W^p_n + \left(\tau_c + \frac{T}{bl}\right)\Delta\gamma \tag{13}$$

where $\boldsymbol{b}$ is the Burgers vector, T is the dislocation line tension or energy per unit length.

This result means that, under conditions of local single slip, *the effect of the dislocation wall energies is to increase the yield strength of the lamella in the amount T/bl.*

5.2. Interpolation boundary layer

A simple estimate of the energy of the boundary layer can be shown to be given by

$$W^{\mathrm{BL}} = \lambda^- W^e(\boldsymbol{F}^{\mathrm{BL}-}) + \lambda^+ W^e(\boldsymbol{F}^{\mathrm{BL}+}) - \bar{W}_n(\boldsymbol{F}) \tag{14}$$

where

$$\boldsymbol{F}^{\mathrm{BL}\pm} = \frac{1}{2}(\boldsymbol{F}^{\pm} + \boldsymbol{F}) \tag{15}$$

5.3. Nonlocal energy

Let the thicknesses of the lamellae be $l^{\pm}$, and the combined thickness of both lamellae $l^c = l^- + l^+$, we have $l^{\pm} = \lambda^{\pm} l^c$.

Combining the above results, it follows that total excess or *nonlocal* energy due to the dislocation walls and the misfit boundary layers contained within a region of the laminate of dimensions $L \times L \times l$ takes the form:

$$E^{\mathrm{NL}} = E^{\mathrm{DW}} + E^{\mathrm{BL}} = L^2 \left\{ \frac{T}{bl^c}(\gamma^- + \gamma^+) + \frac{l_c}{l} W^{\mathrm{BL}} \right\} \tag{16}$$

This excess energy may now be minimized with respect to l^c, with the result:

$$l^c = \sqrt{\frac{T}{b}\frac{(\gamma^- + \gamma^+)l}{W^{\mathrm{BL}}}} \tag{17}$$

which affords an estimate of the lamella size l^c.

6. Comparison with experimental data

The local orientation of the crystal lattice in a plastically deformed crystal is determined by the rotation $\boldsymbol{R}^e$ in the polar decomposition $\boldsymbol{F}^e = \boldsymbol{R}^e \boldsymbol{U}^e$ of the elastic deformation $\boldsymbol{F}^e$. Lamellar microstructures are characterized by the development of uniformly deformed lamellae whose crystal lattice orientations are rotated with respect to one another. Thus, across a lamellar boundary the lattice rotation jumps discontinuously from $\boldsymbol{R}^{e-}$ to $\boldsymbol{R}^{e+}$. The angle θ of the rotation $\boldsymbol{R}^{e+}(\boldsymbol{R}^{e-})^{-1}$ required to realign the lamellar crystal lattices is called the *misorientation angle.*

In order to simulate cold rolling, we exercise the model presented in the foregoing along a macroscopic deformation path:

$$\bar{\boldsymbol{F}} = \begin{pmatrix} \lambda & 0 & 0 \\ 0 & \lambda^{-1} & 0 \\ 0 & 0 & 1 \end{pmatrix}, \tag{18}$$

for λ increasing monotonically in time from an initial value of 1. The elastic constants used in the calculations are: $c_{11} = 168.4$ GPa, $c_{12} = 121.4$ GPa and $c_{44} = 75.4$ GPa. In addition, the critical resolved shear stress is taken to be $\tau_c = 0.1$ GPa, and the dislocation line tension $T = 10$ N. The Burgers vector size is $b = 2.5$ Å. The small number of parameters of the theory is particularly noteworthy. A statistical sample of misorientation angles is generated by applying the macroscopic deformation path to an ensemble of randomly rotated crystals. Thus, each member of this ensemble is characterized by an elastic strain-energy density of the form $W^e(\boldsymbol{F}\boldsymbol{Q})$, and a slip-system set: $\mathcal{S} = \{(\boldsymbol{Q}\boldsymbol{s}^\alpha, \boldsymbol{Q}\boldsymbol{m}^\alpha), \alpha = 1, \dots, N\}$, where $\boldsymbol{Q} \in SO(3)$ and $\{(\boldsymbol{s}^\alpha, \boldsymbol{m}^\alpha), \alpha = 1, \dots, N\}$ is the slip-system set in some reference orientation of the crystal. In all calculations reported subsequently the sample size is $1,000$ crystal orientations. The rotations $\boldsymbol{Q}$ are generated randomly and are taken to be uniformly distributed over $SO(3)$.

The statistical properties of misorientation angles and the lamella width l or GNB spacing have been extensively investigated by Hansen, Hughes et al. (e.g. Hughes and Hansen [4, 9, 3, 11], Hughes et al. [5], Hughes and Nix [6], Zimmer et al. [22]). Their data reveals that the effect of increasing the macroscopic strain is to shift the distribution of both the misorientation angle θ and the GNB spacing l towards larger values. This shift is accompanied by and increase in the width of the distribution, and, correspondingly, by a lowering of its peak. In addition, by scaling the variable by its mean all the data ostensibly collapse onto a single universal curve independent of the macroscopic strain. The re-scaled data are well-approximated by a probability density function of the form:

$$f(\theta, \theta_{\rm av}) = \frac{\alpha^\alpha}{\Gamma(\alpha)} \left(\frac{\theta}{\theta_{\rm av}}\right)^{\alpha-1} {\rm e}^{-\alpha\theta/\theta_{\rm av}} \tag{19}$$

where the exponent $\alpha \approx 2.5$.

Furthermore, the measured average misorientation θ_{av} against the macroscopic von Mises strain bear a power-law relation of the form

$$\theta_{\text{av}} = k(\epsilon_{vM})^{\delta} \tag{20}$$

where

$$\epsilon_{vM} = -\frac{2}{\sqrt{3}} \log \lambda, \tag{21}$$

λ is the upsetting ratio, and k and δ are material dependent constants. For aluminum, $\delta \approx 2/3$. This scaling relation may be regarded as providing a link between a macroscopic measure of strain (ϵ_{vM}) and a property of the microstructure (θ_{av}).

Finally, in cold-rolled nickel (Hughes and Hansen [9]) and cold-rolled aluminum (Hansen and Hughes [3, 11]), the average spacing is observed to decrease monotonically with increasing deformation, often into the micrometer and sub-micrometer range.

We have reproduced numerically these experimental results and found a very good accordance with them. Fig. 1 collects the computed misorientation-angle statistics and compares it against the corresponding experimental fits. Fig. 1b shows the computed misorientation-angle frequencies after scaling by the mean misorientation θ_{av}. Fig. 1c in turn plots the computed average misorientation angle θ_{av} against the macroscopic von Mises strain ϵ_{vM}. Also included in this figure, by way of comparison, is the experimentally calibrated scaling law (20). The exponent which best fits the computed results is $\delta \approx 0.70$, which is in close agreement with the experimentally determined value of $2/3$.

Fig. 2 collects the computed polycrystalline spacing statistics and compares it against the corresponding experimental fit. Fig. 2a displays the computed spacing probability density functions for several upsetting ratios in the range of 0% to 40% of c. r. In these results, calculations have been carried out with a Poisson Voronoi tessellation distribution of grain size (Okabe et al. [18]) representing a variable grain size. As in the misorientation-angle case, the salient features of the experimental recordare in excellent qualitative agreement with observation. Fig. 2b shows the computed spacing frequencies after scaling by the mean spacing l_{av}. Fig. 2c finally plots the computed average spacing l_{av} against the macroscopic von Mises strain ϵ_{vM}. The tendency of the average spacing to decrease with increasing strain is well captured.

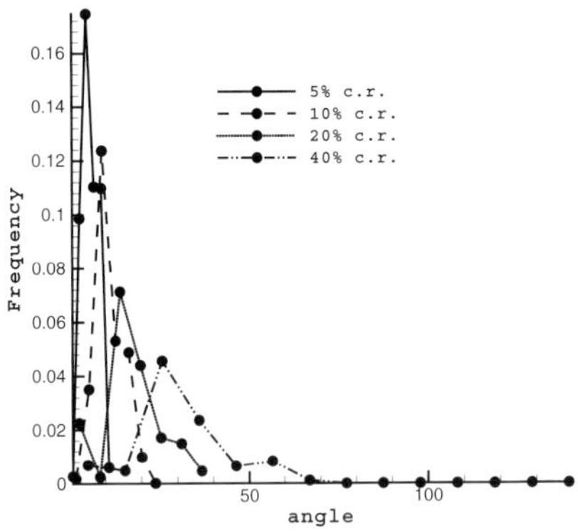

(a)

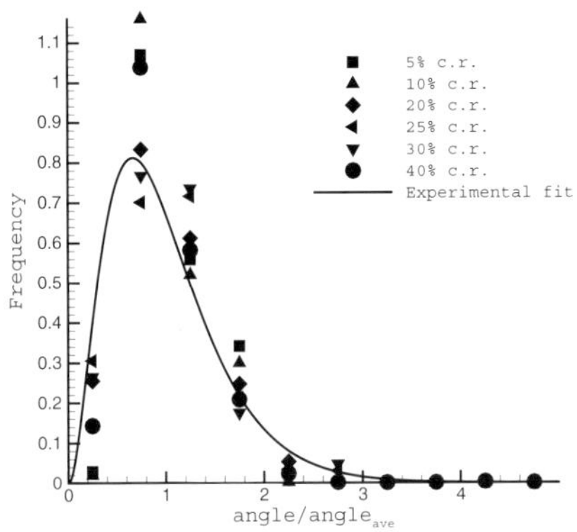

(b)

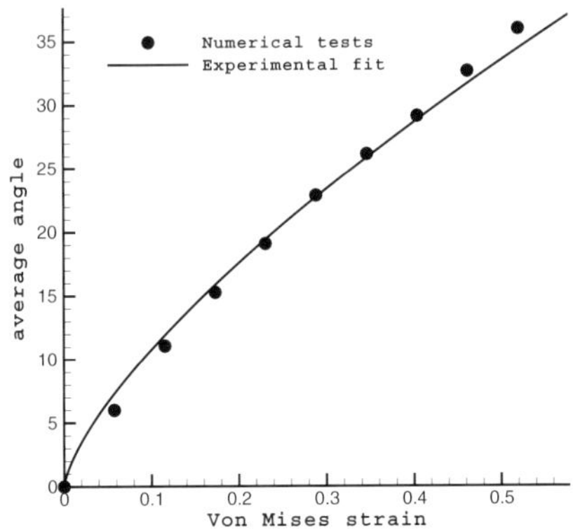

(c)

Figure 1. Computed misorientation-angle statistics. a) Misorientation-angle probability density. b) Misorientation-angle probability density normalized by the average misorientation angle. c) Average misorientation angle vs. von Mises strain. Extracted from Aubry and Ortiz [1].

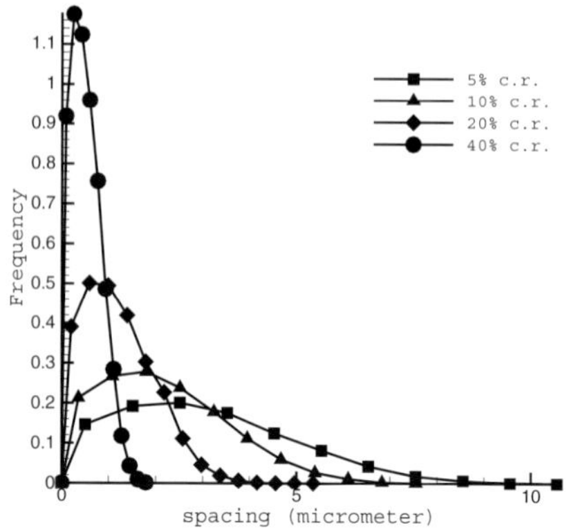

(a)

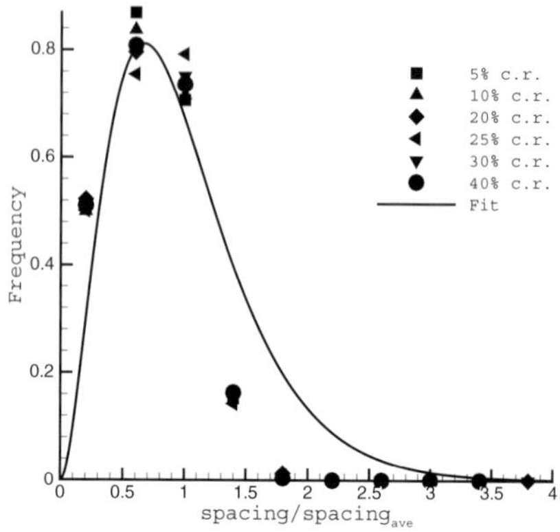

(b)

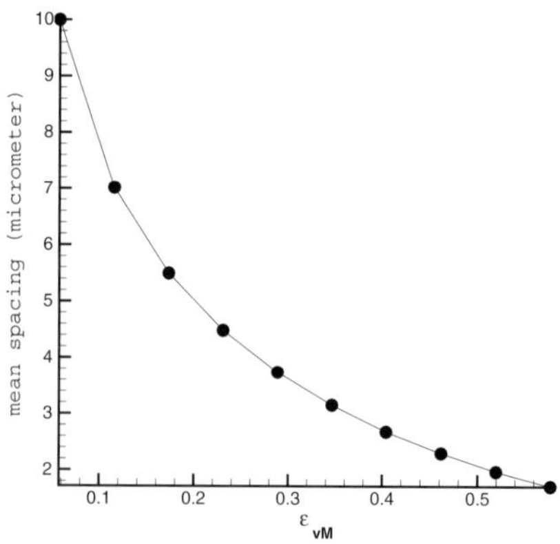

(c)

Figure 2. Computed spacing statistics. a) Spacing probability density. b) Spacing probability density normalized by the average spacing. c) Average spacing vs. von Mises strain. Spacings computed in a random distribution of grains. Extracted from Aubry and Ortiz [1].

Acknowledgments

We are grateful for support provided by the US Department of Energy through Caltech's ASCI/ASAP Center for the Simulation of the Dynamic Behavior of Solids. We are also indebted to Darcy Hughes for many useful discussions and suggestions, and for making her data available to us.

References

[1] Aubry, S. and Ortiz, M. (2002). The mechanics of deformation-induced subgrain dislocation structures in metallic crystals at large strains. *submitted to Proceedings of the royal society of london.*

[2] Godfrey, A. and Hughes, D.A. (2000). Scaling of the spacing of deformation induced dislocation boundaries. *Acta Materialia*, **48(8)**, 1897–1905.

[3] Hansen, N. and Hughes, D.A. (1995). Analysis of large dislocation populations in deformed metals. *Physica Status Solidi A-Applied Research*, **149(1)**, 155–172.

[4] Hughes, D.A. and Hansen, N. (1997). High angle boundaries formed by grain subdivision mechanisms. *Acta Materialia*, **45(9)**, 3871–3886.

[5] Hughes, D.A., Lui, A., Chrzan, D.C. and Hansen, N. (1997). Scaling of microstructural parameters: Misorientations of deformation induced boundaries. *Acta Materialia*, **45(1)**, 105–112.

[6] Hughes, D.A. and Nix, W.D. (1989). Strain hardening and substructural evolution in ni-co solid solutions at large strains. *Materials Science and Engineering*, **A122(2)**, 153–172.

[7] Hughes, D.A. (2001). Deformation microstructures and selected examples of their recrystallization. *Surface and Interface Analysis*, **31(7)**, 560–570.

[8] Hughes, D.A., Chrzan, D.C., Liu, Q. and Hansen, N. (1998). Scaling of misorientation angle distributions. *Physical Review Letters*, **81(21)**, 4664–4667.

[9] Hughes, D.A. and Hansen, N. (1993). Microstructural evolution in nickel during rolling from intermediate to large strains. *Metallurgical Transactions A-Physical Metallurgy and Materials Science*, **24(9)**, 2021–2037.

[10] Hughes, D.A. and Hansen, N. (1995). High-angle boundaries and orientation distributions at large strains. *Scripta Metallurgica et Materialia*, **33(2)**, 315–321.

[11] Hughes, D.A. and Hansen, N. (1997). High angle boundaries formed by grain subdivision mechanisms. *Acta Materialia*, **45(9)**, 3871–3886.

[12] Hughes, D.A. and Hansen, N. (2000). Microstructure and strength of nickel at large strains. *Acta Materialia*, **48(11)**, 2985–3004.

[13] Hughes, D.A. and Hansen, N. (2001). Graded nanostructures produced by sliding and exhibiting universal behavior. *Physical Review Letters*, **8713(13)**, 5503–+.

[14] Hughes, D.A., Liu, Q., Chrzan, D.C. and Hansen, N. (1997). Scaling of microstructural parameters: Misorientations of deformation induced boundaries. *Acta Materialia*, **45(1)**, 105–112.

[15] Kohn, R.V. (1991). The relaxation of a double-well energy. *Continuum Mechanics and Thermodynamics*, **3**, 193–236.

[16] Kohn, R.V. and Strang, G. (1986). Optimal design and relaxation of variational problems i, ii, iii. *Communication on Pure and Applied Mathematics*, **39**, 193–236, 139–182, 353–377.

[17] Lee, E.H. (1969). Elastic-plastic deformation at finite strains. *Journal of Applied Mechanics*, **36(1)**.

[18] Okabe, A., Boots, B. and Sugihara, K. (1992). *Spatial Tessellations: Concepts and Applications of Voronoi Diagrams*. John Wiley and Sons. (Chapter 4).

[19] Ortiz, M. and Repetto, E.A. (1999). Nonconvex energy minimization and dislocation structures in ductile single crystals. *Journal of the Mechanics and Physics of Solids*, **47(2)**, 397–462.

[20] Ortiz, M., Repetto, E.A. and Stainier, L. (2000). A theory of subgrain dislocation structures. *Journal of the Mechanics and Physics of Solids*, **48(10)**, 2077–2114.

[21] Rice, J.R. (1975). Continuum mechanics and thermodynamics of plasticity in relation to microscale deformation mechanisms. In Argon A. (Ed.), editor, *Constitutive Equations in Plasticity*, 23–79. MIT press, Cambridge (Mass).

[22] Zimmer, W.H., Hecker, S.S., Rohr, D.L. and Murr, L.E. (1983). Large strain plastic deformation of commercially pure nickel. *Metal Science*, **17(4)**, 198–206.

Pulling Phase-Transforming Bars: A Three-Dimensional Viewpoint

A. DeSimone
Max Planck Institute for Mathematics in the Sciences
Inselstraße 22–26, 04103 Leipzig, Germany
desimone@mis.mpg.de

Abstract: Phase-transforming bars may respond to imposed elongations by accommodating them through a phase change to a stretched phase, rather than by accumulating internal stresses. The aim of this paper is to contribute to the development of a fully three-dimensional perspective on the essentially one-dimensional geometry of a stretching experiment on a bar, through the case study of stretched sheets of nematic elastomers.

Keywords: Phase-transformations, calculus of variations, quasiconvexity, nonlinear elasticity, elastomers.

1. Introduction

Starting with his seminal papers [5, 6], J.L. Ericksen drew the attention of the Mechanics community to the possibility of exploiting nonlinear hyperelasticity to study some phase transitions occurring in crystalline solids. His one-dimensional analysis is in analogy with the classical treatment of the liquid-vapor phase transformation in a van der Waals fluid. The three-dimensional version of the same idea, however, is hardly a generalization of the one-dimensional cartoon. It requires the entirely new complication of handling the delicate constraints of kinematic compatibility and of frame indifference, both of which are trivial in dimension one.

The idea to model phase transitions of crystalline solids thorough the use of non-convex potentials within nonlinear elasticity has met with considerable success. The one-dimensional case has been studied in great detail, and it has proven a rather rich environment for generating interesting model problems. In fact, the simplicity of the geometrical setting not only leads to explicitly solvable problems, but it also allows one to lift the degeneracies of the original model through the introduction of new ingredients (higher order gradients, kinetic relations, etc.), and to explore in detail the impact of subtle details of these additional constitutive prescriptions.

The three-dimensional problem has been studied even more intensely, see, e.g., Ball and James [1], Müller [7], Šilhavý [9] and the many references quoted therein. From the experimental side, Shields [8] gives a hint of the rich-

C. Miehe (ed.), IUTAM Symposium on Computational Mechanics of Solid Materials at Large Strains, 67-76.

ness of behavior emerging from pulling on a pseudo-elastic crystalline bar. In spite of the many successes, however, the natural question of giving a fully three-dimensional derivation of a physically based stress-strain curve generated through the essentially one-dimensional geometry of a stretching experiment on a bar has proved quite elusive. Besides its practical and conceptual interest, the relevance of this question lies in the fact that one-dimensional models are unable to resolve the rich geometric structure of the phase mixtures which are ubiquitous in these materials, and that (at least without the introduction of a richer kinematic structure) they do not allow to distinguish between material instabilities of constitutive nature, from instabilities of geometric origin. For the study of these issues, the natural three-dimensional setting seems the appropriate one, and this paper is a contribution in this direction.

The paper is organized as follows. Section 2 gives the one-dimensional background. Here some of the original arguments of Ericksen are reviewed, with some slight changes of language motivated by recent results in Continuum Mechanics and in the Calculus of Variations. Section 3 gives a three-dimensional perspective taking advantage of some recent progress on the large deformation behavior of nematic elastomers.

2. Pulling one-dimensional bars

Consider a solid, phase-transforming hyperelastic bar, i.e., a one-dimensional object whose stable equilibrium configurations are obtained by minimizing a stored energy functional I with nonconvex linear density W. We adopt the following notation: $\Omega = (O, L)$ is the reference configuration; $y(x)$ and $u(x) = y(x) - x$ are the deformation and the displacement at x, respectively; $\varepsilon(x) = Du(x) = u'(x)$ is the strain at x; $W(\varepsilon)$ is the elastic energy density, with graph given in Fig. 1; $W'(\varepsilon) = \frac{\partial}{\partial Du} W(Du)$ is the internal "stress". Our goal is to minimize

$$I(u) = \int_0^L W\left(u'(x)\right) dx \tag{1}$$

subject to the displacement boundary conditions (hard loading)

$$u(0) = 0 \quad \text{and} \quad u(L) = \Delta\,. \tag{2}$$

The problem is solved by using the fact that, on a minimizer, I must be stationary with respect to *two* classes of perturbations. The resulting two conditions represent two force balances, namely, the balance of forces and the balance of configurational forces.

The balance of forces is the Euler-Lagrange equation of the given variational problem, obtained by requiring that I be stationary with respect to an arbitrary compactly supported smooth perturbation ϕ in the target space of the candidate minimizer u:

$$O = \frac{d}{dt} I(u + t\phi) \Bigg|_{t=0} = \int_0^L W'\left(u'(x)\right) \phi'(x) dx \,.$$

Since ϕ is arbitrary, Du Bois-Reymond Lemma implies that

$$W'\left(u'(x)\right) \equiv R \,, \tag{3}$$

where the unknown constant R has the physical interpretation of reactive force exerted by the restraint ensuring $u(L) = \Delta$. Assuming enough smoothness to integrate by parts, (3) is usually written as

$$\frac{d}{dx}\left(W'\left(u'(x)\right)\right) = 0 \,. \tag{4}$$

To solve (3), observe from Fig. 1 that, since a horizontal line at height R intersects the graph of W' at most at three points, then either $u'(x)$ is constant, or it belongs to a discrete set containing at most three elements, in which case it is discontinuous.

Non-smooth solutions (broken extremals in the language of classical Calculus of Variations) are dealt with the requirement that I be stationary also with respect to variations of the domain (the only ones that can move discontinuities). This provides an additional equation which enables us to completely solve the problem. Indeed, let $u_t(x) = u(\Phi_t(x))$, where the inverse of Φ_t is the identity plus t times an arbitrary compactly supported smooth perturbation η. Since

$$0 = \frac{d}{dt} I(u_t)\Big|_{t=0} = \int_0^L \left(W(Du(x))1 - W'(Du(x))Du(x)\right) \eta'(x) dx$$

and η is arbitrary, we obtain

$$W(Du(x)) - W'(Du(x))Du(x) \equiv \text{ constant} \,. \tag{5}$$

Integrating by parts, as before, we may write the last equation as

$$\frac{d}{dx}\left(W(Du(x)) - W'(Du(x))Du(x)\right) = 0 \,. \tag{6}$$

Equations (3), (5) solve the problem because they imply that for $x_1 \neq x_2$, the corresponding strains $u'(x_1), u'(x_2)$, must lie on the same tangent line to the graph of W'. A quick inspection of Fig. 1 shows that this can happen with $u'(x_1) \neq u'(x_2)$ only if $u'(x_i)$ is equal to either zero or ε^*.

The arguments above show that minimizers of (1) have either constant strain, or that they are such that the reference configuration is partitioned into subregions χ_0, χ_* in which the strain is identically zero or ε^*, respectively. Whenever the boundary condition leads to a non-uniform strain distribution (i.e., for $\Delta \in (0, \varepsilon_* L)$), minimizers of (1) are not unique. In this case, only the corresponding "volume" fractions

$$\lambda_* = \frac{1}{L} \int_{\chi_*} dx \,, \quad 1 - \lambda_* = \frac{1}{L} \int_{\chi_0} dx$$

are uniquely determined from compatibility with the boundary condition

$$\frac{\Delta}{L} = \frac{1}{L} \int_0^L u'(x) dx = \lambda_* \varepsilon^* \,,$$

which leads to $\lambda_* = \frac{\Delta/L}{\varepsilon^*}$. In spite of lack of complete information on the geometric structure of energy minimizing phase mixtures, the macroscopic response of the system they induce (i.e., the graph of the reaction force R versus the imposed average strain Δ/L) is uniquely determined, and it is obtained as the derivative of the convex envelope W^{**} of W. This envelope plays thus the role of the effective macroscopic energy of the system. The reason why convexity plays an important role becomes clearer if (5) is deduced in a different manner, namely from Weierstrass' necessary condition for (global or strong local) minimizers, stating that if u is a minimizer, then at every x in Ω,

$$W(\varepsilon) - W(u'(x)) - (\varepsilon - u'(x)) W'(u'(x)) \geq 0 \tag{7}$$

for all ε. In other words, only points of global convexity for the graph of W can participate in a minimizer.

We close this section by explicitly noting that indeed (4) and (6) are the one-dimensional versions of the balances of classical and configurational forces. In the higher-dimensional setting, the first variation arguments used above respectively lead to

$$\mathrm{Div}\, S = 0 \quad \text{"balance of forces"}$$
$$\mathrm{Div}\, C = 0 \quad \text{"balance of configurational forces"}$$

where S, the Piola-Kirchhoff stress tensor is given by

$$S = \frac{\partial}{\partial Dy} W(Dy)$$

and C, the Eshelby energy-momentum tensor, is given by

$$C = W(Dy) Id - (Dy)^T \frac{\partial}{\partial Dy} W(Dy)$$

where Id is the identity.

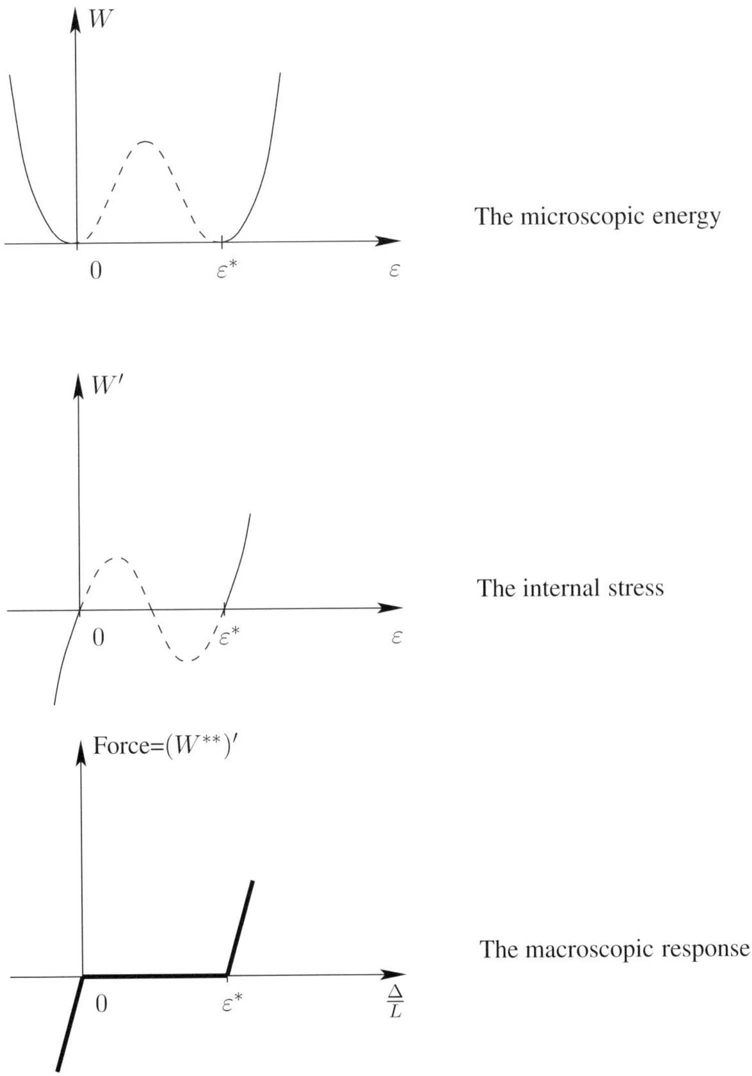

Figure 1. The one-dimensional problem.

3. Stretching sheets of nematic elastomers

We now move to the three-dimensional case. Here, most of the research has focused on crystalline solids. In spite of the many successes, results like the deduction of a stress-strain curve from a fully three-dimensional analysis are still not available. It was recently observed in DeSimone [3] that nematic elastomers, which are amorphous rubbery materials, also exhibit a symmetry breaking phase transformation associated with the establishment of nematic order. Contrary to the crystalline case, where the parent phase only possesses discrete crystallographic symmetry, the parent phase is now amorphous and it possesses full isotropic symmetry. The simplification proved itself decisive, and it led to the explicit determination of the effective energy of the system (i.e., the quasiconvex envelope of the energy density W, which, in dimension one, is just the convex envelope W^{**}), see DeSimone and Dolzmann [4], and to the numerical solution of a realistic boundary value problem simulating experimentally accessible configurations, see Conti et al. [2].

Our goal is now to minimize:

$$I(y) = \int_{\Omega} W\left(Dy(x)\right) dx \tag{8}$$

where Ω is a strip $(0,t) \times (0,L) \times (0,H)$ of thickness t, height H and length L, which is free on the lateral sides, while on the bases $x_2 = 0$, $x_2 = L$, it is subject to rigid clamping and to an imposed stretch s along the basis vector e_2:

$$u\Big|_{x_2=0} = 0 \quad \text{and} \quad u\Big|_{x_2=L} = ((s-1)L)\, e_2\,. \tag{9}$$

W has the expression proposed to model nematic elastomers (Warner and Terentjev [10]):

$$W(F) = \begin{cases} \lambda_1^2(F) + \lambda_2^2(F) + a\lambda_3^2(F) - 3a^{1/3} & \text{if } \det F = 1 \\ +\infty & \text{else} \end{cases} \tag{10}$$

Here $\lambda_1(F) \leq \lambda_2(F) \leq \lambda_3(F)$ are the ordered singular values of the deformation gradient $F = \nabla y$. The assumption of incompressibility, which is described by the constraint $\det F = 1$ on deformations with finite energy, corresponds to $\lambda_1\lambda_2\lambda_3 = 1$. The constant a is a material parameter with typical value $a = 0.5$. A global multiplicative factor, related to the shear modulus of the material, has been set to one in (10) without loss of generality.

The energy (10) is not quasiconvex (see, e.g., Müller [7] for definition and details). This means that it describes a system for which energy minimization may promote the development of fine phase mixtures. It follows that the value of the microscopic energy density $W(F)$ does not give the actual energetic cost of imposing a given macroscopic deformation F, unless the possibility of energy reduction through the development of fine phase mixtures is correctly taken into account. This is precisely the role fulfilled by the quasiconvex envelope of W, see, e.g., Müller [7],

$$W_{\rm qc}(F) = \inf_{y \in W_1^{1,\infty}} \left\{ \frac{1}{|\omega|} \int_\omega W(\nabla y(x))dx : y(x) = Fx \text{ on } \partial\omega, \right\}. \quad (11)$$

Indeed, $W_{\rm qc}$ coarse-grains the energetics of the system: it gives the minimum energy needed to produce the macroscopic deformation F, optimized over all possible admissible microstructures $y(x)$. Here the notation $y \in W_1^{1,\infty}$ means that y is Lipschitz-continuous with $\det \nabla y(x) = 1$. The domain ω, whose volume we denote by $|\omega|$, plays here the role of a representative volume element: it can be verified that $W_{\rm qc}$ does not depend on ω. The use of $W_{\rm qc}$ in the numerical computations allows one to resolve only the length scales associated with macroscopic quantities (e.g., the stresses). The microscopic scales only appear in the definition of $W_{\rm qc}$, hence they are correctly accounted for in the energetics, but averaged out from the kinematics. Clearly, this approach gives only average information on the fine phase mixtures and focuses on the macroscopic response of the system.

An explicit formula for $W_{\rm qc}$ has been derived in DeSimone and Dolzmann [4], which for volume-preserving deformation gradients reads

$$W_{\rm qc}(F) = \begin{cases} 0 & \text{(phase L) if } \lambda_1 \geq a^{1/6} \\ W(F) & \text{(phase S) if } a^{1/2}\lambda_3^2\lambda_1 > 1 \\ \lambda_1^2 + 2a^{1/2}\lambda_1^{-1} - 3a^{1/3} & \text{(phase Sm) else} \end{cases} \quad (12)$$

while it is infinite if $\det F \neq 1$. The three regimes in (12) arising from the collective behavior of energetically optimal fine phase mixtures represent three different modes of macroscopic mechanical response of the system. L stands for liquid, in the sense that in the corresponding region of the phase diagram, the energy is flat (at least, within the approximation that phase-transforming materials evolve in an ideally soft manner). S stands for solid, and it is the region of the phase diagram where, since one singular value is large, the freedom to form microstructure does not allow for energy reduction, and $W_{\rm qc} = W$. Finally, Sm stands for smectic, because here the energy depends only on the smallest singular value. Thus, the material exhibits solid-like response in the direction of the smallest singular value, and liquid-like response in the plane orthogonal to it.

The energy (12) has been used for the numerical simulation of stretching experiments on sheets of nematic elastomers (Conti et al. [2]). A goal of the simulations was to study the role of the clamps in inducing non-uniform stretches and spatially dependent fine phase mixtures. The results show that the graph of the total force at the clamp versus the imposed stretch has a behavior similar to the one emerging from the one-dimensional picture of Section 2, see Fig. 2. The corresponding evolution of the fine phase mixtures can also be monitored in the numerical experiment, by tracing the position of the computed local deformation gradient $F(x)$ on the phase diagram corresponding to (12), see Fig. 3, and by reconstructing the energetically optimal microstructures associated with $F(x)$, see Fig. 4.

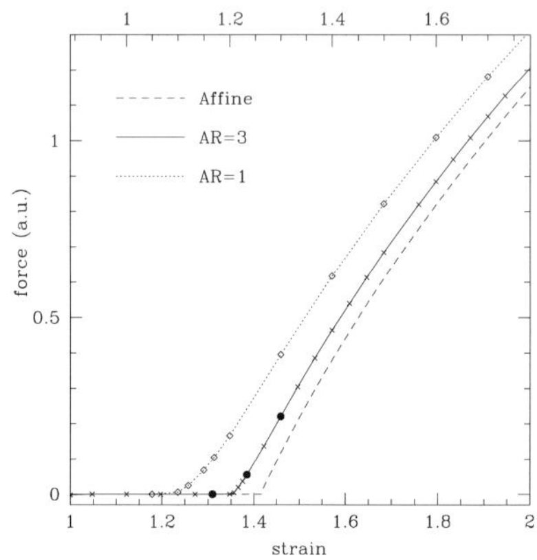

Figure 2. Force versus stretch curves for aspect ratio L/H equal 1, 3, and ∞ (affine approximation). The three black dots mark configurations presented later in more detail. Top axis shows stretch with respect to reference configuration $F \equiv \text{Id}$, bottom axis shows stretch s from initial stress-free configuration.

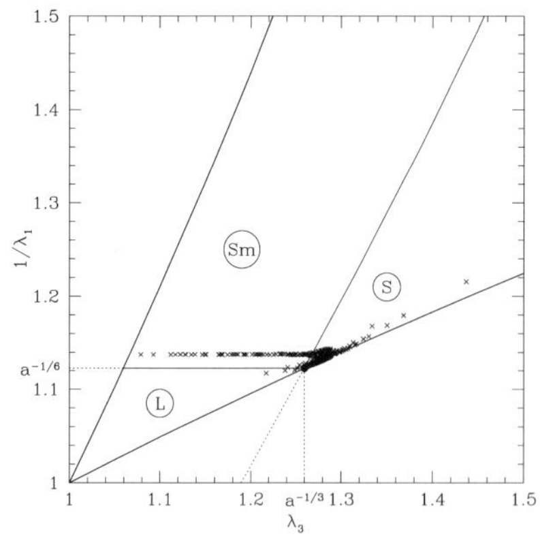

Figure 3. Phase diagram for eq. (12) and computed phase distribution at $s = 1.38$.

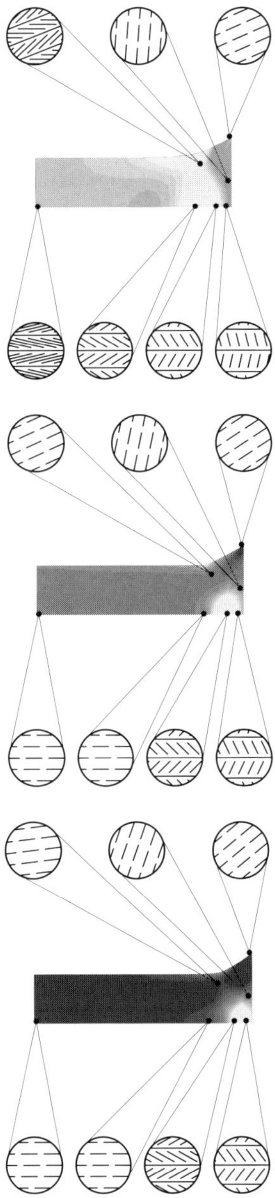

Figure 4. Microstructures at stretches 1.31, 1.38, 1.46. Only one quarter of the sample is shown, due to symmetry. The sticks in the insets indicate the direction of local maximal stretch. Light (dark) color denote presence (absence) of fine phase mixtures.

References

[1] Ball, J.M. and James, R.D. (1992) Proposed experimental tests of a theory of fine microstructures and the two-well problem. *Phil. Trans. Roy. Soc. London*, **A 338**, 389–450.

[2] Conti, S., DeSimone, A. and Dolzmann, G. (2001). Soft elastic response of stretched sheets of nematic elastomers: a numerical study. *J. Mech. Phys. Solids*, in press.

[3] DeSimone, A. (1999). Energetics of fine domain structures. *Ferroelectrics*, **222**, 275–284.

[4] DeSimone, A. and Dolzmann, G. (2001). Macroscopic response of nematic elastomers via relaxation of a class of SO(3)-invariant energies. *Arch. Rat. Mech. Anal.*, in press.

[5] Ericksen, J.L. (1975). Equilibrium of bars. *Journal of Elasticity*, **5**, 191–202.

[6] Ericksen, J.L. (1977). Special topics in elastostatics. *Advances in Applied Mechanics*, **17**, 189–244.

[7] Müller, S. Variational models for microstructure and phase transitions. http://www.mis.mpg.de/cgi-bin/lecturenotes.pl.

[8] Shields, T.W. (1995). Orientation dependence of the pseudo-elastic behavior of single crystals of Cu-Al-Ni in tension. *J. Mech. Phys. Solids*, **43**, 869–895.

[9] Šilhavý, M. this volume.

[10] Warner, M. and Terentjev, E.M. (1996). Nematic elastomers – a new state of matter? *Progress in Polymer Sciences*, **21**, 853–891.

On the Calculation of Microstructures for Inelastic Materials using Relaxed Energies

K. Hackl, U. Hoppe
Institute for Mechanics, University of Bochum
IA 3/126, 44780 Bochum, Germany
hackl@am.bi.ruhr-uni-bochum.de

Abstract: The convexity of energy functionals for inelastic materials is analyzed on the basis of an incremental variational principle. Non-quasiconvex problems give rise to microstructures and often exhibit mesh-dependent results when being solved by standard solution methods, e.g., FEM. A partial rank-one convexification enables a reduction of the mesh-dependency and allows to predict the occurrence and distribution of microstructures independent of the numerical realization.

Keywords: Microstructures, convexity, inelastic materials.

1. Introduction

The formulation of constitutive equations in finite elasticity is confined by several restrictions. Besides the standard principles, e.g., the second law of thermodynamics or the principle of material frame indifference, convexity of the energy function guarantees the existence of solutions of the boundary value problem under consideration and is directly connected to material stability. While there is only one notion of convexity for one-dimensional functions, $f : \mathbb{R} \rightarrow \mathbb{R}$, generalized definitions are needed for the vectorial case, $f : \mathbb{R}^n \rightarrow \mathbb{R}$.

Inelastic materials cannot be directly analyzed with respect to their convexity properties. This requires a variational formulation that is based on an incremental variational principle. For *generalized standard materials* the constitutive equations can be derived from a free energy and a dissipation function. The incremental problem implicitly depends on the internal variables at the beginning and the end of a time step. By means of an incremental variational principle the internal variables can be eliminated, resulting in a so-called reduced energy function, which only depends on the deformation gradient. The reduced energy is used to study the convexity properties of the problem.

As an example a problem from single crystal plasticity is considered. Finite element simulations involving a representative volume element under periodic boundary conditions give mesh-dependent solutions showing laminar microstructures. A consecutive refinement of the mesh size results in a greater number of oscillations of the displacement field. The spatial wavelength of

C. Miehe (ed.), IUTAM Symposium on Computational Mechanics of Solid Materials at Large Strains, 77-86.

these oscillations tends toward zero, exhibiting meaningless results in the limit. By convexification of the energy function the mesh dependency can be substantially reduced, while the occurrence and distribution of microstructures can be predicted in a well defined and physically meaningful manner.

2. Convexity conditions

The analysis of convexity conditions of functions on$\mathbb{R}^n$ requires a generalized definition of convexity. These conditions restrict the possible form of energy functions and guarantee the existence of solutions of boundary value problems in finite elasticity. Moreover, the convexity conditions allow statements on the stability of solutions of the corresponding boundary value problems. For example, the (strong) convexity of the energy function together with a growth condition guarantees the existence of (unique) solutions. The requirement for convexity

$$\Psi(\lambda \mathbf{F}_1 + (1-\lambda)\mathbf{F}_2) \leq \lambda\Psi(\mathbf{F}_1) + (1-\lambda)\Psi(\mathbf{F}_2), \quad \forall\, \mathbf{F}_1, \mathbf{F}_2, \forall\lambda \in [0,1] \quad (1)$$

of functions on $\mathbb{R}^n$ is far too restrictive for the description of a real material behavior (see Marsden and Hughes [8]). Together with a growth condition the notion of quasiconvexity introduced by Morrey [10],

$$\int_\Omega \Psi(\mathbf{F} + \nabla\mathbf{v})\, dV \geq Vol(\Omega)\Psi(\mathbf{F}), \quad \forall\, \mathbf{F}, \forall\, \mathbf{v} = 0 \text{ on } \partial\Omega \qquad (2)$$

is sufficient. (Here Ω is an arbitrary reference volume and$\mathbf{F}$ constant on Ω.) Unfortunately, quasiconvexity is formulated in terms of a nonlocal integral expression that is not suitable for analysis in practice. A necessary condition for quasiconvexity is given by the notion of polyconvexity introduced by Ball [1], which is a local condition. A function Ψ is called polyconvex if there exists a convex function $\tilde{\Psi}$ such that

$$\Psi(\mathbf{F}) = \tilde{\Psi}(\mathbf{F}, (\det \mathbf{F})\mathbf{F}^{-\mathsf{T}}, \det \mathbf{F}). \qquad (3)$$

Polyconvexity implies quasiconvexity. A weaker condition than quasiconvexity is rank-one convexity, which is equivalent to strong (Legendre-Hadamard-) ellipticity and is closely connected to the notion of material instability. A function Ψ is called rank-one convex if

$$\Psi(\mathbf{F} + \lambda\mathbf{a}\otimes\mathbf{b}) \leq \lambda\Psi(\mathbf{F} + \mathbf{a}\otimes\mathbf{b}) + (1-\lambda)\Psi(\mathbf{F}), \quad \forall\, \mathbf{a}, \mathbf{b}, \forall\lambda \in [0,1], \quad (4)$$

where $\mathbf{a} \otimes \mathbf{b}$ is a rank-one matrix. The following implications hold

convexity $\Rightarrow$ polyconvexity $\Rightarrow$ quasiconvexity $\Rightarrow$ rank-one convexity.

For functions defined on $\mathbb{R}$ all four notions reduce to ordinary convexity.

3. Variational formulation

Our analysis of convexity properties of inelastic materials is based on a reduced incremental variational formulation. The necessary steps to obtain such a formulation are now described in brief. More detailed information can be found in Carstensen et al. [3] and Mielke [9].

We consider a rate-independent inelastic material based on the concept of *generalized standard materials* (see Halphen and Nguyen [6], Hackl [5]). The constitutive behavior is characterized by two functions: The specific Helmholtz free energy $\Psi(\mathbf{F}, \mathbf{K})$, which is a function of the deformation gradient $\mathbf{F} = \nabla\phi$ and a set of internal state variables $\mathbf{K}$, and a potential function $J(\mathbf{K}, \mathbf{T})$, which depends on the internal variables and their thermodynamically conjugated forces $\mathbf{T} = -\frac{\partial\Psi}{\partial\mathbf{K}}$. Possible choices for $\mathbf{K}$ comprise variables describing plastic strain, hardening, damage, and phase transformation phenomena. Using a subgradient notation, the evolution equations for the internal variables are given by $\dot{\mathbf{K}} \in \frac{\partial J}{\partial\mathbf{T}}$. Let us define a dissipation functional by the Legendre transform of J

$$\Delta(\mathbf{K}, \dot{\mathbf{K}}) = \sup\left\{ \dot{\mathbf{K}} : \mathbf{T} - J(\mathbf{K}, \mathbf{T}) \;\middle|\; \mathbf{T} \right\},$$

which offers the alternative representation of the conjugated forces $\mathbf{T} \in \frac{\partial\Delta}{\partial\dot{\mathbf{K}}}$ and the evolution equations $0 \in \frac{\partial\Psi}{\partial\mathbf{K}} + \frac{\partial\Delta}{\partial\dot{\mathbf{K}}}$. With the dissipation functional at hand the inelastic problem can be stated in the following variational form

$$\{\phi, \mathbf{K}\} = \arg\min\left\{\mathcal{I}(t, \phi, \mathbf{K}) + \int_0^1 \int_\Omega \Delta(\mathbf{K}(s), \dot{\mathbf{K}}(s))\, dV\, ds \;\middle|\; \phi, \mathbf{K}\right\}, \quad (5)$$

where $\mathcal{I}(t, \phi, \mathbf{K}) = \int_\Omega \Psi(\nabla\phi, \mathbf{K})\, dV - \ell(t, \phi)$ denotes the Gibbs free energy function (ℓ is a potential of external forces). For rate-independent materials the dissipation functional is homogeneous of order one in $\dot{\mathbf{K}}$ and the dissipation (energy release) of a process on a time interval $[t_0, t_1]$ is given by

$$D(\mathbf{K}_0, \mathbf{K}_1) = \inf\left\{ \int_0^1 \Delta(\mathbf{K}(s), \dot{\mathbf{K}}(s))\, ds \;\middle|\; \mathbf{K}(0) = \mathbf{K}_0, \mathbf{K}(1) = \mathbf{K}_1 \right\},$$

which can be interpreted as a dissipation metric on the space of internal variables, see Mielke [9]. An advantage of this variational approach is that it immediately suggests the incremental formulation

$$\{\phi_{n+1}, \mathbf{K}_{n+1}\} = \arg\min\left\{ \int_\Omega \left(\Psi(\nabla\phi, \mathbf{K}) + D(\mathbf{K}_n, \mathbf{K})\right) dV - \ell(t_{n+1}, \phi) \;\middle|\; \phi, \mathbf{K} \right\}. \quad (6)$$

Moreover, the dependency of (6) on the internal variables is local and, hence, the minimization with respect to $\mathbf{K}$ can be carried out separately, leading to a reduced incremental energy function

$$\Psi^{\mathrm{red}}_{\mathbf{K}_n}(\mathbf{F}) = \inf \left\{ \Psi(\mathbf{F},\mathbf{K}) + D(\mathbf{K}_n,\mathbf{K}) \;\middle|\; \mathbf{K} \right\}.$$

Finally, we end up with the so-called *reduced incremental minimization problem* given by the functional

$$\mathcal{I}_{\mathbf{K}_n}(\phi) = \int_\Omega \{\Psi^{\mathrm{red}}_{\mathbf{K}_n}(\mathbf{x}, D\phi) - \ell(t_{n+1},\phi)\}\, dx. \tag{7}$$

The functional (7) is the point of departure for the subsequent analytical and numerical investigation of convexity properties. Note, that it is of the same form as encountered in non-linear elasticity.

4. Example: single slip crystal plasticity

We consider a simple model of finite crystal plasticity with a single slip system. The set of internal variables $\mathbf{K} = \{\gamma, p\}$ consists of the scalar plastic slip γ and a hardening variable p. The plastic deformation is entirely determined by γ, i.e.

$$\mathbf{F_p} = \mathbf{I} + \gamma\, \mathbf{s}_o \otimes \mathbf{n}_o, \tag{8}$$

where $\mathbf{s}_o$ and $\mathbf{n}_o$ denote the referential tangent and normal vector to the slip plane, respectively. We choose a free energy function of a compressible neo-Hookean type

$$\Psi(\mathbf{F},\gamma,p) = U(j) + \frac{\mu}{2}\operatorname{tr}(\mathbf{F}_\mathbf{e}^T\mathbf{F}_\mathbf{e}) + \frac{a}{2}p^2, \quad U(j) = \frac{\Lambda}{4}j^2 - \frac{\Lambda+2\mu}{2}\ln j,$$

where the symbols μ, Λ and a denote material parameters. The set of forces conjugated to $\mathbf{K}$ is $\mathbf{T} = \{\tau, q\}$, with τ denoting the resolved shear stress, Carstensen et al. [3]. The yield function ϕ and its corresponding potential J read

$$\phi(\tau,q) = |\tau| - r - q. \qquad J(\gamma,p,\tau,q) = \begin{cases} 0 & \text{for } \phi(\tau,q) \le 0 \\ \infty & \text{else} \end{cases} \tag{9}$$

Via Legendre transform we obtain the dissipation function (Carstensen et al. [3])

$$\Delta(\gamma,p,\dot\gamma,\dot p) = \begin{cases} r|\dot\gamma| & \text{if } |\dot\gamma| + \dot p \le 0 \\ \infty & \text{else} \end{cases}, \tag{10}$$

for which the energy release function D can explicitly be constructed

$$D(\gamma_0,p_0,\gamma_1,p_1) = \begin{cases} r|\gamma_1-\gamma_0| & \text{if} \quad |\gamma_1-\gamma_0| + p_1 - p_0 \le 0 \\ \infty & \text{else} \end{cases}. \tag{11}$$

Moreover, also the minimization with respect to the internal variables can be carried out explicitly in this example and results in the following expression for the reduced incremental energy

$$\Psi^{\mathrm{red}}_{\gamma_o,p_o}(\mathbf{F}) = U(\det \mathbf{F}) + \frac{\mu}{2}\left[\mathrm{tr}(\mathbf{F}^{\mathsf{T}}\mathbf{F}) - 2\gamma_o\, \mathbf{s}\cdot\mathbf{n} + \gamma_o^2 \mathbf{s}\cdot\mathbf{s} - \frac{(\max\{0, |\mathbf{s}\cdot\mathbf{n} - \gamma_o\, \mathbf{s}\cdot\mathbf{s}| - \frac{r-ap_o}{\mu}\})^2}{\mathbf{s}\cdot\mathbf{s} + a/\mu}\right], \quad (12)$$

with $\mathbf{s} = \mathbf{F}\mathbf{s}_o$ and $\mathbf{n} = \mathbf{F}\mathbf{n}_o$. This function has a form similar to hyperelastic energy density functions in finite elasticity and can be analyzed for its convexity properties. It turns out that $\Psi^{\mathrm{red}}_{\gamma_0,p_0}(\mathbf{F})$ is not rank-one convex and hence is not quasiconvex (Carstensen et al. [3]). This suggests the occurrence of microstructures as minimizers of energy.

4.1. Finite Element Simulation

The occurrence of microstructures can be demonstrated by finite element analysis. We consider a planar simple shear deformation of a representative volume element (RVE) consisting of planar finite elements subjected to periodic boundary conditions, Figure 1. The RVE models the micro-scale behavior of a single material point for a given macro-scale deformation $\mathbf{F}$. The material parameters are chosen as to provide the most pronounced effects of microstructure formation:

Λ	μ	τ_{crit}	a	ϕ
15000 MPa	10000 MPa	10 MPa	1000 MPa	-45.0^o

Figure 2 shows the graph of the reduced energy density function for the above material parameters and for initial conditions $p_o = 0$ and $\gamma_o = 0$. Obviously, the energy function is not rank-one convex. However, microstructures will not arise automatically, even when they represent global minimizers of reduced energy, since the homogeneous solution may still be a local minimizer. To stimulate the formation of microstructures we impose different kinds of imperfections which allow the material to overcome the energy barrier shown in Figure 2 and settle into the global minimum:

- superposition of an initial, static or kinematic imperfection which is successively reduced until it vanishes in the final state. The choice of a useful imperfection requires a large number of parameter studies.
- dynamic analysis with an artificial, successively damped velocity field. For a proper choice of the intensity and orientation of the velocity field the

material will overcome the energy barrier and find a steady state which is a global minimizer of energy.

Generally, both methods finally lead to the same results. Figure 3 shows the distribution of the plastic slip γ for a simulation with a mesh containing 6400 elements. The equilibrium state exhibits a laminar structure composed of twins with reversed plastic slip.

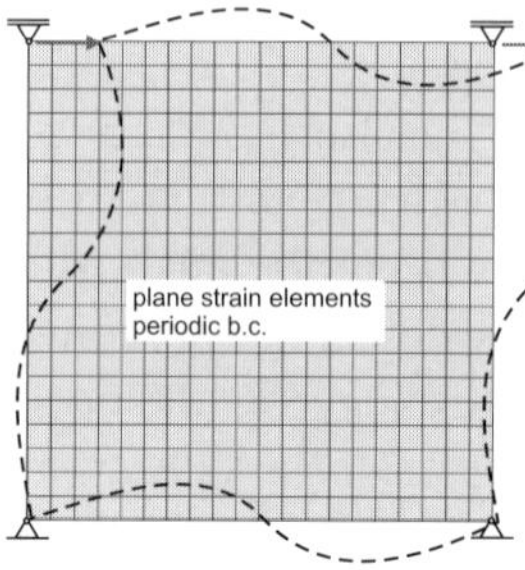

Figure 1. Simple shear problem.

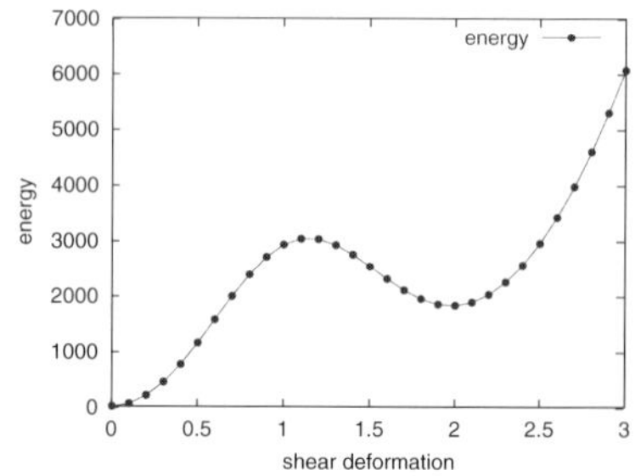

Figure 2. Non-convex energy function.

Figure 4 shows the same distribution for a mesh consisting of 10000 elements. In both cases the same imperfections have been used. We observe that the width of the laminates decreases with the mesh density, although, in particular it strongly depends on the specific imperfection. Figures 5 and 3 show different patterns of laminates for the same mesh consisting of 6400 elements, because different imperfections have been used there.

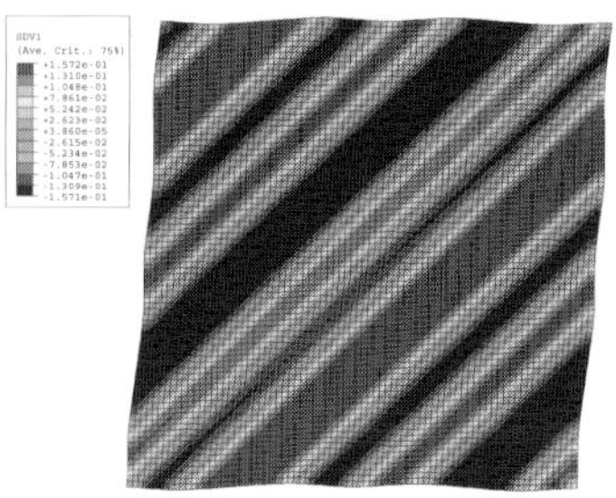

Figure 3. FE simulation with 6400 elements.

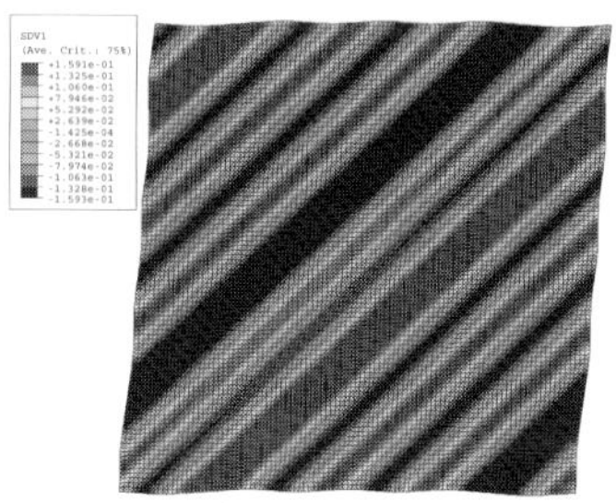

Figure 4. FE simulation with 10000 elements.

Mesh dependency can also be observed in the relationship between total shear force and total shear displacement. Figure 7 shows shear force diagrams for different finite element meshes. We observe a decrease in shear force for finer meshes. The shear force converges to the black curve which depicts the *relaxed* shear force discussed below. Note however, that for very fine meshes one encounters severe stability problems indicated by the rapidly decreasing step-size in Figure 7.

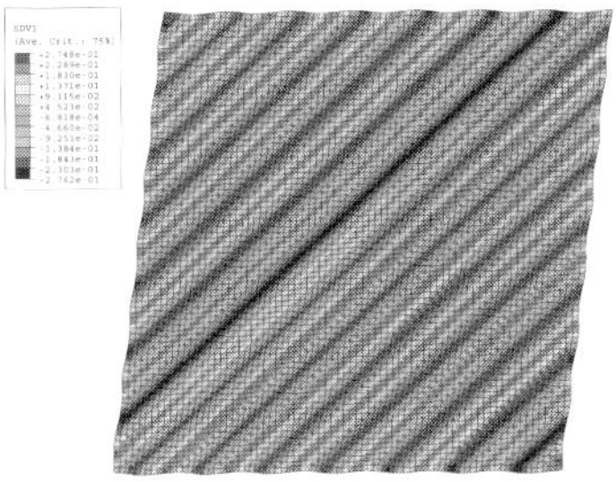

Figure 5. FE simulation with 6400 elements.

Figure 6. FE simulation of a 2nd order laminate.

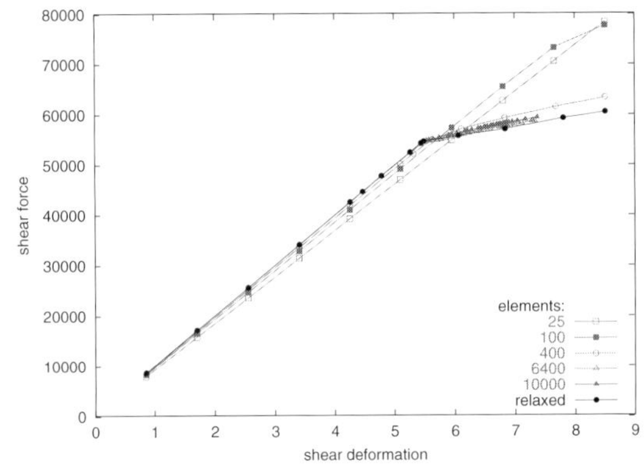

Figure 7. Shear force versus shear deformation.

4.2. RANK-ONE RELAXED ENERGY

The original variational problem does not have a solution for the displacement field on the space of standard finite element functions. Instead, we obtain a series of laminates with finer and finer oscillations. However, when the non-convex energy density function in the original variational problem is replaced by its convex hull, we obtain the so-called relaxed variational problem (Dacorogna [4], Roubicěk [12]). The relaxed variational problem has a solution that can be interpreted as a solution in a generalized sense of the original problem.

Similar to the notion of convexity there exist several notions for the convex envelope of a function Ψ in the vectorial case, e.g., the convex $C\Psi = \sup\{\phi \leq \Psi | \phi \text{ convex}\}$, the polyconvex $(P\Psi)$, the quasiconvex $(Q\Psi)$, and the rank-one convex envelope $(R\Psi)$ (Dacorogna [4]). Corresponding to the definition of quasiconvexity the quasiconvex envelope is defined by the integral identity

$$Q\Psi(\mathbf{F}) = \inf\Big\{\frac{1}{|\Omega|}\int_\Omega \Psi(D\phi)\big|\phi \in W^{1,\infty}(\Omega, \mathbb{R}^3),\ \phi(\mathbf{x}) = \mathbf{F}\mathbf{x} \text{ for } \mathbf{x} \in \partial\Omega\Big\},$$

which is non-local and not suitable for numerical calculations. The envelope with respect to first-order laminates

$$R_1\Psi(\mathbf{F}) = \inf\big\{(1-\lambda)\Psi(\mathbf{F}-\lambda\,\mathbf{a}\otimes\mathbf{b}) \\ +\lambda\,\Psi(\mathbf{F}+(1-\lambda)\,\mathbf{a}\otimes\mathbf{b})\big|\lambda\in[0,1],\quad \mathbf{a},|\mathbf{b}|=1\big\}. \quad (13)$$

is an approximation of the rank-one convex and hence the quasi-convex envelope and is applicable to many problems in solid mechanics.

The convexification (13) of a non-convex function can be formulated as an optimization problem. In case of the example considered above the rank-one convexification of the reduced energy density function (12) leads to a restricted minimization problem with objective function

$$\tilde{\Psi}^{\mathrm{red}}(\mathbf{z},\mathbf{F}) = (1-\lambda)\Psi^{\mathrm{red}}(\underbrace{\mathbf{F}-\lambda\mathbf{a}\otimes\mathbf{b}}_{\mathbf{F}^-}) + \lambda\Psi^{\mathrm{red}}(\underbrace{\mathbf{F}+(1-\lambda)\mathbf{a}\otimes\mathbf{b}}_{\mathbf{F}^+}). \quad (14)$$

In the planar (two-dimensional) case the optimization variables can be represented as

$$\mathbf{z} = (\lambda,\rho,\alpha,\beta),\quad \mathbf{a} = \rho\begin{pmatrix}\cos(\alpha)\\ \sin(\alpha)\end{pmatrix},\quad \mathbf{b} = \begin{pmatrix}\cos(\beta)\\ \sin(\beta)\end{pmatrix},\quad \rho\geq 0. \quad (15)$$

The vectors $\mathbf{a}$ and $\mathbf{b}$ form the rank-one matrix $\mathbf{a}\otimes\mathbf{b}$ and the parameter λ describes the volume ratio of the two phases with deformation gradients $\mathbf{F}^+$ and $\mathbf{F}^-$, respectively. The rank-one convexified reduced energy function is computed by solving the minimization problem

$$R_1\Psi^{\mathrm{red}}(\mathbf{F}) = \min_{\mathbf{z}\in\mathcal{S}} \tilde{\Psi}^{\mathrm{red}}(\mathbf{z},\mathbf{F}) \quad (16)$$

for a given $\mathbf{F}$, where the domain $\mathcal{S}$ of $\mathbf{z}$ is

$$\mathcal{S} = \{\mathbf{z}\in\mathbb{R}^4 \mid 0\leq\lambda\leq 1,\ 0\leq\rho,\ 0\leq\alpha,\beta\leq\pi,\ \det(\mathbf{F}^\pm)>0\}. \quad (17)$$

The global restricted optimization problem (16) has been solved using different numerical procedures, e.g. a sequential quadratic programming method by Spellucci [13]. The search for a global minimum is cumbersome, because the domain $\mathcal{S}$ is not convex and the objective function (14) usually possesses a large number relative minima. Therefore, a large number of starting points have to be used to find the global minimum. Figure 5 shows the results of the convexification (16) applied to the reduced energy (12). It gives the original (upper curve) and relaxed (lower curve) energy for a deformation gradient of the form $\mathbf{F} = \mathbf{I} + \gamma\left(\begin{smallmatrix}0&1\\0&0\end{smallmatrix}\right)$, i.e. a variation of $\mathbf{F}$ in a rank-one direction. One notices that the curve is strictly convex, hence does not agree with the straight Maxwell-line one might expect.

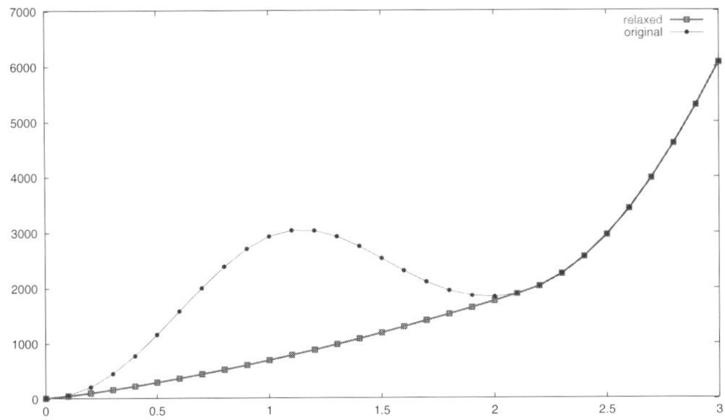

Figure 8. Rank-one convexification.

Figure 9. 1st order laminate.

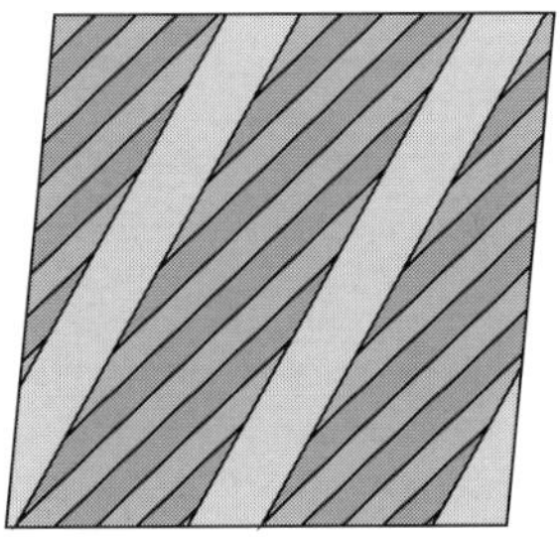

Figure 10. 2nd order laminate.

Figure 9 shows the result of a numerical relaxation for the simple shear problem. The corresponding finite element solution is given in Fig. 5. Note that Fig. 5 was computed with a finite element mesh consisting of 6400 finite element elements with 4 integration points each, whereas Fig. 9 is the result of the relaxation procedure at a single material point. The direction and distribution of the laminates can be computed from the values of the optimization variables λ, ρ, α, and β at the global minimum. We like to point out that the relaxed energy approach only predicts the phase ratio in a statistical sense but does not determine certain details of the specific microsturcture present as for example the width of single laminates. In Fig. 9 the width of the laminates has been chosen to coincide with the finite element simulation. Figure 10 shows a laminate of 2nd order and corresponds to the finite element solution in Fig. 6. The computation of higher order laminates requires a dynamic memory allocation technique, because the number of solution variables grows exponentially with the order of the laminate. In most solid mechanical problems, however, only laminates of order one or two will occur, or will at least represent a sufficiently accurate approximation of higher order laminates.

5. Conclusion

This work presents an approach to the calculation of microstructures for inelastic materials based on a reduced incremental variational principle. In essence, the reduced functional is of the same form as in nonlinear elasticity and can be analyzed by standard methods from the direct calculus of variations and convex analysis. It is demonstrated, that finite element simulations of non-quasiconvex problems may result in mesh-dependent solutions possibly showing a laminar microstructure. By rank-one relaxation of the reduced energy the formation of microstructures can be predicted in an essentially mesh-independent manner.

Further work will be devoted to more advanced material models, e.g., von-Mises plasticity, damage, and phase transition models, and will also be directed to the development of efficient numerical algorithms.

References

[1] Ball, J.M. (1977). Convexity conditions and existence theorems in nonlinear elasticity. *Arch. Rational Mech. Analysis*, **63**, 337–403.

[2] Carstensen, C. and Hackl, K. (2000). On microstructures occurring in a model of finite-strain elastoplasticity involving a single slip-system. *Z. Angew. Math. Mech.*, **80**, S421–S422.

[3] Carstensen, C., Hackl, K. and Mielke, A. (2001). Nonconvex potentials and microstructures in finite-strain plasticity. *Proc. Roy. Soc. London*, (accepted).

[4] Dacorogna, B. (1989). *Direct methods in the calculus of variations*, Springer Verlag.

[5] Hackl, K. (1997). Generalized standard media and variational principles in classical and finite strain elastoplasticity. *J. Mech. Phys. Solids*, **45**, 667–688.

[6] Halphen, B. and Nguyen, Q. (1975). Sur les matériaux standards géneralises. *J. Mechanique*, **14**, 39–63.

[7] Luskin, M. (1996). On the computation of crystalline microstructure. *Acta Numerica*, 191–256.

[8] Marsden, J. and Hughes, T. (1983). *Mathematical Foundation of Elasticity*. Prentice-Hall.

[9] Mielke, A. (2002). Finite plasticity, lie groups and geodesics on sl(d). *Geometry, Dynamics, and Mechanics*, P. Newton, A. Weinstein, P. Holmes (eds.), to appear.

[10] Morrey, C. (1952). Quasi-convexity and the lower semicontinuity of multiple integrals. *Pacific Journal Math.*, **2**, 25–53.

[11] Ortiz, M. and Repetto, E. (1999). Nonconvex energy minimisation and dislocation in ductile single crystals. *J. Mech. Physics Solids*, **47**, 397–462.

[12] Roubicěk, T. (1997). *Relaxation in optimization theory and variational calculus*, De Gruyter.

[13] Spellucci, P. (1998). An SQP method for general nonlinear programs using only equality constrained subproblems. *Math. Prog.*, **82**, 413–448.

Computational Homogenization of Materials with Microstructures based on Incremental Variational Formulations

C. Miehe, M. Lambrecht, J. Schotte
Institute of Applied Mechanics (CE), University of Stuttgart
Pfaffenwaldring 7, 70569 Stuttgart, Germany
cm@mechbau.uni-stuttgart.de

Abstract: The paper presents incremental minimization principles for the broad class of standard dissipative materials at finite strains. Starting with a minimization principle for the local constitutive material response we define a minimization problem for the incremental boundary-value problem of standard dissipative solids. The existence of this principle allows the determination of micro-structure developments in non-stable dissipative materials based on a convexification analysis. Finally, we propose a minimization principle for the boundary-value problem of homogenization in heterogeneous microstructures.

Keywords: Dissipative materials, minimization principles, non-convex response, convexification, energy relaxation, micro-structures, homogenization.

1. Introduction

The paper presents new variational formulations for the homogenization analysis of inelastic solid materials with microstructures undergoing finite strains. The point of departure is a general internal variable formulation for the constitutive response of standard dissipative materials in terms of an energy storage and a dissipation function. Consistent with this type of finite inelasticity we develop an *incremental variational formulation of the local constitutive response*, where a quasi-hyperelastic stress potential is obtained from a local minimization problem with respect to the internal variables. It is shown that this local minimization problem determines the internal state of the material for finite increments of time.

The existence of this variational formulation allows the definition of the *material stability of an inelastic solid based on weak convexity conditions* of the incremental stress potential in analogy to treatments of finite elasticity. Furthermore, localization phenomena are interpreted as microstructure developments on multiple levels associated with non-convex incremental stress potentials in analogy to elastic phase decomposition problems. These micro-structures can be resolved by the relaxation of non-convex energy functionals based on a convexification of the stress potential. The relaxed problem provides a well-posed

C. Miehe (ed.), IUTAM Symposium on Computational Mechanics of Solid Materials at Large Strains, 87-100.

formulation for a mesh-objective analysis of localizations as close as possible to the non-convex original problem.

Furthermore, the existence of the quasi-hyperelastic incremental stress potentials for standard dissipative materials allows the extension of homogenization approaches of elasticity to the incremental setting of inelasticity. Focussing on macro-deformation-driven micro-structures of heterogeneous materials, we develop a new *incremental variational formulation of the homogenization problem* for standard dissipative materials at finite strains, where a quasi-hyperelastic macro-stress potential is obtained from a global minimization problem with respect to the fine-scale displacement fluctuation field. It is shown that this global minimization problem determines the state of the heterogeneous micro-structure for finite increments of time. A representative application is the texture analysis in polycrystalline aggregates of single crystal grains. Table 1 overviews and summarizes the subsequent developments.

Table 1. Minimization principles for standard dissipative materials.

(M) Constitutive Model. $\boldsymbol{F} \in GL(3)$ is the local deformation of the material and $\boldsymbol{\mathcal{I}}$ a generalized vector of internal variables. A normal-dissipative set of local material equations has the structure

stress equation $\quad \boldsymbol{P} = \partial_{\boldsymbol{F}} \psi(\boldsymbol{F}, \boldsymbol{\mathcal{I}})$

evolution equation $\quad \partial_{\boldsymbol{\mathcal{I}}} \psi(\boldsymbol{F}, \boldsymbol{\mathcal{I}}) + \partial_{\dot{\boldsymbol{\mathcal{I}}}} \phi(\dot{\boldsymbol{\mathcal{I}}}, \boldsymbol{\mathcal{I}}) = 0\ , \ \boldsymbol{\mathcal{I}}(0) = \boldsymbol{\mathcal{I}}_0$

defined in terms of energy storage function ψ and dissipation function ϕ.

(V) Incremental Variational Formulation of the Constitutive Model. In a finite time increment $[t_n, t_{n+1}]$, the *minimization problem of constitutive response*

stress equation $\quad \boldsymbol{P}_{n+1} = \partial_{\boldsymbol{F}} W(\boldsymbol{F}_{n+1})$

stress potential $\quad W(\boldsymbol{F}_{n+1}) = \inf_{\boldsymbol{\mathcal{I}}} \int_{t_n}^{t_{n+1}} [\dot{\psi} + \phi]\, dt\ , \ \boldsymbol{\mathcal{I}}(t_n) = \boldsymbol{\mathcal{I}}_n$

determines the current internal state $\boldsymbol{\mathcal{I}}_{n+1}$ and provides a potential for the stresses at t_{n+1}.

(S) Stability of the Incremental Constitutive Response. In $[t_n, t_{n+1}]$ the material is locally stable at $\boldsymbol{F}_{n+1}$ if the *incremental stress potential* W *is quasi-convex*

quasi-convexity $\quad \frac{1}{|D|} \int_D W(\boldsymbol{F}_{n+1} + \nabla \boldsymbol{w}(\boldsymbol{y}))\, dV \geq W(\boldsymbol{F}_{n+1})$

(R) Relaxation of a Non-Convex Incremental Response. For an instable non-convex incremental response, the *minimization problem of quasi-convexification*

macro-stresses $\quad \boldsymbol{P}_{Qn+1} = \partial_{\boldsymbol{F}} W_Q(\boldsymbol{F}_{n+1})$

convexification $\quad W_Q(\boldsymbol{F}_{n+1}) = \inf_{\boldsymbol{w} \in \mathcal{W}} \frac{1}{|D|} \int_D W(\boldsymbol{F}_{n+1} + \nabla \boldsymbol{w}(\boldsymbol{y}))\, dV$

determines the current microstructure-fluctuations $\boldsymbol{w}(\boldsymbol{y})$.

(H) Homogenization of Heterogeneous Microstructures. $\bar{\boldsymbol{F}} \in GL(3)$ is the macro-deformation of a heterogeneous microstructure at t_{n+1}. The *minimization problem of homogenization*

macro-stresses $\quad \bar{\boldsymbol{P}}_{n+1} = \partial_{\bar{\boldsymbol{F}}} \bar{W}(\boldsymbol{F}_{n+1})$

homogenization $\quad \bar{W}(\boldsymbol{F}_{n+1}) = \inf_{\boldsymbol{w}} \frac{1}{|\mathcal{B}|} \int_{\mathcal{B}} W(\boldsymbol{F}_{n+1} + \nabla \boldsymbol{w}; \boldsymbol{X})\, dV$

determines the fluctuation field $\boldsymbol{w}(\boldsymbol{X})$ on the heterogeneous microstructure.

---

2. Minimization problem for standard dissipative materials

2.1. INTERNAL VARIABLE FORMULATION OF INELASTIC MATERIALS

Let $\boldsymbol{F} \in GL_+(3)$ with $\det[\boldsymbol{F}] > 0$ be the homogeneous deformation of a material at time $t \in \mathcal{R}_+$. Focussing on purely mechanical problems, the homogeneous stress response of the material is physically constrained by the so-called Clausius-Planck inequality for the dissipation of the material

$$\mathcal{D} := \boldsymbol{P} : \dot{\boldsymbol{F}} - \dot{\psi} \geq 0 \,, \tag{1}$$

where $\boldsymbol{P}$ denotes the nominal (first Piola) stress tensor. ψ is an objective *energy storage function* that is assumed to depend on the deformation $\boldsymbol{F}$ and a generalized vector $\boldsymbol{\mathcal{I}} \in \mathcal{G}$ of internal variables. $\mathcal{G}$ indicates a vector space $\mathcal{R}^n$ of n scalar functions of internal variables which may be constrained to a manifold, e.g. the Lie group $SL(3)$ of unimodular tensors in isochoric finite inelasticity. Insertion into (1) yields by a standard argument the constitutive equation for the stresses

$$\boxed{\boldsymbol{P} = \partial_{\boldsymbol{F}} \psi(\boldsymbol{F}, \boldsymbol{\mathcal{I}})} \tag{2}$$

and the reduced dissipation inequality

$$\mathcal{D} = \boldsymbol{\mathcal{F}} \cdot \dot{\boldsymbol{\mathcal{I}}} \geq 0 \qquad \text{with} \qquad \boldsymbol{\mathcal{F}} := -\partial_{\boldsymbol{\mathcal{I}}} \psi(\boldsymbol{F}, \boldsymbol{\mathcal{I}}) \,, \tag{3}$$

where $\boldsymbol{\mathcal{F}} \in \mathcal{R}^n$ is a generalized vector of internal forces conjugate to the internal variables $\boldsymbol{\mathcal{I}}$. A broad spectrum of materials is covered by the so-called standard dissipative materials where the evolution $\dot{\boldsymbol{\mathcal{I}}}$ of the internal variables on $\mathcal{G}$ is governed by a scalar *dissipation function* ϕ, depending on the flux $\dot{\boldsymbol{\mathcal{I}}}$ of the internal variables and the internal variables $\boldsymbol{\mathcal{I}}$ themselves, see for example Ziegler [25] and Halphen and Nguyen [6]. It governs the evolution of $\boldsymbol{\mathcal{I}}$ in time by the constitutive differential equation

$$\boxed{\partial_{\boldsymbol{\mathcal{I}}} \psi(\boldsymbol{F}, \boldsymbol{\mathcal{I}}) + \partial_{\dot{\boldsymbol{\mathcal{I}}}} \phi(\dot{\boldsymbol{\mathcal{I}}}, \boldsymbol{\mathcal{I}}) = \boldsymbol{0} \quad \text{with} \quad \boldsymbol{\mathcal{I}}(0) = \boldsymbol{\mathcal{I}}_0} \tag{4}$$

We denote this equation Biot's equation of standard dissipative systems, referring to Biot [2]. The two constitutive equations (2) and (4) determine the stress response of a *smooth* normal-dissipative material in a deformation-driven process where $\boldsymbol{F}$ is prescribed. Plasticity and dry friction are time-independent or non-viscous irreversible processes, governed by *non-smooth* dissipation functions. These functions are positively homogeneous of degree one with respect to the flux $\dot{\boldsymbol{\mathcal{I}}}$, i.e. $\phi(\epsilon \dot{\boldsymbol{\mathcal{I}}}, \boldsymbol{\mathcal{I}}) = \epsilon \phi(\dot{\boldsymbol{\mathcal{I}}}, \boldsymbol{\mathcal{I}})$ for all $\epsilon \in \mathcal{R}_+$. Such a function has a cone-like graph and is not differentiable at the point $\dot{\boldsymbol{\mathcal{I}}} = \boldsymbol{0}$. A generalization of the above formulation needs a generalization of the differential operator of smooth functions to the notion of a *sub-differential* operator ∂ of non-smooth convex functions as follows, see Moreau [17], Halphen and Nguyen [6] and references therein. Using the definition of the sub-differential, in what follows we formally understand Biot's equation (4) to be generalized to $0 \in \partial_{\boldsymbol{\mathcal{I}}}\psi(\boldsymbol{F},\boldsymbol{\mathcal{I}}) + \partial_{\dot{\boldsymbol{\mathcal{I}}}}\phi(\dot{\boldsymbol{\mathcal{I}}},\boldsymbol{\mathcal{I}})$.

2.2. Incremental variational formulation of inelastic materials

We now proceed with the construction of an integrated version of constitutive equations giving a consistent approximation of the continuous differential equation (4) in a finite increment $[t_n, t_{n+1}] \in \mathcal{R}_+$ of time. Following conceptually the works Ortiz and Repetto [22] and Carstensen et al. [4] on plasticity, the key point is the definition of an *incremental stress potential function* W depending of the deformation $\boldsymbol{F}_{n+1} := \boldsymbol{F}(t_{n+1})$ at time t_{n+1} that determines the stresses at t_{n+1} by the quasi-hyperelastic function evaluation

$$\boxed{\boldsymbol{P}_{n+1} = \partial_{\boldsymbol{F}} W(\boldsymbol{F}_{n+1})} \tag{5}$$

Clearly, this function must cover characteristics of the storage function ψ and the dissipation function ϕ introduced above. To this end, consider the *minimization problem*

$$\boxed{W(\boldsymbol{F}_{n+1}) = \inf_{\boldsymbol{\mathcal{I}}\in\mathcal{G}} \int_{t_n}^{t_{n+1}} [\dot{\psi} + \phi]\, dt \quad \text{with} \quad \boldsymbol{\mathcal{I}}(t_n) = \boldsymbol{\mathcal{I}}_n} \tag{6}$$

for dissipative standard materials proposed in Miehe [10]. The principle provides an alternative to the formulation proposed by Ortiz and Repetto [22] whose variational principle minimizes the incremental work $\int_{t_n}^{t_{n+1}} \boldsymbol{P} : \dot{\boldsymbol{F}} dt = \int_{t_n}^{t_{n+1}} [\dot{\psi} + \boldsymbol{\mathcal{F}} \cdot \dot{\boldsymbol{\mathcal{I}}}] dt$. Starting with the given initial condition $\boldsymbol{\mathcal{I}}(t_n) = \boldsymbol{\mathcal{I}}_n$, the minimum problem defines an optimal path of the internal variables $\boldsymbol{\mathcal{I}}(t)$ for $t \in [t_n, t_{n+1}]$, including the right boundary value $\boldsymbol{\mathcal{I}}_{n+1} := \boldsymbol{\mathcal{I}}(t_{n+1})$. We refer to Martin [9] and the recent work of Mielke [16] for a discussion of extremum paths in linear and nonlinear plasticity. The two equations (5) and (6) provide an approximative variational counterpart of the continuous setting (2) and (4) of the constitutive equations in the finite time step $[t_n, t_{n+1}]$ under consideration. In order to show the consistency, we at first recast (6) into the form

$$W(\boldsymbol{F}_{n+1}) = \inf_{\boldsymbol{\mathcal{I}}\in\mathcal{G}} \left\{ [\psi(\boldsymbol{F},\boldsymbol{\mathcal{I}})]_{t_n}^{t_{n+1}} + \int_{t_n}^{t_{n+1}} \phi(\dot{\boldsymbol{\mathcal{I}}},\boldsymbol{\mathcal{I}})\, dt \right\} . \tag{7}$$

The necessary condition for the minimum problem is that the variation with respect to the internal variables of the term in brackets vanishes. For smooth functions, integration by parts yields the expression

$$[(\partial_{\boldsymbol{\mathcal{I}}}\psi + \partial_{\dot{\boldsymbol{\mathcal{I}}}}\phi)\cdot\delta\boldsymbol{\mathcal{I}}]_{t_n}^{t_{n+1}} + \int_{t_n}^{t_{n+1}} [-\frac{d}{dt}(\partial_{\dot{\boldsymbol{\mathcal{I}}}}\phi) + \partial_{\boldsymbol{\mathcal{I}}}\phi]\cdot\delta\boldsymbol{\mathcal{I}}\, dt = 0\,. \quad (8)$$

Thus the variational problem (6) yields Biot's equation (4)

$$\partial_{\boldsymbol{\mathcal{I}}}\psi + \partial_{\dot{\boldsymbol{\mathcal{I}}}}\phi = \boldsymbol{0} \quad \text{for} \quad t = t_{n+1} \quad (9)$$

at the discrete right boundary of the interval $[t_n, t_{n+1}]$. The minimizing path of the internal variables inside the interval is determined by the Euler equation

$$-\frac{d}{dt}(\partial_{\dot{\boldsymbol{\mathcal{I}}}}\phi) + \partial_{\boldsymbol{\mathcal{I}}}\phi = \boldsymbol{0} \quad \text{for} \quad t \in [t_n, t_{n+1}]\,. \quad (10)$$

For the limit $t_{n+1} \to t_n$, the form of the minimization path becomes irrelevant, because the time increment degenerates to a discrete time t_{n+1}. Because (9) still holds in this case, it is shown that the proposed variational formulation (6) represents a consistent point-wise approximation of Biot's normal-dissipative evolution equation (4). Furthermore, taking the derivative of the incremental potential function with respect to the deformation $\boldsymbol{F}_{n+1}$, we have

$$\partial_{\boldsymbol{F}} W(\boldsymbol{F}_{n+1}) = \partial_{\boldsymbol{F}}\psi(\boldsymbol{F}_{n+1}, \boldsymbol{\mathcal{I}}_{n+1})\,, \quad (11)$$

where $\boldsymbol{\mathcal{I}}_{n+1}$ is considered to be given by the minimization problem (6). A comparison with (2) then shows the consistency of the potential equation (5) with the continuous setting.

3. Minimization problem for standard dissipative solids

3.1. Incremental variational formulation of inelastic solids

The existence of the constitutive minimization problem (6) allows the introduction of an incremental minimization formulation of the boundary-value problem of finite inelasticity for standard dissipative solids. Let $\boldsymbol{\varphi} : \mathcal{B} \times \mathcal{R}_+ \to \mathcal{R}^3$ denote the nonlinear deformation map of an inelastic continuum $\mathcal{B} \subset \mathcal{R}^3$ at a material point $\boldsymbol{X} \in \mathcal{B}$ and time $t \in \mathcal{R}^+$. Then $\boldsymbol{F} := \nabla\boldsymbol{\varphi}$ is the deformation gradient with Jacobian $J := \det[\boldsymbol{F}] > 0$. Now consider a functional of the current deformation field $\boldsymbol{\varphi}_{n+1}$ at the right boundary of the increment $[t_n, t_{n+1}]$

$$\Pi(\boldsymbol{\varphi}_{n+1}) = \int_{\mathcal{B}} W(\boldsymbol{F}_{n+1})\, dV - \Pi_{ext}(\boldsymbol{\varphi}_{n+1}) \quad (12)$$

with the external part $\Pi_{ext}(\boldsymbol{\varphi}_{n+1}) = \int_{\mathcal{B}} \boldsymbol{\varphi}_{n+1}\cdot\boldsymbol{\gamma}_{n+1}\, dV - \int_{\partial\mathcal{B}_t} \boldsymbol{\varphi}_{n+1}\cdot\boldsymbol{t}_{n+1}\, dA$ of dead body forces $\boldsymbol{\gamma}(\boldsymbol{X}, t)$ in $\mathcal{B}$ and surface tractions $\boldsymbol{t}(\boldsymbol{X}, t)$ on $\partial\mathcal{B}_t$. W is the incremental stress potential function defined in (6). The current deformation map of inelastic standard dissipative materials can then be determined by a *principle of minimum incremental energy for standard dissipative solids*

$$\boxed{\Pi(\varphi^*_{n+1}) = \inf_{\varphi_{n+1} \in \mathcal{W}} \Pi(\varphi_{n+1})} \tag{13}$$

subject to $\varphi_{n+1} \in \mathcal{W} := \{\varphi \in \mathcal{W}^{1,p}(\mathcal{B}) | \varphi = \bar{\varphi}(\boldsymbol{X}) \text{ on } \partial\mathcal{B}_\varphi\}$ associated with prescribed deformations $\bar{\varphi}$ at $\boldsymbol{X} \in \partial\mathcal{B}_\varphi$ on the boundary. The minimization problem has a structure identical to the principle of minimum potential energy in finite elasticity.

3.2. INCREMENTAL STABILITY OF STANDARD DISSIPATIVE MATERIALS

With the incremental minimization problem (13) at hand, existence results of finite elasticity as outlined in Ball [1] and Dacorogna [5] can straightforwardly be applied to the incremental response of standard dissipative materials. The internal part $I(\varphi_{n+1}) := \int_\mathcal{B} W(\boldsymbol{F}_{n+1})\, dV$ of the functional (12) is sequentially weakly lower semicontinuous (s.w.l.s.), if the incremental stress potential $W(\boldsymbol{F}_{n+1})$ defined in (6) is quasiconvex, see Morrey [18] and Dacorogna [5]. This is considered to be the key property for the existence of sufficiently regular minimizer of the variational problem (13). Hence, the quasiconvexity

$$\frac{1}{|D|} \int_D W(\boldsymbol{F}_{n+1} + \nabla \boldsymbol{w}(\boldsymbol{y}))\, dV \geq W(\boldsymbol{F}_{n+1}) \tag{14}$$

of the incremental stress potential W is considered to be the fundamental *criterion for a local material stability of the incremental response of standard dissipative solids.* The condition states that for all fluctuations $\boldsymbol{w}$ at $\boldsymbol{y} \in D$ on an arbitrarily chosen part D of the inelastic solid with support $\boldsymbol{w} = \boldsymbol{0}$ on ∂D the homogeneous deformation given by $\boldsymbol{F}_{n+1}$ provides an absolute minimizer of the incremental potential. Thus (14) rules out internal buckling, the development of local fine-scale microstructures and phase decomposition of a homogeneous local deformation state.

3.3. MICROSTRUCTURE DEVELOPMENT OF NON-STABLE MATERIALS

The variational formulation of the incremental boundary value problem for standard dissipative solids outlined above opens up the opportunity to apply the concept of relaxation of non-convex problems where an energy-minimizing *microstructure* develops. A relaxation is associated with a *convexification* of a non-convex function W by constructing its *convex envelopes* W_Q as discussed below, see Kohn and Strang [8], Dacorogna [5] and Müller [20] for a sound mathematical basis. The convexification procedure determines the developing microstructure. We consider the *quasi-convexified incremental energy functional of standard dissipative solids*

$$\Pi_Q(\boldsymbol{\varphi}_{n+1}) = \int_{\mathcal{B}} W_Q(\boldsymbol{F}_{n+1})\, dV - \Pi_{ext}(\boldsymbol{\varphi}_{n+1})\,, \tag{15}$$

where the internal part of the relaxed energy functional is obtained by replacing the non-convex integrand by its*quasiconvex envelope.* The current deformation field of the elastic-plastic solid is then determined by the relaxed incremental variational principle

$$\boxed{\Pi_Q(\boldsymbol{\varphi}^*_{n+1}) = \inf_{\boldsymbol{\varphi}_{n+1}\in\mathcal{W}} \Pi_Q(\boldsymbol{\varphi}_{n+1})} \tag{16}$$

that *minimizes the relaxed incremental potential energy.* This relaxed problem (16) is considered to be a *well-posed problem as close as possible to the ill-posed problem* (13). In (16), the quasi-convexified incremental stress potential is defined by the minimization problem of relaxation with respect to the microscopic fluctuation field $\boldsymbol{w}$

$$\boxed{W_Q(\boldsymbol{F}_{n+1}) = \inf_{\boldsymbol{w}\in\mathcal{W}_0} \frac{1}{|D|}\int_D W(\boldsymbol{F}_{n+1} + \nabla\boldsymbol{w}(\boldsymbol{y}))\, dV} \tag{17}$$

with $\boldsymbol{y} \in D$ subject to the constraint $\boldsymbol{w} \in \mathcal{W}_0 := \{\boldsymbol{w} \in \mathcal{W}^{1,\infty}(D) | \boldsymbol{w} = \boldsymbol{0}$ on $\partial D\}$ providing a support on ∂D. The first and second derivatives of this function define relaxed stresses and tangent moduli in the form

$$\boldsymbol{P}_{Qn+1} := \partial_{\boldsymbol{F}} W_Q(\boldsymbol{F}_{n+1}) \quad \text{and} \quad \mathbb{C}_{Qn+1} := \partial^2_{\boldsymbol{F}\boldsymbol{F}} W_Q(\boldsymbol{F}_{n+1})\,. \tag{18}$$

It determines a micro-fluctuation field $\boldsymbol{w}$ for an arbitrarily chosen domain $D \subset \mathcal{R}^3$.

4. Minimization problem of homogenization for dissipative solids

4.1. Fluctuation fields on heterogeneous microstructures

Let $\boldsymbol{\varphi} : \mathcal{B} \times \mathcal{R}_+ \to \mathcal{R}^3$ denote the deformation map of a heterogeneous micro-structure $\mathcal{B} \subset \mathcal{R}^3$ at $\boldsymbol{X} \in \mathcal{B}$ and time $t \in \mathcal{R}_+$. The deformation is assumed to be driven by a prescribed macro-deformation $\bar{\boldsymbol{F}}(t)$

$$\boldsymbol{\varphi}(\boldsymbol{X}, t) = \bar{\boldsymbol{F}}(t)\boldsymbol{X} + \boldsymbol{w}(\boldsymbol{X}, t) \tag{19}$$

consisting of a linear part $\bar{\boldsymbol{F}}(t)\boldsymbol{X}$ and a superimposed fine-scale fluctuation field $\boldsymbol{w} : \mathcal{B} \times \mathcal{R}_+ \to \mathcal{R}^3$. Following Hill [7], the macro-deformation $\bar{\boldsymbol{F}}(t)$ is assumed to be governed by surface data of the microscopic deformation field $\bar{\boldsymbol{F}}(t) = \frac{1}{|\mathcal{B}|}\int_{\partial\mathcal{B}} \boldsymbol{\varphi}(\boldsymbol{X}, t) \otimes \boldsymbol{n}(\boldsymbol{X})\, dA$. Then, as a consequence of the decomposition (19), the fluctuation field $\boldsymbol{w}$ has to satisfy the constraint $\frac{1}{|\mathcal{B}|}\int_{\partial\mathcal{B}} \boldsymbol{w}(\boldsymbol{X},t)\otimes\boldsymbol{n}(\boldsymbol{X})\,dA = \boldsymbol{0}$. This constraint is satisfied for the three classes of constraints

$$\left.\begin{array}{lll} \bar{C}_D(\boldsymbol{w};\boldsymbol{t}) & := -\dfrac{1}{|\mathcal{B}|}\displaystyle\int_{\partial\mathcal{B}} \boldsymbol{t}\cdot\boldsymbol{w}\,dA & = 0 \\ \bar{C}_P(\boldsymbol{w};\boldsymbol{t}^+) & := -\dfrac{1}{|\mathcal{B}|}\displaystyle\int_{\partial\mathcal{B}^+} \boldsymbol{t}^+\cdot(\boldsymbol{w}^+ - \boldsymbol{w}^-)\,dA & = 0 \\ \bar{C}_S(\boldsymbol{w};\bar{\boldsymbol{P}}) & := -\dfrac{1}{|\mathcal{B}|}\displaystyle\int_{\partial\mathcal{B}} (\bar{\boldsymbol{P}}\boldsymbol{n})\cdot\boldsymbol{w}\,dA & = 0 \end{array}\right\} \tag{20}$$

associated with the surface $\partial\mathcal{B}$ of the micro-structure. The first constraint $\bar{C}_D$ in (20) demands *homogeneous deformation* with zero fluctuations $\boldsymbol{w} = \boldsymbol{0}$ on the boundary $\partial\mathcal{B}$ of the micro-structure, where the traction field $\boldsymbol{t}(\boldsymbol{X},t)$ is considered as a Lagrangian multiplier. The second constraint $\bar{C}_P$ states a *non-trivial periodicity* $\boldsymbol{w}^+ = \boldsymbol{w}^-$ of the superimposed fluctuation field on $\partial\mathcal{B}$. Here, the boundary $\partial\mathcal{B}$ is understood to be decomposed into two parts $\partial\mathcal{B} = \partial\mathcal{B}^-\cup\partial\mathcal{B}^+$ with outward normals $\boldsymbol{n}^+ = -\boldsymbol{n}^-$ at two associated points $\boldsymbol{X}^-\in\partial\mathcal{B}^-$ and $\boldsymbol{X}^+\in\partial\mathcal{B}^+$. $\boldsymbol{t}^+(\boldsymbol{X},t)$ is a Lagrangian multiplier field that characterizes the antiperiodic tractions on $\partial\mathcal{B}^+$. The third constraint $\bar{C}_S$ in (20) is associated with *homogeneous stresses* $\boldsymbol{t} = \bar{\boldsymbol{P}}\boldsymbol{n}$ on the boundary $\partial\mathcal{B}$, where the Lagrangian multiplier $\bar{\boldsymbol{P}}(t)$ is the macro-stress dual to the macro-deformation $\bar{\boldsymbol{F}}(t)$.

4.2. Incremental Variational Formulation of Homogenization

Based on the constitutive minimization problem for the incremental response of standard dissipative materials defined in (6) we obtain the current stresses by the quasi-hyperelastic function evaluation (5). Hence, the micro-stress $\boldsymbol{P}_{n+1}$ and micro-moduli $\mathbb{C}_{n+1}$ at a local point $\boldsymbol{X}\in\mathcal{B}$ of the heterogeneous micro-structure at time t_{n+1} are defined by

$$\boldsymbol{P}_{n+1} = \partial_{\boldsymbol{F}} W(\boldsymbol{F}_{n+1};\boldsymbol{X}) \quad\text{and}\quad \mathbb{C}_{n+1} = \partial^2_{\boldsymbol{F}\boldsymbol{F}} W(\boldsymbol{F}_{n+1};\boldsymbol{X})\,. \tag{21}$$

The analogy to finite elasticity induces the following incremental variational formulation of the homogenization problem for normal-dissipative inelastic solids. As the key homogenization condition, we consider the *principle of minimum average incremental energy* proposed in Miehe [10]

$$\boxed{\bar{W}(\bar{\boldsymbol{F}}_{n+1}) = \inf_{\boldsymbol{w}_{n+1}} \frac{1}{|\mathcal{V}|}\int_{\mathcal{B}} W(\bar{\boldsymbol{F}}_{n+1} + \nabla\boldsymbol{w}_{n+1};\boldsymbol{X})\,dV} \tag{22}$$

for heterogeneous micro-structures with constituents of standard dissipative materials. The minimization problem is understood to be subject to the three alternative constraints defined in (20). The minimization problem determines the fluctuation field $\boldsymbol{w}_{n+1}$ on the micro-structure at discrete time t_{n+1}. The macro-stress potential $\bar{W}$ is assumed to define the macro-stresses $\bar{\boldsymbol{P}}_{n+1}$ and the macro-moduli $\bar{\mathbb{C}}_{n+1}$ by the quasi-hyperelastic function evaluations

$$\bar{\boldsymbol{P}}_{n+1} = \partial_{\bar{\boldsymbol{F}}} \bar{W}(\bar{\boldsymbol{F}}_{n+1}) \quad \text{and} \quad \bar{\mathbb{C}}_{n+1} = \partial^2_{\bar{\boldsymbol{F}}\bar{\boldsymbol{F}}} \bar{W}(\bar{\boldsymbol{F}}_{n+1}) \tag{23}$$

in complete analogy to (21). Thus the minimization problem (22) provides a shift of associated variables from the micro-scale to the macro-scale, often denoted as micro-to-macro transition. It extends the so-called *average variational principle* of nonlinear elasticity, outlined by Müller [19] and Ponte Castañeda and Suquet [24], to the incremental formulation of finite inelasticity. The attractive feature of the proposed formulation (22) is that the structure of the average variational principle is preserved, i.e. the energy storage function ψ of finite elasticity is replaced by the incremental stress potential W of finite inelasticity.

5. Characteristic numerical examples

5.1. Microstructure development in single-slip plasticity

This first example applies a two-phase rank-one relaxation analysis developed in Miehe and Lambrecht [14] and Miehe et al. [15] to a model problem of single-slip crystal plasticity and investigate the relaxed and non-relaxed stress response of a specimen in shear. The main goals are the demonstration of the mesh-invariance of the technique and the analysis of the developing microstructures. The dimensions of the specimen are 16 mm (width) and 32 mm (height). The upper and lower bounds are fixed in both the horizontal and vertical directions. The upper edge is moved horizontally in constant increments $\Delta u = 0.05$ mm to its final displacement $u = 40$ mm. The plastic slip system is orientated with a counterclockwise angle $\alpha = 135^o$ against the horizontal line. The specimen is discretized with 6×12, 8×16, 12×24, 16×32 and 20×40 finite elements. For a detailed representation of the constitutive functions including the material parameters we refer to Miehe et al. [15]. The evolution of the first-order laminates is documented in Fig. 1 for the 8×16-element mesh and four different displacements $u = 10$ mm, $u = 20$ mm, $u = 30$ mm and $u = 40$ mm. The magnified microstructures correspond to the center Gauss point of the red marked finite elements. With increasing deformation the bifurcation of the homogeneous deformation intensifies and the volume fraction $1-\xi$ of the micro-phase $(-)$ augments until a stable homogeneous deformation state is recovered ($\xi = 0$). The resolution of the microstructures as first-order laminates results from the rank-one convexification of the non-convex incremental stress potentials W. Fig. 2a shows the shape of a non-convex incremental stress potential W associated with an integration point of a finite

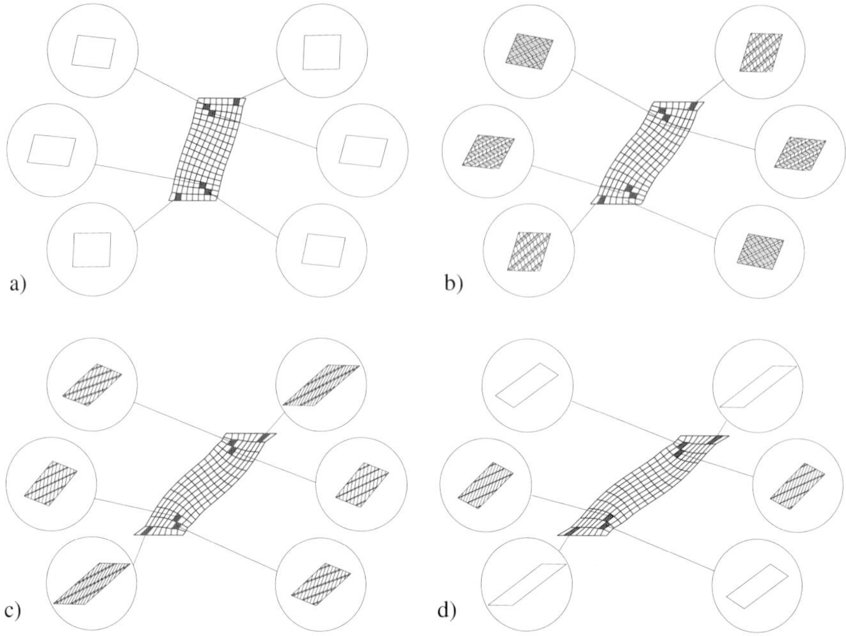

Figure 1. Specimen in shear. Development of microstructures at the colored integration points for the top displacements a) $u = 10\,\mathrm{mm}$, b) $u = 20\,\mathrm{mm}$, c) $u = 30\,\mathrm{mm}$, d) $u = 40\,\mathrm{mm}$.

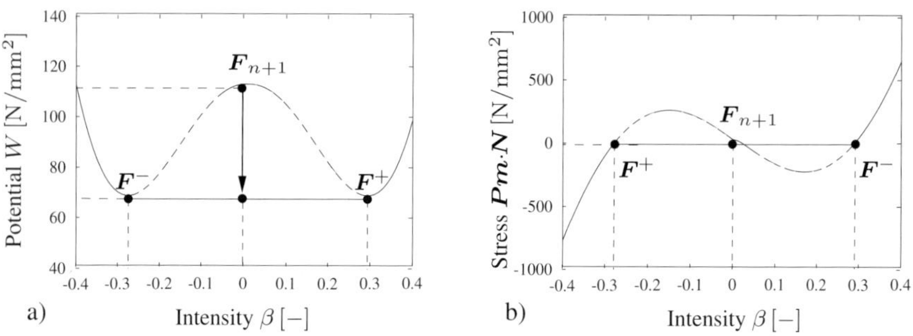

Figure 2. Specimen in shear. β parameterizes the intensity of the laminate $\boldsymbol{F}^{\pm} = \boldsymbol{F}_{n+1} + \beta^{\pm}\boldsymbol{m} \otimes \boldsymbol{N}$. a) At $\boldsymbol{F}_{n+1}$ the potential is not rank-one convex (dashed line). $\boldsymbol{F}_{n+1}$ decomposes into micro-phases $\boldsymbol{F}^{\pm}$ (solid line). b) The relaxed stress-strain relation characterizes a snap-through Maxwell-line behavior between the micro-phases $\boldsymbol{F}^{\pm}$.

element. Obviously, the incremental stress potential $W(\boldsymbol{F}_{n+1})$ is greater than the interpolation of potentials $W(\boldsymbol{F}^{+})$ and $W(\boldsymbol{F}^{-})$ corresponding to micro-phases $(+)$ and $(-)$. As a consequence, the homogeneous deformation state is not stable and decomposes into the micro-deformations $\boldsymbol{F}^{\pm}$. The relaxed stress-deformation relation plotted in Fig. 2b characterizes a snap-through behavior (Maxwell line) between the two micro-phases $\boldsymbol{F}^{\pm}$ due to the constant slope of the rank-one convex envelope. In order to point out the mesh-objectivity of the relaxed formulation we compare the associated load-deflection curves plotted in Fig. 3 for the five different mesh-sizes considered. The zigzagging of the non-relaxed solution results in Fig. 3a from the abrupt activation of the slip-

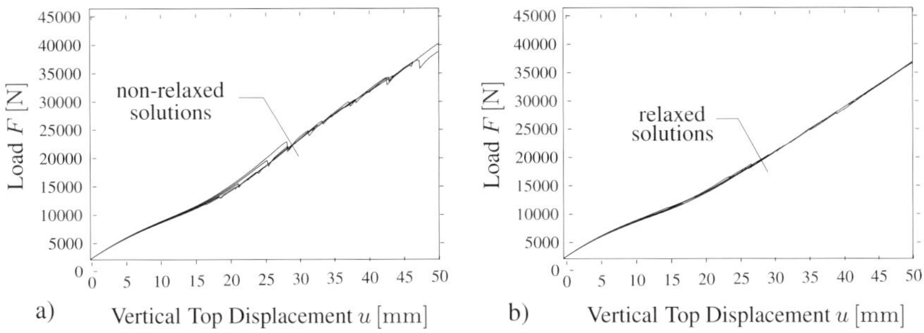

Figure 3. Specimen in shear. Load-displacement curves for five finite element meshes in terms of a) non-relaxed and b) objective relaxed formulation.

systems that strongly depends on the coarseness of the mesh-discretizations. As a consequence the incremental boundary-value problem is ill-posed and the non-convexified stress response not objective. In contrast to the non-relaxed formulation, application of the proposed relaxation technique yields a mesh-invariant response. The load-deflection curves do not depend on the mesh-size, but are identical for all different mesh-densities, see Fig. 3b. It turns out that the resolution of the microstructure as a first-order laminate is sufficient with regard to an objective stress response. Note that the objectivity of the material behavior is obtained without the introduction of an internal length scale parameter.

5.2. HOMOGENIZATION OF A POLYCRYSTALLINE MICROSTRUCTURE

In the second example we investigate the effect of different boundary constraints on the fluctuation field for the prediction of the texture evolution in heterogeneous polycrystals, using the homogenization analysis outlined in Miehe et al. [13]. The micro-structure considered is a copper unit cube discretized with $7^3 = 343$ so-called Q1P0 mixed finite elements. Individual crystal orientations are assigned to each integration point to model a polycrystalline aggregate of 2744 single crystal grains of equal volume. These grains are governed by a model of finite single crystal plasticity described in Miehe et al. [13]. The micro-structure is driven by a prescribed macro-deformation of plane strain compression with $\bar{\boldsymbol{F}}(t) = 1(\boldsymbol{e}_1 \otimes \boldsymbol{e}_1) + \exp[vt](\boldsymbol{e}_2 \otimes \boldsymbol{e}_2) + \exp[-vt](\boldsymbol{e}_3 \otimes \boldsymbol{e}_3)$ and $v = 1.0s^{-1}$. During the simulation the specimen is compressed down to 21% of its initial height, see Fig. 4 for plot sequences of the deformed geometry for three different fluctuation constraints. The homogeneous deformation of the boundary in Fig. 4a provides the stiffest response, the homogeneous stress

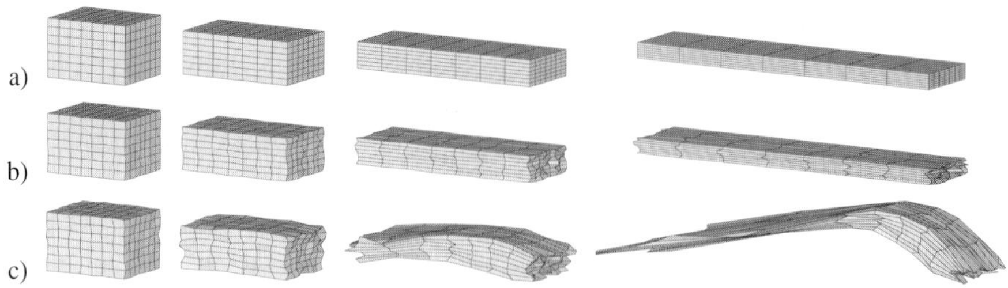

Figure 4. Plane strain compression test. Deformed micro–structures after a reduction of 18%, 41%, 63% and 79%. a) Homogeneous deformation. b) Periodic deformation. c) Homogeneous stresses on the boundary.

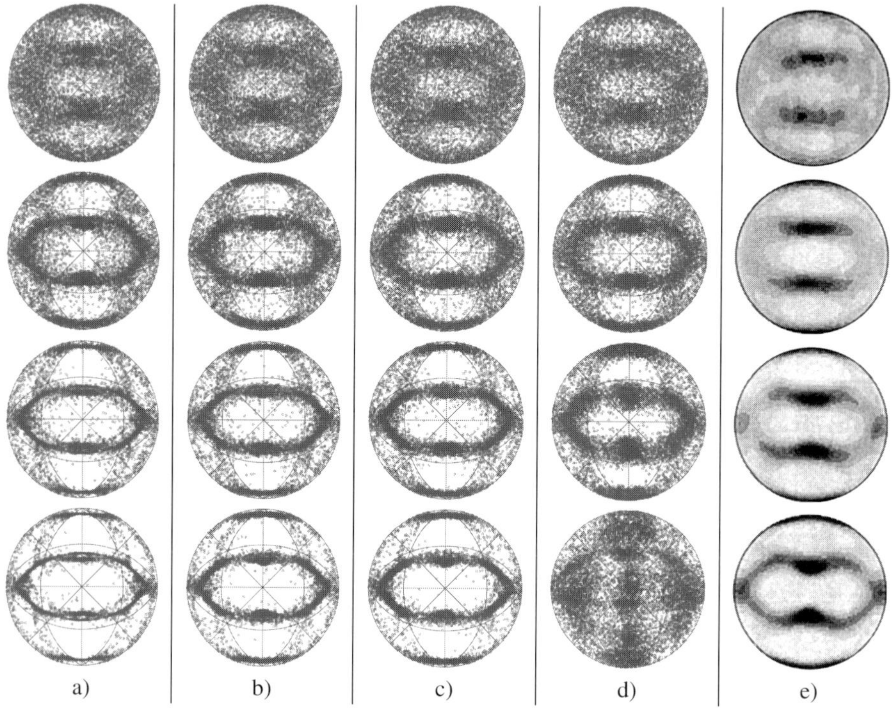

Figure 5. Plane strain compression test. Sequences of $\langle 111\rangle$-pole figures in the e_1e_2-plane corresponding to the deformation states of Figure 4. a) Taylor calculation using 2744 grains. b) Homogeneous deformation. c) Periodic deformation. d) Homogeneous stresses on the boundary. e) Experimental results of Bronkhorst et al. [3]

boundary condition of Fig. 4c the softest response of the micro-structure, resulting in heavily distorted finite element meshes. The texture development of the micro-structure is compared with experimental data and the prediction of a corresponding Taylor-type analysis. Fig. 5 presents the $\langle 111\rangle$-pole figures associated with the deformation states of Fig. 4. All pole figures plotted here are not extracted from a stereographic but an equal-area projection. For an initial quasi-isotropic orientation distribution of the single crystal grains this yields a constant point density in the undeformed state. The results of all simulations are in good agreement with the textures observed in experiments. As expected, the Taylor-type approach in Fig. 5a yields the sharpest orientation distributions due to the stiff constraint of uniform deformation of the micro-structure. The simulations in Fig. 5b–c relax this assumption. With respect to the sharpness of the pole figures, the solutions in Fig. 5c with periodic constraints are bounded by the solutions Fig. 5b and 5d associated with homogeneous deformation and homogeneous stresses on the boundary of the micro-structure. The textures obtained by the homogeneous and periodic constraints differ little. Compared to the Taylor-approach, due to the softer response of the micro-structure, they are more diffuse with a somewhat wider elliptical band. Fig. 5d reports on the development of crystal orientations in case of homogeneous stresses on the boundary of the aggregate. This approach simulates the softest micro-structure and, accordingly, the texture plots are more diffuse. The last pole figure should be ignored, since it results from the heavily distorted mesh of Fig. 4c.

References

[1] Ball, J.M. (1977). Convexity Conditions and Existence Theorems in Nonlinear Elasticity. *Archive of Rational Mechanics and Analysis*, **63**, 337–403.

[2] Biot, M.A. (1965). *Mechanics of Incremental Deformations*, Wiley & Sons, New York.

[3] Bronkhorst, C.A., Kalidindi, S.R. and Anand, L. (1992). Polycrystalline Plasticity and the Evolution of Crystallographic Texture in f.c.c. Metals. *Philosophical Transactions Royal Society London, Series A*, **341**, 443–477.

[4] Carstensen, C., Hackl, K. and Mielke, A. (2002). Nonconvex potentials and microstructures in finite-strain plasticity. *Royal Society London, Proc. Series A*, **458**, 299–317.

[5] Dacorogna, B. (1989). *Direct Methods in the Calculus of Variations*, Springer-Verlag, Berlin, Heidelberg.

[6] Halphen, B. and Nguyen, Q.S. (1975). Sur les Matériaux Standards Généralisés. *Journal de Mécanique*, **40**, 39–63.

[7] Hill, R. (1972). On Constitutive Macro-Variables for Heterogeneous Solids at Finite Strain. *Proceedings of the Royal Society London, Series A*, **326**, 131–147.

[8] Kohn, R.V. and Strang, G. (1983). Explicit Relaxation of a Variational Problem in Optimal Design. *Bulletin of the American Mathematical Society*, **9**, 211–214.

[9] Martin, J.B. (1975). *Plasticity. Fundamentals and General Results*, MIT press, Cambridge, Massachusetts.

[10] Miehe, C. (2002). Strain-Driven Homogenization of Inelastic Microstructures and Composites Based on an Incremental Variational Formulation. *International Journal for Numerical Methods in Engineering*, **55**, 1285–1322.

[11] Miehe, C., Schröder, J. and Schotte, J. (1999). Computational Homogenization Analysis in Finite Plasticity. Simulation of Texture Development in Polycrystalline Materials. *Computer Methods in Applied Mechanics and Engineering*, **171**, 387–418.

[12] Miehe, C., Schröder, J. and Becker, M. (2002). Computational Homogenization Analysis in Finite Elasticity. Material and Structural Instabilities on the Micro- and Macro-Scales of Periodic Composites and their Interaction. *Computer Methods in Applied Mechanics and Engineering*, **191**, 4971–5005.

[13] Miehe, C., Schotte, J. and Lambrecht, M. (2002). Homogenization of Inelastic Solid Materials at Finite Strains Based on Incremental Minimization Principles. *Journal of the Mechanics and Physics of Solids*, **50**, 2123–2167.

[14] Miehe, C. and Lambrecht, M. (2000). Analysis of Microstructure Development in Shearbands by Energy Relaxation of Incremental Stress Potentials: Large-Strain Theory for Generalized Standard Solids. *International Journal for Numerical Methods in Engineering*, in press.

[15] Miehe, C., Lambrecht, M. and Gürses, E. (2002). Analysis of Material Instabilities of Inelastic Solids based on Incremental Variational and Relaxation Methods: Application to Single-Slip Crystal Plasticity. *Journal of the Mechanics and Physics of Solids*, submitted.

[16] Mielke, A. (2002). *Finite elastoplasticity, Lie groups and geodesics on SL(d)*, In: *Geometry, Dynamics, and Mechanics*, P. Newton, A. Weinstein and P. Holmes (eds.), Springer-Verlag, Berlin, 61–90.

[17] Moreau, J.J. (1976). Application of Convex Analysis to the Treatment of Elastoplastic Systems. In: *Application of Methods of Functional Analysis to Problems in Mechanics*, P. Germain and B. Nayroles (eds.), Springer Verlag.

[18] Morrey, C.B. (1952). Quasiconvexity and the Semicontinuity of Multiple Integrands. *Pacific Journal of Mathematics*, **2**, 25–53.

[19] Müller, S. (1987). Homogenization of Nonconvex Integral Functionals and Cellular Elastic Materials. *Archive of Rational Mechanics and Analysis*, **99**, 189–212.

[20] Müller, S. (1999). Variational models for microstructure and phase transitions. In: *Lectures given at the 2nd Session of the Centro Internazionale Matematico Estivo held in Cetaro, Italy, June 15–22, 1996*, F. Bethuel, G. Huisken, S. Müller, K. Steffen, S. Hildebrandt and M. Struwe (eds.), Springer, Berlin.

[21] Nguyen, Q.S. (2000). *Stability and Nonlinear Solid Mechanics*, Wiley & Sons, Chichester.

[22] Ortiz, M. and Repetto, E.A. (1999). Nonconvex Energy Minimization and Dislocation Structures in Ductile Single Crystals. *Journal of the Mechanics and Physics of Solids*, **47**, 397–462.

[23] Ortiz, M., Repetto, E.A. and Stainier, L. (2000). A Theory of Subgrain Dislocation Structures. *Journal of the Mechanics and Physics of Solids*, **48** (10), 2077–2114.

[24] Ponte Castañeda, P. and Suquet, P. (1998). Nonlinear Composites. *Advances in Applied Mechanics*, **34**, 171–303.

[25] Ziegler, H. (1963). Some Extremum Principles in Irreversible Thermodynamics with Application to Continuum Mechanics. In: *Progress in Solid Mechanics*, I.N. Sneddon and R. Hill (eds.), **IV**, North-Holland Publishing Company, Amsterdam.

Modeling of Complex Material Response

Superelasticity in Shape-Memory Materials

L. Anand, P. Thamburaja
Department of Mechanical Engineering, Massachusetts Institute of Technology
77 Massachusetts Avenue, Cambridge, MA 02139, U.S.A.
anand@mit.edu

Abstract: The recently developed crystal-mechanics-based constitutive model of Thamburaja and Anand [8] for superelasticity of shape-memory materials is reviewed and shown to quantitatively predict the superelastic response of an initially-textured Ti-Ni alloy in a variety of reasonably complex deformation modes.

Keywords: Shape-memory, crystal mechanics, finite elements, mechanical testing.

1. Introduction

The individual grains in polycrystalline shape-memory materials can abruptly change their lattice structure in the presence of suitable thermo-mechanical loading. The high-temperature phase in shape-memory materials is called *austenite*, and the low-temperature phase is called *martensite*. The austenitic phase typically has a higher crystallographic symmetry than the martensitic phase. Under zero stress, shape-memory materials are distinguished by the following four temperatures: θ_{ms}, martensite start temperature; θ_{mf}, martensite finish temperature; θ_{as}, austenite start temperature; and θ_{af}, austenite finish temperature. This capability of undergoing a solid-solid, diffusionless, displacive phase transformation leads to the technologically-important properties of *superelasticity* (also called *pseudoelasticity*), and *shape-memory*. In this paper we focus on the superelastic response of shape-memory materials. This response is a consequence of a stress-induced transformation from austenite to martensite and back when a sample is tested between zero and a finite but small strain, ≈ 5 percent, at a constant ambient temperature above its austenite finish temperature, θ_{af}. There is little or no permanent deformation experienced by the specimen in such a strain cycle; this gives an impression that the material has only undergone elastic deformation, and hence the term *superelastic*. However, there is dissipation; the mechanism responsible for the dissipation is the motion of sharp interfaces between the two material phases.

We have recently formulated a crystal-mechanics-based constitutive model (Thamburaja and Anand [8]) for superelasticity of shape-memory materials. The purpose of this paper is to briefly summarize this model and present the results of some superelastic experiments conducted on initially-textured Ti-Ni

C. Miehe (ed.), IUTAM Symposium on Computational Mechanics of Solid Materials at Large Strains, 103-114.

alloy, together with corresponding numerical simulations. The numerical simulations are shown to be in good accord with the experiments.

2. Single crystal constitutive model

Although the underlying physics of inelastic deformation due to phase transformations is quite different from that for dislocation-based plasticity of metals, we formulate a constitutive model for deformation of a single crystal of a shape-memory alloy by austenite-martensite phase transformations, by modifying the widely used framework for crystal plasticity by crystallographic slip.

The overall inelastic deformation of a crystal is always inhomogeneous at length scales associated with the *fine-scale microstructures* accompanying the austenite-martensite phase transformations. We emphasize that our model shall not account for the typical fine-scale microstructure, but only the volume fractions of different types of fine structure. Further, for the continuum level of interest here, the inelastic deformation will be defined as an average over a volume element that contains enough plate-like transformed regions to result in an acceptably smooth process; the smallest such volume element above which the inelastic response may be considered smooth, is labeled a representative-volume element.

Using the standard notation of modern continuum mechanics, we recall that the deformation gradient $\mathbf{F}$ maps *referential* segments $\mathbf{dX}$ to segments $\mathbf{dx} = \mathbf{F}\mathbf{dX}$ in the *deformed* configuration. We base the theory on the following multiplicative decomposition of the deformation gradient: $\mathbf{F} = \mathbf{F}^e\mathbf{F}^{tr}$. Here, (i) $\mathbf{F}^{tr}$ represents the local deformation of *referential* segments $\mathbf{dX}$ to segments $\mathbf{dl} = \mathbf{F}^{tr}\mathbf{dX}$ in a *relaxed lattice configuration* due to "inelastic transformations" which are based on the generation, growth and annihilation of transformed plate-like regions in a microscopic representative volume element (RVE) representing an infinitesimal "macroscopic" material neighborhood. (ii) $\mathbf{F}^e(\mathbf{X})$ represents the mapping of segments $\mathbf{dl}$ in the relaxed lattice configuration into segments $\mathbf{dx} = \mathbf{F}^e\mathbf{dl}$ in the *deformed* configuration due the "elastic mechanisms" of stretching and rotation of the lattice.

The phase transformations are assumed to take place by the nucleation and growth of plate-like regions whose boundaries are prescribed by transformation fronts $i = 1, 2, \cdots, N$ in the lattice, with each front i defined by a *transformation direction* $\mathbf{b}_0^i$ and a *transformation-front normal* $\mathbf{m}_0^i$, where $\mathbf{b}_0^i$ and $\mathbf{m}_0^i$ are *constant* (not necessarily orthogonal) unit vectors in the lattice; the tensor $\mathbf{b}_0^i \otimes \mathbf{m}_0^i$ is referred to as the ith *transformation tensor*. The presumption that phase transformation takes place through nucleation and growth of plate-like volume elements leads to the requirement that the evolution of $\mathbf{F}^{tr}$ be governed by *transformation rates*, $\dot{\xi}^i$, which correspond to *rate of change of transformed volume fractions* ξ^i *in a microscopic representative volume element* (RVE) via the *flow rule* $\dot{\mathbf{F}}^{tr} = \mathbf{L}^{tr}\mathbf{F}^{tr}$, with $\mathbf{L}^{tr} = \sum_{i=1}^{N} \dot{\xi}^i \, \gamma_T^i \, \mathbf{b}_0^i \otimes \mathbf{m}_0^i$, where γ_T^i is a strictly positive constant "transformation strain" associated with each system i. A phase transformation from austenite to martensite occurs when $\dot{\xi}^i > 0$; we call this the *forward transformation*. A transformation from martensite to austenite occurs when $\dot{\xi}^i < 0$, and we call this the *reverse transformation*. The volume fraction ξ^i of martensite per unit volume of the relaxed lattice configuration in an RVE corresponding to the ith system is restricted to lie in the range $0 \leq \xi^i \leq 1$. The total volume fraction of transformed regions in the RVE denoted by $\xi = \sum_{i=1}^{N} \xi^i$, is also restricted to lie in the range $0 \leq \xi \leq 1$.

With this brief background, and using the variables

$\mathbf{T}$	$\mathbf{T} = \mathbf{T}^\top$	Cauchy stress,
$\mathbf{F}$	$\det \mathbf{F} > 0$	deformation gradient,
θ	$\theta > 0$	absolute temperature,
θ_0		reference temperature,
$\mathbf{F}^{tr}$		transformation deformation gradient,
$(\mathbf{b}_0^i, \mathbf{m}_0^i)$		transformation systems,
γ_T^i	$\gamma_T^i > 0$	transformation strains,
ξ^i	$0 \leq \xi^i \leq 1$	transf. vol. fraction in an RVE for system i,
ξ	$0 \leq \xi \leq 1$	total transformed volume fraction in an RVE,
f_c^i	$f_c^i > 0$	critical transformation resistances,

together with the definitions

$\mathbf{F}^e = \mathbf{F}\mathbf{F}^{tr-1}, \det \mathbf{F}^e > 0$	elastic deformation gradient,
$\mathbf{C}^e = \mathbf{F}^{e\top}\mathbf{F}^e$	elastic right Cauchy-Green strain,
$\mathbf{E}^e = \frac{1}{2}(\mathbf{C}^e - \mathbf{1})$	elastic strain,
$\mathbf{T}^e = \det(\mathbf{F}^e)\mathbf{F}^{e\,-1}\mathbf{T}\mathbf{F}^{e-\top}$	stress conjugate to the elastic strain $\mathbf{E}^e$,
$\tau^i = \mathbf{b}_0^i \cdot \mathbf{T}^e\mathbf{m}_0^i$	resolved stress,

our constitutive model is summarized below:

1 **Equation for the stress**:

$$\mathbf{T}^e = \boldsymbol{\mathcal{C}}(\theta_0, \xi)[\mathbf{E}^e - \mathbf{A}(\theta_0, \xi)(\theta - \theta_0)], \tag{1}$$

where $\boldsymbol{\mathcal{C}}(\theta_0, \xi)$ is the elasticity tensor, and $\mathbf{A}(\theta_0, \xi)$ is the thermal expansion tensor.

2 **Transformation conditions**:
Let

$$f^i = \gamma_T^i \left\{\tau^i - C(\theta - \theta_T)\right\}, \tag{2}$$

define a *driving force* for phase transformation. Here C is a positive-valued parameter with dimensions of stress, and θ_T is the phase equilibrium temperature. Then, the transformation conditions are taken as

$$(f^i - f_c^i) = 0 \quad \text{for austenite to martensite,} \tag{3}$$

$$(f^i + f_c^i) = 0 \quad \text{for martensite to austenite.} \tag{4}$$

3 **Flow rule**:
The evolution equation for $\mathbf{F}^{tr}$ is

$$\dot{\mathbf{F}}^{tr} = \mathbf{L}^{tr}\mathbf{F}^{tr}, \tag{5}$$

$$\mathbf{L}^{tr} = \sum_{\alpha=1}^{N} \dot{\xi}^i \, \gamma_T^i \, \mathbf{b}_0^i \otimes \mathbf{m}_0^i, \tag{6}$$

with the transformation rates $\dot{\xi}^i$ restricted as follows:

$$\dot{\xi}^i \geq 0, \quad (f^i - f_c^i) \leq 0, \quad \text{and} \quad \dot{\xi}^i \, (f^i - f_c^i) = 0,$$

for forward transformation, and

$$\dot{\xi}^i \leq 0, \quad (f^i + f_c^i) \geq 0, \quad \text{and} \quad \dot{\xi}^i \, (f^i + f_c^i) = 0,$$

for reverse transformation.

4 **Evolution equation for the transformation resistances** f_c^i:
As a first approximation, the transformation resistances f_c^i are taken to be constant, so that

$$\dot{f}_c^i = 0. \tag{7}$$

5 **Consistency conditions**:
During forward transformations the following consistency conditions

$$\dot{\xi}^i \, \overline{(f^i - f_c^i)}^{\boldsymbol{\cdot}} = 0 \quad \text{if} \quad (f^i - f_c^i) = 0, \tag{8}$$

must be satisfied, and the conditions

$$\dot{\xi}^i \, \overline{(f^i + f_c^i)}^{\boldsymbol{\cdot}} = 0 \quad \text{if} \quad (f^i + f_c^i) = 0, \tag{9}$$

must be satisfied during reverse transformations.

To complete the constitutive model for a particular material the constitutive parameter/functions that need to be specified are

$$\left\{ \boldsymbol{\mathcal{C}}(\theta_0, \xi), \mathbf{A}(\theta_0, \xi), \mathbf{b}_0^i, \mathbf{m}_0^i, \gamma_T^i, \theta_T, C, f_c^i \right\}.$$

We have implemented our constitutive model in the finite-element program ABAQUS/Explicit [1]. Algorithmic details of the time-integration procedure used to implement the model in the finite-element code are given in Thamburaja and Anand [8]. This computational capability allows us to model a *polycrystalline aggregate* by considering a suitably large number of elements, with each element[1] representing a single grain whose constitutive response is given by our single-crystal constitutive model.

3. Application to polycrystalline Ti-Ni alloy

In this section we consider application of the constitutive model to the technologically important Ti-Ni polycrystalline shape-memory alloy. For Ti-Ni the high-temperature austenitic phase has a body-centered-cubic superlattice (B2) crystal structure, while the low-temperature martensitic phase is monoclinic. The crystallographic theory of martensitic transformation shows that phase transformation in Ti-Ni can occur on 192 possible transformation systems (Hane and Shield [3]). However, it is not clear whether all the 192 possible systems are actually operative during thermo-mechanical loading. Here, we shall only consider the 24 transformation systems used by a variety of recent researchers (e.g. Matsumoto et al. [6], Lu and Weng [5], Gall and Sehitoglu [2]). The components of the 24 transformation systems $(\mathbf{m}_0^i, \mathbf{b}_0^i)$ with respect to an orthonormal basis associated with the parent cubic austenite crystal lattice are given in Table 1. The transformation strains γ_T^i are the same for all these 24 transformation systems,$\gamma_T^i \equiv \gamma_T = 0.1308$. We shall assume that the values of the critical transformation resistances for these systems are also identical, $f_c^i \equiv f_c$.

A suitably thermo-mechanically-processed and heat-treated Ti-Ni alloy intended for superelastic applications was obtained from a commercial source in rod form. Its crystallographic texture was determined by using X-ray methods. The measured $\{111\}$, $\{100\}$, and $\{110\}$ pole figures are shown in Figure 1. This figure also shows a numerical representation of this measured texture using 729 unweighted grain orientations; the numerical representation was obtained by using the Preferred Orientation Package, PoPLa (Kallend et al. [4]).

The transformation temperatures for the Ti-Ni, determined by using differential scanning calorimetric (DSC) techniques, are: $\theta_{ms} = 251.3\,\mathrm{K}, \theta_{mf} = 213.0\,\mathrm{K}, \theta_{as} = 260.3\,\mathrm{K}, \theta_{af} = 268.5\,\mathrm{K}$.

Details of the methodology to estimate the other single-crystal material parameters for Ti-Ni may be found in Thamburaja and Anand [8]. Briefly, these were estimated by a curve-fitting procedure in which the stress-strain curve from a finite element calculation for simple tension was fit to a superelasticity experiment in tension. In our finite-element model of a polycrystalline aggregate, each finite-element represents one crystal, and a set of crystal orientations which approximate the initial crystallographic texture of the shape-memory alloy are randomly assigned to the elements. The macroscopic stress-strain responses are calculated as volume averages over the entire aggregate. Calculations for simple tension on an aggregate of 729 unweighted different grain orientations representing a polycrystalline material with the initial crystallographic texture, Figure 1, were carried out to estimate the material parameters. The quality of the curve-fit is shown in Figure 2c. The residual strain exhibited in the experiments after complete unloading of the specimen is probably due to plastic

Table 1. Transformation systems.

i	$m^i_{0,1}$	$m^i_{0,2}$	$m^i_{0,3}$	$b^i_{0,1}$	$b^i_{0,2}$	$b^i_{0,3}$
1	-0.8888	-0.4045	0.2153	0.4343	-0.4878	0.7576
2	-0.8888	0.4045	0.2153	0.4343	0.4878	0.7576
3	-0.8888	0.2153	-0.4045	0.4343	0.7576	-0.4878
4	-0.8888	0.2153	0.4045	0.4343	0.7576	0.4878
5	-0.8888	-0.2153	0.4045	0.4343	-0.7576	0.4878
6	-0.8888	-0.2153	-0.4045	0.4343	-0.7576	-0.4878
7	-0.8888	0.4045	-0.2153	0.4343	0.4878	-0.7576
8	-0.8888	-0.4045	-0.2153	0.4343	-0.4878	-0.7576
9	0.4045	-0.8888	0.2153	0.4878	0.4343	0.7576
10	-0.4045	-0.8888	0.2153	-0.4878	0.4343	0.7576
11	0.2153	-0.8888	-0.4045	0.7576	0.4343	-0.4878
12	0.2153	-0.8888	0.4045	0.7576	0.4343	0.4878
13	-0.2153	-0.8888	0.4045	-0.7576	0.4343	0.4878
14	-0.2153	-0.8888	-0.4045	-0.7576	0.4343	-0.4878
15	0.4045	-0.8888	-0.2153	0.4878	0.4343	-0.7576
16	-0.4045	-0.8888	-0.2153	-0.4878	0.4343	-0.7576
17	0.2153	0.4045	-0.8888	0.7576	0.4878	0.4343
18	0.2153	-0.4045	-0.8888	0.7576	-0.4878	0.4343
19	0.4045	0.2153	-0.8888	0.4878	0.7576	0.4343
20	-0.4045	0.2153	-0.8888	-0.4878	0.7576	0.4343
21	0.4045	-0.2153	-0.8888	0.4878	-0.7576	0.4343
22	-0.4045	-0.2153	-0.8888	-0.4878	-0.7576	0.4343
23	-0.2153	0.4045	-0.8888	-0.7576	0.4878	0.4343
24	-0.2153	-0.4045	-0.8888	-0.7576	-0.4878	0.4343

deformation of the austenite and some stabilized martensite; these effects are not taken into account in our constitutive model, and thus are not predicted. The single-crystal material parameters[2] used to obtain this fit are: elastic moduli for austenite: $C^a_{11} = 130\,\text{GPa}$, $C^a_{12} = 98\,\text{GPa}$, $C^a_{44} = 34\,\text{GPa}$; elastic moduli for martensite: $C^m_{11} = 65\,\text{GPa}$, $C^m_{12} = 49\,\text{GPa}$, $C^m_{44} = 17\,\text{GPa}$; coefficients of thermal expansion: $\alpha^a = 11 \times 10^{-6}/\text{K}$, $\alpha^m = 6.6 \times 10^{-6}/\text{K}$; transformation strain: $\gamma_T = 0.1308$; phase equilibrium temperature: $\theta_T = 256\,\text{K}$; stress-temperature coefficient: $C = 3.0\,\text{MPa/K}$; critical transformation resistance: $f_c = 8.2\,\text{MJ/m}^3$.

Next, the model is used to calculate the superelastic response in torsion and compare it against a corresponding experimental result. The initial mesh used in the simulation for tubular torsion is shown in Figure 3a. The deformed mesh

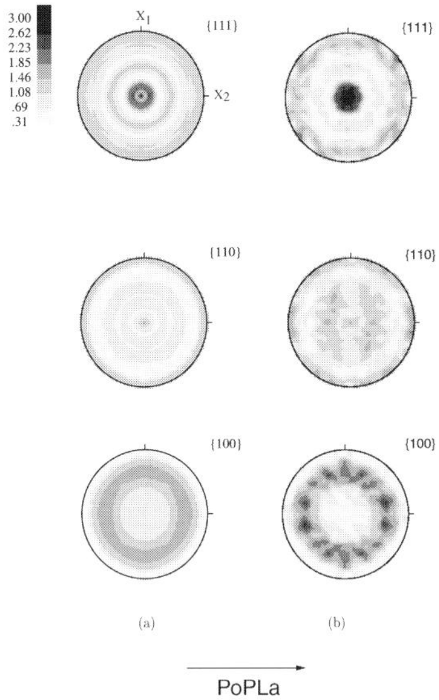

Figure 1. (a) Experimentally-measured texture in the as-received Ti-Ni rod. (b) Its numerical representation using 729 unweighted discrete crystal orientations.

after a shear strain of 9% is shown in Figure 3b, together with the contours for the martensite volume fraction. The predicted nominal shear stress-strain curve is shown in Figure 3c, where it is compared against the corresponding experimentally-measured curve. The experiment was performed under isothermal conditions at room temperature. The numerical prediction is in good accord with the experiment.

We have also evaluated the predictive capability of our model for (i) a proportional-loading, tension-torsion experiment, and (ii) a path-change, tension-torsion experiment (Thamburaja and Anand [7]). These experiments were also performed under isothermal testing conditions at room temperature.

3.1. PROPORTIONAL-LOADING TENSION-TORSION EXPERIMENT

The strain-history imposed on a thin-walled tubular specimen during the proportional-loading tension-torsion experiment is shown in Figure 4a. The axial strain ϵ is increased at a constant rate from 0% to 3% in 500 seconds, while the shear strain γ is increased at a constant rate from 0% to 4.5% in the same time period. These strains are then linearly reduced to zero, in another 500 seconds.

The initial finite-element mesh for the combined tension-torsion simulation is shown in Figure 4b, and the finite-element mesh after 3% strain in tension and 4.5% strain in shear is shown in Figure 4c, along with the contours of the martensite volume fraction.

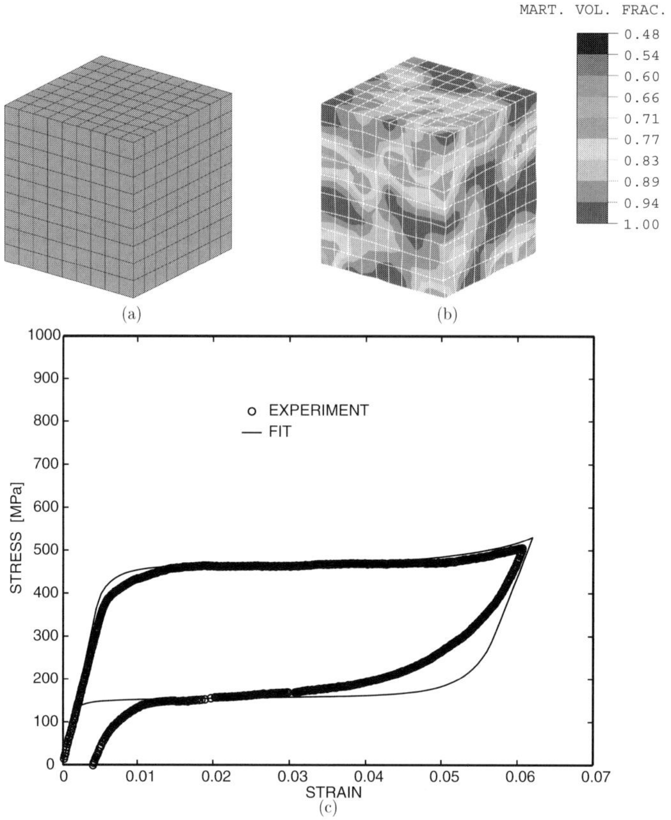

Figure 2. (a) Undeformed mesh of 729 ABAQUS C3D8R elements. (b) Deformed mesh at a tensile strain of 6%. Contours of martensite volume fraction are shown. (c) Superelastic stress-strain curve in tension. The experimental data from this test were used to estimate the constitutive parameters. The curve-fit using the full finite element model of the polycrystal is also shown.

Predictions of the axial-stress versus axial-strain, and shear-stress versus shear-strain from the constitutive model are compared against corresponding experimental measurements in Figures 4d and 4e, respectively. The measured superelastic stress-strain response in the tension-torsion experiment is well-approximated by the predictions from the constitutive model.

3.2. PATH-CHANGE TENSION-TORSION EXPERIMENT

The strain-history imposed on a thin-walled tubular specimen during the path-change tension-torsion experiment is shown in Figure 5a. First, the axial strain is increased at a constant rate from 0% to 3% in 150 seconds, while the shear strain is maintained at 0%. After this, the shear strain is linearly increased

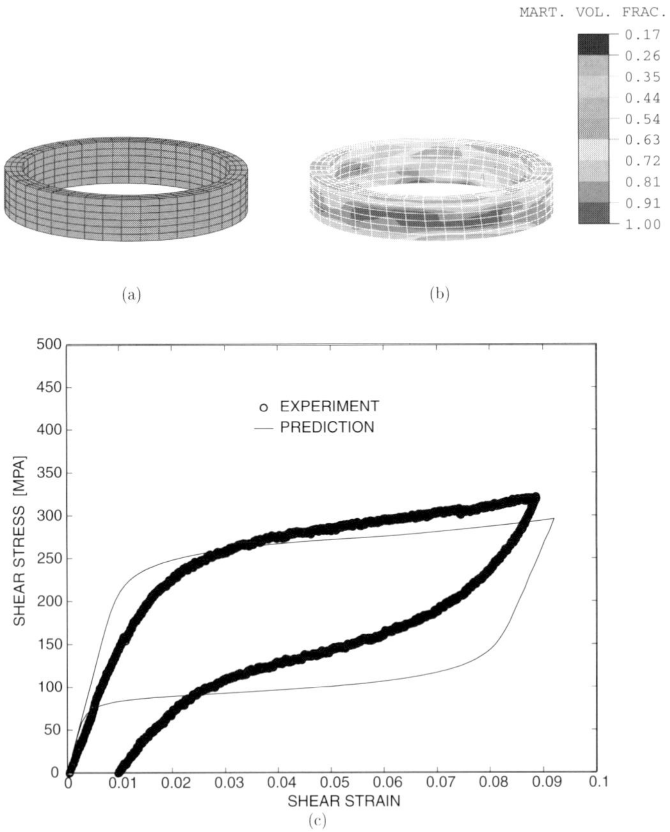

Figure 3. (a) Undeformed mesh of 720 ABAQUS C3D8R elements. (b) Deformed mesh at a shear strain of 9%. Contours of martensite volume fraction are shown. (c) Superelastic stress-strain curve in torsion. The prediction from the full finite element model for the polycrystal is also shown.

from 0% to 7% in 350 seconds while the tensile strain is maintained at 3%. Next, the shear strain is linearly reduced from 7% to 0% in 350 seconds, while still maintaining the tensile strain at 3%. Finally, the tensile strain is linearly reduced from 3% to 0% in another 150 seconds, while maintaining the shear strain at 0%.

The initial finite-element mesh for the path-change tension-torsion simulation is shown in Figure 5b, and the finite-element mesh after 3% strain in tension and 7% strain in shear is shown in Figure 5c, along with the contours of the martensite volume fraction.

Predictions of the axial-stress versus axial-strain, and shear-stress versus shear-strain from the constitutive model are compared against corresponding experimental measurements in Figures 5d and 5e, respectively. Again, the measured

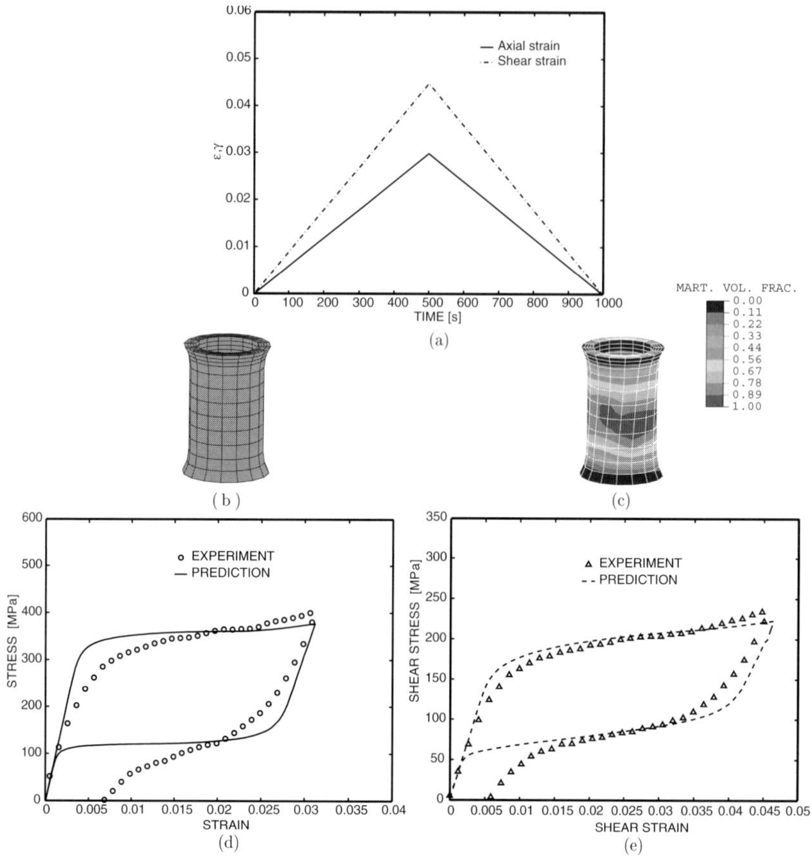

Figure 4. (a) Loading program for combined tension-torsion experiment. (b) Undeformed mesh of 768 ABAQUS C3D8R elements. (c) Deformed mesh at an axial strain of 3% and a shear strain of 4.5%. Contours of martensite volume fraction are shown. (d) Superelastic stress-strain curve in tension. (e) Superelastic stress-strain curve in shear. The prediction from the full finite element model for the polycrystal is also shown.

superelastic stress-strain response in the tension-torsion experiment is well-approximated by the predictions from the constitutive model.

Of particular note is the result that when the axial strain reaches 3% and is maintained at this level, the axial stress drops when the specimen is subsequently sheared. The axial stress rises again when the direction of shearing is reversed. As shown by the arrows in Figure 5d, this response is remarkably well-captured by the simulation.

Further, note from Figure 5e, that the shear stress has a negative value after the shear strain has completed its cycle back to zero. This shear stress decreases

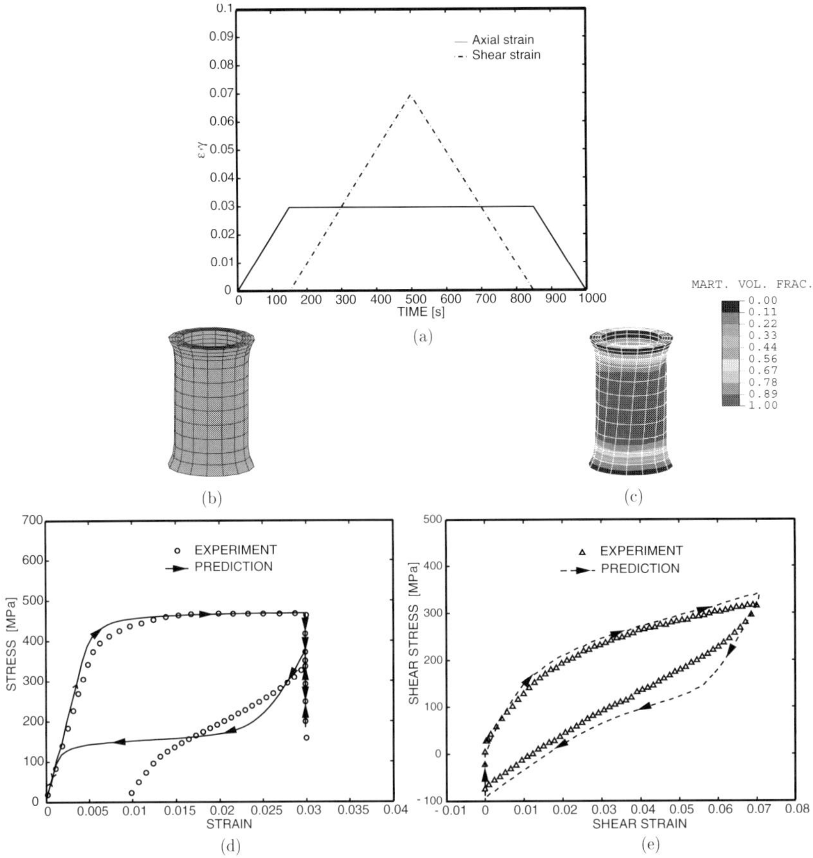

Figure 5. (a) Loading program for the path-change tension-torsion experiment. (b) Undeformed mesh of 768 ABAQUS C3D8R elements. (c) Deformed mesh at a tensile strain of 3% and a shear strain of 7%. Contours of martensite volume fraction are shown. (d) Superelastic stress-strain curve in tension. (e) Superelastic stress-strain curve in shear. The prediction from the full finite element model for the polycrystal is also shown.

back to zero only when the axial strain is finally reduced back to zero. This response is also well-predicted by the numerical simulation.

4. Conclusion

A crystal-mechanics-based constitutive model to describe the superelastic response of polycrystalline shape-memory alloys has been developed and shown (to within the idealization of no plastic deformation of austenite and of no detwinning of martensite) to *quantitatively* predict the superelastic response of an initially-textured Ti-Ni alloy in a variety of tension, torsion and combined tension-torsion experiments.

Acknowledgments

The financial support for this work was provided by the National Science Foundation under Grant CMS-002930.

Notes

1. More refined calculations which divide each grain into numerous finite elements are also possible.
2. For simplicity, the values of the elastic constants are taken to follow the rule of mixtures. Also, we have been unable to find experimentally-measured values of the anisotropic elastic moduli of single-crystal monoclinic martensite of Ti-Ni. Typically, the Young's Modulus of austenite is about two to three times that of martensite, while the Poisson's ratios for the material in the two different phases are approximately equal to each other. Guided by this information, we have assumed that anisotropic elastic constants of the monoclinic martensite may be approximately treated as those of a cubic material, and that the corresponding values of the stiffnesses, C_{ij}, are one half as large as those for the austenite.

References

[1] ABAQUS Reference Manuals. (2001). Providence, R.I.

[2] Gall, K. and Sehitoglu, H. (1999). The role of texture in tension-compression asymmetry in polycrystalline Ni-Ti. *International Journal of Plasticity*, **15**, 69–92.

[3] Hane, K.F. and Shield, T.W. (1999). Microstructure in the cubic to monoclinic transition in titanium-nickel shape memory alloys. *Acta Materialia*, **47**, 2603–2617.

[4] Kallend, J.S., Kocks, U.F., Rollett, S. and Wenk, H.R. (1989). *The Preferred Orientation Package from Los Alamos (PoPLa)*, Los Alamos National Laboratory, TMS, Warrendale, PA.

[5] Lu, Z.K. and Weng, G.J. (1998). A self-consistent model for the stress-strain behavior of shape-memory alloy polycrystals, *Acta Materialia*, **46**, 5423–5433.

[6] Matsumoto, O., Miyazaki, S., Otsuka, K. and Tamura, H. (1987). Crystallography of martensitic transformation in Ti-Ni single crystals. *Acta Metallurgica*, **35**, 2137–2144.

[7] Thamburaja, P. and Anand, L. Superelastic behavior in tension-torsion of an initially-textured Ti-Ni shape memory alloy. *International Journal of Plasticity*, in press.

[8] Thamburaja, P. and Anand, L., (2001). Polycrystalline shape-memory materials : effect of crystallographic texture. *Journal of the Mechanics and Physics of Solids*, **49**, 709–737.

Physically-Based Single and Polycrystal Plasticity Models and their Experimental Verification

S. Nemat-Nasser
Center of Excellence for Advanced Materials, University of California, San Diego
9500 Gilman Drive, La Jolla, CA 92093-0416, U.S.A.

Abstract: Physically-based constitutive relations for single crystals are formulated on the basis of slip-induced plastic deformation, and the accompanying elastic lattice distortion. The slip is produced through the motion of dislocations. Due account is taken of the crystal structure and the barriers which the dislocations must overcome in their motion through the lattice. A slip model that directly accounts for both the temperature- and strain-rate effects, is presented. The short- and long-range barriers that the dislocations must overcome in their motion are identified. The hardening issue, associated with the long-range athermal resistance to the motion of the dislocations, is examined. Illustrative physically-based models of the flow stress associated with slip, are developed. Explicit results for bcc and fcc crystals are produced, taking into account the temperature, strain rate, and the long-range hardening effects. As an illustration, explicit results for a commercially pure tantalum (bcc), and OFHC copper (fcc), are given. Using the results of novel experiments, performed to obtain the isothermal and adiabatic stress-strain relations of a number of polycrystalline metals over broad ranges of the strain-rates and temperatures, the resulting constitutive relations are verified. The references listed at the end of this paper provide the necessary background data for the modeling and the corresponding experimental assessment.

Keywords: Dislocation motion, slip, single and polycrystals, modeling, experiments.

1. Introduction

The kinematics of crystal plasticity is now fairly well understood; see Rice [24], Hill and Rice [6], Hill and Havner [5], and Havner [4] who provides a comprehensive account and references. On the other hand, the constitutive relations for the slip-induced plastic deformation of the crystals, are generally phenomenological, based on a power law and its variants; see, e.g., Hutchinson [7], Pan and Rice [20], Peirce et al. [21], Nemat-Nasser and Obata [17], Rashid et al. [22], Zikry and Nemat-Nasser [26], and Anand and Kalidindi [1].

Crystallographic slip occurs through the motion of dislocations. In this event, the dislocations must overcome their short-range and long-range barriers, as well as any velocity-dependent drag force that may hinder their motion. The short-range barriers may include the Peierls stress (the primary short-range resistance for bcc metals), point defects (vacancies and self-interstitials), other dislocations which intersect the slip plane (the primary short-range resistance for

C. Miehe (ed.), IUTAM Symposium on Computational Mechanics of Solid Materials at Large Strains, 115-130.

fcc metals), or substitutional atoms. The long-range barriers may correspond to the elastic stress fields of other dislocations, point defects, and grain boundaries. This resistance is not dependent on the strain rate. It is affected by temperature through the temperature-dependence of the elastic constants.

Recently, Nemat-Nasser et al. [13, 16, 19] have used dislocation-based models to represent rate- and temperature-dependent crystalline slip in (bcc) single-crystal tantalum and (fcc) copper, and then have simulated experimental results with considerable success. Here, some of these results are summarized together with some comparison with experimental data. Direct polycrystal modeling and experimental verification of the results are presented elsewhere; see references cited at the end of this paper.

2. Fundamentals

2.1. Crystal kinematics

The total deformation gradient, $\mathbf{F}$, measured relative to the undeformed configuration of the crystal, is divided into a non-plastic (elastic plus rigid-body rotation) deformation gradient, $\mathbf{F}^*$, and an inelastic (plastic) deformation gradient, $\mathbf{F}^p$, as follows:

$$\mathbf{F} = \mathbf{F}^* \mathbf{F}^p. \tag{1}$$

The total velocity gradient is defined by

$$\mathbf{L} = \dot{\mathbf{F}} \mathbf{F}^{-1}, \tag{2}$$

where dot and the superscript -1 stand for the time derivative and the inverse, respectively. Similarly, the non-plastic velocity gradient, $\mathbf{L}^*$, and the velocity gradient due to plastic deformation, $\mathbf{L}^p$, are given by

$$\mathbf{L}^* = \dot{\mathbf{F}}^* \mathbf{F}^{*^{-1}}, \tag{3}$$

and

$$\hat{\mathbf{L}}^p = \dot{\mathbf{F}}^p \mathbf{F}^{p^{-1}}, \tag{4}$$

where hat is used when the velocity gradient is measured relative to the*initial configuration* of the crystal lattice. Substitution of (1), (3), and (4) into (2) yields

$$\mathbf{L} = \mathbf{L}^* + \mathbf{F}^* \hat{\mathbf{L}}^p \mathbf{F}^{*^{-1}}. \tag{5}$$

We focus on metals whose plastic deformation is caused essentially by crystalline slip that in turn is produced by the motion of dislocations on slip planes. Hence, the plastic velocity gradient, $\hat{\mathbf{L}}^p$, is expressed by the summation of slips associated with all slip systems. It is often assumed that the primary slip systems in bcc single crystals are $\{110\}\langle 111\rangle$, $\{112\}\langle 111\rangle$, and $\{123\}\langle 111\rangle$, whereas those in fcc single crystals are $\{111\}\langle 110\rangle$. It thus follows that there are 48 primary slip systems in the bcc, and 12 in the fcc single crystals. To deal with both cases at once, we write

$$\hat{\mathbf{L}}^p = \sum_{\alpha=1}^{N} \dot{\gamma}^{(\alpha)} \mathbf{l}_0^{(\alpha)}, \tag{6}$$

where $\dot{\gamma}^{(\alpha)}$ is a slip rate, N is 48 for bcc and 12 for fcc crystals, and $\mathbf{l}_0^{(\alpha)}$ is a slip system tensor defined by

$$\mathbf{l}_0^{(\alpha)} = \mathbf{s}_0^{(\alpha)} \otimes \mathbf{n}_0^{(\alpha)}. \tag{7}$$

Here, $\mathbf{s}$ and $\mathbf{n}$ are the slip direction and the slip plane normal, respectively, and the subscript 0 stands for the initial configuration of the lattice.

For most metals, the elastic strains are very small. Thus, when (6) is substituted into (5), we obtain

$$\mathbf{L} = \mathbf{L}^* + \sum_{\alpha=1}^{N} \dot{\gamma}^{(\alpha)} \mathbf{l}^{(\alpha)}, \tag{8}$$

where

$$\mathbf{l}^{(\alpha)} = \mathbf{R}^* \mathbf{l}_0^{(\alpha)} \mathbf{R}^{*^T}, \tag{9}$$

and $\mathbf{R}^*$ is the non-plastic rotation tensor. For small elastic strains, the non-plastic velocity gradient is written as

$$\mathbf{L}^* = \dot{\boldsymbol{\varepsilon}} + \boldsymbol{\varepsilon}\boldsymbol{\Omega}^* - \boldsymbol{\Omega}^*\boldsymbol{\varepsilon} + \boldsymbol{\Omega}^*, \tag{10}$$

where ε is the elastic strain measured in the rotated lattice, and $\boldsymbol{\Omega}^*$ is the spin rate defined by

$$\boldsymbol{\Omega}^* = \dot{\mathbf{R}}^* \mathbf{R}^{*^T}. \tag{11}$$

The symmetric, $\mathbf{D}$, and antisymmetric, $\mathbf{W}$, parts of the total velocity gradient, $\mathbf{L}$, respectively, are

$$\mathbf{D} = \mathbf{D}^* + \sum_{\alpha=1}^{N} \dot{\gamma}^{(\alpha)} \mathbf{p}^{(\alpha)}, \tag{12}$$

and

$$\mathbf{W} = \mathbf{W}^* + \sum_{\alpha=1}^{N} \dot{\gamma}^{(\alpha)} \mathbf{w}^{(\alpha)}, \tag{13}$$

where $\mathbf{p}^{(\alpha)}$ is the symmetric, and $\mathbf{w}^{(\alpha)}$ is the anti-symmetric part of $\mathbf{l}^{(\alpha)}$.

2.2. On nature of slip systems

For bcc crystals, the forty-eight slip systems are divided into four families according to the slip directions, $[111]$, $[\bar{1}11]$, $[1\bar{1}1]$, and $[11\bar{1}]$. Hence, twelve slip systems belong to each family, and the slip-plane normal of slip systems in each family lie on the plane perpendicular to the slip direction. Thus, only two of twelve slip systems in each family are linearly independent. Therefore, ten conditions for $\mathbf{p}^{(\alpha)}$'s are obtained in each of the families, leading to forty conditions for $\mathbf{p}^{(\alpha)}$'s. These forty conditions are given explicitly by Nemat-Nasser et al. [19].

In addition, there are three other conditions which arise since$\mathbf{p}^{(\alpha)}$ is the symmetric part of the corresponding slip system, and hence, the slip-plane normal, s, and the slip direction, $\mathbf{n}$, play the same role to produce$\mathbf{p}^{(\alpha)}$. Hence it follows that only five of the forty-eight slip systems are linearly independent.

In the same manner, it can be shown that only 5 of the 12 slip systems in fcc crystals, are independent. Indeed, on each slip plane, only two slip directions can be independent. Since there are four slip planes in an fcc crystal, this leaves eight potentially independent systems. Here, again, three additional linear conditions relate slip systems of various slip planes, similarly to those for bcc crystals, leaving only five independent slip systems. In general, it can be concluded that no more than five slip systems in any crystal structure can be linearly independent. This stems from the tensorial nature of the slip systems which are defined by traceless second-order tensors formed by the diadic product of orthogonal unit vectors. In a three-dimensional space, only five such tensors can possibly be linearly independent. For further details, see Nemat-Nasser et al. [16, 19].

3. Physically-based model of crystalline slip

In the process of plastic deformation through crystalline slip by dislocation motion, dislocations have to overcome their short-range and long-range obstacles, as well as any resulting viscous drag (Kocks et al. [10]). As pointed out before, the short-range barriers may include the Peierls stress, point defects, other dislocations which intersect the slip plane, or substitutional atoms. The long-range barriers are generally caused by the elastic stress field due to grain boundaries, far-field forests of dislocations, and far-field defects. The resistance due to the short-range barriers may be overcome by thermal activation, while that due to the long-range barriers is essentially independent of temperature (except through the dependence of elastic moduli on temperature), and can be overcome with the aid of the applied resolved shear stress. The classical rate-independent plasticity is based on only the long-range barriers to the dislocation motion.

The flow stress, $\tau^{(\alpha)}$, of the αth slip system may therefore be written as

$$\tau^{(\alpha)} = \tau^{*(\alpha)} + \tau_a{}^{(\alpha)} + \tau_d{}^{(\alpha)}, \tag{14}$$

where $\tau^{*(\alpha)}$, $\tau_a{}^{(\alpha)}$, and $\tau_d{}^{(\alpha)}$ are the thermal, athermal, and the drag parts of the resisting (resolved in the direction of motion) shear stress corresponding to the motion of dislocations which sustain the (α)th slip rate. In line with the analysis presented by Nemat-Nasser and Li [15] for polycrystalline metals, the thermal part of the flow stress, here $\tau^{*(\alpha)}$, is viewed as a decreasing function of temperature of the crystal, T, and an increasing function of the plastic strain rate $\dot{\gamma}^{(\alpha)}$. The athermal part of the flow stress, $\tau_a{}^{(\alpha)}$, is essentially independent of temperature, T, and the plastic strain rate, $\dot{\gamma}^{(\alpha)}$, but it is affected by them in certain indirect ways (Nemat-Nasser and Li [15]), e.g., through the dependence of the elastic moduli on T.

3.1. THERMAL PART OF THE FLOW STRESS

For a dislocation moving on a typical slip plane in a given slip direction, the frequency of the occurrence of a jump over a short-range barrier is given by

$$\omega^{(\alpha)} = \omega_0^{(\alpha)} \exp\left(-\frac{\Delta G^{(\alpha)}}{kT}\right), \tag{15}$$

where $\omega_0^{(\alpha)}$ and k are the attempt frequency and the Boltzmann constant, respectively, and $\Delta G^{(\alpha)}$ is the activation energy, associated with the (α)th slip system. For convenience, the superscript (α) is used for all parameters associated with the αth slip system, even though some of these parameters may have values common to several or even all slip systems. The total average time, $t^{(\alpha)}$, that the dislocation takes to travel between two neighboring sets of barriers, consists of two parts,

$$t^{(\alpha)} = t_w{}^{(\alpha)} + t_r{}^{(\alpha)}, \tag{16}$$

where $t_w{}^{(\alpha)}$ is the waiting period before a dislocation crosses a barrier, and $t_r{}^{(\alpha)}$ is the running period that the dislocation moves until meeting the next barrier. Noting that the average travel distance is the sum of the average distance between barriers, $d^{(\alpha)}$, and the average effective barrier width, $\lambda^{(\alpha)}$, the average velocity of the motion of the dislocations is estimated by

$$\bar{v}^{(\alpha)} \approx \frac{d^{(\alpha)} + \lambda^{(\alpha)}}{t_w{}^{(\alpha)} + t_r{}^{(\alpha)}} \approx \frac{d^{(\alpha)}}{t_w{}^{(\alpha)}}, \tag{17}$$

where $d^{(\alpha)} \gg \lambda^{(\alpha)}$ when intersecting dislocations are the barriers, and $t_w{}^{(\alpha)} \gg t_r{}^{(\alpha)}$, in general. For the (α)th slip system, the waiting period, $t_w{}^{(\alpha)}$, is approximately given by

$$t_w^{(\alpha)} \approx 1/\omega^{(\alpha)}. \tag{18}$$

Hence, using the frequency given by (15), from (18) the average dislocation velocity for the (α)th slip system may be approximated by

$$\bar{v}^{(\alpha)} \approx d^{(\alpha)}\omega^{(\alpha)} = d^{(\alpha)}\,\omega_0^{(\alpha)}\,\exp\left(-\frac{\Delta G^{(\alpha)}}{kT}\right). \tag{19}$$

The slip rate, $\dot{\gamma}^{(\alpha)}$, of the (α)th slip system can therefore be expressed as

$$\dot{\gamma}^{(\alpha)} = b^{(\alpha)}\rho_m^{(\alpha)}\bar{v}^{(\alpha)}, \tag{20}$$

where $b^{(\alpha)}$ is the magnitude of the corresponding Burgers vector, and $\rho_m^{(\alpha)}$ is the density of the mobile dislocations associated with the (α)th slip system. Combining the last equation with (19), the slip rate, $\dot{\gamma}^{(\alpha)}$, is given by

$$\dot{\gamma}^{(\alpha)} = \dot{\gamma}_r^{(\alpha)}\exp\left(-\frac{\Delta G^{(\alpha)}}{kT}\right), \tag{21}$$

where

$$\dot{\gamma}_r^{(\alpha)} \equiv b^{(\alpha)}d^{(\alpha)}\rho_m^{(\alpha)}\omega_0^{(\alpha)}, \tag{22}$$

is the reference slip rate.

Following Kocks et al. [10], the activation energy, $\Delta G^{(\alpha)}$, is represented by

$$\Delta G^{(\alpha)} = G_0^{(\alpha)}\left\{1 - \left(\frac{\tau^{*(\alpha)}}{\hat{\tau}^{(\alpha)}}\right)^p\right\}^q, \qquad \hat{\tau}^{(\alpha)} = \tau^{*(\alpha)}\,|_{T=0}, \tag{23}$$

where $\hat{\tau}^{(\alpha)}$ is the threshold shear stress which is the stress required by a dislocation to overcome its short-range barrier without any assistance from the thermal activation;

$$G_0^{(\alpha)} = \hat{\tau}^{(\alpha)}V^{*(\alpha)} \equiv \hat{\tau}^{(\alpha)}b^{(\alpha)}\lambda^{(\alpha)}d^{(\alpha)}, \tag{24}$$

is the total free energy needed to overcome the barrier without the aid of external work; and $V^{*(\alpha)} \equiv b^{(\alpha)}\lambda^{(\alpha)}d^{(\alpha)}$ is the activation volume. The constants p and q in (23) are the parameters defining the shape of the energy-barrier profile. From (21) and (23), it follows that

$$\tau^{*(\alpha)} = \hat{\tau}^{(\alpha)}\left[1 - \left(-\frac{kT}{G_0^{(\alpha)}}\ln\frac{|\dot{\gamma}^{(\alpha)}|}{\dot{\gamma}_r^{(\alpha)}}\right)^{\frac{1}{q}}\right]^{\frac{1}{p}}, \qquad T \le T_c^{(\alpha)}, \tag{25}$$

$$\tau^{*(\alpha)} = 0, \qquad \text{for } T > T_c^{(\alpha)}, \qquad T_c^{(\alpha)} = -G_0^{(\alpha)}/\{k\,\ln[|\dot{\gamma}^{(\alpha)}|/\dot{\gamma}_r^{(\alpha)}]\}. \tag{26}$$

Here, $T_c^{(\alpha)}$ is the critical temperature at which a dislocation can overcome its short-range barriers without the assistance of the applied resolved shear stress.

Examine (24), where $G_0^{(\alpha)}$ and $b^{(\alpha)}$ are constants, and $\lambda^{(\alpha)}$ is dependent on the profile of the barrier. The average distance between barriers, $d^{(\alpha)}$, depends on the type of barrier. For Peierls barriers, which are the primary short-range obstacles of bcc metals, we may set $d^{(\alpha)} = b^{(\alpha)}$. For fcc crystals, the short-range barriers to the motion of dislocations primarily are the dislocation forests which intersect the slip plane. In this case, the average distance between barriers, $d^{(\alpha)}$, is the same as the average spacing, $l^{(\alpha)}$, of the dislocations which intersect the slip plane. Hence, $d^{(\alpha)} = l^{(\alpha)}$ is a function of the current density of dislocations, $\rho_c^{(\alpha)}$, that intersect the (α)th slip plane, and one has

$$d^{(\alpha)} = l^{(\alpha)} \approx [\rho_c^{(\alpha)}]^{-1/2}. \tag{27}$$

The current density of dislocations, $\rho_c^{(\alpha)}$, associated with the (α)th slip system, is defined by the accumulated dislocations which are on the "intersecting" slip planes, where an intersecting slip plane is a plane whose dislocations pierce the considered slip plane.

It is necessary to develop an appropriate constitutive relation for the average dislocation spacing, $l^{(\alpha)}$, in terms of the temperature and deformation history. Indeed, the $l^{(\alpha)}$'s are the natural *length scales* in physics-based plasticity of metals. This suggests that the length scale in plasticity is a *constitutive* rather than a *kinematical* quantity. Quantities such as macroscopic plastic strain and its gradients do not qualify as *thermodynamic structural parameters*, whereas those that characterize the dislocation density and its distribution do.

Therefore, we follow Nemat-Nasser and Li [15], view $l^{(\alpha)}$'s as fundamental constitutive parameters characterizing the microstructure, and seek to obtain constitutive relations for them. Based on the physics of the phenomenon, it is reasonable to expect that $l^{(\alpha)}$ increases with increasing temperature (annealing) and decreases with increasing density of the dislocations of the intersecting slip planes. Since the *effective plastic strain rate*, $\dot{\gamma} = (2/3\,\mathbf{D}^p : \mathbf{D}^p)^{1/2}$ is by definition positive, the total accumulated effective plastic strain, γ, is a monotonically increasing parameter and hence may be used as an independent variable to express the evolution of all other parameters which characterize the microstructure and hence affect the the resistance to dislocation motion. Here, $\mathbf{D}^p$ is the plastically-induced part of the crystal's deformation-rate tensor. In a polycrystal, one may use instead the overall polycrystals plastic deformation-rate tensor, as suggested by Nemat-Nasser et al. [16, 19].

We now write, for fcc crystals,

$$d^{(\alpha)} = l^{(\alpha)} = l_0^{(\alpha)} / f^{(\alpha)}(\gamma, T), \tag{28}$$

$$f^{(\alpha)}(\gamma, T) > 0, \qquad f^{(\alpha)}(0, T_0) = 1, \tag{29}$$

$$\frac{\partial f^{(\alpha)}(\gamma, T)}{\partial \gamma} \geq 0, \qquad \frac{\partial f^{(\alpha)}(\gamma, T)}{\partial T} \leq 0, \tag{30}$$

where $l_0^{(\alpha)}$ is the initial average spacing of the dislocations intersecting the slip plane, T_0 is the initial temperature, and γ is the total effective plastic strain produced by all slip systems. As an illustration, consider the simple model

$$f^{(\alpha)} = 1 + a^{(\alpha)}(T)\gamma^{n_0}, \tag{31}$$

where for copper, for example, it may be assumed that

$$a^{(\alpha)}(T) = a_0^{(\alpha)}[1 - (T/T_m)^2], \tag{32}$$

with T_m being the melting temperature (≈ 1350 K for copper). In calculations, we may then use a common value for all $a_0^{(\alpha)}$'s, which are non-negative constants that depend on the initial dislocation (barrier) spacing, and view this common value and n_0 as adjustable constitutive parameters.

We emphasize that γ should be viewed as a "loading parameter" used to express the dislocation density. It is *not* a parameter to be used to define the microstructure, since plastic strain is not an appropriate state variable.

Now, from the relation (24), noting that $G_0^{(\alpha)}$ is constant, it follows that

$$\hat{\tau}^{(\alpha)} = \hat{\tau}_0^{(\alpha)} f^{(\alpha)}(\gamma, T), \qquad \hat{\tau}_0^{(\alpha)} = G_0^{(\alpha)}/[b^{(\alpha)}\lambda^{(\alpha)} l_0^{(\alpha)}]. \tag{33}$$

To obtain the plastic strain rate, use (28) in (21),

$$\dot{\gamma}_r^{(\alpha)} = \dot{\gamma}_0^{(\alpha)}/f^{(\alpha)}(\gamma, T), \qquad \dot{\gamma}_0^{(\alpha)} \equiv b^{(\alpha)}\rho_m^{(\alpha)}\omega_0^{(\alpha)} l_0^{(\alpha)}. \tag{34}$$

Consequently, when the dislocation forests intersecting the slip plane are the primary short-range obstacles, as for fcc crystals, the thermal part of the flow stress is obtained as

$$\tau^{*(\alpha)} = \hat{\tau}_0^{(\alpha)} f^{(\alpha)}(\gamma, T) \left[1 - \left[-\frac{kT}{G_0^{(\alpha)}}\left(\ln \frac{|\dot{\gamma}^{(\alpha)}|\, f^{(\alpha)}(\gamma, T)}{\dot{\gamma}_0^{(\alpha)}}\right)\right]^{\frac{1}{q}}\right]^{\frac{1}{p}}, \tag{35}$$

for $T \leq T_c$, and $\tau^{*(\alpha)} = 0$, for $T > T_c^{(\alpha)}$, where the critical temperature $T_c^{(\alpha)}$ is determined by the following equation

$$T_c^{(\alpha)} = -G_0^{(\alpha)}/\{k\, \ln[|\dot{\gamma}^{(\alpha)}| f^{(\alpha)}(\gamma, T_c^{(\alpha)})/\dot{\gamma}_0^{(\alpha)}]\}. \tag{36}$$

3.2. Athermal Part of the Flow Stress

The athermal part of the flow stress, $\tau_a^{(\alpha)}$, represents the resistance to the mobile dislocations due to the elastic stress field of other dislocations, grain boundaries, and defects of various kinds. A dislocation must overcome this resistance (with the aid of the applied resolved shear stress) before it can overcome its short-range obstacles. In general, $\tau_a^{(\alpha)}$ is a function of the density and distribution of the dislocations (moble and stationary), grain sizes and their distribution, as well as the density and distribution of precipitates, substitutional atoms, and other impurities and defects. Each of these densities corresponds to a*length scale* in this model. We thus write,

$$\tau_a^{(\alpha)} = \hat{g}^{(\alpha)}(\rho, d_G, ...)\tau_{a1}^{(\alpha)}, \tag{37}$$

where $\tau_{a1}^{(\alpha)} = \tau_{a1}^{(\alpha)}(T)$ is a material parameter that depends on temperature through the elastic constants. The arguments in the function$g^{(\alpha)}$ may be expressed as functions ofγ. Hence, we may simplify (37) and write

$$\tau_a^{(\alpha)} = g^{(\alpha)}(\gamma)\tau_{a1}^{(\alpha)}. \tag{38}$$

3.3. DRAG RESISTANCE

Figure 1, from Kapoor and Nemat-Nasser [9], shows the flow stress vs. the strain rate for a commercially pure polycrystal tantalum, for the indicated temperature. The stress increases with increasing strain rate, achieving a plateau at strain rates exceeding 3,000/s. This increased strain-rate sensitivity is usually attributed to the viscous-drag effects on the mobile dislocations; see Follansbee and Weertman [3], Zerilli and Armstrong [25], Chiem [2], and Regazzoni et al. [23]. The viscous-drag stress, $\tau_d^{(\alpha)}$, is usually related linearly to the average dislocation velocity, $\bar{v}^{(\alpha)}$, by $\tau_d^{(\alpha)} \approx B\bar{v}^{(\alpha)}/b^{(\alpha)}$, where B is the drag coefficient. Based on these and other experimental results (Nemat-Nasser and Guo [12]), it may now be assumed that

$$\tau_d^{(\alpha)} = m_0 \left[1 - exp\left(-\alpha_0\dot{\gamma}^{(\alpha)}\right)\right], \tag{39}$$

where m_0 is a material constant which can be measured directly at a very high strain rate, and α_0 represents an effective damping coefficient affecting the dislocation motion. α_0 is of the order of 10^{-4} when $\dot{\gamma}$ is measured per second, and its actual value can be estimated empirically using high-temperature data at several strain rates, as suggested by Nemat-Nasser and Guo [12]. The drag resistance may be incorporated into the expression for the constitutive relation of the (α)th slip rate, as show by Nemat-Nasser [11]. For the present application, we ignore this part of the resistance to dislocation motion, and focus only on the activation and athermal parts.

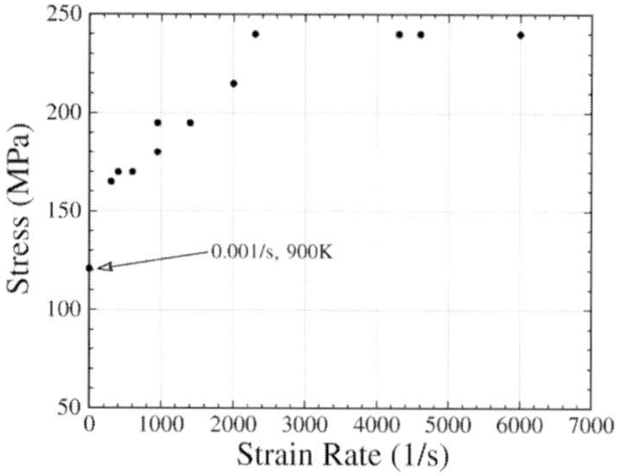

Figure 1. Flow stress at 0.05 strain at 900K as a function of strain rate.

3.4. THE CONSTITUTIVE RELATION

From (35) and (38), and since the slip rate, $\dot{\gamma}^{(\alpha)}$, and the driving shear stress, $\tau^{(\alpha)}$, have the same sign, the resistance to slip of the (α)th slip system is given by

$$\tau^{(\alpha)} = \mathrm{sgn}(\dot{\gamma}^{(\alpha)}) \Bigg[g^{(\alpha)}(\gamma)\tau_{a1}^{(\alpha)} + + \hat{\tau}_0^{(\alpha)} f^{(\alpha)}(\gamma, T) \left[1 - \left(-\frac{kT}{G_0^{(\alpha)}} \ln \frac{|\dot{\gamma}^{(\alpha)}|\, f^{(\alpha)}(\gamma, T)}{\dot{\gamma}_0^{(\alpha)}} \right)^{\frac{1}{q}} \right]^{\frac{1}{p}} \Bigg], \tag{40}$$

for $T \leq T_c^{(\alpha)}$, and

$$\tau^{(\alpha)} = \mathrm{sgn}(\dot{\gamma}^{(\alpha)}) g^{(\alpha)}(\gamma)\tau_{a1}^{(\alpha)}, \tag{41}$$

for $T > T_c^{(\alpha)}$. Therefore, the plastic strain rate of the (α)th slip system is,

$$\dot{\gamma}^{(\alpha)} = \frac{\mathrm{sgn}(\tau^{(\alpha)})\dot{\gamma}_0^{(\alpha)}}{f^{(\alpha)}(\gamma, T)} \exp\left[-\frac{G_0^{(\alpha)}}{kT} \left\{ 1 - \left(\frac{|\tau^{(\alpha)}| - \tau_a^{(\alpha)}}{\hat{\tau}_0^{(\alpha)} f^{(\alpha)}(\gamma, T)} \right)^p \right\}^q \right], \tag{42}$$

when $|\tau^{(\alpha)}| \geq \tau_a^{(\alpha)}$, and $\dot{\gamma}^{(\alpha)} = 0$ otherwise.

We assume that the increment of temperature is proportional to the increment of the plastic work, so that the temperature, T, is given by

$$T = T_0 + \eta \Delta w^p, \tag{43}$$

where Δw^p is the increment of the plastic work, and η is the conversion factor, including a factor which defines the portion of the plastic work used to heat the material; see Kapoor and Nemat-Nasser [8]. For a single crystal, the increment of the plastic work, Δw^p, can be calculated by

$$\Delta w^p = \sum_{\alpha=1}^{N} \int_0^{\gamma^{(\alpha)}} \tau^{(\alpha)}\, d\gamma^{(\alpha)}. \tag{44}$$

As explained before, the primary barriers to the motion of dislocations in fcc and bcc crystals are different: the dominant barriers in fcc crystals are the dislocation forests which intersect the slip plane, whereas those in bcc crystals are the Peierls lattice stresses, although both barriers influence the dislocation motion in essentially all crystal systems. For fcc crystals, the energy required to cross the Peierls barrier is generally small (less than 0.2eV), and only at very low temperatures does the Peierls barrier become important. For bcc crystals, the average spacing of the barriers, $d^{(\alpha)}$, equals the magnitude of the Burgers vector, $b^{(\alpha)}$, which is a constant, and $l^{(\alpha)} = l_0^{(\alpha)} = b^{(\alpha)}$ is also a constant. Then, for bcc crystals, one may assume $f^{(\alpha)} \approx 1$.

In the case of fcc crystals, on the other hand, the intersecting dislocations are the barriers. The core of a dislocation involves several surrounding atoms. Therefore, the magnitude of the energy barrier is expected to be considerably greater than the lattice resistance. Thus, the constitutive parameter $k/G_0^{(\alpha)}$, for fcc OFHC copper is estimated to be 4.9×10^{-5}/K by Nemat-Nasser and Li [15], which is about half of the value estimated for bcc tantalum, 8.62×10^{-5}/K, by Nemat-Nasser and Isaacs [13]. We use the same values in our crystal plasticity model calculations.

3.5. Annealed versus As-Received FCC Samples

The initial average dislocation (barrier) spacing for annealed fcc samples is expected to be greater than that for the as-received samples. Hence, the value of $\hat{\tau}_0^{(\alpha)}$ for annealed samples is expected to be smaller than that for the as-received samples. The parameter $a_0^{(\alpha)}$ is also expected to be greater for the annealed samples than that for the as-received ones. For the continuum model of OFHC copper, Nemat-Nasser and Li [15] choose $a_0^{(\alpha)} \equiv a_0 = 20$ for the annealed samples, and $a_0^{(\alpha)} \equiv a_0 = 1.8$ for the as-received samples. In our crystal plasticity model calculation for fcc OFHC copper, we also assume $a_0^{(\alpha)} = a_0$, for all the slip systems, and use those same values of a_0 as used for the continuum model in Nemat-Nasser and Li [15]. In view of (34), $\dot{\gamma}_0^{(\alpha)}$ depends on the initial dislocation spacing, $l_0^{(\alpha)}$. As is seen, $\dot{\gamma}_0^{(\alpha)}$ also depends on the density of the mobile dislocations, $\rho_m^{(\alpha)}$, and the frequency $\omega_0^{(\alpha)}$, both of which are affected by $l_0^{(\alpha)}$. We assume that the product $\rho_m^{(\alpha)} \omega_0^{(\alpha)} l_0^{(\alpha)}$ remains essentially the same whether the sample is annealed or not. This may be a crude approximation, but it seems to be adequate for the present application.

4. Computational techniques and numerical results

Nemat-Nasser et al. [19, 16] have successfully employed a computational method to solve finite-deformation problems of bcc and fcc single crystals and polycrystals, using the physically-based constitutive model of Nemat-Nasser [11]. Consider an fcc or bcc polycrystal composed of M single crystals. Assign an average grain size to all crystals, and let their initial orientations be uniformly distributed. Use the Taylor averaging method, i.e., assume that each grain of the polycrystal undergoes the same deformation rate, and hence the same deformation. Hence, in all grains, $\mathbf{L} = \bar{\mathbf{L}}$, where $\mathbf{L}$ and $\bar{\mathbf{L}}$ denote the local and overall velocity gradients, respectively. The equivalent plastic strain rate and the equivalent stress of the polycrystal must be calculated by averaging over all the grains. The overall temperature is then computed using the average equivalent plastic strain rate and stress.

As discussed in Nemat-Nasser and Okinaka [18], for each single crystal, the deformation history is divided into three regimes according to the number of active slip systems. A slip system is called *active*, if its resolved shear stress satisfies $|\tau^{(\alpha)}| \geq \tau_a^{(\alpha)}$. From the constitutive relation (42), $\dot{\gamma}^{(\alpha)} = 0$ for the non-active slip system. Denoting the number of active systems in the single crystal by N_a, the three regimes are:

1 Rapidly-changing regime, where the elastic deformation plays an important role,

2 Transition regime, for which $0 < N_a \leq N$, and

3 Steady-state regime, for which $N_a > N$.

When a material is in the transition regime, the resolved shear stresses may increase with time, leading to activation of additional slip systems. When the loading starts, or when the loading conditions are suddenly changed, then the elastic deformation plays an essential role, and the material is in the rapidly-changing regime.

Solutions are obtained incrementally, applying different numerical methods for different deformation regimes. For the transition and the steady-state regimes, the *plastic-predictor, elastic-corrector method* proposed by Nemat-Nasser and Li [14], is used. We start with the tentative assumption that the total deformation increment is solely due to plastic slip, and then correct the results by including the accompanying elastic contribution. Since, in general, the elastic contribution is much smaller than the plastic one, the method efficiently yields accurate results. Indeed, in many cases the slip rate due to elastic distortion (but not the associated stress rate) may be neglected in the transition and steady-state regimes. In the rapidly-changing regime, the explicit Euler method is employed. The time increment used for the explicit Euler method is much smaller than that used for the plastic-predictor, elastic-corrector method. The time increments are thus, in general, not uniform for all grains. Hence, in the calculation of the polycrystal, it is necessary to assign a *time period*, at the end of which the averaging is performed to obtain the equivalent overall plastic strain rate and the equivalent stress of the polycrystal. The overall temperature is then calculated from the average plastic work rate per unit volume. This average work rate is assumed to remain unchanged within each averaging time interval. Hence, the volume average of the Cauchy stress at the end of the averaging interval, t_j^{av}, $j = 1, 2, .., J$, is given by

$$\mathbf{\Sigma}(t_j^{av}) = \frac{1}{V}\int_V \boldsymbol{\sigma}(t_j^{av})dV = \frac{1}{M}\sum_{n=1}^{M} \boldsymbol{\sigma}_n(t_j^{av}), \tag{45}$$

where σ_n is the Cauchy stress for the nth grain of the polycrystal. The equivalent stress, $\Sigma_{eq}(t_j^{av})$, is then calculated as

$$\Sigma_{eq}(t_j^{av}) = (3/2\mathbf{\Sigma}(t_j^{av}) : \mathbf{\Sigma}(t_j^{av}))^{1/2}. \tag{46}$$

The plastic part of the strain rate for the nth grain of the polycrystal is given by

$$\mathbf{D}_n^p = \sum_{\alpha=1}^{N} \dot{\gamma}_n^{(\alpha)} \mathbf{p}_n^{(\alpha)}, \tag{47}$$

where $n = 1, ..., M$ are the grain indices. The volume average of the plastic strain rate, $\bar{\mathbf{D}}^p$, and the equivalent plastic strain rate, $\dot{\gamma}_{eq}$, at t_j^{av}, $j = 1, 2, .., J$, are defined by

$$\bar{\mathbf{D}}^p(t_j^{av}) = \frac{1}{M}\sum_{n=1}^{M} \mathbf{D}_n^p(t_j^{av}), \tag{48}$$

and

$$\dot{\gamma}_{eq}(t_j^{av}) = (2/3\bar{\mathbf{D}}^p(t_j^{av}) : \bar{\mathbf{D}}^p(t_j^{av}))^{1/2}, \tag{49}$$

respectively.

The plastic work rate per unit volume is approximated by

$$\dot{W}^p(t_j^{av}) \approx \Sigma_{eq}(t_j^{av})\dot{\gamma}_{eq}(t_j^{av}). \tag{50}$$

The approximate temperature of the polycrystal material element (the test sample) is calculated from

$$T_A(t) = T(t_{j-1}^{av}) + \eta \dot{W}^p(t_{j-1}^{av})[t - t_{j-1}^{av}], \tag{51}$$

where t is in the jth averaging interval and satisfies $t_{j-1}^{av} < t \leq t_j^{av}$. Again, the conversion factor η includes a factor which defines the fraction of the plastic work converted into heat.

Typical simulation results are given in Figures 2 and 3 for annealed OFHC copper and commercially pure tantalum. In these simulations, the values of all parameters are taken to be the same for all slip systems. For the copper (Figure 2) we have set $\hat{\tau}_0^{(\alpha)} = \hat{\tau}_0 = 9$ MPa and $\tau_0^{(\alpha)} = \tau_0 = 50$ MPa. The values of all the remaining constitutive parameters are taken to be the same as those estimated by Nemat-Nasser and Li [15] in their continuum model. These parameters are: $k/G_0^{(\alpha)} = 4.9 \times 10^{-5}$ /K, $\dot{\gamma}_0^{(\alpha)} = 2 \times 10^{10}$ /s, $\tau_{a0}^{(\alpha)} = \tau_{a0} = 0$, $a_0 = 20$ for annealed copper; and $a_0 = 1.8$ for as-received copper, $n_0 = 0.5$, $n_1 = 0.3$, $p = 2/3$, and $q = 2$. For tantalum, the material constants are as follows: $p = \frac{2}{3}$, $q = 2$, $\dot{\gamma}_r^{(\alpha)} = 5.46 \times 10^8$, $k/G_0 = 8.62 \times 10^{-5}$, which are the same as those used in the continuum model by Nemat-Nasser and Isaacs [13]. In addition, we have set $\alpha = 0.433$, $\tau_{a0}^{(\alpha)} = 75$ MPa, $\tau_0^{(\alpha)} = 65$ MPa, $\hat{\tau}^{(\alpha)} = 440$ MPa, and $n = 0.5$. Additional results are found in the references.

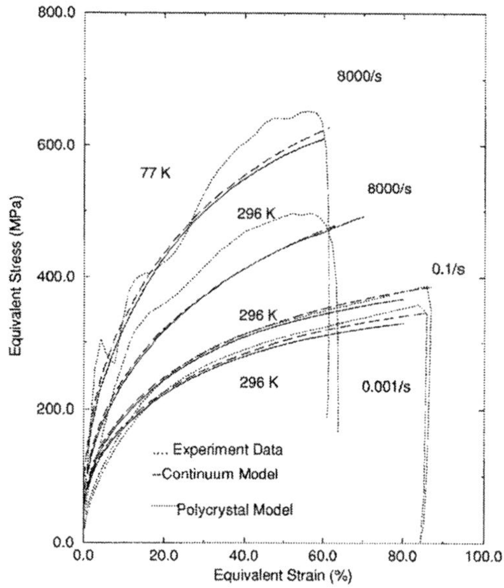

Figure 2. Comparison between the experimental data, the continuum model calculation, and the crystal plasticity model calculation of polycrystalline annealed OFHC copper at strain rates of 8000, 0.1, and 0.001 s-1 and initial temperatures of 77 and 296K. Dotted curves represent the experimental data, long dash curves represent the continuum model calculation, and solid curves are the model calculation using the crystal plasticity constitutive model, and the Taylor averaging method.

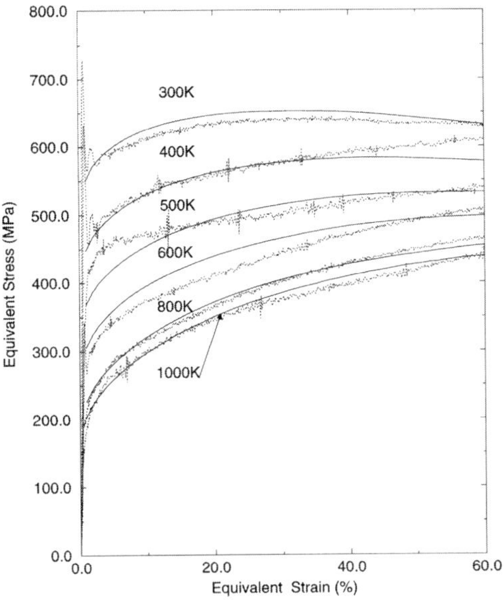

Figure 3. Comparison between the experimental data and the model calculation for polycrystalline tantalum at 5000/s strain rate and initial temperatures of 300, 400, 500, 600, 800, and 1000K. Dotted curves represent the experimental data, and solid curves are given by the model calculation, using the Taylor averaging method.

Acknowledgments

This research was supported by the Army Research Office under contract number ARO MURI DAAH04-96-1-0376, to the University of California, San Diego.

References

[1] Anand, L. and Kalidindi, S.R. (1994). The process of shearband formation in plane strain compression of fcc metals: effects of crystallographic texture. *Mech. Mater.*, **17**, 223–243.

[2] Chiem, C.Y. (1992). Material deformation at high strain rates. In: *Shock-Wave and High-Strain-rate Phenomena in Materials*, Marcel Dekker, New Your, 69–85.

[3] Follansbee, P.S. and Weertman, J. (1982). On the question of flow stress at high strain rated controlled by dislocation viscous flow. *Mech. Mater.*, **1**, 345–350.

[4] Havner, K.S. (1992). *Finite Plastic Deformation of Crystalline Solids*, Cambridge Press, New York.

[5] Hill, R. and Havner, K.S. (1982). Perspectives in the mechanics of elastoplastic crystals. *J. Mech. Phys. Solids*, **30**, 5–22.

[6] Hill, R. and Rice, J.R. (1972). Constitutive analysis of elastic-plastic crystals at arbitrary starin. *J. Mech. Phys. Solids*, **20**, 401–413.

[7] Hutchinson, J.W. (1976). Bounds and self-consistent estimates for creep of polycrystalline materials. *Proc. R. Soc. Lond.*, **A348**, 101–127.

[8] Kapoor, R. and Nemat-Nasser, S. (1998). Determination of temperature rise during high strain rate. *Mech. Mater.*, **27**, 1–12.

[9] Kapoor, R. and Nemat-Nasser, S. (2000). Comparision between high and low strain-rate deformation of tantalum. *Metall. Mat. Trans. A*, **31A**, 815–823.

[10] Kocks, U.F., Argon, A.S. and Ashby, M.F. (1975). Thermodynamics and Kinematics of Slip. *Progress in Materials Science*, Pergamon Press, **19**.

[11] Nemat-Nasser, S. (1997). Plasticity: Inelastic Flow of Heterogeneous Solids at Finite Strains and Rotations. In: *Proceeding of the ICTAM, Kyoto, Japan, August 25-31, 1996*, Elseveri Science B.V., 25–31.

[12] Nemat-Nasser, S. and Guo, W. G. (2000). High-strain-rate response of commercially pure vanadium. *Mech. Mater.*, **32**, 243–260.

[13] Nemat-Nasser, S. and Isaacs, J.B. (1997). Direct measurement of isothermal flow stress of metals at elevated temperatures and high strain rates with application to Ta and Ta-W alloys. *Acta Mater.*, **45**, 907–919.

[14] Nemat-Nasser, S. and Li, Y.-F. (1994). An Algorithm for Large-Scale Computational Finite-Deformation Plasticity. *Mech. Mater.*, **18**, 231–264.

[15] Nemat-Nasser, S. and Li, Y.L. (1998). Flow stress of fcc polycrystals with application to OFHC CU. *Acta Mater.*, **46**, 565–577.

[16] Nemat-Nasser, S., Ni, L. and Okinaka, T. (1998). A constitutive model for fcc crystals with application to polycrystalline OFHC copper. *Mech. Mater.*, **30**, 4, 325–341.

[17] Nemat-Nasser, S. and Obata, M. (1986). Rate-Dependent, Finite Elasto-Plastic Deformation of Polycrystals. *Proc. R. Soc. Lond*, **A407**, 343–375.

[18] Nemat-Nasser S. and Okinaka, T. (1996). A New Computational Approach to Crystal Plasticity: FCC Single Crystal. *Mech. Mater.*, **24**, 43–58.

[19] Nemat-Nasser, S., Okinaka, T. and Ni, L. (1998). A physically-based constitutive model for bcc crystals with application to polycrystalline tantalum. *J. Mech. Phys. Solids*, **46**, 6, 1009–1038.

[20] Pan, J. and Rice, J.R. (1983). Rate sensitivity of plastic flow and implications for yield-surface vertices. *Int. J. Solids Struct.*, **19**, 973–987.

[21] Peirce, D., Asaro, R.J. and Needleman, A. (1983). Material rate dependence and localized deformation in crystalline solids. *Acta. Metall.*, **31**, 1951–1976.

[22] Rashid, M.M., Gray, G.T. and Nemat-Nasser, S. (1992). Heterogeneous deformations in copper single crystals at high and low strain rates. *Phil. Mag.*, **65**, 707–735.

[23] Regazzoni, G., Kocks, U.F. and Follansbee, P.S. (1987). Dislocation kinetics at high strain rates. *Acta. Mater.*, **46**, 565–577.

[24] Rice, J.R. (1971). Inelastic constitutive relations for solids: an internal-variable theory and its application to metal plasticity. *J. Mech. Phys. Solids*, **19**, 433–455.

[25] Zerilli, F.J. and Armstrong, R.W. (1992). The effect of dislocation drag on the stress-strain behavior of fcc metals. *Acta. Mater.*, **40**, 1803–1808.

[26] Zikry, M.A. and Nemat-Nasser, S. (1990). High strain-rate localization and failure of crystalline materials. *Mech. Mater.*, **10**, 215–237.

Localized Plastic Flow in Single Crystals: A Nonlocal Analysis

A. Needleman
Division of Engineering, Brown University
Barus & Holley Building, Providence, RI 02912, U.S.A
needle@engin.brown.edu

Abstract: Plane strain tension of a single crystal is analyzed numerically. The crystal geometry is idealized in terms of a planar double slip model and the material is characterized by the nonlocal crystal plasticity theory of Acharya and Bassani [1]. This nonlocal theory preserves the classical structure of boundary value problems. Hence, implementation into a conventional single crystal finite element code is straightforward. The plane strain tension boundary value problem is formulated accounting for finite strains and rotations, and material rate sensitivity. The implications of a particular characterization of the gradient hardening for the growth of diffuse necking modes and for the development of localized plastic flow are explored.

Keywords: Plasticity, localization, nonlocal theory, finite elements.

1. Introduction

In recent years, there has been considerable interest in developing nonlocal theories of plasticity to account for experimental observations of the size dependence of plastic flow processes. Various phenomenological plasticity theories have been proposed that give rise to such size dependence, for example, Aifantis [2], Fleck and Hutchinson [7], Acharya and Bassani [1] and Gurtin [8]. Different phenomenological theories give rise to different boundary value problem formulations. One way to investigate the implications of the various theories is by obtaining boundary value problem solutions using them, as for example in Begley and Hutchinson [5], Shu and Fleck [12] and Bassani et al. [4].

In this paper, the numerical solution to a finite deformation boundary value problem for a single crystal characterized by a nonlocal plasticity constitutive relation is presented. The boundary value problem is plane strain tension of a single crystal. This is the same boundary value problem analyzed in Peirce et al. [11] except that here the material is characterized by a nonlocal constitutive relation. The calculations account for finite strains and rotations, and for material rate dependence.

The nonlocal plasticity theory employed is that of Acharya and Bassani [1], which is particularly simple to implement in that the nonlocal terms only appear in the hardening matrix. As a consequence, the classical structure of the

C. Miehe (ed.), IUTAM Symposium on Computational Mechanics of Solid Materials at Large Strains, 131-140.

rate boundary value problem is preserved. This makes incorporating the theory of Acharya and Bassani [1] into a conventional local finite element code straightforward. While the formulation precludes capturing certain nonlocal effects such as boundary layers, the results in Bassani et al. [4] suggest that there may be a class of nonlocal effects that this simple theory can model.

As a consequence of maintaining the classical boundary value problem structure, when the rate independent idealization is used and when bifurcation occurs from a homogeneous deformation state, the bifurcation predictions of the nonlocal theory of Acharya and Bassani [1] coincide with those of the corresponding local plasticity theory. However, in the calculations, diffuse necking precedes the onset of localized plastic flow so that the predictions of the nonlocal theory and conventional continuum slip plasticity theory for the development of localized plastic flow do not necessarily coincide. The numerical solutions illustrate an effect of nonlocal material behavior on the interaction between diffuse necking and localized shearing deformation modes.

2. Crystal plasticity

The formulation and numerical implementation follow that in Peirce et al. [11], where further details and references are given. The deformation gradient is written as

$$\mathbf{F} = \mathbf{F}^* \cdot \mathbf{F}^p \tag{1}$$

The deformation $\mathbf{F}^p$ consists solely of crystallographic slipping along specific slip systems $(\mathbf{s}^{(\alpha)}, \mathbf{m}^{(\alpha)})$, where $\mathbf{s}^{(\alpha)}$ is the slip direction and $\mathbf{m}^{(\alpha)}$ is the slip plane normal for slip system α. The vectors $\mathbf{s}^{(\alpha)}$ and $\mathbf{m}^{(\alpha)}$ are regarded as lattice vectors so that they stretch and rotate by

$$\mathbf{s}^{(\alpha)*} = \mathbf{F}^* \cdot \mathbf{s}^{(\alpha)} \qquad \mathbf{m}^{(\alpha)*} = \mathbf{m}^{(\alpha)} \cdot \mathbf{F}^{*-1} \tag{2}$$

Differentiating (1), with respect to time gives

$$\dot{\mathbf{F}} \cdot \mathbf{F}^{-1} = (\mathbf{D}^* + \boldsymbol{\Omega}^*) + (\mathbf{D}^p + \boldsymbol{\Omega}^p) \tag{3}$$

where $\dot{( )} = \partial( )/\partial t$, $(\mathbf{D}^* + \boldsymbol{\Omega}^*)$, is the elastic part of the rate of deformation tensor and the second term is the plastic part. The plastic stretching, $\mathbf{D}^p$, and spin tensors, $\boldsymbol{\Omega}^p$, are given by

$$\mathbf{D}^p = \sum_\alpha \dot{\gamma}^{(\alpha)} \mathbf{P}^{(\alpha)} \qquad \boldsymbol{\Omega}^p = \sum_\alpha \dot{\gamma}^{(\alpha)} \mathbf{W}^{(\alpha)} \tag{4}$$

with $\dot{\gamma}^{(\alpha)}$ the rate of shearing on slip system α and

$$\mathbf{P}^{(\alpha)} = \frac{1}{2}(\mathbf{s}^{(\alpha)*}\mathbf{m}^{(\alpha)*} + \mathbf{m}^{(\alpha)*}\mathbf{s}^{(\alpha)*})$$

$$\mathbf{W}^{(\alpha)} = \frac{1}{2}(\mathbf{s}^{(\alpha)*}\mathbf{m}^{(\alpha)*} - \mathbf{m}^{(\alpha)*}\mathbf{s}^{(\alpha)*}) \tag{5}$$

The lattice Jaumann rate of Kirchhoff stress and the elastic rate of stretching are related by

$$\hat{\boldsymbol{\tau}}^* = \mathbf{L} : \mathbf{D}^* \tag{6}$$

where

$$\hat{\boldsymbol{\tau}}^* = \dot{\boldsymbol{\tau}} + \boldsymbol{\tau} \cdot \boldsymbol{\Omega}^* - \boldsymbol{\Omega}^* \cdot \boldsymbol{\tau} \tag{7}$$

Elasticity is assumed to be unaffected by slip and isotropic, and the elastic strains remain small so that,

$$\mathbf{L} = 2G\left[\mathbf{I}' + \frac{\nu}{1-2\nu}\mathbf{II}\right] \tag{8}$$

Here, $G = E/(2(1+\nu))$ is the shear modulus, with E being Young's modulus and ν Poisson's ratio, $\mathbf{I}$ is the second order identity tensor and $\mathbf{I}'$ is the fourth order identity tensor.

The resulting expression for the material convected rate of Kirchhoff stress, $\dot{\boldsymbol{\tau}}^c$, is

$$\dot{\boldsymbol{\tau}}^c = \mathbf{L} : \mathbf{D} - \sum_\alpha \dot{\gamma}^{(\alpha)}\mathbf{R}^{(\alpha)} - \mathbf{D} \cdot \boldsymbol{\tau} - \boldsymbol{\tau} \cdot \mathbf{D} \tag{9}$$

with

$$\mathbf{R}^{(\alpha)} = \mathbf{L} : \mathbf{P}^{(\alpha)} + \mathbf{W}^{(\alpha)} \cdot \boldsymbol{\tau} - \boldsymbol{\tau} \cdot \mathbf{W}^{(\alpha)} \tag{10}$$

Strain rate hardening is described by the power law relation,

$$\dot{\gamma}^{(\alpha)} = \dot{a}\left[\frac{\tau^{(\alpha)}}{g^{(\alpha)}}\right]\left[\left|\frac{\tau^{(\alpha)}}{g^{(\alpha)}}\right|\right]^{(1/m)-1} \tag{11}$$

Here, $\dot{a}$ is a reference strain rate, m is the strain rate hardening exponent, $g^{(\alpha)}$ is the slip system hardness, and the slip system resolved shear stress, $\tau^{(\alpha)}$, is given by

$$\tau^{(\alpha)} = \mathbf{s}^{(\alpha)*} \cdot \boldsymbol{\tau} \cdot \mathbf{m}^{(\alpha)*} \tag{12}$$

The hardness evolves according to

$$\dot{g}^{(\alpha)} = \sum_\beta h_{\alpha\beta}\dot{\gamma}^{(\beta)} \tag{13}$$

where $g^{(\alpha)}(0) = \tau_0$ and $h_{\alpha\beta}$ is the slip-system hardening matrix.

In the local theory, the hardening matrix can be taken to be some function of the accumulated slips $\gamma^{(\beta)}$. The nonlocal theory of Acharya and Bassani [1] is particularly simple in that the only difference from the local theory is that the $h_{\alpha\beta}$

are functions of a measure of the lattice incompatibility. Various measures have been proposed for measuring lattice incompatibility at finite strain as discussed by Cermelli and Gurtin [6]. For volume preserving plastic flow, one such measure is

$$\boldsymbol{\alpha} = \mathbf{F}^p \cdot \nabla \times \mathbf{F}^p \tag{14}$$

where $\mathbf{F}^p$ can be calculated from the evolution equation

$$\dot{\mathbf{F}}^p \cdot \mathbf{F}^{p-1} = \sum_\alpha \dot{\gamma}^{(\alpha)} \mathbf{s}^{(\alpha)} \mathbf{m}^{(\alpha)} \tag{15}$$

A bar subject to plane strain tension is analyzed, with the plane of deformation being the x_1-x_2-plane and the tensile axis is aligned with the x_2-axis. Symmetry about both the x_1 and x_2-axes is assumed. For a tensile specimen having initial length $2L_0$ and width $2w_0$, the boundary conditions for the rate problem on the quadrant analyzed numerically are

$$\dot{u}_2 = \dot{\epsilon}_0 (U + L_0) \quad \dot{T}_1 = 0 \quad \text{on } x_2 = L_0 \tag{16}$$

$$\dot{u}_2 = 0 \quad \dot{T}_1 = 0 \quad \text{on } x_2 = 0 \tag{17}$$

$$\dot{T}_1 = 0 \quad \dot{T}_2 = 0 \quad \text{on } x_1 = w_0 \tag{18}$$

$$\dot{u}_1 = 0 \quad \dot{T}_2 = 0 \quad \text{on } x_1 = 0 \tag{19}$$

where $U = u_2(x_1, L_0)$.

The bars analyzed have an initial geometric imperfection of the form

$$w_0 = \bar{w}_0 \left[1 - \xi \cos\left(\frac{\pi x_2}{L_0} \right) \right] \tag{20}$$

As in Peirce et al. [11], the crystal geometry considered is the double slip planar model crystal of Asaro [3] in which the two slip systems are symmetrically oriented about the tensile axis. For this model double slipping crystal, $h_{\alpha\beta}$ in (13) is a 2×2 matrix. Here, attention is confined to a hardening matrix for which $h_{11} = h_{22} = h_{12} = h_{21} = h$ with h depending on the slips and on the plastic incompatibility.

Acharya and Bassani [1] do not propose a form for the dependence of the hardening on $\boldsymbol{\alpha}$ in cases of multiple slip. To specify such a dependence for use in the numerical calculations, the tensor $\boldsymbol{\alpha}$ in (14) is written in Cartesian tensor notation as

$$\alpha_{ij} = F^p_{im} e_{mrs} F^p_{js,r} \tag{21}$$

where e_{mrs} is the alternating symbol and $(\ )_{,i}$ denotes differentiation with respect to x_i. For the planar double slip crystal model, all field quantities are independent of x_3, $F^p_{13} = F^p_{23} = 0$ and $F^p_{33} = 1$. Hence, (21) implies that the only non-vanishing components of α_{ij} are

$$\alpha_{3j} = \left(F^p_{j2,1} - F^p_{j1,2}\right) \quad j = 1,2 \tag{22}$$

We explore one simple form for incorporating a measure of plastic incompatibility into the hardening description. In the local theory, the components of $h_{\alpha\beta}$ are taken to depend on the slip through the sum

$$\gamma_e = |\gamma^{(1)}| + |\gamma^{(2)}| \tag{23}$$

In the nonlocal theory, (23) is used together with a similar dependence on the non-vanishing components of α_{ij}, i.e.

$$\alpha_e = |\alpha_{31}| + |\alpha_{32}| \tag{24}$$

Based on the expression in Bassani et al. [4], the hardening function h is written as

$$h(\gamma_e, \alpha_e) = h_0 \left(\frac{\gamma_e}{\gamma_0} - 1\right)^{N-1} \left[1 + \ell^2 \left(\frac{\alpha_e}{\gamma_0}\right)^2\right]^p \tag{25}$$

where ℓ is the material length scale and p is a parameter.

For a perfect bar, $\xi = 0$, an equilibrium state of homogeneous plane strain is one equilibrium solution available for all values of applied displacement U. For rate independent material behavior bifurcation into both localized and diffuse modes are possible, see Peirce et al. [10], Hill and Hutchinson [9]. Bifurcation first becomes possible after the maximum load into the diffuse necking mode with the longest wavelength consistent with the specimen geometry. Subsequently, shorter wavelength modes become available, approaching a limiting bifurcation point for arbitrarily short wavelength surface modes. At a larger value of U, bifurcation into a localized shear band mode becomes possible. Unless the specimen is particularly stubby, bifurcation into the diffuse necking mode depends only on the plane strain tangent stiffness and the specimen geometry. On the other hand, bifurcations into the short wavelength surface mode and into the localized shearing mode are sensitive to the overall constitutive behavior, particularly to the response to a change in loading path. Since the prebifurcation state is one of homogeneous plane strain tension, the bifurcation predictions of the nonlocal plasticity theory of Acharya and Bassani [1] are the same as for the local theory. For the slightly rate dependent material behavior considered here, these bifurcations do not occur but the rate independent bifurcation behavior gives a good indication of the available instability modes and the order in which they are activated.

The solution of the boundary value problem for an imperfect bar will lead to more complex behavior. Subsequent to the maximum load point, the deformation pattern reflects the diffuse necking mode. Then, the deformation in the specimen is non-homogeneous. Therefore, the development of the short wavelength surface mode and the localized shearing mode will be affected by the nonlocal material behavior.

3. Numerical implementation and results

Since the theory of Acharya and Bassani [1] preserves the classical boundary value problem structure, the virtual work statement of equilibrium is

$$\int_V \mathbf{s} : \delta\mathbf{F} dV = \int_S \mathbf{T} \cdot \delta\mathbf{u} dS \tag{26}$$

where $\mathbf{s}$ is the non-symmetric nominal stress tensor, V and S are the volume and external surface, respectively, of the body in the reference configuration, and $\mathbf{T}$ is the nominal traction vector. Standard kinematic relations are used to relate $\dot{\tau}^c$ in (9) to $\dot{\mathbf{s}}$ and (26) serves as the basis for the finite element discretization.

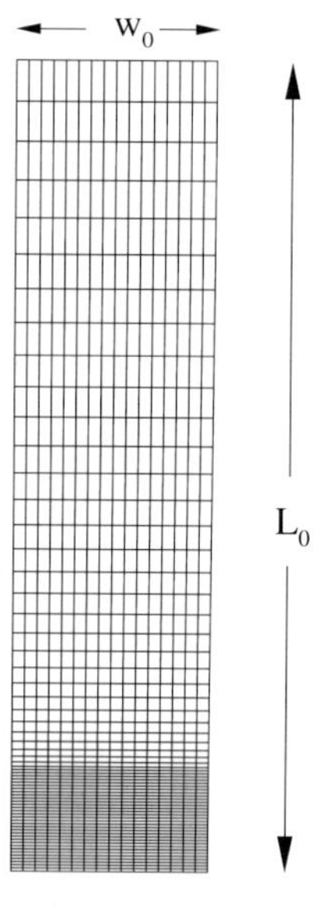

Figure 1. Finite element mesh.

The calculations are carried out for a bar having $\bar{w}_0/L_0 = 0.25$ and with an initial geometric imperfection $\xi = 0.01$ in (20). The deformation history is calculated in a linear incremental manner and in order to increase the stable time step, a rate tangent modulus time stepping procedure is used. The finite

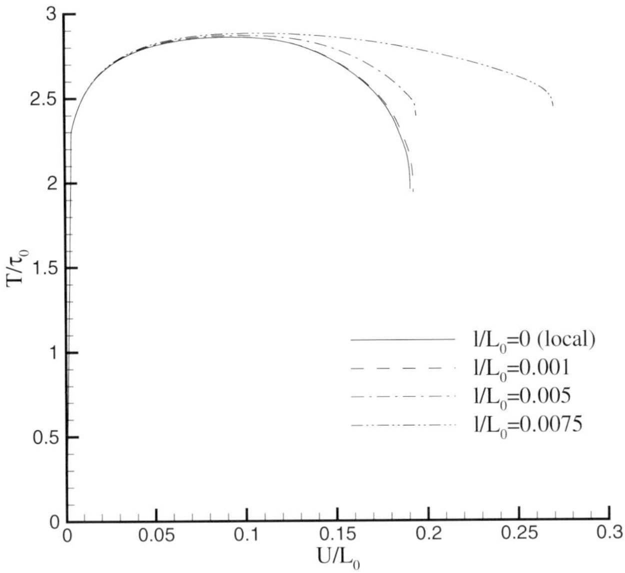

Figure 2. Load versus displacement curves.

element discretization is based on linear displacement triangles arranged in "crossed triangle" rectangles. The finite element mesh used in the computations consists of 16×64 rectangles, graduated in the x_2−direction as shown in Fig. 1. The gradients of $\mathbf{F}^p$ in (22) are calculated as described in Bassani et al. [4]: (i) the components of $\mathbf{F}^p$, which are stored at the centroids of each triangular element, are extrapolated to the nodal points of each rectangular element using bilinear shape functions; (ii) the nodal values are taken to be the average value of those of each rectangle connected to a node; and (iii) the derivatives are defined as differences across the rectangle so that these derivatives have the same value for each triangle making up a rectangle.

The two slip systems are oriented at $\pm 30°$ to the tensile axis and the material properties are given by $\tau_0/G = 0.005$, $\nu = 1/3$, $h_0/\tau_0 = 10$, $\gamma_0 = 0.01$, $N = 0.1$, $m = 0.005$ and $p = 1/2$. Various values of the material length ℓ are used.

Figure 2 shows traction displacement curves for four calculations: $\ell/L_0 = 0$, $\ell/L_0 = 0.001$, $\ell/L_0 = 0.005$ and $\ell/L_0 = 0.0075$. Here, the traction T is the force per unit original area. Increasing the ratio ℓ/L_0 can be interpreted as increasing the material length scale for a fixed specimen size or as decreasing the specimen size with a fixed material length scale. For the purposes of discussion, the interpretation of a fixed material length and decreasing specimen size will be used. There is very little difference in the four curves of load versus

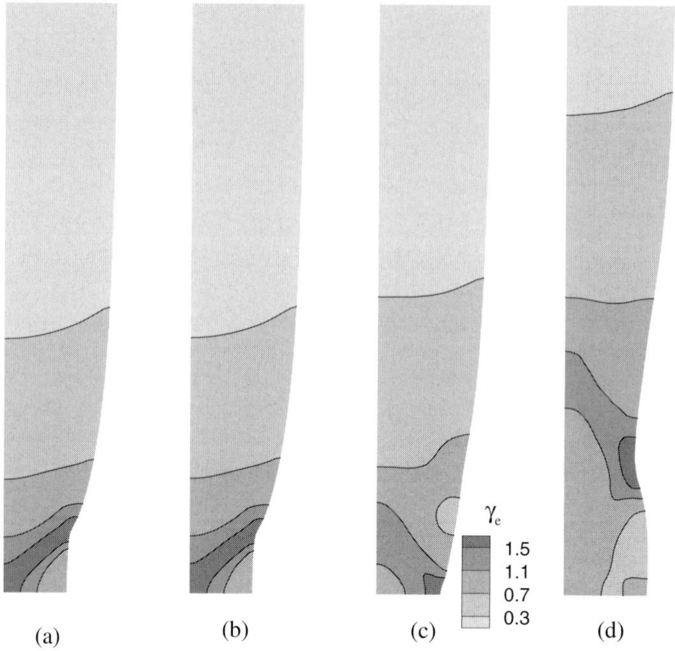

Figure 3. Contours of accumulated slip γ_e. (a) $\ell/L_0 = 0$ (local). (b) $\ell/L_0 = 0.001$. (c) $\ell/L_0 = 0.005$. (d) $\ell/L_0 = 0.0075$. In each case, the stage of deformation shown corresponds to the last point on the load-displacement curves in Fig. 2.

displacement prior to the load maximum and the predictions of the local theory and the nonlocal theory with $\ell/L_0 = 0.001$ are close over the full range calculated. A significant post maximum load stiffening is seen for $\ell/L_0 = 0.005$ that increases further for $\ell/L_0 = 0.0075$. For these two larger values of ℓ/L_0, the main effects are the more gradual load shedding beyond the load maximum and the delay in the sharp load drop.

Contours of γ_e are shown in Fig. 3 for the four values of ℓ/L_0. For $\ell/L_0 = 0$ and $\ell/L_0 = 0.001$, the deformation patterns are quite similar. However, when $\ell/L_0 = 0.005$ the shear band that develops is qualitatively different than for the cases with the two smaller values of ℓ/L_0; the shear band is broader, the slip in it is smaller (even though the overall strain of the specimen is larger), and the shear band now intersects the $x_2 = 0$ plane at the free surface, rather than in the center of the specimen. With $\ell/L_0 = 0.0075$, the diffuse neck is centered away from the $x_2 = 0$ plane (so that for the full specimen there are two necks) and a rather broad shear band forms between the free surface at the diffuse neck and the mid-plane. With $\ell/L_0 = 0.01$ (not shown), the diffuse neck develops at $x_2 = L_0$.

At least for the hardening law used, the nonlocal plasticity theory of Acharya and Bassani [1] predicts that with decreasing specimen size both diffuse necking and localization into a shear band are delayed. However, the net effect is that, with decreasing specimen size, the diffuse necking mode of deformation becomes more pronounced.

Presuming that a representative value of the material characteristic length, ℓ, is of the order of 1 μm, $\ell/L_0 = 0.005$ corresponds to $L_0 = 0.2$ mm. This gives an indication of the order of magnitude of the size scale at which nonlocal effects become significant in the calculations here. These calculations were carried out using an arbitrarily chosen measure, α_e, of incompatibility in the hardening law. The extent to which the qualitative features of the response depend on this choice, or on the particular parameter values used in the calculations, remains to be determined. There is also the larger issue of whether or not the simple theory of Acharya and Bassani [1] appropriately models the nonlocal effects in ductile single crystals arising from lattice incompatibility.

Shear band localization is a common occurrence in ductile metal crystals, in single crystal tensile specimens and in grains of a polycrystal. Determining the implications of various nonlocal plasticity theories for such localization phenomena offers one possibility for finding experimentally observable qualitative differences in predicted behavior between the various nonlocal crystal plasticity theories, e.g. Acharya and Bassani [1], Gurtin [8], a finite deformation version of the theory in Shu and Fleck [12]. Since such shear localization occurs at relatively large strains and lattice rotation is an integral aspect of the phenomenon, a finite strain nonlocal framework is required. Also, regardless of the nonlocal framework, apportioning the incompatibility measure among multiple slip systems remains an open issue.

Acknowledgments

Support from the Materials Research Science and Engineering Center on *On Micro-and-Nano-Mechanics of Electronic and Structural Materials* at Brown University (NSF Grant DMR-0079964) is gratefully acknowledged.

References

[1] Acharya, A. and Bassani, J.L. (2000). Incompatibility and crystal plasticity. *J.Mech. Phys. Solids*, **48**, 1565–1595.

[2] Aifantis, E.C. (1984). On the microstructural origin of certain inelastic models. *J. Eng. Mat. Tech.*, **106**, 326–330.

[3] Asaro, R.J. (1979). Geometrical effects in the inhomogeneous deformation of ductile single crystals. *Acta. Metall.*, **27**, 445–453.

[4] Bassani, J.L., Needleman, A. and Van der Giessen, E. (2001). Plastic flow in a composite: a comparison of nonlocal continuum and discrete dislocation predictions. *Int. J. Solids Structs.*, **38**, 833–853.

[5] Begley, M.R. and Hutchinson, J.W. (1998). The mechanics of size-dependent indentation. *J. Mech. Phys. Solids*, **46**, 2049–2068.

[6] Cermelli, P. and Gurtin, M.E. (2001). On the characterization of geometrically necessary dislocations in finite plasticity. *J. Mech. Phys. Solids*, **49**, 1539–1568.

[7] Fleck, N.A. and Hutchinson, J.W. (1997). Strain gradient plasticity. *Adv. Appl. Mech.*, **33**, 295–361.

[8] Gurtin, M.E. (2000). On the plasticity of single crystals: free energy, microforces plastic-strain gradients. *J. Mech. Phys. Solids*, **48**, 989–1036.

[9] Hill, R. and Hutchinson, J.W. (1975). Bifurcation phenomena in the plane tensile test. *J. Mech. Phys. Solids*, **23**, 239–264.

[10] Peirce, D., Asaro, R.J. and Needleman, A. (1982). An analysis of nonuniform and localized deformation in ductile single crystals. *Acta Metall.*, **30**, 1087–1119.

[11] Peirce, D., Asaro, R.J. and Needleman, A. (1983). Material rate dependence and localized deformation in crystalline solids. *Acta Metall.*, **31**, 1951–1976.

[12] Shu, J.Y. and Fleck, N.A. (1999). Strain gradient crystal plasticity: size-dependent deformation of bicrystals. *J. Mech. Phys. Solids*, **47**, 297–324.

Continuum Thermodynamic Modeling and Simulation of Additional Hardening due to Deformation Incompatibility

B. Svendsen
Department of Mechanical Engineering, University of Dortmund
Leonhard-Euler-Straße 5, 44221 Dortmund, Germany
bob.svendsen@uni-dortmund.de

S. Reese
Department of Civil Engineering, Ruhr University of Bochum
IA 01/128, 44780 Bochum, Germany
reese@nm.ruhr-uni-bochum.de

Abstract: The purpose of this work is the formulation and investigation of a phenomenological constitutive model for the inelastic material behavior of single crystals and polycrystals including the possible effects of incompatibility in the local inelastic deformation on this behavior. In particular, attention is focused here on additional hardening caused by the production of additional dislocations (i.e., geometrically-necessary dislocations) accommodating such incompatibility. In the context of crystal plasticity, such local inelastic deformation is represented by the glide-system (scalar) slips. As it turns out, this quantity can be used as a basis on which to formulate a non-local continuum thermodynamic generalization of standard crystal plasticity in order to account for the effects of such incompatibility on the material behavior. The corresponding continuum thermodynamic formulation yields a generalized Ginzburg-Landau or Cahn-Allen-type evolution/field relation for the scalar glide-system slip whose form is determined by that of the free energy, a dissipation potential, and that of the evolution relations for the inelastic state variables. In the last part of the work, the behavior of the model is elucidated in the context of the constrained simple shear of a crystalline strip containing one or two glide systems.

Keywords: Thermodynamics, crystal plasticity, dislocations, hardening, incompatibility.

1. Introduction

Standard micromechanical modeling of the inelastic material behavior of metallic single crystals and polycrystals (see Hill and Rice [10], Asaro [2] and Cuitiño and Ortiz [5]) is commonly based on the premise that resistance to glide is due mainly to the random trapping of mobile dislocations during locally homogeneous deformation. Such trapped dislocations are commonly referred to as statistically-stored dislocations (SSDs), and act as obstacles to further dislocation motion, resulting in hardening. As anticipated in the work of Nye [16] and Kröner [11], and discussed by Ashby [3], an additional contribution to

C. Miehe (ed.), IUTAM Symposium on Computational Mechanics of Solid Materials at Large Strains, 141-150.

the density of immobile dislocations and so to hardening can arise when the continuum lengthscale (e.g. grain size) approaches that of the dominant microstructural features (e.g. mean spacing between precipitates relative to the precipitate size, or mean spacing between glide planes). Indeed, in this case, the resulting deformation incompatibility between, e.g. "hard" inclusions and a "soft" matrix, is accomodated by the development of so-called geometrically-necessary dislocations (GNDs). Experimentally-observed effects in a large class of materials such as increasing material hardening with decreasing (grain) size (i.e., the Hall-Petch effect) are commonly associated with the development of such GNDs.

Such considerations and results have motivated a number of workers over the last few years to formulate various extensions of phenomenological and crystal plasticity (e.g., Fleck and Hutchinson [7], Naghdi and Srinivasa [15], Le and Stumpf [12], Steinmann [21], Dai and Parks [6], Forest et al. [8], Shizawa and Zbib [18], Shu and Fleck [19], Menzel and Steinmann [14], Acharya and Bassani [1], Gurtin [9], Ortiz et al. [17], Svendsen [23]) in order to account in various ways for the dependence of the material behavior on deformation inhomogeneity, or in the crystal plasticity context, deformation incompatibility. It is not the purpose of the current work to compare and contrast all of these approaches; rather, the focus here is on the application of one approach from Svendsen [23] to the case of single and double slip in a crystalline strip (e.g., Shu et al. [20]). In particular, attention is focused in this regard on additional hardening due to local deformation incompatibility.

2. Basic constitutive and field relations

In the case of phenomenological crystal plasticity, any material point is endowed with a set of n glide systems. The geometry and orientation of each such glide system is described as usual by an orthonormal basis$(\boldsymbol{s}_\mathfrak{a}, \boldsymbol{n}_\mathfrak{a}, \boldsymbol{t}_\mathfrak{a})$ $(\mathfrak{a} = 1, \dots, n)$. Here, $\boldsymbol{s}_\mathfrak{a}$ represents the direction of glide in the plane,$\boldsymbol{n}_\mathfrak{a}$ the glide-plane normal, and $\boldsymbol{t}_\mathfrak{a} := \boldsymbol{s}_\mathfrak{a} \times \boldsymbol{n}_\mathfrak{a}$ the direction transverse to $\boldsymbol{s}_\mathfrak{a}$ in the glide plane. Since we neglect in this work the effects of any processes involving a change in or evolution of either the glide direction $\boldsymbol{s}_\mathfrak{a}$ or the glide-system orientation $\boldsymbol{n}_\mathfrak{a}$ (e.g. texture development), these referential unit vectors, and so$\boldsymbol{t}_\mathfrak{a}$ as well, are assumed constant with respect to the reference placement. Assuming then that inelastic slip in the glide planes is the only source of local inelastic deformation $\boldsymbol{F}_\mathrm{P}$ in the material, the usual form

$$\dot{\boldsymbol{F}}_\mathrm{P} = \sum\nolimits_{\mathfrak{a}=1}^{m} (\boldsymbol{s}_\mathfrak{a} \otimes \boldsymbol{n}_\mathfrak{a}) \boldsymbol{F}_\mathrm{P}\, \dot{\gamma}_\mathfrak{a} \tag{1}$$

for the evolution of the local inelastic deformation $\boldsymbol{F}_\mathrm{P}$ in the material holds. Here, $\gamma_\mathfrak{a}$ represents the (scalar inelastic) slip deformation in glide plane$\mathfrak{a}$, and

$m \leq n$ the set of active glide systems. Beyond $\boldsymbol{F}_{\mathrm{P}}$, we require a measure of (local) deformation incompatibility relevant to the case of crystal plasticity. As recognized in essence by Nye [16] and many others, the simplest such possibility is represented by the incompatibilitycurl $\boldsymbol{F}_{\mathrm{P}}$ of $\boldsymbol{F}_{\mathrm{P}}$ (other possibilities are discussed in Svendsen [23]). From (1), we have the evolution relation

$$\overline{\operatorname{curl} \boldsymbol{F}_{\mathrm{P}}}^{\cdot} = \sum\nolimits_{\mathfrak{a}=1}^{m} (\boldsymbol{s}_{\mathfrak{a}} \otimes \boldsymbol{n}_{\mathfrak{a}})(\operatorname{curl} \boldsymbol{F}_{\mathrm{P}})\,\dot{\gamma}_{\mathfrak{a}} + \boldsymbol{s}_{\mathfrak{a}} \otimes (\nabla\dot{\gamma}_{\mathfrak{a}} \times \boldsymbol{F}_{\mathrm{P}}^{\mathrm{T}} \boldsymbol{n}_{\mathfrak{a}}) \tag{2}$$

for curl $\boldsymbol{F}_{\mathrm{P}}$ via the fact that $\boldsymbol{s}_{\mathfrak{a}}$, $\boldsymbol{n}_{\mathfrak{a}}$ and $\boldsymbol{t}_{\mathfrak{a}}$ are constant. Lastly, beyond $\boldsymbol{F}_{\mathrm{P}}$ and curl $\boldsymbol{F}_{\mathrm{P}}$, we assume in this work that the inelastic state of any material point, and in particular its hardening behavior, is characterised in the current crystal plasticity context by the set $\boldsymbol{\nu} := (\nu_1, \ldots, \nu_n)$ of glide-system accumulated slip deformations, with

$$\dot{\nu}_{\mathfrak{a}} := |\dot{\gamma}_{\mathfrak{a}}| \quad . \tag{3}$$

Note that the evolution relation (1) for $\boldsymbol{F}_{\mathrm{P}}$ depends (linearly) on $\boldsymbol{F}_{\mathrm{P}}$ and the set $\dot{\boldsymbol{\gamma}} := (\dot{\gamma}_1, \ldots, \dot{\gamma}_m)$. Similarly, that (2) for curl $\boldsymbol{F}_{\mathrm{P}}$ depends (linearly) on curl $\boldsymbol{F}_{\mathrm{P}}$, $\boldsymbol{F}_{\mathrm{P}}$, $\dot{\boldsymbol{\gamma}}$, and the set $\nabla\dot{\boldsymbol{\gamma}} := (\nabla\dot{\gamma}_1, \ldots, \nabla\dot{\gamma}_m)$. Lastly, that (3) for $\nu_{\mathfrak{a}}$ depends non-linearly on $\dot{\gamma}_{\mathfrak{a}}$. In the current thermomechanical context, then, (1)–(3) imply the general material frame-indifferent constitutive form

$$\mathcal{C} = \mathcal{C}(\theta, \boldsymbol{C}, \nabla\theta, \boldsymbol{\nu}, \boldsymbol{F}_{\mathrm{P}}, \operatorname{curl} \boldsymbol{F}_{\mathrm{P}}, \dot{\boldsymbol{\gamma}}, \nabla\dot{\boldsymbol{\gamma}}) \tag{4}$$

for all dependent constitutive quantities. Here, θ represents the temperature, $\boldsymbol{C} = \boldsymbol{F}^{\mathrm{T}}\boldsymbol{F}$ the right Cauchy-Green deformation, $\boldsymbol{F} := \nabla\boldsymbol{\xi}$ the deformation gradient, and $\boldsymbol{\xi}$ the motion of the material with respect to the reference configuration under consideration.

Applying methods developed in previous work (Svendsen [22, 24]), Svendsen [23] showed that the set $\boldsymbol{\gamma}$ of glide-system slips can be interpreted from a phenomenological, continuum thermodynamic point of view as either (i), generalized internal variables (GIVs), or as (ii), internal degrees of freedom (DOFs). In particular, when modeled as internal variables, these are refered to as "generalized" since, in contrast to classical internal variables, the material behavior depends on the gradients of these as well. In the context of crystal plasticity, one could also refer to these as activation "order" parameters or phase fields in the sense that their evolution reflects a non-local activation process, i.e., dislocation motion and interaction in and between glide systems. On the other hand, the modeling of the $\boldsymbol{\gamma}$ as degrees of freedom is meant in the sense of "generalized" Lagrangian degrees of freedom, e.g. of a multicomponent system, the $\boldsymbol{\gamma}$ being in addition "internal" to the system in the sense of internal variables. For simplicity and brevity, the GIV approach and results are reviewed briefly in what follows; these results will then be used for the modeling and simulation of additional hardening in a crystalline strip in the next section. For more details, the reader is referred to Svendsen [23].

In the GIV approach, the invariance of the total energy balance yields in particular the standard referential forms

$$
\begin{aligned}
\dot{\varrho} &= 0\,, \\
\varrho\ddot{\boldsymbol{\xi}} &= \operatorname{div}(\boldsymbol{F}\boldsymbol{S}) + \boldsymbol{f}\,, \\
\boldsymbol{0} &= \boldsymbol{S} - \boldsymbol{S}^{\mathrm{T}}\,,
\end{aligned}
\tag{5}
$$

for (local) mass, continuum momentum, and moment of momentum, balance, respectively. Here, ϱ represents the referential mass density, $\boldsymbol{S}$ the second Piola-Kirchhoff stress, and $\boldsymbol{f}$ the momentum supply rate density. As in the standard continuum case, $\boldsymbol{S}$ represents a dependent constitutive quantity of the form (4). In this framework, the thermodynamic analysis in the GIV approach is based on the entropy balance and Clausius-Duhem form for the entropy supply rate, resulting in the expression

$$
\delta = \tfrac{1}{2}\,\boldsymbol{S}\cdot\dot{\boldsymbol{C}} - \dot{\psi} - \eta\,\dot{\theta} + \operatorname{div}\left(\theta\boldsymbol{\phi} - \boldsymbol{q}\right) - \boldsymbol{\phi}\cdot\nabla\theta
\tag{6}
$$

(e.g., Maugin [13]) for the dissipation rate density δ in terms of the entropy flux $\boldsymbol{\phi}$, heat flux $\boldsymbol{q}$ and referential free energy density ψ, all dependent constitutive quantities of the form (4). Assuming then the existence of a dissipation potential d_{V} of the form (4) for the current constitutive class, exploitation of the corresponding Coleman-Noll dissipation principle results in particular in the evolution/field relation

$$
d_{\mathrm{V},\dot{\gamma}_{\mathfrak{a}}} = \varpi_{\mathrm{V}\mathfrak{a}} + \operatorname{div}\left(\boldsymbol{\varphi}_{\mathrm{V}\mathfrak{a}} + d_{\mathrm{V},\nabla\dot{\gamma}_{\mathfrak{a}}}\right)
\tag{7}
$$

for $\gamma_{\mathfrak{a}}$ as based in part on the reduced form $\psi = \psi(\theta, \boldsymbol{C}, \boldsymbol{F}_{\mathrm{P}}, \boldsymbol{\nu}, \operatorname{curl}\boldsymbol{F}_{\mathrm{P}})$ for ψ. Here, $\varpi_{\mathrm{V}\mathfrak{a}} := \tau_{\mathfrak{a}} - x_{\mathfrak{a}} - \operatorname{sgn}(\dot{\gamma}_{\mathfrak{a}})\, r_{\mathfrak{a}}$ and $\boldsymbol{\varphi}_{\mathrm{V}\mathfrak{a}} := \boldsymbol{F}_{\mathrm{P}}^{\mathrm{T}}\boldsymbol{n}_{\mathfrak{a}} \times (\psi_{,\operatorname{curl}\boldsymbol{F}_{\mathrm{P}}})^{\mathrm{T}}\boldsymbol{s}_{\mathfrak{a}}$ represent an effective production-rate density and thermodynamic flux, respectively, for local inelastic deformation/dislocation motion, with $\tau_{\mathfrak{a}} := -\boldsymbol{s}_{\mathfrak{a}} \otimes \boldsymbol{n}_{\mathfrak{a}} \cdot \psi_{,\boldsymbol{F}_{\mathrm{P}}}\boldsymbol{F}_{\mathrm{P}}^{\mathrm{T}}$ and $x_{\mathfrak{a}} := \boldsymbol{s}_{\mathfrak{a}} \otimes \boldsymbol{n}_{\mathfrak{a}} \cdot \psi_{,\operatorname{curl}\boldsymbol{F}_{\mathrm{P}}}(\operatorname{curl}\boldsymbol{F}_{\mathrm{P}})^{\mathrm{T}}$ and $r_{\mathfrak{a}} := \psi_{,\nu_{\mathfrak{a}}}$ the glide-system Schmid stress $\tau_{\mathfrak{a}}$, a contribution $x_{\mathfrak{a}}$ to back stress, and yield stress $r_{\mathfrak{a}}$, respectively. In addition, the thermodynamic analysis yields the usual "thermoelastic" constitutive forms $\boldsymbol{S} = 2\,\psi_{,\boldsymbol{C}}$ and $\eta = -\psi_{,\theta}$ for the second Piola-Kirchhoff stress and the entropy density, respectively. In addition, an evolution/field relation for θ is obtained from the entropy balance. In summary, then, one obtains the evolution/field relations $(5)_2$ and (7) for the basic unknowns $\boldsymbol{\xi}$ and $\gamma_{\mathfrak{a}}$ ($\mathfrak{a} = 1, \ldots, m$), respectively, in the GIV approach. In particular, as discussed in more detail elsewhere (Svendsen [23]), the second of these represents a generalized Ginzburg-Landau or Cahn-Allen-type (phase-field-type) relation for $\gamma_{\mathfrak{a}}$ (e.g., Cahn and Allen [4].

3. Simple shear of a crytalline strip

In order to investigate the consequences of the approach and formulation discussed in the previous section further, consider its application to the case of the (isothermal) simple shear of a crystalline strip containing one or two glide planes. This example has been used in the recent work of Shu et al. [20] in order to compare the predictions of discrete dislocation computer simulation of Van der Giessen and Needleman [25] and non-local strain-gradient approach of Fleck and Hutchinson [7] and applied to crystal plasticity (e.g., Shu and Fleck [19]). As dicusssed by them, it represents a model problem for the type of plastic constraint found at grain boundaries of a polycrystal, or the surface of a thin film, or at interfaces in a composite. To be comparable with the formulation of Shu et al. [20], we simplify the results of the previous section to the special case of (i), isothermal and quasi-static conditions, (ii), small deformation, (iii), elastic isotropy, and (iv), linear self-hardening alone. In particular, the momentum balance $(5)_2$ then reduces to

$$\mathbf{0} = \operatorname{div} \boldsymbol{T} \tag{8}$$

(i.e., neglecting body forces) in terms of the (small-strain) stress tensor $\boldsymbol{T} = \psi_{,\boldsymbol{E}}$ and small-strain measure $\boldsymbol{E} := \operatorname{sym}(\nabla \boldsymbol{u})$, $\boldsymbol{u}$ being the displacement field as usual. In addition, these assumptions result in the form

$$\begin{aligned} \psi = {} & \tfrac{1}{2}\,\lambda\,(\boldsymbol{I}\cdot\boldsymbol{E}_{\mathrm{E}})^2 + \mu\,\boldsymbol{E}_{\mathrm{E}}\cdot\boldsymbol{E}_{\mathrm{E}} \\ & + \sum\nolimits_{\mathfrak{a}=1}^{n} r_0\,\nu_{\mathfrak{a}} + \tfrac{1}{2}\,h_0\,\nu^2 + g^{-1}\mu\,\ell^g\,|\operatorname{curl}\boldsymbol{H}_{\mathrm{P}}|^g \end{aligned} \tag{9}$$

for the free energy density ψ, with $\nu := \sqrt{\sum_{\mathfrak{a}=1}^{n} \nu_{\mathfrak{a}}^2}$. Here, $\boldsymbol{E}_{\mathrm{E}} := \boldsymbol{E} - \boldsymbol{E}_{\mathrm{P}}$ represents the (small) elastic strain, and $\boldsymbol{E}_{\mathrm{P}} := \operatorname{sym}(\boldsymbol{H}_{\mathrm{P}})$ the (small) inelastic strain, with $\boldsymbol{H}_{\mathrm{P}} = \sum_{\mathfrak{a}=1}^{n}(\boldsymbol{s}_{\mathfrak{a}}\otimes\boldsymbol{n}_{\mathfrak{a}})\,\gamma_{\mathfrak{a}}$. In addition, λ and μ represent the Lamé constants, r_0 a characteristic initial glide-system yield stress, and h_0 a characteristic hardening modulus for linear self-hardening. Further, ℓ and g represent a characteristic length and exponent, respectively, for energy stored in the material due to local deformation incompatibility. Lastly, we work with the simple power-law form

$$d_{\mathrm{V}} = \frac{n}{n+1}\,\varsigma\,\dot{\nu}_0\,\sum\nolimits_{\mathfrak{a}=1}^{m}\left(\frac{|\dot{\gamma}_{\mathfrak{a}}|}{\dot{\nu}_0}\right)^{(n+1)/n} \tag{10}$$

for the dissipation potential d_{V}. Here, ς represents a characteristic energy scale for activation of dislocation glide motion with units of J m^{-3} or Pa, and $\dot{\nu}_0$ a characteristic value of $|\dot{\gamma}_{\mathfrak{a}}|$. In particular, (9) and (10) imply the evolution/field relation

$$\varsigma\,\dot{\nu}_0^{-1}\dot{\gamma}_{\mathfrak{a}} = \mu\,\ell^2 \sum\nolimits_{\mathfrak{b}=1}^{m} \boldsymbol{A}_{\mathfrak{ab}} \cdot \nabla(\nabla\gamma_{\mathfrak{b}}) \\ +\; 2\mu\,\boldsymbol{s}_{\mathfrak{a}} \cdot \boldsymbol{E}_{\mathrm{E}}\boldsymbol{n}_{\mathfrak{a}} - \mathrm{sgn}(\dot{\gamma}_{\mathfrak{a}})\,(r_0 + h_0\,\nu_{\mathfrak{a}}) \tag{11}$$

for the glide system slip $\gamma_{\mathfrak{a}}$ via (the small-deformation forms of) (7) on the basis of the choices $g = 2$ and $n = 2$ in (9), with $\boldsymbol{A}_{\mathfrak{ab}} := (\boldsymbol{s}_{\mathfrak{a}}\cdot\boldsymbol{s}_{\mathfrak{b}})\,[(\boldsymbol{n}_{\mathfrak{a}}\cdot\boldsymbol{n}_{\mathfrak{b}})\,\boldsymbol{I}-\boldsymbol{n}_{\mathfrak{a}}\otimes\boldsymbol{n}_{\mathfrak{b}}]$. In particular, note that $\boldsymbol{A}_{\mathfrak{aa}}\cdot\nabla(\nabla\gamma_{\mathfrak{a}}) = [\boldsymbol{I}-\boldsymbol{n}_{\mathfrak{a}}\otimes\boldsymbol{n}_{\mathfrak{a}}]\cdot\nabla(\nabla\gamma_{\mathfrak{a}}) = [\boldsymbol{s}_{\mathfrak{a}}\otimes\boldsymbol{s}_{\mathfrak{a}}+\boldsymbol{t}_{\mathfrak{a}}\otimes\boldsymbol{t}_{\mathfrak{a}}]\cdot\nabla(\nabla\gamma_{\mathfrak{a}})$ represents the divergence of $\nabla\gamma_{\mathfrak{a}}$ projected onto the $\mathfrak{a}^{\mathrm{th}}$ glide plane. It is worth emphasizing that the form of this projection results from the dependence of ψ on curl $\boldsymbol{H}_{\mathrm{P}}$. For comparison, note that $\boldsymbol{A}_{\mathfrak{ab}} := (\boldsymbol{s}_{\mathfrak{a}}\cdot\boldsymbol{s}_{\mathfrak{b}})\,(\boldsymbol{n}_{\mathfrak{a}}\cdot\boldsymbol{n}_{\mathfrak{b}})\boldsymbol{I}$, and so $\boldsymbol{A}_{\mathfrak{ab}}\cdot\nabla(\nabla\gamma_{\mathfrak{b}}) = (\boldsymbol{s}_{\mathfrak{a}}\cdot\boldsymbol{s}_{\mathfrak{b}})\,(\boldsymbol{n}_{\mathfrak{a}}\cdot\boldsymbol{n}_{\mathfrak{b}})\,\mathrm{div}(\nabla\gamma_{\mathfrak{b}})$, would hold if ψ depended on the inhomogeneity $\nabla\boldsymbol{H}_{\mathrm{P}}$ instead of on the incompatibility curl $\boldsymbol{H}_{\mathrm{P}}$ of $\boldsymbol{H}_{\mathrm{P}}$. In the crystal plasticity and current context, at least, the distinction is significant in the sense that no additional hardening results in the current context when ψ depends directly on $\nabla\boldsymbol{H}_{\mathrm{P}}$.

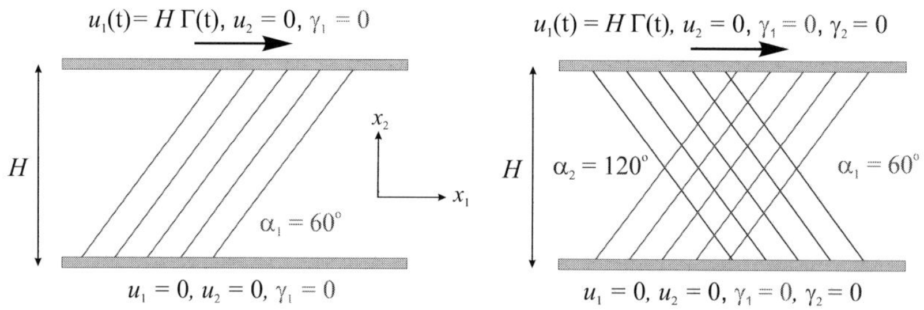

Figure 1. Geometry and boundary conditions for single slip (left) and double slip (right).

The specific geometry and corresponding boundary-value problem for single and double slip are depicted in Figure 1. In particular, the discrete dislocation simulations of this case discussed in Shu et al. [20] are carried out under the assumption that dislocation motion is blocked at the lower ($x_2 = 0$) and upper ($x_2 = H$) boundaries of the strip. Since the development of $\gamma_{\mathfrak{a}}$ is due to dislocation mobility, such boundary conditions are realized here by assuming that $\gamma_{\mathfrak{a}}$ vanishes there. The simulations discussed in Shu et al. [20], as well as those to be discussed here in what follows, were carried out for plain-strain conditions at an average strain rate of $\dot{\Gamma} = 10^3\ \mathrm{s}^{-1}$, with $\Gamma(t) = \dot{\Gamma}t$. In addition, they modeled the strip as unbounded in the x_1 and x_3 directions via the use of periodic boundary conditions. For simplicity, in the current preliminary study, this situation was approximated by working with a 10x100 finite-element mesh of thickness or height H and by focusing on conditions in the middle of the strip in order to minimize the effects of the vertical boundaries. The parameter values used by Shu et al. [20] are representative for aluminium, implying in particular $\lambda = 51053$ MPa and $\mu = 26300$ MPa for the elasticity constants.

Because the respective model relations are quite different, the remaining parameters were chosen for the preliminary investigation here in order to obtain approximately the same effective yield stress and hardening level as obtained by them for double slip and $H \gg \ell$, yielding for the choices $g = 2$ and $n = 2$ the values $\varsigma = 30$ MPa, $\dot{\nu}_0 = 2 \times 10^3\ \mathrm{s}^{-1}$, $r_0 = 0.5$ MPa, and $h_0 = 50$ MPa. Consider first the results for the shear displacement gradient profiles across the strip (Figure 2). The non-uniformity of the deformation is due here to slip being

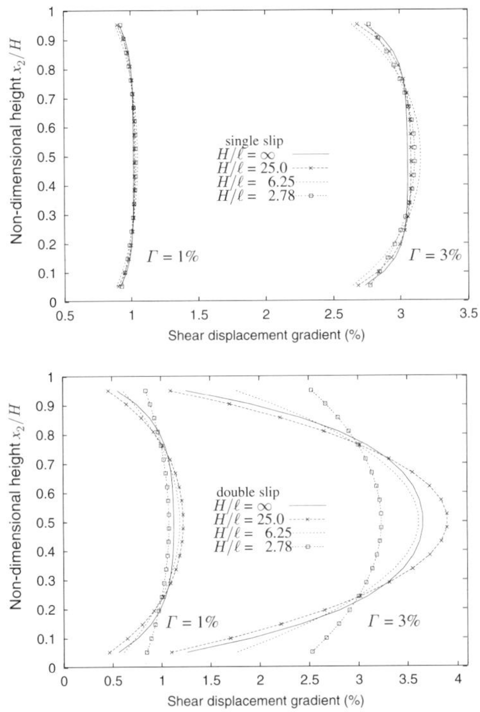

Figure 2. Profiles of the shear displacement gradient $u_{1,2}$ for single and double slip.

blocked at the upper and lower boundaries of the strip, resulting in dislocation pile-up. In both cases, the thickness of the boundary layer increases, and the deformation inhomogeneity becomes more pronounced, in the range as H/ℓ decreases from infinity down to a value of about 6.25 in the case of single slip, and about 25 for double slip. Interestingly, a further decrease of H/ℓ results in a reverse of this trend, leading to a progressive disappearance of the boundary layer and to the development of uniform shear strain across the layer. The increase of the boundary-layer thickness with decreasing strip thickness is consistent with the discrete dislocation simulation; the subsequent reversal of the boundary layer development and trend toward uniform shear strain across the layer with decreasing H/ℓ, however, is not observed in these, at least for the range of parameters investigated. Also evident in Figure 2 is a thickening of the boundary layer with increasing deformation, again for values of H/ℓ larger than about 6.25 for single slip, and about 25 for double slip, after which it begins to disappear. Such a thickening of the boundary layer with increasing deformation is also observed in the discrete dislocation simulations, but not predicted by the non-local strain-gradient-based theory.

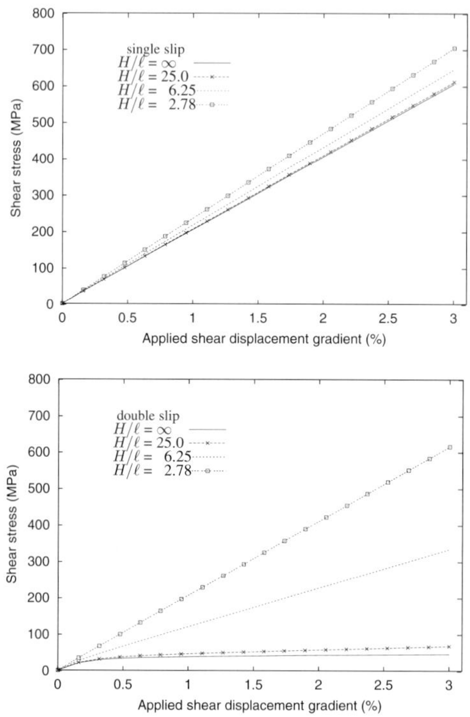

Figure 3. Shear stress T_{12} vs. applied shear displacement gradient Γ for single and double slip.

Turning next to the dependence of the material behavior on specimen size, consider the stress-strain behavior predictions for single and double slip shown in Figure 3. In the case of single slip, for example, dislocation pile-up at the boundaries results in substantial hardening which is rather insensitive to other model details, in agreement with discrete dislocation results. As for the double slip case, the results imply that, at larger strip thicknesses, the additional slip resulting from a second active slip system is able to relax dislocation pile-up and counteract the blockage of slip at the boundaries, resulting in a substantial reduction in the level of hardening relative to the single slip case. As the strip thickness decreases further, however, dislocation pile-up again becomes dominate, resulting again in substantial hardening analogous to the single slip case. Note that this effect correlates with the reduction of boundary-layer thickness with decreasing strip thickness discussed above, and like this latter effect, is not seen in the discrete dislocation simulation for the range of parameters investigated.

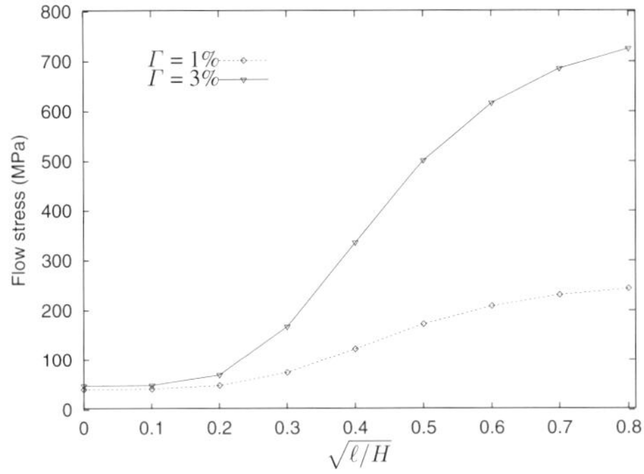

Figure 4. Flow stress as a function of the Hall-Petch ratio for double slip.

Consider finally the dependence of the flow stress on the Hall-Petch ratio $\sqrt{\ell/H}$ for two different values of the total applied shear Γ and double slip. As shown in Figure 4, for values of this ratio between 0.3 and 0.5, the dependence of the flow stress on this ratio is linear, reminiscent of the Hall-Petch effect. Below and above this region, however, the scaling becomes non-linear. In addition, it appears to depend on the level of deformation. Investigation of these and other aspects and consequences of the current formulation, including three-dimensional effects (e.g. parabolic hardening, screw dislocations) and those at large deformation, represent work in progress.

Acknowledgments

The authors would like to thank Alan Needleman for useful discussions on basic aspects of this work.

References

[1] Acharya, A. and Bassani, J.L. (2000). Lattice incompatibility and a gradient theory of crystal plasticity. *J. Mech. Phys. Solids*, **48**, 1565–1595.

[2] Asaro, R.J. (1983). Micromechanics of crystals and polycrystals. *Adv. Appl. Mech.*, **23**, 1–115.

[3] Ashby, M.F. (1970). The deformation of plastically non-homogeneous materials. *Phil. Mag.*, **21**, 399–424.

[4] Cahn, J.W. and Allen, S.M. (1977). A microscopic theory for domain wall motion and its experimental verification in Fe-Al alloy domain growth kinetics. *J. de Physique C7*, **38**, 51–54.

[5] Cuitiño, A.M. and Ortiz, M. (1992). Computational modeling of single crystals. *Modelling Simul. Mater. Sci. Eng.*, **1**, 225–263.

[6] Dai, H. and Parks, D.M. (1997). Geometrically-necessary dislocation density and scale-dependent crystal plasticity. *Proceedings of Plasticity '97*, A.S. Khan (ed.), Neat Press, 17–18.

[7] Fleck, N.A. and Hutchinson, J.W. (1993). A phenomenological theory for strain gradient effects in plasticity. *J. Mech. Phys. Solids*, **41**, 1825–1857.

[8] Forest, S., Cailletaud, G. and Sievert, R. (1997). A Cosserat theory for elastoviscoplastic single crystals at finite deformation. *Arch. Mech.*, **49**, 705–736.

[9] Gurtin, M.E. (2000). On the plasticity of single crystals: free energy, microforces, plastic-strain gradients. *J. Mech. Phys. Solids*, **48**, 989–1036.

[10] Hill, R. and Rice, J.R. (1972). Constitutive analysis of elastic-plastic crystals at arbitrary strain. *J. Mech. Phys. Solids*, **20**, 401–413.

[11] Kröner, E. (1960). Allgemeine Kontinuumstheorie der Versetzungen und Eigenspannungen. *Arch. Rat. Mech. Anal.*, **4**, 273–334.

[12] Le, K.C. and Stumpf, H. (1996). Non-linear continuum theory of dislocations. *Int. J. Engng. Sci.*, **34**, 339–358.

[13] Maugin, G. (1990). On internal variables and dissipative structures. *Int. J. Non-equil. Thermo.*, **15**, 173–192.

[14] Menzel, A. and Steinmann, P. (2000). On the continuum formulation of higher gradient plasticity for single and polycrystals. *J. Mech. Phys. Solids*, **48**, 1777–1796.

[15] Naghdi, P. and Srinivasa, A.R. (1994). Characterization of dislocations and their influence on plastic deformation in single crystals. *Int. J. Solids Structs.*, **7**, 1157–1182.

[16] Nye, J.F. (1953). Some geometric relations in dislocated crystals. *Acta Metall.*, **1**, 153–162.

[17] Ortiz, M., Repetto, E.A. and Stainier, L. (2000). A theory of dislocation structures. *J. Mech. Phys. Solids*, **48**, 2077–2114.

[18] Shizawa, K. and Zbib, H.M. (1999). A thermodynamical theory of gradient elastoplasticity with dislocation density tensor, I. Fundamentals. *Int. J. Plast.*, **15**, 899–938.

[19] Shu, J.Y. and Fleck, N.A. (1999). Strain gradient crystal plasticity: size-dependent deformation of bicrystals. *J. Mech. Phys. Solids*, **47**, 297–324.

[20] Shu, J.Y., Fleck, N.A., Van der Giessen, E. and Needleman, A. (2001). Boundary layers in constrained plastic flow: comparison of nonlocal and discrete dislocation plasticity. *J. Mech. Phys. Solids*, **49**, 1361–1395.

[21] Steinmann, P. (1996). Views on multiplicative elastoplasticity and the continuum theory of dislocations. *Int. J. Engng. Sci.*, **34**, 1717–1735.

[22] Svendsen, B. (1999). On the thermodynamics of isotropic thermoelastic materials with scalar internal degrees of freedom. *Cont. Mech. Thermodyn.*, **11**, 247–262.

[23] Svendsen, B. (2001). Continuum thermodynamic models for crystal plasticity including the effects of geometrically-necessary dislocations. *J. Mech. Phys. Solids*, in press.

[24] Svendsen, B. (2001). On the formulation of balance relations and configurational fields for materials with microstructure via invariance. *Int. J. Solids Structs.*, **38**, 700–740.

[25] Van der Giessen, E. and Needleman, A. (1995). Discrete dislocation plasticity: a simple planar model. *Model. Simul. Mater. Sci. Engng.*, **3**, 689–735.

Objective Rates in Finite Elastoplasticity

O.T. Bruhns
Institute of Mechanics, Ruhr-University Bochum
Universitätsstraße 150, 44780 Bochum, Germany
bruhns@tm.bi.ruhr-uni-bochum.de

Abstract: An Eulerian rate formulation of finite elastoplasticity is a composite one, in which objective Eulerian tensor rates are used. Among a large variety of objective rates, how to choose suitable ones has been one of the crucial points in finite elastoplasticity. It is realized that the foregoing composite formulation of elastoplasticity should fulfill certain criteria in order to avoid inconsistencies or contradictions.

Keywords: Finite elastoplasticity, Eulerian description, logarithmic strain, logarithmic rate.

1. Introduction and Motivation

Traditional formulations of elastoplasticity in the presence of finite strain and large rotation are Eulerian type and widely used; they are based upon, among other things, the additive decomposition of the stretching or the Eulerian strain-rate into elastic and plastic parts, refer to, e.g., Hill [5], Lehmann [10], and Nemat-Nasser [12]. In such formulations, yield functions and objective rate constitutive equations are expressed in terms of objective Eulerian tensor quantities, including the stretching, the Kirchhoff stress, internal state variables, etc. Each of the latter transforms in a corotational manner under a change of the observing frame. According to the principle of material frame-indifference or objectivity, each constitutive function should be form-invariant, whenever the observing frame is changed by any given rotation. On the other hand, the principle of material symmetry requires that each constitutive function should fulfill invariance restrictions under the action of the initial material symmetry group.

Eulerian rate type formulations of the given kind consist of a rate equation for elastic behavior and a flow rule for plastic behavior as well as an evolution equation for hardening behavior, in which objective Eulerian tensor rates are used. It is well known that the material time derivative of an objective quantity does not need to be objective, too. Therefore, a variety of objective time derivatives has been proposed. However, the question still remains whether the requirement of objectivity is sufficient for the formulation of such rates, since unexpected results have been observed in specific computations (see e.g. Dienes [3], Nagtegaal and De Jong [11]).

C. Miehe (ed.), IUTAM Symposium on Computational Mechanics of Solid Materials at Large Strains, 151-160.

Among a large variety of objective rates, how to choose suitable ones has been one of the crucial points in finite elastoplasticity. It is realized that the foregoing composite formulation of elastoplasticity should fulfill certain *consistency criteria* in order to avoid inconsistencies or contradictions. These criteria narrow the choice of possible objective rates. Recently, Bruhns et al. [1, 19] have introduced the *self-consistency criterion*: in the above description of finite elastoplasticity, with the additive decomposition of the stretching, the rate equation intended for characterization of elastic behavior should be exactly integrable to deliver an elastic relation. It has been demonstrated that, if a composite formulation of elastoplasticity with objective rates is required to fulfill the just-stated criterion, then the newly discovered logarithmic rate (refer to, e.g. Xiao et al. [16]) is the only possible choice among all objective corotational rates, including the Zaremba-Jaumann rate and the Green-Naghdi rate.

Moreover, it has been passionately discussed in the past whether in finite elastoplasticity the use of an additive decomposition of the deformation rate (Green and Naghdi [4]) or the multiplicative decomposition of the deformation gradient (Lee [8]) has to be used, since both descriptions seemed to be incompatible. With the introduction of the logarithmic rate, both descriptions may be related to each other.

We will restrict our considerations here on the same vector space. Let $\mathbf{a}$, $\mathbf{b}$, $\mathbf{c}$ and $\mathbf{d}$ be first order tensors (vectors) and $\mathbf{A}$, $\mathbf{B}$ second order tensors. Throughout, the following notation will be used

$$(\mathbf{a}\otimes\mathbf{b}):(\mathbf{c}\otimes\mathbf{d}) = (\mathbf{a}\cdot\mathbf{c})(\mathbf{b}\cdot\mathbf{d}) \quad \rightarrow \quad \mathbf{A}:\mathbf{B} = \mathrm{tr}(\mathbf{A}\mathbf{B}^{\mathrm{T}})$$

$$(\mathbf{a}\otimes\mathbf{b}\otimes\mathbf{c}\otimes\mathbf{d})^{\mathrm{T}} = \mathbf{c}\otimes\mathbf{d}\otimes\mathbf{a}\otimes\mathbf{b}$$

$$\mathrm{sym}\mathbf{A} = \tfrac{1}{2}(\mathbf{A}+\mathbf{A}^{\mathrm{T}}), \quad \mathrm{skw}\mathbf{A} = \tfrac{1}{2}(\mathbf{A}-\mathbf{A}^{\mathrm{T}})$$

$$\mathbf{Q}\star\mathbf{a} = \mathbf{Q}\mathbf{a}, \quad \mathbf{Q}\star\mathbf{A} = \mathbf{Q}\mathbf{A}\mathbf{Q}^{\mathrm{T}},$$

where $\mathbf{A}^{\mathrm{T}}$ and $\mathrm{tr}\mathbf{A}$ are the transposed and the trace of $\mathbf{A}$, and $\mathbf{Q}$ is an orthogonal tensor.

2. Some fundamental relations

Consider a body $\mathcal{B}$ in the state of motion. Let $\mathbf{x}$ be the position vector in R^3 occupied by a typical particle $\mathbf{X}$ in the current configuration $\mathcal{C}$. A motion of the body is then a mapping χ defined by $\mathbf{x} = \chi(\mathbf{X}, t)$, with the deformation gradient

$$\mathbf{F} = \mathrm{Grad}\chi = \frac{\partial\chi}{\partial\mathbf{X}}, \qquad \det\mathbf{F} > 0\,. \tag{1}$$

The particle velocity and the velocity gradient are denoted, respectively, by

$$\mathbf{v} = \dot{\mathbf{x}}, \qquad \mathbf{L} = \mathrm{grad}\mathbf{v} = \frac{\partial\mathbf{v}}{\partial\mathbf{x}} = \dot{\mathbf{F}}\mathbf{F}^{-1}\,. \tag{2}$$

Throughout, the superposed dot designates the material time derivative, and $\mathbf{A}^{-1}$ represents the inverse of a second order tensor $\mathbf{A}$.

According to the polar decomposition theorem, the deformation gradient $\mathbf{F}$ has the unique left and right multiplicative decompositions:

$$\mathbf{F} = \mathbf{V}\mathbf{R} = \mathbf{R}\mathbf{U}, \quad \mathbf{R}^{\mathrm{T}} = \mathbf{R}^{-1}, \quad \det\mathbf{R} = 1\,, \tag{3}$$

where $\mathbf{V}$ and $\mathbf{U}$ are the left and right stretch tensors, and $\mathbf{R}$ is the rotation tensor. Both $\mathbf{V}$ and $\mathbf{U}$ are positive definite symmetric tensors and determined uniquely by

$$\mathbf{B} = \mathbf{V}^2 = \mathbf{F}\mathbf{F}^{\mathrm{T}}, \quad \mathbf{C} = \mathbf{U}^2 = \mathbf{F}^{\mathrm{T}}\mathbf{F}\,, \tag{4}$$

where $\mathbf{B}$ and $\mathbf{C}$ are the left and right Cauchy-Green tensors. Moreover, the rotation $\mathbf{R}$ is also uniquely determined by $\mathbf{F}$.

On the other hand, the velocity gradient $\mathbf{L}$ has the unique additive decomposition:

$$\mathbf{L} = \mathbf{D} + \mathbf{W}\,, \quad \mathbf{D} = \mathrm{sym}\mathbf{L}, \quad \mathbf{W} = \mathrm{skw}\mathbf{L}\,, \tag{5}$$

where $\mathbf{D}$ and $\mathbf{W}$ are the stretching and the vorticity tensor, respectively.

The stretch tensor $\mathbf{V}$ may be expressed by its spectral decomposition

$$\mathbf{V} = \sum_{i=1}^{3} \lambda_i \mathbf{v}_i \otimes \mathbf{v}_i\,, \tag{6}$$

where λ_i and $\mathbf{v}_i$, $(i = 1, 2, 3)$ are used to denote the eigenvalues and related eigendirections. If $\mathbf{V}$ has multiple eigenvalues, the corresponding eigendirections are not unique. It is therefore more convenient to use eigenprojections. Let λ_σ be the m distinct eigenvalues of $\mathbf{V}$, and $\mathbf{V}_\sigma$ $(\sigma = 1, \ldots, m)$ the corresponding subordinate eigenprojections, then

$$\mathbf{V} = \sum_{\sigma=1}^{m} \lambda_\sigma \mathbf{V}_\sigma\,, \quad \sum_{\sigma=1}^{m} \mathbf{V}_\sigma = \mathbf{I}\,, \quad \mathbf{V}_\sigma \mathbf{V}_\tau = \begin{cases} 0, & \sigma \neq \tau \\ \mathbf{V}_\sigma, & \sigma = \tau \end{cases}\,. \tag{7}$$

Here, $\mathbf{I}$ is used to denote the second order identity tensor. The eigenprojections can be expressed as function of $\mathbf{V}$ and its eigenvalues (Sylvester's formula)

$$\mathbf{V}_\sigma = \delta_{1m}\mathbf{I} + \prod_{\sigma \neq \tau}^{m} \frac{\mathbf{V} - \lambda_\tau \mathbf{I}}{\lambda_\sigma - \lambda_\tau}\,. \tag{8}$$

Let now $\mathbf{A}$ be a time-differentiable objective symmetric second order Eulerian tensor and $\mathbf{\Omega}^*$ a skewsymmetric spin. The latter can determine a proper orthogonal tensor $\mathbf{Q}$ to within a constant proper orthogonal tensor through

$$\dot{\mathbf{Q}} = -\mathbf{Q}\mathbf{\Omega}^*. \tag{9}$$

Then we have

$$\mathbf{A} \quad \rightarrow \quad \mathbf{Q}\,\mathbf{A}\mathbf{Q}^{\mathrm{T}} = \mathbf{Q} \star \mathbf{A}$$

and

$$\overline{(\mathbf{Q} \star \mathbf{A})}^{\cdot} = \mathbf{Q}\,\dot{\mathbf{A}}\mathbf{Q}^{\mathrm{T}} + \mathbf{Q}\,\mathbf{A}\dot{\mathbf{Q}}^{\mathrm{T}} + \dot{\mathbf{Q}}\,\mathbf{A}\mathbf{Q}^{\mathrm{T}} = \mathbf{Q} \star \overset{\circ}{\mathbf{A}}{}^{*}, \tag{10}$$

where the symmetric second order tensor

$$\overset{\circ}{\mathbf{A}}{}^{*} = \dot{\mathbf{A}} + \mathbf{A}\boldsymbol{\Omega}^{*} - \boldsymbol{\Omega}^{*}\mathbf{A} \tag{11}$$

is the corotational rate of $\mathbf{A}$ defined by the spin $\boldsymbol{\Omega}^{*}$. It represents the rate-of-change of $\mathbf{A}$ observed in a rotating frame with the spin $\boldsymbol{\Omega}^{*}$. A corotational rate of an objective Eulerian tensor need not be objective. The objectivity of a corotational rate depends on its defining spin. There are infinitely many objective corotational rates. The well-known Zaremba-Jaumann rate defined by the spin $\boldsymbol{\Omega}^{*} = \mathbf{W}$ provides one classical example.

Xiao et al. [17] showed that the most general form of objective corotational rates is related to the spin

$$\boldsymbol{\Omega}^{*} = \mathbf{W} + \boldsymbol{\Upsilon}(\mathbf{B},\mathbf{D}) = \mathbf{W} + \sum_{\sigma \neq \tau}^{m} h\left(\frac{\chi_\sigma}{I_1}, \frac{\chi_\tau}{I_1}\right) \mathbf{B}_\sigma \mathbf{D} \mathbf{B}_\tau . \tag{12}$$

Here, $\chi_\sigma = \lambda_\sigma^2$ denote the distinct eigenvalues of the Cauchy-Green tensors $\mathbf{C}$ and $\mathbf{B}$, and $\mathbf{C}_\sigma$ and $\mathbf{B}_\sigma$ are the corresponding subordinate eigenprojections. Moreover, I_i $(i = 1, 2, 3)$ are the basic invariants of $\mathbf{B}$, and the spin function has the property $h(x, y) = -h(y, x)$. A basis-free expression for the antisymmetric function $\boldsymbol{\Upsilon}$ has been given in Xiao et al. [17]. We note, however, that this expression is in general not continuous for repeated eigenvalues.

Some well-known examples of the generalized expression (12) are:

(i) The spin of the Eulerian triad $\dot{\mathbf{n}}_i = \boldsymbol{\Omega}^{\mathrm{E}}\,\mathbf{n}_i$ $(i = 1, 2, 3)$

$$\boldsymbol{\Omega}^{\mathrm{E}} = \mathbf{W} + \sum_{\sigma \neq \tau}^{m} \frac{1 + (\chi_\sigma/\chi_\tau)}{1 - (\chi_\sigma/\chi_\tau)} \mathbf{B}_\sigma \mathbf{D} \mathbf{B}_\tau \,. \tag{13}$$

(ii) The polar spin

$$\boldsymbol{\Omega}^{\mathrm{R}} = \dot{\mathbf{R}}\mathbf{R}^{\mathrm{T}} = \mathbf{W} + \sum_{\sigma \neq \tau}^{m} \frac{1 - \sqrt{\chi_\sigma/\chi_\tau}}{1 + \sqrt{\chi_\sigma/\chi_\tau}} \mathbf{B}_\sigma \mathbf{D} \mathbf{B}_\tau \,. \tag{14}$$

(iii) The logarithmic spin

$$\begin{aligned}\boldsymbol{\Omega}^{\mathrm{Log}} &= \dot{\mathbf{R}}^{\mathrm{Log}}(\mathbf{R}^{\mathrm{Log}})^{\mathrm{T}} \\ &= \mathbf{W} + \sum_{\sigma \neq \tau}^{m} \left(\frac{1 + (\chi_\sigma/\chi_\tau)}{1 - (\chi_\sigma/\chi_\tau)} + \frac{2}{\ln(\chi_\sigma/\chi_\tau)} \right) \mathbf{B}_\sigma \mathbf{D} \mathbf{B}_\tau \,,\end{aligned} \tag{15}$$

with the logarithmic rotation $\mathbf{R}^{\mathrm{Log}}$ determined by

$$\dot{\mathbf{R}}^{\mathrm{Log}} = \mathbf{\Omega}^{\mathrm{Log}}\mathbf{R}^{\mathrm{Log}}, \quad \mathbf{R}^{\mathrm{Log}}|_{t=0} = \mathbf{I}. \tag{16}$$

The following formula will be useful

$$\overline{(\mathbf{R}^{\mathrm{Log}} \star \mathbf{A})}^{\cdot} = \mathbf{R}^{\mathrm{Log}} \star \overset{\circ}{\mathbf{A}}{}^{\mathrm{Log}}. \tag{17}$$

For computational purposes a continuous basis-free expression for the spin $\mathbf{\Omega}^{\mathrm{Log}}$ may be given in unified form

$$\mathbf{\Omega}^{\mathrm{Log}} = \begin{cases} \mathbf{W}, & m = 1 \\ \mathbf{W} + \gamma_1[\mathbf{BD}], & m = 2 \\ \mathbf{W} + \gamma_1[\mathbf{BD}] + \gamma_2[\mathbf{GD}] + \gamma_3[\mathbf{GD}\bar{\mathbf{B}}], & m = 3 \end{cases} \tag{18}$$

where $\bar{\mathbf{B}}$ is the deviator of $\mathbf{B}$, and the square brackets designate $[\mathbf{A}] = 2\,\mathrm{skw}\mathbf{A}$. Moreover, $\gamma_i = \gamma_i(I_1, I_2, I_3)$, and

$$\mathbf{G} = \left(I_2 - \frac{I_1^2}{3}\right)\mathbf{B}^2 + \left(\frac{I_1 I_2}{3} - I_3\right)\mathbf{B} + \frac{1}{3}(I_1 I_3 - I_2^2)\,\mathbf{I}. \tag{19}$$

Other kinds of objective corotational rates are possible, see, e.g., those used by Lee, Mallett and Wertheimer [9] and Dafalias [2]. Associated with these rates are the spin tensors that are dependent upon some internal state variables, such as the back stress $\boldsymbol{\alpha}$ and the plastic spins, etc.

On the other hand, among non-corotational objective rates are e.g. the well-known Oldroyd rates, and the Truesdell rate:

$$\overset{\circ}{\mathbf{A}}{}^{\mathrm{uO}} = \dot{\mathbf{A}} - \mathbf{A}\mathbf{L}^{\mathrm{T}} - \mathbf{L}\mathbf{A},$$

$$\overset{\circ}{\mathbf{A}}{}^{\mathrm{lO}} = \dot{\mathbf{A}} + \mathbf{A}\mathbf{L} + \mathbf{L}^{\mathrm{T}}\mathbf{A},$$

$$\overset{\circ}{\mathbf{A}}{}^{\mathrm{T}} = \dot{\mathbf{A}} - \mathbf{A}\mathbf{L}^{\mathrm{T}} - \mathbf{L}\mathbf{A} + (\mathrm{tr}\mathbf{D})\mathbf{A}.$$

It may be observed that each known objective tensor rate $\overset{\circ}{\mathbf{A}}{}^{*}$ cited above is defined by the material time derivative $\dot{\mathbf{A}}$ added by a term related to $\mathbf{A}$, both gradients $\mathbf{F}$ and $\mathbf{L}$, and some internal variables in some cases. This observation motivates a fully general form of tensor rates

$$\overset{\circ}{\mathbf{A}}{}^{*} = \dot{\mathbf{A}} + \varphi(\mathbf{A}, \mathbf{L}, \mathbf{F}, \mathbf{\Lambda}), \tag{20}$$

where φ is a second order tensor-valued function of several tensor arguments, and $\mathbf{\Lambda}$ may be a set of internal variables.

According to Hill [6], a general class of Eulerian and Lagrangean strain measures may be defined through a single scale function. Their forms can be given by

$$\mathbf{E} = \mathbf{g}(\mathbf{C}) = \sum_{\sigma=1}^{m} g(\chi_\sigma)\mathbf{C}_\sigma\,, \quad \mathbf{e} = \mathbf{g}(\mathbf{B}) = \sum_{\sigma=1}^{m} g(\chi_\sigma)\mathbf{B}_\sigma\,, \tag{21}$$

where the scale function $g(\cdot)$ is a smooth increasing function with normalized property $g(1) = g'(1) - \frac{1}{2} = 0$. In particular, the natural logarithmic scale function $g(\chi) = \frac{1}{2}\ln\chi$ yields Hencky's logarithmic strain measures:

$$\mathbf{H} = \frac{1}{2}\ln\mathbf{C} = \frac{1}{2}\sum_{\sigma=1}^{m}(\ln\chi_\sigma)\mathbf{C}_\sigma\,, \quad \mathbf{h} = \frac{1}{2}\ln\mathbf{B} = \frac{1}{2}\sum_{\sigma=1}^{m}(\ln\chi_\sigma)\mathbf{B}_\sigma\,. \tag{22}$$

The following relations hold

$$\mathbf{h} = \mathbf{R}\star\mathbf{H}, \qquad \mathbf{H} = \mathbf{R}^{\mathrm{T}}\star\mathbf{h}\,. \tag{23}$$

Following Hill [6], the symmetric second order tensor $\mathbf{S}$ determined through the work-conjugacy relation

$$\dot{w} = \mathbf{S} : \dot{\mathbf{E}} = \boldsymbol{\tau} : \mathbf{D}, \quad \boldsymbol{\tau} = (\det\mathbf{F})\boldsymbol{\sigma}, \tag{24}$$

is called the work-conjugated stress measure of the Lagrangean strain measure $\mathbf{E}$. In the above, $\boldsymbol{\tau}$ is used to designate the Kirchhoff stress, where $\boldsymbol{\sigma}$ is the Cauchy stress.

Hill's work-conjugacy relation may be generalized to give

$$\dot{w} = \mathbf{S} : \dot{\mathbf{E}} = \mathbf{s} : \overset{\circ}{\mathbf{e}}{}^{*}, \tag{25}$$

where now $\mathbf{s}$ is the work-conjugated stress of the Eulerian strain $\mathbf{e}$, and $\overset{\circ}{\mathbf{e}}{}^{*}$ is an objective corotational rate of $\mathbf{e}$ defined by the spin $\boldsymbol{\Omega}^*$, i.e. defined in a rotating frame. Thus $(\mathbf{S},\mathbf{E})$ and $(\mathbf{s},\mathbf{e})$ are Lagrangean and Eulerian work-conjugated pairs related to the spin $\boldsymbol{\Omega}^*$. In particular, we denote the work-conjugated stress of the Lagrangean logarithmic strain measure $\mathbf{H}$ by $\boldsymbol{\Pi}$, and the Eulerian counterpart of $\boldsymbol{\Pi}$ by $\boldsymbol{\pi}$, with relations

$$\boldsymbol{\pi} = \mathbf{R}\star\boldsymbol{\Pi}, \qquad \boldsymbol{\Pi} = \mathbf{R}^{\mathrm{T}}\star\boldsymbol{\pi}\,. \tag{26}$$

An explicit expression of $\boldsymbol{\Pi}$ in terms of the right stretch $\mathbf{U}$ and the rotated Cauchy stress $\mathbf{R}^{\mathrm{T}}\star\boldsymbol{\sigma}$ was first derived by Hoger [7]. An alternative expression in terms of $\mathbf{C}$ and $\mathbf{R}^{\mathrm{T}}\star\boldsymbol{\tau}$ has been derived in Xiao et al. [18]. From this and Eq. $(26)_1$ we obtain, e.g.

(i) $\boldsymbol{\Omega}^{\mathrm{R}}$-conjugated:

$$\dot{w} = \boldsymbol{\pi} : \overset{\circ}{\mathbf{h}}{}^{\mathrm{R}} \quad \rightarrow \quad \boldsymbol{\pi} = \sum_{\sigma,\tau=1}^{m} \sqrt{\chi_\sigma\chi_\tau}^{\,-1}\,\frac{\chi_\sigma - \chi_\tau}{\ln\chi_\sigma - \ln\chi_\tau}\,\mathbf{B}_\sigma\boldsymbol{\tau}\mathbf{B}_\tau\,. \tag{27}$$

(ii) $\mathbf{\Omega}^{\text{Log}}$-conjugated:

$$\dot{w} = \boldsymbol{\pi} : \overset{\circ}{\mathbf{h}}{}^{\text{Log}} \quad \rightarrow \quad \boldsymbol{\pi} = \boldsymbol{\tau} \,. \tag{28}$$

We now want to compare the different objective rates introduced above, and reduced the variety of reasonable rates described with Eq. (20).

Prager [13] was the first who has pointed out the advantage of corotational rates compared with non-corotational ones. For a composite elastic, perfectly plastic material he argued that to avoid contradictions, the yield function should be stationary when the stress rate vanishes. This *stationarity criterion* was generalized in Xiao et al. [20]:

If the simultaneous vanishing of the rates $\overset{\circ}{\boldsymbol{\tau}}{}^*$ and $\overset{\circ}{\boldsymbol{\alpha}}{}^*$ implies that any possible yield function $f = f(\boldsymbol{\tau}, \boldsymbol{\alpha}, k)$ is stationary, then the rates $\overset{\circ}{\boldsymbol{\tau}}{}^*$ and $\overset{\circ}{\boldsymbol{\alpha}}{}^*$ must be of the same kind of corotational rates defined by the same spin tensor.

Let us consider a description of finite elastoplasticity with a yield function $f(\boldsymbol{\tau}, \boldsymbol{\alpha}, k)$ separating the domains of elastic and (elasto-)plastic deformations. From the theory of invariants, we know that each possible yield function, which is an isotropic invariant of both $\boldsymbol{\tau}$ and $\boldsymbol{\alpha}$, may be expressed as a function of k and a set of basic invariants and simultaneous invariants of $\boldsymbol{\tau}$ and $\boldsymbol{\alpha}$

$$\begin{cases} \text{tr}\boldsymbol{\tau},\ \text{tr}\boldsymbol{\tau}^2,\ \text{tr}\boldsymbol{\tau}^3;\ \text{tr}\boldsymbol{\alpha},\ \text{tr}\boldsymbol{\alpha}^2,\ \text{tr}\boldsymbol{\alpha}^3\,; \\ \text{tr}(\boldsymbol{\tau}\boldsymbol{\alpha}),\ \text{tr}(\boldsymbol{\tau}^2\boldsymbol{\alpha}),\ \text{tr}(\boldsymbol{\tau}\boldsymbol{\alpha}^2),\ \text{tr}(\boldsymbol{\tau}^2\boldsymbol{\alpha}^2)\,. \end{cases} \tag{29}$$

We thus infer that each possible yield function $f(\boldsymbol{\tau}, \boldsymbol{\alpha}, k)$ is stationary whenever the basic invariants and the simultaneous invariants listed above are stationary.

First, we consider the requirement that the vanishing of $\overset{\circ}{\boldsymbol{\tau}}{}^*$ implies that the three basic invariants of $\boldsymbol{\tau}$ are stationary, which is equivalent to the fact that the eigenvalues of $\boldsymbol{\tau}$ are stationary. Hence, the foregoing statement means that there is a proper orthogonal tensor $\mathbf{Q}$ and a constant $\boldsymbol{\tau}_0$ such that

$$\overset{\circ}{\boldsymbol{\tau}}{}^* = \mathbf{0} \quad \rightarrow \quad \boldsymbol{\tau} = \mathbf{Q} \star \boldsymbol{\tau}_0, \quad \dot{\boldsymbol{\tau}}_0 = \mathbf{0}\,. \tag{30}$$

Introducing this into Eq. (20), we may deduce

$$\overline{(\mathbf{Q} \star \boldsymbol{\tau}_0)}^{\boldsymbol{\cdot}} + \boldsymbol{\varphi}(\mathbf{Q} \star \boldsymbol{\tau}_0, \mathbf{L}, \mathbf{F}, \boldsymbol{\Lambda}) = \mathbf{0},$$

and hence we have

$$\boldsymbol{\varphi}(\mathbf{Q} \star \boldsymbol{\tau}_0, \mathbf{L}, \mathbf{F}, \boldsymbol{\Lambda}) = (\mathbf{Q} \star \boldsymbol{\tau}_0)\mathbf{\Omega}^* - \mathbf{\Omega}^*(\mathbf{Q} \star \boldsymbol{\tau}_0)\,, \tag{31}$$

with

$$\mathbf{\Omega}^* = \dot{\mathbf{Q}}\mathbf{Q}^{\mathrm{T}} = -\mathbf{Q}\dot{\mathbf{Q}}^{\mathrm{T}}\,. \tag{32}$$

Generally, the three eigenvalues of $\boldsymbol{\tau}_0$ may be three arbitrary constants within the yield limit, and the proper orthogonal tensor $\mathbf{Q}$ in Eq. (32) may be replaced by $\mathbf{Q}\mathbf{Q}_0$ with $\mathbf{Q}_0$ an arbitrary constant proper orthogonal tensor. These facts mean that the tensor $\mathbf{Q} \star \boldsymbol{\tau}_0$ in Eq. (31) may be any given symmetric second order tensor. Thus, we infer

$$\varphi(\boldsymbol{\tau}, \mathbf{L}, \mathbf{F}, \boldsymbol{\Lambda}) = \boldsymbol{\tau}\boldsymbol{\Omega}^* - \boldsymbol{\Omega}^*\boldsymbol{\tau}. \tag{33}$$

Expression (33) indicates that the rate $\overset{\circ}{\boldsymbol{\tau}}{}^*$ (20) must be a corotational rate defined by a spin $\boldsymbol{\Omega}^*$. Next, from similar considerations related with the vanishing of the rate $\overset{\circ}{\boldsymbol{\alpha}}{}^*$, and the simultaneously vanishing of the rates $\overset{\circ}{\boldsymbol{\tau}}{}^*$ and $\overset{\circ}{\boldsymbol{\alpha}}{}^*$, we conclude that both rates must be of the same kind of corotational rates defined by the same spin $\boldsymbol{\Omega}^*$.

Thus, Prager's stationarity criterion implies that the suitable choice of the rates $\overset{\circ}{\boldsymbol{\tau}}{}^*$ and $\overset{\circ}{\boldsymbol{\alpha}}{}^*$ in Eulerian rate constitutive formulations of finite elastoplasticity should be the same kind of corotational rates defined by the same spin tensor, and thus non-corotational rates should be excluded. However, this does not mean that any given kind of corotational rates may be suitable, since Prager's stationarity criterion is necessary but not sufficient. To further narrow the choice of objective rates, new consistency criteria are needed.

3. Elasticity and the logarithmic rate

It is commonly accepted that the stretching $\mathbf{D}$ is a well-defined fundamental kinematic quantity measuring the rate of change of local deformation in a deforming body. However, it has long been unknown whether or not the stretching can be really written as a rate of a strain measure. The pertinent question is: whether or not a strain measure $\mathbf{e}$ and a spin $\boldsymbol{\Omega}^*$ can be found such that the objective corotational rate of $\mathbf{e}$ is exactly identical with the stretching tensor $\mathbf{D}$

$$\overset{\circ}{\mathbf{e}}{}^* = \dot{\mathbf{e}} + \mathbf{e}\boldsymbol{\Omega}^* - \boldsymbol{\Omega}^*\mathbf{e} = \mathbf{D}\,. \tag{34}$$

Using eigenprojections, Xiao et al. [16] were able to show that the above relation holds iff the Eulerian strain measure is the logarithmic strain $\mathbf{h}$ given by Eq. $(22)_2$, and the spin is the logarithmic spin $\boldsymbol{\Omega}^{\mathrm{Log}}$ (Eq. 15)

$$\overset{\circ}{\mathbf{h}}{}^{\mathrm{Log}} = \dot{\mathbf{h}} + \mathbf{h}\boldsymbol{\Omega}^{\mathrm{Log}} - \boldsymbol{\Omega}^{\mathrm{Log}}\mathbf{h} = \mathbf{D}\,. \tag{35}$$

Introducing this into Eq. (17), we arrive at

$$\overline{(\mathbf{R}^{\mathrm{Log}} \star \mathbf{h})}^{\boldsymbol{\cdot}} = \mathbf{R}^{\mathrm{Log}} \star \overset{\circ}{\mathbf{h}}{}^{\mathrm{Log}} = \mathbf{R}^{\mathrm{Log}} \star \mathbf{D}\,. \tag{36}$$

Hypoelastic materials have a constitutive relation of the form

$$\overset{\circ}{\boldsymbol{\tau}}{}^* = \mathbb{H}(\boldsymbol{\tau}) : \mathbf{D}\,, \tag{37}$$

where the fourth order hypoelasticity tensor $\mathbb{H}$ has the minor and major symmetry properties. The fundamental structure of this hypoelastic relation does not depend on the time derivative used. Therefore, in a general sense, Truesdell and Noll [15] are correct, when they assert that: "... Despite claims and whole papers to the contrary, any advantage claimed for one such rate over another is pure illusion." However, depending on the time derivative used, fundamental differences may occur in modeling material behavior (refer to, e.g., Dienes [3], Nagtegaal and De Jong [11]).

Simó and Pister [14] showed that for any of the commonly known objective stress rates, the corresponding rate type model for the elastic response is in general not integrable and thus inconsistent with the notion of elasticity, in particular hyperelasticity.

Let us now assume a hypoelastic model based upon the logarithmic rate, i.e.

$$\overset{\circ}{\boldsymbol{\tau}}{}^{\mathrm{Log}} = \mathbb{H} : \mathbf{D}, \quad \boldsymbol{\tau}|_{t=0} = \mathbf{0}, \quad \mathbf{h}|_{t=0} = \mathbf{0}. \tag{38}$$

Sequentially, we can derive

$$\mathbf{R}^{\mathrm{Log}} \star \overset{\circ}{\boldsymbol{\tau}}{}^{\mathrm{Log}} = \mathbb{H} : (\mathbf{R}^{\mathrm{Log}} \star \mathbf{D}),$$

and with the help of Eq. (36), Eq. $(38)_1$ may be integrated

$$\overline{\mathbf{R}^{\mathrm{Log}} \star \boldsymbol{\tau}}^{\,\cdot} = \mathbb{H} : (\overline{\mathbf{R}^{\mathrm{Log}} \star \mathbf{h}}^{\,\cdot}), \quad \rightarrow \quad \boldsymbol{\tau} = \mathbb{H} : \mathbf{h}\,. \tag{39}$$

We thus conclude that the hypoelastic model based upon the logarithmic rate is exactly integrable to deliver a linear, isotropic relation between the Kirchhoff stress and the spatial logarithmic strain.

Moreover, let $\Sigma(\boldsymbol{\tau})$ be any given differentiable isotropic scalar function. In Bruhns et al. [1] it has been proved that the rate equation

$$\mathbf{D} = \frac{\partial^2 \Sigma}{\partial \boldsymbol{\tau} \partial \boldsymbol{\tau}} : \overset{\circ}{\boldsymbol{\tau}}{}^{*} \tag{40}$$

is exactly integrable to deliver an isotropic elastic relation if and only if the stress rate $\overset{\circ}{\boldsymbol{\tau}}{}^{*}$ is the logarithmic rate. The unique integrable-exactly rate equation defines the hyperelastic relation $\mathbf{h} = \partial \Sigma / \partial \boldsymbol{\tau}$.

4. Conclusion

In his textbook *Plasticity Theory*, with reference to a linear hypoelastic model (37) based on the Jaumann rate, J. Lubliner has stated: "However, ... such an equation with constant coefficients cannot be derived from an energy-based elastic (hyperelastic) stress strain relation ..., and the result holds when the Jaumann rate is replaced by any other objective rate."

It has been shown that this is no longer true. A "self-consistent" hypoelastic equation with the logarithmic rate is exactly integrable to deliver a hyperelastic relation. The strain that has to be used is the logarithmic strain $\mathbf{h}$. Under these conditions the equivalence of the additive and the multiplicative decompositions has been demonstrated.

References

[1] Bruhns, O.T., Xiao, H. and Meyers, A. (1999). Self-consistent Eulerian rate type elasto-plasticity models based upon the logarithmic stress rate. *Int. J. Plasticity*, **15**, 479–520.

[2] Dafalias, Y.F. (1985). The plastic spin. *J. Appl. Mech.*, **52**, 865–871.

[3] Dienes, J.K. (1979). On the analysis of rotation and stress rate in deforming bodies. *Acta Mechanica*, **32**, 217-232.

[4] Green, A.E. and Naghdi, P.M. (1965) A general theory of an elastic-plastic continuum. *Arch. Rat. Mech. Anal.*, **18**, 251–281.

[5] Hill, R. (1958). A general theory of uniqueness and stability in elastic-plastic solids. *J. Mech. Phys. Solids*, **6**, 236–249.

[6] Hill, R. (1978). Aspects of invariance in solid mechanics. *Adv. Appl. Mech.*, **18**, 1–75.

[7] Hoger, A. (1987). The stress conjugate to the logarithmic strain. *Int. J. Solids Struct.*, **23**, 1645–1656.

[8] Lee, E.H. (1969). Elastic-plastic deformation at finite strains. *J. Appl. Mech.*, **36**, 1–6.

[9] Lee, E.H., Mallett, R.L. and Wertheimer, T.B. (1983). Stress analysis for anisotropic hardening in finite deformation plasticity. *J. Appl. Mech.*, **50**, 554–560.

[10] Lehmann, Th. (1962). Zur Beschreibung großer plasticher Formänderungen unter Berücksichtigung der Werkstoffverfestigung. *Rheologica Acta*, **2**, 247–254.

[11] Nagtegaal, J.C. and De Jong, J.E. (1982). Some aspects of non-isotropic work-hardening in finite strain plasticity. In: *Plasticity of Metals at Finite Strain: Theory, Computation and Experiment*, E.H. Lee and R.L. Mallet (eds.), Stanford University, pp.65–102.

[12] Nemat-Nasser, S. (1982). On finite deformation elastoplasticity. *Int. J. Solids Struct.*, **18**, 857–872.

[13] Prager, W. (1960). An elementary discussion of definitions of stress rate. *Q. Appl. Math.*, **18**, 403–407.

[14] Simó, J.C. and Pister, K.S. (1984). Remarks on rate constitutive equations for finite deformation problem: computational implications. *Comp. Meth. Appl. Mech. Engng.*, **46**, 201–215.

[15] Truesdell, C.A. and Noll, W. (1965). The nonlinear field theories of mechanics. In: *Handbuch der Physik*, S. Flügge (ed.), vol. III/3, Berlin: Springer-Verlag.

[16] Xiao, H., Bruhns, O.T. and Meyers, A. (1997). Logarithmic strain, logarithmic spin and logarithmic rate. *Acta Mechanica*, **124**, 89–105.

[17] Xiao, H., Bruhns, O.T. and Meyers, A. (1998). On objective corotational rates and their defining spin tensors. *Int. J. Solids Struct.*, **35**, 4001–4014.

[18] Xiao, H., Bruhns, O.T. and Meyers, A. (1998). Objective corotational rates and unified work-conjugacy relation between Lagrangean and Eulerian stress and strain measures. *Arch. Mech.*, **50**, 913–943.

[19] Xiao, H., Bruhns, O.T. and Meyers, A. (2000). A consistent finite elastoplasticity theory combining additive and multiplicative decomposition of the stretching and the deformation gradient. *Int. J. Plasticity*, **16**, 143–177.

[20] Xiao, H., Bruhns, O.T. and Meyers, A. (2000). The choice of objective rates in finite elastoplasticity: general results on the uniqueness of the logarithmic rate. *Proc. Roy. Soc. London*, **A 456**, 1865–1882.

Finite Deformation Plasticity with Void Growth and Asymmetric Compression-Tension Behavior

R. Mahnken
Alstom (Switzerland) Ltd
Baden, Switzerland
rolf.mahnken@power.alstom.com

E. Stein
Institute for Structural Mechanics and Numerical Mechanics, University of Hannover
Appelstraße 9A, 30167 Hannover, Germany
stein@ibnm.uni-hannover.de

Abstract: A generalized single surface yield function is presented for void growth effects, asymmetric tension-compression behavior and pressure dependence. The constitutive equations are formulated within the framework of multiplicative elasto-plasticity, where two different conceptions are considered. Representative examples including comparisons with experimental data illustrate the capability of the proposed approach.

Keywords: Void growth, asymmetric effects, pressure dependence, finite strains.

1. Introduction

Classical macroscopic phenomena of polycrystalline metals are plastic deformations above the yield stress which is identical in compression and tension tests. At higher loading metallographic observations reveal the three stages of nucleation, growth and coalescence of microvoids (see e.g. Le Roy et al. [9]). Further experimental – non-classical – results of high-strength materials, mainly martensite steels, are reported by Spitzig et al. [12] and references therein, where the yield stresses are greater in uniaxial compression than in uniaxial tension, and where hydrostatic pressure raises the stress-strain curve. In this contribution a generalized single surface yield function for simulating the above mentioned experimental effects is presented. The constitutive equations are formulated within the framework of multiplicative elasto-plasticity, where two different conceptions are considered. Representative examples illustrate the capability of the proposed approach.

Notations. Tensor calculus on manifolds as advocated by Giessen and Kollmann [2] and references therein is used, which, as a main advantage, allows

C. Miehe (ed.), IUTAM Symposium on Computational Mechanics of Solid Materials at Large Strains, 161-170.

for unambiguous transformation rules between different vector spaces. Square brackets $[\bullet]$ used throughout denote "function of".

2. Formulation of the constitutive equations

2.1. Thermodynamic framework

Thermodynamic formulations for isothermal processes are based on the principle of positive dissipation

$$\mathcal{D} = \mathcal{P} - \partial_t \Psi \geq 0. \tag{1}$$

Here, relative to the reference configuration $\mathcal{B}_0$ the stress-power is given as $\mathcal{P} = \mathbf{S} : \mathbf{D}$ in terms of the contra-variant 2. Piola Kirchhoff stress tensor $\mathbf{S}$ and the co-variant rate of deformation tensor $\mathbf{D} = \partial_t \mathbf{E}$, where $\mathbf{E}$ is defined in Eq. (A.1.1-a) of the Appendix. Relative to $\bar{\mathcal{B}}$, the intermediate configuration, $\mathcal{P} = \bar{\mathbf{S}} : \bar{\mathbf{D}}$ holds, where $\bar{\mathbf{S}} = {}_*\phi_p^\sharp[\mathbf{S}]$ and $\bar{\mathbf{D}} = {}_*\phi_p^\flat[\mathbf{D}]$ are plastic push-forwards of $\mathbf{S}$ and $\mathbf{D}$, respectively. Upon assuming $\Psi = \Psi[\bar{\mathbf{C}}_e, \boldsymbol{q}]$ for the free energy function, with $\bar{\mathbf{C}}_e$ defined in Eq. (A.3) and $\boldsymbol{q}$ representing internal variables, the dissipation (1) results into

$$\mathcal{D} = \left(\bar{\mathbf{S}} - 2\frac{\partial \Psi}{\partial \bar{\mathbf{C}}_e}\right) : \frac{1}{2}\partial_t \bar{\mathbf{C}}_e + \bar{\mathbf{S}} : (\bar{\mathbf{C}}_e \cdot \bar{\mathbf{L}}_p) - \frac{\partial \Psi}{\partial \boldsymbol{q}} \star \partial_t \boldsymbol{q} \geq 0 \tag{2}$$

and where $\bar{\mathbf{L}}_p$ is defined in Eq. (A.2). The above relation holds for all processes $\bar{\mathbf{C}}_e$ and $\bar{\mathbf{L}}_p$, which implies $\bar{\mathbf{S}} = 2\partial \Psi^{el} / \partial \bar{\mathbf{C}}_e$. Furthermore, upon defining the Mandel stress tensor $\bar{\mathbf{T}} = \bar{\mathbf{C}}_e \cdot \bar{\mathbf{S}}$ and the thermodynamic force $\boldsymbol{Q} = \partial \Psi^p / \partial \boldsymbol{q}$ the dissipation (2) reduces to

$$\mathcal{D}^{red} = \bar{\mathbf{T}} : \bar{\mathbf{L}}_p - \boldsymbol{Q} \star \partial_t \boldsymbol{q} \geq 0. \tag{3}$$

2.2. Flow rule

A general flow rule is formulated for the plastic velocity tensor $\bar{\mathbf{L}}_p$ occurring in Eq. (3) in a nine dimensional space. Applying the concept of a non-associated flow rule the following approach is considered:

$$\bar{\mathbf{L}}_p = \lambda \frac{\partial \Phi^*}{\partial \bar{\mathbf{T}}}. \tag{4}$$

Here $\Phi^* = \Phi^*[\bar{\mathbf{T}}, \boldsymbol{Q}]$ is the plastic potential, formulated in terms of $\bar{\mathbf{T}}$, which appears as the work conjugate (dual) tensor to $\bar{\mathbf{L}}_p$ in Eq. (3). Furthermore, the scalar λ is the plastic multiplier obtained from the loading and unloading conditions $\lambda \geq 0, \quad \Phi \leq 0, \quad \lambda\Phi = 0$, and where $\Phi = \Phi[\bar{\mathbf{T}}, \boldsymbol{Q}]$ is the yield function generally different from Φ^*.

From now on the case of *isotropy* is considered, thus ensuing that $\bar{\mathbf{C}}_e$ and $\bar{\mathbf{S}}$ commute. Consequently $\bar{\mathbf{G}} \cdot \partial\Phi^*/\partial\bar{\mathbf{T}} = \bar{\mathbf{G}} \cdot \bar{\mathbf{C}}_e^{-1} \cdot \partial\Phi^*/\partial\bar{\mathbf{S}} = \partial\Phi^*/\partial\bar{\mathbf{S}} \cdot \bar{\mathbf{C}}_e^{-1} \cdot \bar{\mathbf{G}}$, thus implying skew$[\bar{\mathbf{L}}_p^\flat] = \mathbf{0}$, which yields for $\bar{\mathbf{D}}_p$ defined in Eq. (A.1.3.c) $\bar{\mathbf{D}}_p = \mathbf{G} \cdot \bar{\mathbf{L}}_p$. Thus Eq. (4) rewrites as

$$\bar{\mathbf{D}}_p = \lambda\bar{\mathbf{G}} \cdot \bar{\mathbf{C}}_e^{-1} \cdot \frac{\partial\Phi^*}{\partial\bar{\mathbf{S}}}. \tag{5}$$

Furthermore the scenario of small elastic strains allows the approximations $\mathbf{G} \cdot \bar{\mathbf{C}}_e^{-1} \cong \mathbf{G} \cdot \mathbf{G}^\sharp = \bar{\mathbf{1}}$ and additionally $\mathbf{G}^\sharp \cdot \bar{\mathbf{T}} = \mathbf{G}^\sharp \cdot \bar{\mathbf{C}}_e \cdot \bar{\mathbf{S}} \cong \mathbf{G}^\sharp \cdot \mathbf{G} \cdot \bar{\mathbf{S}} = \bar{\mathbf{S}}$. Thus Eq. (5) approximates as

$$\bar{\mathbf{D}}_p \cong \lambda\frac{\partial\Phi^*}{\partial\bar{\mathbf{S}}}. \tag{6}$$

Upon using the relations $\bar{\mathbf{L}}_p = -1/2\mathcal{L}_p^\sharp[\bar{\mathbf{G}}^\sharp] \cdot \bar{\mathbf{G}}$ and $\mathbf{G}^\sharp \cdot \bar{\mathbf{D}}_p \cdot \mathbf{G}^\sharp = -1/2\mathcal{L}_p^\sharp[\bar{\mathbf{G}}^\sharp]$ the above flow rules (4) and (6) can be reformulated as

$$\boxed{\begin{array}{llcll} 1. & -\frac{1}{2}\mathcal{L}_p^\sharp[\bar{\mathbf{G}}^\sharp] \cdot \bar{\mathbf{G}} & = & \lambda\dfrac{\partial\Phi^*}{\partial\bar{\mathbf{T}}}, & \text{where } \bar{\mathbf{T}} = \bar{\mathbf{C}}_e \cdot \bar{\mathbf{S}} \\ 2. & -\frac{1}{2}\mathcal{L}_p^\sharp[\bar{\mathbf{G}}^\sharp] & \cong & \lambda\dfrac{\partial\Phi^*}{\partial\bar{\mathbf{S}}^\flat}, & \text{where } \bar{\mathbf{S}}^\flat = \bar{\mathbf{G}} \cdot \bar{\mathbf{S}} \cdot \bar{\mathbf{G}} \end{array}} \tag{7}$$

and where the plastic Lie derivative operator $\mathcal{L}_p^\sharp[\bar{\bullet}] = {}_*\phi_p^\sharp\left[\partial_t[{}^*\phi_p^\sharp[\bar{\bullet}]]\right]$ has been used. Furthermore an elastic push-forward ${}_*\phi_e$ of the relations (7) yields representations for the flow rules w.r.t. the spatial configuration

$$\boxed{\begin{array}{llcll} 1. & -\frac{1}{2}\mathcal{L}^\sharp[\mathbf{b}_e] \cdot \mathbf{b}_e^{-1} & = & \lambda\dfrac{\partial\Phi^*}{\partial\boldsymbol{\tau}'}, & \text{where } \boldsymbol{\tau}' = \mathbf{g} \cdot \boldsymbol{\tau} \\ 2. & -\frac{1}{2}\mathcal{L}^\sharp[\mathbf{b}_e] & \cong & \lambda\dfrac{\partial\Phi^*}{\partial\bar{\boldsymbol{\tau}}^\flat}, & \text{where } \bar{\boldsymbol{\tau}}^\flat = \mathbf{b}_e^{-1} \cdot \boldsymbol{\tau} \cdot \mathbf{b}_e^{-1}. \end{array}} \tag{8}$$

Here the Kirchhoff stress tensor is obtained from $\boldsymbol{\tau} = {}_*\phi_e^\sharp[\bar{\mathbf{S}}]$, and $\mathbf{b}_e = {}_*\phi_e^\sharp[\bar{\mathbf{G}}^\sharp]$ is the elastic left Cauchy-Green tensor, respectively. Furthermore the Lie derivative operator $\mathcal{L}^\sharp[\bullet] = {}_*\phi^\sharp\left[\partial_t[{}^*\phi^\sharp[\bullet]]\right]$ has been used.

It is noteworthy, that the l.h.s. of the mixed-variant formulation (8.1) reflects the approach in Simo [11], Simo and Miehe [10] and Miehe [8], whereas the l.h.s. of the contra-variant formulation (8.2) reflects the approach considered by Lührs et al. [3]. An exponential map integration scheme for the flow rule (8.1) is described e.g. in Simo [11], which, as a main practical advantage, leads to an update formula in principal directions with the same structure as in the geometrically linear case. This feature cannot be exploited from the Euler backward rule described in Lührs et al. [3].

2.3. YIELD FUNCTION AND PLASTIC POTENTIAL

For the case of isotropy the yield function reduces to $\Phi_Y = \Phi_Y[\,^{\bar{G}\bar{S}}I_1,\ ^{\bar{G}\bar{S}}I_2,\ ^{\bar{G}\bar{S}}I_3]$, where the three basic invariants $^{\bar{G}\bar{S}}I_i, i = 1,2,3$ are determined according to Eq. (A.4). In Ehlers [1] a yield function of the form $\Phi_Y[\,^{\bar{G}\bar{S}}I_1,\ ^{\bar{G}\bar{S}_{dev}}I_2,\ ^{\bar{G}\bar{S}_{dev}}I_3]$ is introduced for geomaterials, where $\bar{\mathbf{S}}^{dev}$ denotes the deviatoric part of $\bar{\mathbf{S}}$. Upon introducing a process vector $\mathbf{p} = [f, e_v, ...]$, which collectively summarizes the void volume fraction f, the effective plastic strain e_v and possible further process variables, a more general yield function is considered (see Mahnken [4, 5]).

$$\boxed{\begin{array}{llll} 1. & \Phi_Y & = & ^{\bar{G}\bar{S}_{dev}}I_2 + \mathcal{H}_N\,\mathcal{D}_\phi, \quad \text{where} \\ 2. & \mathcal{H}_N & = & \sum_{i=0}^{N} a_i[\mathbf{p}]\ ^{\bar{G}\bar{S}}I_1^i \\ 3. & \mathcal{D}_\phi & = & (1+\gamma[\mathbf{p}]\vartheta)^{-m}, \qquad \vartheta = {}^{\bar{G}\bar{S}_{dev}}I_3 / {}^{\bar{G}\bar{S}_{dev}}I_2^{3/2} \\ 4. & \mathbf{p} & = & [e_v, f, ...]. \end{array}} \tag{9}$$

In the relations (9) the coefficients $a_i[\mathbf{p}]$, $i = 0, ..., N$ are dependent on the vector of process variables $\mathbf{p}$, whereas in the original formulation of Ehlers [1] they appear as constants a_i, $i = 0, ..., N = 4$ and thus can directly be interpreted as material parameters. As noted by Ehlers [1] the function $\mathcal{H}_N$ in Eq. (9.2) represents the yield function in the hydrostatic plane, whereas the function $\mathcal{D}_\phi$ in Eq. (9.3) controls the deviation from a circle towards the compression and extension radii in the octahedral plane. Next, the general yield function (9) is specified as

$$\boxed{\begin{array}{llll} 1. & \Phi_Y & = & ^{\bar{G}\bar{S}_{dev}}I_2 + \mathcal{H}_2\,\mathcal{D}_\phi, \quad \text{where} \\ 2. & \mathcal{H}_2 & = & a_0[f,e_v] + a_1[f,e_v]\ ^{\bar{G}\bar{S}}I_1 + a_2[f,e_v]\ ^{\bar{G}\bar{S}}I_1^2 \\ 3. & a_0[f,e_v] & = & -\frac{1}{3}Y^2[e_v](1-f)^2 \\ 4. & a_1[f,e_v] & = & r_1 + q_1 f^{m_1} + p_1 e_v^{k_1} \\ 5. & a_2[f,e_v] & = & r_2 + q_2 f^{m_2} + p_2 e_v^{k_2} \\ & & & + \begin{cases} 0 & \text{if} \quad f \le f_c \\ k(f-f_c)^2 & \text{if} \quad f > f_c \end{cases} \\ 6. & \mathcal{D}_\phi & = & (1+\gamma[f,e_v]\vartheta)^{-m_D} \\ 7. & \gamma[f,e_v] & = & r_\gamma + q_\gamma f^{m_\gamma} + p_\gamma e_v^{k_\gamma} \end{array}} \tag{10}$$

where r_i, m_i, k_i, $i = 1, 2, \gamma$, k, f_c and m_D are constants. In order to account for a non-associated flow rule, furthermore, a plastic potential Φ^* is introduced

$$
\begin{array}{llcl}
1. & \Phi^* & = & {}^{\bar{G}\bar{S}_{dev}}I_2 + \mathcal{H}_2^* \, \mathcal{D}_\phi^*, \quad \text{where} \\
2. & \mathcal{H}_2^* & = & a_0^*[f, e_v] + a_1^*[f, e_v] \; {}^{\bar{G}\bar{S}}I_1 + a_2^*[f, e_v] \; {}^{\bar{G}\bar{S}}I_1^2 \\
3. & a_0^*[f, e_v] & = & a_0[f, e_v] = -\frac{1}{3} Y^2[e_v](1 - f)^2 \\
4. & a_1^*[f, e_v] & = & r_1^* + q_1^* f^{m_1^*} + p_1^* e_v^{k_1^*} \\
5. & a_2^*[f, e_v] & = & r_2^* + q_2^* f^{m_2^*} + p_2^* e_v^{k_2^*} \\
 & & & + \begin{cases} 0 & \text{if} \quad f \leq f_c \\ k^*(f - f_c)^2 & \text{if} \quad f > f_c \end{cases} \\
6. & \mathcal{D}_\phi^* & = & (1 + \gamma^*[f, e_v]\vartheta)^{-m_D^*} \\
7. & \gamma^*[f, e_v] & = & r_\gamma^* + q_\gamma^* f^{m_\gamma^*} + p_\gamma^* e_v^{k_\gamma^*}
\end{array}
\tag{11}
$$

with additional constants r_i^*, m_i^*, k_i^*, $i = 1, 2, \gamma$, k^*, f_c and m_D^*. The relations (10.4-7) and (11.4-7) do account for ductile damage and plastic deformation represented by the process variables f and e_v, respectively. Material hardening for the yield function (10) is incorporated through the yield stress

$$
Y[e_v] = Y_0 + R[e_v], \qquad \text{where} \qquad R[e_v] = q\,(1 - \exp(-b e_v)) + H e_v \tag{12}
$$

and where Y_0, q, b, H are material constants.

The above plastic potential (11) has been combined in Mahnken [4, 6] with the flow rules (7) and (8), respectively. Additionally an evolution equation for the void growth rate is assumed to be proportional to the plastic volume dilatation rate, however not specified here. A rate equation for the effective plastic strain rate e_v is obtained by equating the plastic work rate with the matrix dissipation, see Mahnken [4, 6]. Furthermore, examination of the dissipation inequality (1) in Mahnken [6] yields a limit for the void growth as

$$
f[e_v] \leq \frac{Y_0}{Y_0 + R[e_v]}. \tag{13}
$$

Consequently the hardening capability influences directly the dissipation property of the macroscopic material model.

The above constitutive equations have been verified on the basis of various experimental results from the literature. For this purpose it became also necessary to determine the associated material parameters. Details on the gradient-based optimization procedure, originating from the developments in Mahnken and Stein [7], can be found in Mahnken [4, 6].

3. Representative examples

3.1. SIMULATION OF THREE CARBON STEELS

In this section the experimental results of Le Roy et al. [9] for three types of carbon steel (1090, 1045, 1015) are simulated. Details on the experimental procedure and the chemical composition of the materials are documented in Le Roy et al. [9] . The simulation of the experimental results is based on the yield function and plastic potential Eq. (10) and Eq. (11), respectively, where, by setting $\mathcal{D}_\phi = 1$ no tension-compression asymmetry is considered. Two formulations for the flow rule, i.e. the mixed-variant form and the contra-variant form specified in Eq. (7) and (8), respectively, have been compared.

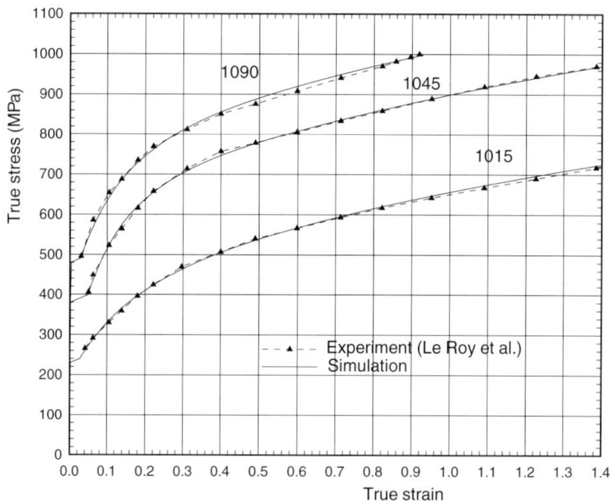

Figure 1. Simulation of three carbon steels: Comparison of experiment by Le Roy et al. [9] with simulation (mixed-variant formulation) for the true stress versus the true strain.

A comparison between experimental and simulated results for the true stress versus the true strain is plotted in Figure 1, whereas the void volume fraction against the true strain is compared in Figure 2 for all three materials. In both diagrams it becomes apparent, that the proposed model is able to give a very good agreement with the experimental data. The difference between the the mixed-variant formulation and the contra-variant formulation is negligible. This is due to the occurrence of small elastic strains which has been a major assumption for derivation of the contra-variant formulation.

3.2. SIMULATION OF QUENCHED-AND-TEMPERED MARTENSITE STEEL

This section intends to simulate the experimental results of Spitzig et al. [12] for a quenched-and-tempered martensite steel AISI 4310. The loadings of the

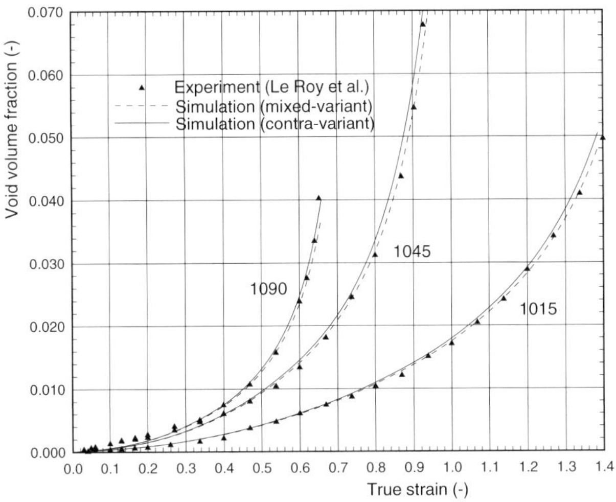

Figure 2. Simulation of three carbon steels: Comparison of experiment by Le Roy et al. [9] with simulation (mixed-variant and contra-variant formulation) for the void volume fraction versus true strain.

specimens are tension and compression where additionally, hydrostatic pressure is applied. Further details on the experimental procedure, on the composition of the steels and the preparation of the specimen including the quenching procedure are described in Spitzig et al. [12].

Due to the occurrence of small strains the simulation of the experimental data is performed within a geometrically linear formulation. Both, the yield function (10) and the plastic potential (11), respectively, are used to obtain a non-associated flow rule. Since according to the observation by Spitzig et al. [12], damage effects due to microscopic void growth are not regarded as a source for the observed effects the corresponding terms in the formulations (10) and (11) are left out. More details on the model formulation and determination of the material parameters can be found in Mahnken [5].

Figure 3 shows the results of the uniaxial tension and compression tests, and clearly illustrates the asymmetry effect, in which the flow stress is greater in uniaxial compression than in uniaxial tension. Additionally it becomes apparent that testing under hydrostatic pressure raises the stress-strain curve but does not significantly affect the work hardening characteristics. The experimental results in Figure 4 reveal a dependence of the von Mises stress $\sigma_v = \sqrt{3I_2[\boldsymbol{\sigma}^{dev}]}$ on the first invariant $I_1[\boldsymbol{\sigma}]$. Also the asymmetry in tension and compression is observed.

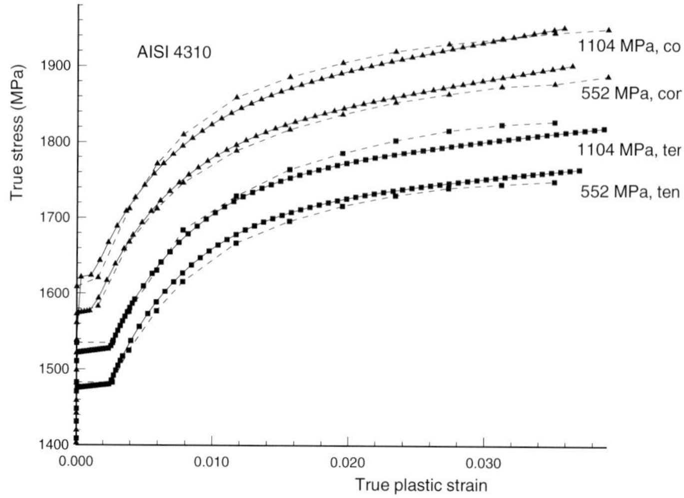

Figure 3. Simulation of martensite steel AISI 4310: Effect of hydrostatic pressure on the compressive and tension stress strain curves, comparison of experiment (dashed lines) with simulation(solid lines). The numbers on the r.h.s refer to the applied pressure.

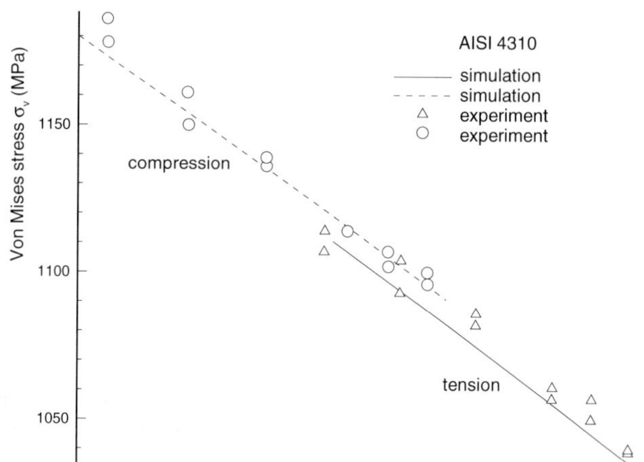

Figure 4. Simulation of martensite steel AISI 4310: von Mises stress versus first invariant in tension and compression, comparison of experiment and simulation.

The results of the model formulation are also depicted in Figure 3 and Figure 4, respectively, thus revealing its capability to simulate both the asymmetry effects and the pressure dependence of yielding.

4. Final remarks

A generalized yield function has been presented for simulating void growth effects, asymmetric tension-compression behavior and furthermore a pressure dependence of yielding. The constitutive equations are formulated within the framework of multiplicative elasto-plasticity, where two different conceptions have been considered. Whereas the algorithmic differences are rather significant, the deviations in the final results become negligible for the case of small elastic strains. From a practical point of view the authors favourize the mixed-variant approach, since it attains an identical algorithmic structure in principal directions as in the geometrically linear theory. Representative examples demonstrate the successful verification of the approach. The extension to anisotropy is regarded as the most important aspect in future work.

Appendix: Kinematics in multiplicative plasticity

Upon introducing an intermediate configuration $\bar{\mathcal{B}}$ in addition to a reference configuration $\mathcal{B}_0$ and a spatial configuration $\mathcal{B}$, the underlying concept of multiplicative elasto-plasticity assumes the decomposition of the deformation gradient $\mathbf{F}$ into elastic and plastic parts, respectively, i.e. $\mathbf{F} = \mathbf{F}_e \cdot \mathbf{F}_p$. The three associated tangent spaces $T\mathcal{B}_0, T\mathcal{B}$ and $T\bar{\mathcal{B}}$ are equipped with co-variant Riemannian metric tensors $\mathbf{G}$, $\mathbf{g}$, $\bar{\mathbf{G}}$, respectively, and analogously the associated dual spaces $T\mathcal{B}_0^*, T\mathcal{B}^*$ and $T\bar{\mathcal{B}}^*$ with contra-variant Riemannian metric tensors $\mathbf{G}^\sharp = (\mathbf{G})^{-1}$, $\mathbf{g}^\sharp = (\mathbf{g})^{-1}$, $\bar{\mathbf{G}}^\sharp = (\bar{\mathbf{G}})^{-1}$, respectively. Upon using a symbolic notation the following quantities related to the co-variant Green strain tensor $\mathbf{E} = 1/2\left(\mathbf{F}^t \cdot \mathbf{g} \cdot \mathbf{F} - \mathbf{G}\right)$ can be defined:

$$
\begin{array}{llll}
\text{1.a,b,c)} & \mathbf{E} = \frac{1}{2}\left({}^*\phi^\flat[\mathbf{g}] - \mathbf{G}\right) & \bar{\mathbf{E}} = \frac{1}{2}\left({}^*\phi_e^\flat[\mathbf{g}] - {}_*\phi_p^\flat[\mathbf{G}]\right) & \bar{\mathbf{D}} = \mathcal{L}_p^\flat\left[\bar{\mathbf{E}}\right] \\
\text{2.a,b,c)} & \mathbf{E}_e = \frac{1}{2}\left({}^*\phi^\flat[\mathbf{g}] - {}^*\phi_p^\flat[\bar{\mathbf{G}}]\right) & \bar{\mathbf{E}}_e = \frac{1}{2}\left({}^*\phi_e^\flat[\mathbf{g}] - \bar{\mathbf{G}}\right) & \bar{\mathbf{D}}_e = \mathcal{L}_p^\flat\left[\bar{\mathbf{E}}_e\right] \\
\text{3.a,b,c)} & \mathbf{E}_p = \frac{1}{2}\left({}^*\phi_p^\flat[\bar{\mathbf{G}}] - \mathbf{G}\right) & \bar{\mathbf{E}}_p = \frac{1}{2}\left(\bar{\mathbf{G}} - {}_*\phi_p^\flat[\mathbf{G}]\right) & \bar{\mathbf{D}}_p = \mathcal{L}_p^\flat\left[\bar{\mathbf{E}}_p\right]
\end{array}
\quad \text{(A.1)}
$$

where the operators ${}^*\phi^\flat$, ${}^*\phi_e^\flat$, ${}^*\phi_p^\flat$, ${}_*\phi_p^\flat$ denote full pull-back, elastic pull back, plastic pull-back and plastic push-forward operations of co-variant tensor objects, respectively, defined in Mahnken [6]. In this way all six representations a)-b) in Eq. (A.1) reveal the strain tensors as the differences of two co-variant metric tensors transformed onto the same configuration. The three representations c) in Eq. (A.1) define the total, elastic and plastic rate of deformation tensors $\bar{\mathbf{D}}$, $\bar{\mathbf{D}}_e$ and $\bar{\mathbf{D}}_p$ relative to the intermediate configuration. These are obtained by application of the plastic Lie derivative operator $\mathcal{L}_p^\flat[\bar{\bullet}] = {}_*\phi_p^\flat\left[\partial_t[{}^*\phi_p^\flat[\bar{\bullet}]]\right]$ to $\bar{\mathbf{E}}$, $\bar{\mathbf{E}}_e$ and $\bar{\mathbf{E}}_p$, where $\partial_t[\bullet]$ denotes the time derivative. The tensor $\bar{\mathbf{D}}_p$ of Eq. (A.1.3-c) can be specified as

$$
\bar{\mathbf{D}}_p = \mathcal{L}_p^\flat\left[\bar{\mathbf{E}}_p\right] = \mathrm{sym}[\bar{\mathbf{G}} \cdot \bar{\mathbf{L}}_p], \qquad \text{where } \bar{\mathbf{L}}_p := \dot{\mathbf{F}}_p \cdot \mathbf{F}_p^{-1} = -\mathbf{F}_p \cdot \dot{\mathbf{F}}_p^{-1}. \quad \text{(A.2)}
$$

Furthermore the elastic right Cauchy-Green tensor $\bar{\mathbf{C}}_e$ is defined as

$$\bar{\mathbf{C}}_e := {}^*\phi_e^\flat[\mathbf{g}] = \mathbf{F}_e^t \cdot \mathbf{g} \cdot \mathbf{F}_e \tag{A.3}$$

which occurs in Eq. (A.1.1-b) and Eq. (A.1.2-b). Lastly we define all three *basic invariants* of a contra-variant second order tensor $\bar{\mathbf{A}}^\sharp = \bar{A}^{ij}\bar{\mathbf{G}}_i \otimes \bar{\mathbf{G}}_j$, related to the actual configuration, as dual pairings of mixed-variant tensors

$${}^{\bar{G}\bar{A}^\sharp}I_i \quad := \quad \frac{1}{i}\bar{\mathbf{1}} : \left(\bar{\mathbf{G}} \cdot \bar{\mathbf{A}}^\sharp\right)^i, \quad i = 1, 2, 3. \tag{A.4}$$

Here $\bar{\mathbf{G}} = G_{ij}\bar{\mathbf{G}}^i \otimes \bar{\mathbf{G}}^j$ is a co-variant metric tensor, $\bar{\mathbf{1}} = \bar{\mathbf{G}}_i \otimes \bar{\mathbf{G}}^i$ is a (mixed-variant) second order unit tensor, and $\mathbf{G}_i$ and $\mathbf{G}^i$ are basis vectors.

References

[1] Ehlers, W. (1995). A single-surface yield function for geomaterials. *Archive of Applied Mechanics*, **65**, 246–259.

[2] Giessen, E. van der and Kollmann, F.G. (1996). On Mathematical Aspects of Dual Variables in Continuum Mechanics. Part 1: Mathematical Principles. *ZAMM*, **76**, 8, 447–462.

[3] Lührs, G., Hartmann, S. and Haupt, P. (1997). On the numerical treatment of finite deformations in elasto-viscoplasticity. *Comp. Meths. Appl. Mech. Eng.*, **144**, 1–21.

[4] Mahnken, R. (2001). Theoretical, numerical and identification aspects of a new model class for ductile damage. Accepted for publication in: *Int. J. Plast.*, **15**, 1111–1137.

[5] Mahnken, R. (2001). Strength difference in compression and tension and pressure dependence of yielding in elasto-plasticity. *Comp. Meths. Appl. Mech. Eng.*, **190**, 5057–5080.

[6] Mahnken, R. (2001). Void growth in finite deformation elasto-plasticity due to hydrostatic stress states. Submitted to: *Comp. Meths. Appl. Mech. Eng.*, **19**, 5057–5080.

[7] Mahnken R. and Stein, E. (1997). Parameter Identification for Finite Deformation Elasto-Plasticity in Principal Directions. *Comp. Meths. Appl. Mech. Eng.*, **147**, 17–39.

[8] Miehe, C. (1994). On the Representation of Prandlt-Reuss Tensors within the Framework of Multiplicative Elastoplasticity. *Int. J. Plast.*, Vol. 10, **6**, 609–621.

[9] Le Roy, G., Embury, J.D., Edward. G. and Ashby, M.F. (1981). A model of ductile fracture based on the nucleation and growth of voids. *Acta Metallurgica*, **29**, 1509–1522.

[10] Simo, J.C. and Miehe, C. (1992). Associative Coupled Thermoplasticity at Finite Strains: Formulation, Numerical Analysis and Implementation. *Comp. Meths. Appl. Mech. Eng.*, **98**, 41–104.

[11] Simo, J.C. (1992). Algorithms for static and dynamic multiplicative plasticity that preserve the classical return mapping schemes of the infinitesimal theory. *Comp. Meths. Appl. Mech. Eng.*, **99**, 61–112.

[12] Spitzig, W.A., Sober, R.J. and Richmond, O. (1975). Pressure dependence of yielding and associated volume expansion in tempered martensite. *Acta Metallurgica*, **23**, 885–893.

On the Construction of Polyconvex Anisotropic Free Energy Functions

J. Schröder
Institut für Mechanik, Universität Essen
Universitätsstraße 15, 45117 Essen, Germany
j.schroeder@uni-essen.de

P. Neff
Fachbereich Mathematik, Technische Universität Darmstadt
Schloßgartenstraße 7, 64289 Darmstadt, Germany
neff@mathematik.tu-darmstadt.de

Abstract: In this paper we discuss the formulation of polyconvex, anisotropic free energy functions within the framework of the invariant theory. In order to guarantee the existence of solutions in large strain elasticity, the free energy function has to be quasiconvex. This condition is rather complicated to handle. Therefore, polyconvex functions are usually considered which are always quasiconvex. For isotropic materials, there exists a wide range of constitutive functions which satisfy the polyconvexity requirement. However, to the authors knowledge, there exists no systematic treatment of anisotropic, polyconvex free energies in the literature. The goal of this contribution is the construction of transversely isotropic free energy functions which automatically satisfy the polyconvexity condition.

Keywords: Anisotropic hyperelasticity, polyconvexity.

1. Introduction

The existence of minimizers of the energy in hyperelasticity is based on the concept of quasiconvexity, introduced by Morrey [7]. This integral inequality is rather complicated to handle. Thus, the sufficient condition for quasiconvexity, the polyconvexity condition in the sense of Ball [1], is a more important concept for practical applications, see also Ciarlet [4] and Dacorogna [5]. In the case of isotropy there exist some models which satisfy this condition. The present work is concerned with the construction of polyconvex anisotropic free energy functions. The main goal of this contribution is the representation of the governing constitutive equations in the framework of the invariant theory which automatically fulfill the polyconvexity condition. For an introduction to the invariant theory see Smith and Rivlin [10] , Wang [13, 14], Spencer [12], Smith [11] and the references therein. The main idea is the introduction of additional argument tensors, the so-called structural tensors. The free energies

C. Miehe (ed.), *IUTAM Symposium on Computational Mechanics of Solid Materials at Large Strains*, 171-180.

and the anisotropic stress response functions are represented by scalar-valued and tensor-valued isotropic tensor functions in terms of principal and mixed invariants of a suitable deformation tensor and the structural tensors. For the construction of anisotropic constitutive equations see e.g. Boehler [3] and Betten [2]. In the following we focus on the construction of transversely isotropic polynominal invariants which fulfill the polyconvexity condition a priori and we present a representative numerical example.

2. Foundations

The body of interest in the reference configuration is denoted by $\mathcal{B} \subset \mathbb{R}^3$, parameterized in $\boldsymbol{X}$, and the current configuration by $\mathcal{S} \subset \mathbb{R}^3$, parameterized in $\boldsymbol{x}$. The nonlinear deformation map $\varphi_t : \mathcal{B} \to \mathcal{S}$ at time $t \in \mathbb{R}_+$ maps points $\boldsymbol{X} \in \mathcal{B}$ onto points $\boldsymbol{x} \in \mathcal{S}$. The deformation gradient $\boldsymbol{F}$ is defined by $\boldsymbol{F}(\boldsymbol{X}) := \nabla\varphi_t(\boldsymbol{X})$ with the Jacobian $J(\boldsymbol{X}) := \det \boldsymbol{F}(\boldsymbol{X}) > 0$. An important strain measure, the right Cauchy-Green tensor, is defined by $\boldsymbol{C} := \boldsymbol{F}^T\boldsymbol{F}$. Let $\boldsymbol{N}\ dA$ and $\boldsymbol{n}\ da$ denote the infinitesimal area vectors and dV and dv denote the infinitesimal volume elements defined in the reference and the current configuration, respectively, then $\boldsymbol{n}\ da = \mathrm{Cof}[\boldsymbol{F}]\boldsymbol{N}\ dA$ and $dv = \det[\boldsymbol{F}]\ dV$ hold. It should be mentioned that the argument $(\boldsymbol{F}, \mathrm{Adj}\boldsymbol{F}, \det \boldsymbol{F})$, with $\mathrm{Adj}\boldsymbol{F} = (\mathrm{Cof}\boldsymbol{F})^T$, plays an important role in the definition of polyconvexity.

We consider hyperelastic materials which postulate the existence of a so-called Helmholtz free-energy function $\psi = \hat{\psi}(\boldsymbol{F}, \bullet)$. The argument $(\bullet)$ denotes additional tensor arguments, which characterize the anisotropy of the material. The principle of material frame indifference leads to well known reduced constitutive equations for the second Piola-Kirchhoff stresses, e.g. $\boldsymbol{S} = 2\partial_{\boldsymbol{C}}\hat{\psi}(\boldsymbol{C})$. In the case of anisotropy we introduce a material symmetry group $\mathcal{G}_k \subset \mathrm{SO}(3)$ with respect to a local reference configuration, which characterizes the anisotropy class of the material. $\mathrm{SO}(3)$ characterizes the special orthogonal group. The concept of material symmetry leads to the invariance condition

$$\hat{\psi}(\boldsymbol{F}\boldsymbol{Q}) = \hat{\psi}(\boldsymbol{F}) \quad \forall\, \boldsymbol{Q} \in \mathcal{G}_k, \boldsymbol{F}\ . \tag{1}$$

We say that the function ψ in Equation (1) is a $\mathcal{G}_k$-invariant function.

3. Representation of anisotropic tensor functions

The main idea to fulfill the above mentioned invariance condition in a coordinate-free representation is based on the extension of $\mathcal{G}_k$-invariant functions into functions which are invariant under a larger group, here the special orthogonal group. This implies that it is in principal possible to transform an anisotropic constitutive function into an isotropic function through the application of some tensors, the so-called structural tensors, which reflect the symmetry group of the considered material. The symmetry group of a material is defined by (1). Here we consider transverse isotropic materials which can be characterized by one second-order tensor $\boldsymbol{M}$ defined in the reference configuration. Let $\mathcal{G}_M$ be the invariance group of the structural tensor, i.e

$$\mathcal{G}_M := \{\boldsymbol{Q} \in \mathrm{SO}(3),\ \boldsymbol{Q}^T\boldsymbol{M}\boldsymbol{Q} = \boldsymbol{M}\}\ . \tag{2}$$

If $\mathcal{G}_M \equiv \mathcal{G}_K$, where $\mathcal{G}_K$ is defined by (1), then the invariance group preserves the characteristics of the anisotropic solid. In this case we can transform (1) into a function which is invariant under the special orthogonal group. This leads to a scalar-valued isotropic tensor function of the form

$$\psi = \hat{\psi}(\boldsymbol{C}, \boldsymbol{M}) = \hat{\psi}(\boldsymbol{Q}^T\boldsymbol{C}\boldsymbol{Q}, \boldsymbol{Q}^T\boldsymbol{M}\boldsymbol{Q})\ \forall\, \boldsymbol{Q}\ \in\ \mathrm{SO}(3)\ . \tag{3}$$

It should be noted, that $\boldsymbol{Q}^T\boldsymbol{S}(\boldsymbol{C}, \boldsymbol{M})\boldsymbol{Q} = \boldsymbol{S}(\boldsymbol{Q}^T\boldsymbol{C}\boldsymbol{Q}, \boldsymbol{M})$ holds only for transformations $\boldsymbol{Q} \in \mathcal{G}_k$ and only for these transformations, thus the structural tensor $\boldsymbol{M}$ characterizes the material symmetry as pointed out above. Transverse isotropy is characterized by one preferred unit direction $\boldsymbol{a}$, and we set $||\boldsymbol{a}|| = 1$. The material symmetry group is defined by

$$\mathcal{G}_{ti} := \{\boldsymbol{I};\ \boldsymbol{Q}(\alpha, \boldsymbol{a})\ |\ 0 < \alpha < 2\,\pi\}\ , \tag{4}$$

where $\boldsymbol{Q}(\alpha, \boldsymbol{a})$ are all rotations about the $\boldsymbol{a}$-axis. The structural tensor $\boldsymbol{M}$, whose invariance group preserves the material symmetry group $\mathcal{G}_{ti}$, is given by $\boldsymbol{M} := \boldsymbol{a} \otimes \boldsymbol{a}$. The integrity bases consist of the so-called principal invariants and the mixed invariants. For the considered symmetry class a possible basis is given by

$$\mathcal{P} := \{I_1, I_2, I_3, I_4, I_5\}\ . \tag{5}$$

The principal and mixed invariants entering (5) are as follows

$$I_1 = \mathrm{tr}\,\boldsymbol{C},\ I_2 = \mathrm{tr}[\mathrm{Cof}\boldsymbol{C}],\ I_3 = \mathrm{det}\boldsymbol{C},\ I_4 = \mathrm{tr}[\boldsymbol{C}\boldsymbol{M}],\ I_5 = \mathrm{tr}[\boldsymbol{C}^2\boldsymbol{M}]\ . \tag{6}$$

For the free energy function we assume the general form $\psi = \hat{\psi}(I_i|i = 1\ldots5) + c$, where we have introduced the constant $c \in \mathbb{R}$ in order to fulfill the non-essential normalization condition $\psi(\boldsymbol{1}, \boldsymbol{M}) = 0$.

4. Representation of polyconvex free energy functions

As mentioned before we focus on polyconvex functions. Let $W \in C^2(M^{3\times3}, \mathbb{R})$ be a scalar-valued energy density.

Definition 4.1 (Polyconvexity) $\boldsymbol{F} \mapsto W(\boldsymbol{F})$ *is polyconvex if and only if there exists a function* $P : M^{3\times3} \times M^{3\times3} \times \mathbb{R} \mapsto \mathbb{R}$ *such that*

$$W(\boldsymbol{F}) = P(\boldsymbol{F}, \mathrm{Adj}\boldsymbol{F}, \det \boldsymbol{F})$$

and $\mathbb{R}^{19} \mapsto \mathbb{R},\ (X, Y, Z) \mapsto P(X, Y, Z)$ *is convex for all points* $\boldsymbol{X} \in \mathbb{R}^3$.

A consequence of the above definition for a more restrictive class of energy densities is the following corollary for additive polyconvex functions: Let $W(\boldsymbol{F}) = W_1(\boldsymbol{F}) + W_2(\mathrm{Adj}\boldsymbol{F}) + W_3(\det \boldsymbol{F})$. If $W_i,\ i = 1,2$ are convex in the associated variable respectively and $W_3 : \mathbb{R}^+ \mapsto \mathbb{R}$ is convex in the associated variable as well, then W is altogether polyconvex. This will be one of our primary tools in constructing polyconvex strain energy functions.

Definition 4.2 (Ellipticity) *We say that the elastic free energy* $W(\boldsymbol{F}) = \psi(\boldsymbol{C}) \in C^2(M^{3\times 3}, \mathbb{R})$ *leads to a uniformly elliptic equilibrium system whenever the so-called uniform Legendre-Hadamard condition*

$$\exists\, c^+ > 0\, \forall\, \boldsymbol{F} \in M^{3\times 3} : \forall\, \xi, \eta \in \mathbb{R}^3 : D_F^2 W(\boldsymbol{F}).(\xi \otimes \eta, \xi \otimes \eta) \geq c^+ \cdot \|\xi\|^2 \|\eta\|^2$$

holds. W *is elliptic if and only if the Legendre-Hadamard condition*

$$\forall\, \boldsymbol{F} \in M^{3\times 3} : \forall\, \xi, \eta \in \mathbb{R}^3 : \quad D_F^2 W(\boldsymbol{F}).(\xi \otimes \eta, \xi \otimes \eta) \geq 0$$

holds. We say that W *is rank-one convex if the function* $f : \mathbb{R} \mapsto \mathbb{R},\ f(t) = W(\boldsymbol{F} + t \cdot (\xi \otimes \eta))$ *is convex for all* $\boldsymbol{F} \in M^{3\times 3}$ *and all* $\xi, \eta \in \mathbb{R}^3$.

To obtain various strain energy terms we assume an additive structure, that means $\psi = \sum_{j=1}^{n} \psi_j$. The stresses are given by the chain rule as follows

$$\boldsymbol{S} = 2 \frac{\partial \psi}{\partial \boldsymbol{C}} = 2 \sum_{j=1}^{n} \sum_{i=1}^{5} \frac{\partial \psi_j}{\partial I_i} \frac{\partial I_i}{\partial \boldsymbol{C}} \,. \tag{7}$$

The tensor generators $\partial_C I_i$ are independent of the specific form of the free energy function. They only depend on the symmetry of the material characterized by the introduced structural tensors.

5. Polyconvex anisotropic free energy functions

The free energy function consists of several isotropic ψ_i^{iso} and anisotropic terms ψ_j^{ti} which leads to $\psi = \sum_i \psi_i^{iso} + \sum_j \psi_j^{ti}$. For the isotropic ones we choose some well-known formulations from the literature, see e.g. Ciarlet [4]. In this context see also Hartmann and Neff [6] who proposed several new isotropic polyconvex functions. The anisotropic terms are discussed in the following, see also Schröder and Neff [9]. It should be noted that direct extensions of the small strain theory to large strain formulations by replacing the linear strain tensor with the Green-Lagrange strain tensor $\boldsymbol{E} := \frac{1}{2}(\boldsymbol{C} - \boldsymbol{1})$ are a priori not polyconvex, see Raoult [8].

First we analyze the simple quadratic polynominal function in $\boldsymbol{C}$

$$\psi_C := c_1\, I_1^2 + c_2\, J_2 + c_3\, I_1\, I_4 + c_4\, I_4^2 + c_5\, I_5 + c_6\, f_G + c_7\, f_M\,, \tag{8}$$

with $J_2 := I_1^2 - 2I_2 = \mathrm{tr}\,[\boldsymbol{C}^2]$. Here f_G and f_M are non-specified functions which we introduce in order to fulfill the condition of a stress-free reference configuration with respect to the isotropic and anisotropic contributions. The first and second terms in (8) are polyconvex; the proof is omitted here. Let us now analyze the remaining anisotropic terms.

The quadratic function $I_1 I_4$ in the variable $\boldsymbol{C}$ does not fulfill the polyconvexity condition, i.e. the expression

$$\boldsymbol{F} \mapsto \mathrm{tr}\,(\boldsymbol{F}^T\boldsymbol{F}\boldsymbol{M})\mathrm{tr}\,(\boldsymbol{F}^T\boldsymbol{F}) = \mathrm{tr}\,[\boldsymbol{C}\boldsymbol{M}]\mathrm{tr}\,\boldsymbol{C} = I_1\, I_4 \tag{9}$$

is not polyconvex, since it is not even elliptic. The outline of the proof is as follows: Equation (9) can be expressed in the form

$$\mathrm{tr}\,[\boldsymbol{F}^T\boldsymbol{F}\boldsymbol{M}]\mathrm{tr}\,[\boldsymbol{F}^T\boldsymbol{F}] = \|\boldsymbol{F}\|^2\|\boldsymbol{F}\boldsymbol{a}\|^2_{\mathrm{R}^3}\,.$$

Calculating the second differential with respect to $\boldsymbol{F}$ yields

$$\begin{aligned} &D_F^2\left(\|\boldsymbol{F}\|^2\cdot\|\boldsymbol{F}\boldsymbol{a}\|^2_{\mathrm{R}^3}\right).(\boldsymbol{H},\boldsymbol{H}) = \\ &\qquad 8\langle\boldsymbol{F},\boldsymbol{H}\rangle\langle\boldsymbol{F}\boldsymbol{a},\boldsymbol{H}\boldsymbol{a}\rangle_{\mathrm{R}^3} + 2\|\boldsymbol{F}\boldsymbol{a}\|^2_{\mathrm{R}^3}\|\boldsymbol{H}\|^2 + 2\|\boldsymbol{F}\|^2\|\boldsymbol{H}\boldsymbol{a}\|^2_{\mathrm{R}^3}. \end{aligned} \tag{10}$$

We see that this expression is in general non-positive (take $\boldsymbol{F}, \boldsymbol{H}$ in diagonal form), which excludes convexity. Furthermore, it is possible to show the non-ellipticity as well. Evaluating the second derivative of $I_1\, I_4$ for some special choice of $\boldsymbol{F}$ leads to

$$D_F^2\,(I_1\, I_4) = D_F^2\left(\|\boldsymbol{F}\|^2\cdot\|\boldsymbol{F}\boldsymbol{a}\|^2\right).(\xi\otimes\eta,\xi\otimes\eta) < 0.$$

Powers of the basic invariant I_4 are polynominal functions of the form

$$\boldsymbol{F} \mapsto (\mathrm{tr}\,[\boldsymbol{F}^T\boldsymbol{F}\boldsymbol{M}])^k = (\mathrm{tr}\,[\boldsymbol{C}\boldsymbol{M}])^k = I_4^k \qquad \text{with } k \geq 1 \tag{11}$$

which are polyconvex. This statement can be shown by checking the condition $D_F^2\left(\langle\boldsymbol{F},\boldsymbol{F}\boldsymbol{M}\rangle^k\right).(\boldsymbol{H},\boldsymbol{H}) \geq 0$, see Schröder and Neff [9].

The fifth term in (8) is characterized by the nonlinear mapping

$$\boldsymbol{F} \mapsto \mathrm{tr}\,[\boldsymbol{F}^T\boldsymbol{F}\boldsymbol{F}^T\boldsymbol{F}\boldsymbol{M}] = \mathrm{tr}\,[\boldsymbol{C}^2\boldsymbol{M}] = I_5\,. \tag{12}$$

It can be shown by evaluating the ellipticity condition with

$$\boldsymbol{F} = \begin{pmatrix} 1 & 0 & 0 \\ 0 & 2 & 0 \\ 0 & 0 & \frac{1}{n} \end{pmatrix}, \quad \xi = \begin{pmatrix} 0 \\ 0 \\ 1 \end{pmatrix}, \quad \boldsymbol{a} = \begin{pmatrix} \alpha \\ \alpha \\ \alpha \end{pmatrix}, \quad \eta = \begin{pmatrix} \alpha \\ -\alpha \\ \alpha \end{pmatrix}$$

and $\alpha = \frac{1}{\sqrt{3}}$ that for some reasonable n the expression

$$D_F^2\,(I_5) = D_F^2\left(\|\boldsymbol{F}^T\boldsymbol{F}\boldsymbol{a}\|^2_{\mathrm{R}^3}\right).(\xi\otimes\eta,\xi\otimes\eta) \leq (-2+\frac{2}{3n^2})\frac{1}{3}+\frac{4}{n} < 0$$

holds. Thus the considered term is not elliptic and hence not polyconvex.

We can conclude that the terms I_5 and $I_1 I_4$ have to be excluded as ansatz functions for the free energy. In order to construct further quadratic ansatz functions we recall the fact that $\mathrm{Cof}[\boldsymbol{C}]$ is a quadratic function in the right Cauchy-Green tensor. Furthermore, it seems reasonable to construct a polynominal mixed invariant, which reflects the deformation of a preferred area element of an infinitesimal volume element.

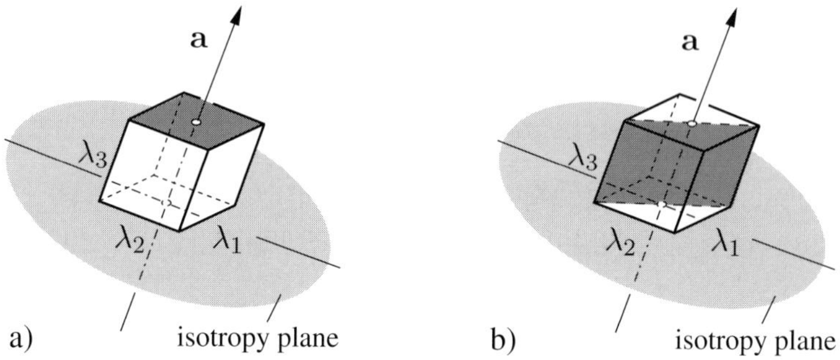

Figure 1. Geometric interpretation of the the polynominal invariants a) K_1 and b) K_2.

Based on this geometric motivation we start with the Cayley-Hamilton theorem and multiply the characteristic polynominal in $\boldsymbol{C}$ with $\boldsymbol{C}^{-1}\boldsymbol{M}$. This leads with $\mathrm{Cof}\boldsymbol{C} = \mathrm{Adj}\boldsymbol{C}$ to the expression

$$\mathrm{Cof}[\boldsymbol{C}]\boldsymbol{M} = \boldsymbol{C}^2\boldsymbol{M} - I_1\boldsymbol{C}\boldsymbol{M} + I_2\boldsymbol{M} \ . \tag{13}$$

The trace of Equation (13) is the polyconvex polynominal invariant

$$K_1 := \mathrm{tr}\left[\mathrm{Cof}[\boldsymbol{C}]\boldsymbol{M}\right] = I_5 - I_1\, I_4 + I_2 \ . \tag{14}$$

The proof of the convexity of the powers of K_1 is straightforward by replacing $\boldsymbol{F}$ with $\mathrm{Cof}\boldsymbol{F}$ in (11). Expressing the invariant K_1 in the form

$$K_1 = \mathrm{tr}\left[\mathrm{Cof}[\boldsymbol{C}]\boldsymbol{M}\right] = \mathrm{Cof}[\boldsymbol{F}^T\boldsymbol{F}] : \boldsymbol{a}\otimes\boldsymbol{a} = ||\mathrm{Cof}[\boldsymbol{F}]\boldsymbol{a}||^2 \ , \tag{15}$$

we can give a rather simple geometric interpretation of this polynominal invariant. $\sqrt{K_1} = ||\mathrm{Cof}[\boldsymbol{F}]\boldsymbol{a}||$ controls the deformation of the area element with unit normal $\boldsymbol{a}$. Consider the deformation of a unit cube with $\boldsymbol{F} = \mathrm{diag}(\lambda_1, \lambda_2, \lambda_3)$ where λ_3 represents the stretch in preferred direction, then we arrive at $\sqrt{K_1} = \lambda_1\lambda_2$. An illustration of this simple example is given in Figure 1a where $\lambda_1\lambda_2$ represents the shaded area.

With the same physical motivation as before we are now constructing another polyconvex polynominal invariant by

$$K_2 := \mathrm{tr}\left[\mathrm{Cof}[\boldsymbol{C}](\boldsymbol{1} - \boldsymbol{M})\right] = I_1\, I_4 - I_5 \ . \tag{16}$$

The proof is omitted here. After some algebraic manipulations we obtain

$$K_2 = \mathrm{tr}\,[\mathrm{Cof}[\boldsymbol{C}]\boldsymbol{D}] = ||\mathrm{Cof}[\boldsymbol{F}]||^2 - ||\mathrm{Cof}[\boldsymbol{F}]\boldsymbol{a}||^2 \,. \tag{17}$$

Thus $\sqrt{K_2}$ controls the deformation of an area element with a normal in the isotropy plane, i.e. with a normal perpendicular to $\boldsymbol{a}$. Consider again the deformation of a unit cube with $\boldsymbol{F} = \mathrm{diag}(\lambda_1, \lambda_2, \lambda_3)$ with λ_3 being the stretch in the preferred direction. Thus λ_1, λ_2 represent the stretches of the edges of the unit cube in the isotropy plane and we arrive at $\sqrt{K_2} = \lambda_3\sqrt{\lambda_2^2 + \lambda_1^2}$. Figure 1b shows the controlled area element, which is shaded.

6. Numerical example: 3D-analysis of a perforated plate

In this section we analyze a three-dimensional perforated plate with centered hole and discuss a specific free energy function in detail. The dimensions of the plate are width/ height/ thickness $=20/20/2$ and the radius is given by 2 units. The system is discretized with 3200 eight-noded standard displacement elements, where we have used 4 elements in thickness direction, see Figure 2. The chosen free energy function ψ consists of seven additive terms

$$\psi = \sum_{j=1}^{7} \psi_j(I_1, I_2, I_3, I_4, I_5) \,. \tag{18}$$

Here we have chosen a minimum number of seven ansatz functions five are necessary to capture the independent constants of transversely isotropic materials and two are necessary in order to fulfill the condition of a stress free reference configuration. In this context see also the interpretation of Equation (8). The first ansatz function

$$\psi_1(I_1) = \alpha_1 \,\mathrm{tr}\,\boldsymbol{C} \tag{19}$$

is linear in the invariant I_1. As such it only effects the second Piola-Kirchhoff stresses, i.e. it leads to a positive volumetric stress contribution. The remaining polyconvex ansatz functions are

$$\left.\begin{aligned}
\psi_2(I_3) &= \alpha_2\; 1/\det \boldsymbol{C} \\
\psi_3(I_4) &= \alpha_3\; (\mathrm{tr}\,[\boldsymbol{MC}])^2 \\
\psi_4(I_1, I_2, I_3, I_4, I_5) &= \alpha_4\; \mathrm{tr}\,[\boldsymbol{M}\mathrm{Cof}\boldsymbol{C}]/(\det \boldsymbol{C})^{1/3} \\
\psi_5(I_1, I_4, I_5) &= \alpha_5\; (\mathrm{tr}\,[(\boldsymbol{1} - \boldsymbol{M})\mathrm{Cof}\boldsymbol{C}])^2 \\
\psi_6(I_1, I_3, I_4, I_5) &= \alpha_6\; \mathrm{tr}\,[(\boldsymbol{1} - \boldsymbol{M})\boldsymbol{C}]/(\det \boldsymbol{C})^{1/3} \\
\psi_7(I_1, I_3, I_4, I_5) &= \alpha_7\; (\mathrm{tr}\,[(\boldsymbol{1} - \boldsymbol{M})\boldsymbol{C}])^2/(\det \boldsymbol{C})^{1/3}
\end{aligned}\right\} \,. \tag{20}$$

Here we have used more complicated polyconvex ansatz functions than those derived in the previous section; the proofs of the polyconvexity are given in

Schröder and Neff [9]. With the free energy function defined in Equation (18) with (19) and (20) we arrive with the requirement of a stress free reference configuration $\boldsymbol{S}(\mathbf{1},\boldsymbol{M}) = \mathbf{0}$ at

$$\begin{aligned}(2\,\alpha_1 + \frac{4}{3}\,\alpha_4 + 8\,\alpha_5 + \frac{2}{3}\,\alpha_6 + \frac{16}{3}\,\alpha_7 - 2\,\alpha_2)\,\mathbf{1}\\ +(4\,\alpha_3 - 2\,\alpha_4 + 8\,\alpha_5 - 2\,\alpha_6 - 8\,\alpha_7)\,\boldsymbol{M} = \mathbf{0}\end{aligned} \tag{21}$$

The remaining equations for the determination of the material parameters are given by the values of the moduli $\mathbb{C}_0$ at the reference configuration. They are obtained by linearization of the stress response functions (7) at$\boldsymbol{C} = \mathbf{1}$. Thus we obtain with the Green-Lagrange strain tensor the equation

$$\mathrm{Lin}[\boldsymbol{S}] = \boldsymbol{S}(\mathbf{1},\boldsymbol{M}) + \mathbb{C}_0 : \mathrm{Lin}[\boldsymbol{E}] \quad \text{with } \mathbb{C}_0 := 2\partial_C \boldsymbol{S}|_{\mathbf{1}} = 4\partial^2_{CC}\psi|_{\mathbf{1}} \,. \tag{22}$$

The moduli appear in the form

$$\begin{aligned}\mathbb{C}_0^{ABCD} &= c_1\,\boldsymbol{G}^{AB}\boldsymbol{G}^{CD} + c_2\,[\boldsymbol{G}^{AB}\boldsymbol{G}^{CD} - \boldsymbol{G}^{AC}\boldsymbol{G}^{BD}]\\ &+ c_3\,\boldsymbol{M}^{AB}\boldsymbol{M}^{CD} + c_4\,[\boldsymbol{G}^{AB}\boldsymbol{M}^{CD} + \boldsymbol{M}^{AB}\boldsymbol{G}^{CD}]\\ &+ c_5\,[\boldsymbol{G}^{AC}\boldsymbol{M}^{BD} + \boldsymbol{M}^{AC}\boldsymbol{G}^{BD}]\,,\end{aligned} \tag{23}$$

where $\boldsymbol{G}^{AB}$ denotes the index representation of the contravariant metric tensors $\boldsymbol{G}^{-1}$. In (23) the $c_i|i = 1\ldots 5$ are abbreviations of sums of the derivatives of the free energy function with respect to the elements of the polynominal basis, e.g.

$$c_2 := 4\sum_{j=1}^{n}\left(\frac{\partial\psi_j}{\partial I_2} + \frac{\partial\psi_j}{\partial I_3}\right) \,. \tag{24}$$

For the considered free energy function the parameters are given by

$$\left.\begin{aligned}c_1 &= 8\,\alpha_5 + \frac{40}{9}\,\alpha_7 + 8\,\alpha_2 - \frac{8}{9}\,\alpha_4 + \frac{8}{9}\,\alpha_6\\ c_2 &= \frac{8}{3}\,\alpha_4 - 4\,\alpha_2 - \frac{8}{3}\,\alpha_6 - \frac{16}{3}\,\alpha_7\\ c_3 &= 8\,\alpha_3 + 8\,\alpha_5 + 8\,\alpha_7\\ c_4 &= -\frac{8}{3}\,\alpha_4 + 24\,\alpha_5 - \frac{8}{3}\,\alpha_7 + \frac{4}{3}\,\alpha_6\\ c_5 &= 4\,\alpha_4 - 16\,\alpha_5\end{aligned}\right\} \,. \tag{25}$$

Let the X_3-axis be the preferred direction, i.e. $\boldsymbol{a} = (0,0,1)^T$. Then we can identify the coefficients in (23) with the values of the linearized moduli with respect to this local coordinate system. We obtain

$$\left.\begin{aligned}\mathbb{C}_0^{1111} = \mathbb{C}_0^{2222} = c_1, \quad \mathbb{C}_0^{1122} = c_1 + c_2\\ \mathbb{C}_0^{3333} = c_1 + c_3 + 2(c_4 + c_5)\\ \mathbb{C}_0^{2323} = \mathbb{C}_0^{1313} = (c_5 - c_2)/2, \quad \mathbb{C}_0^{1133} = \mathbb{C}_0^{2233} = c_1 + c_2 + c_4\end{aligned}\right\} \,. \tag{26}$$

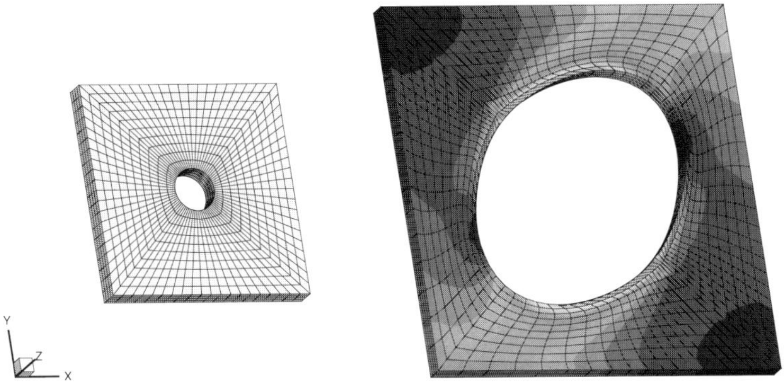

Figure 2. Perforated plate: reference configuration (left) and final configuration (right).

For the simulation of the boundary value problem we choose the values

$$\left.\begin{array}{c}\mathbb{C}_0^{1111} = 1215.0, \quad \mathbb{C}_0^{1122} = 445.0, \quad \mathbb{C}_0^{1133} = 581.0, \\ \mathbb{C}_0^{3333} = 2900.0, \quad \mathbb{C}_0^{1313} = 500.0\end{array}\right\} . \tag{27}$$

From (21), (26) and (25) we get the material parameters of the polyconvex free energy terms in the invariant setting to

$$\left.\begin{array}{c}\alpha_1 \approx 2.6875, \quad \alpha_2 \approx 136.41, \quad \alpha_3 \approx 112.21, \quad \alpha_4 \approx 80.393 \\ \alpha_5 \approx 5.7233, \quad \alpha_6 \approx 162.15, \quad \alpha_7 \approx 1.1920\end{array}\right\} . \tag{28}$$

The plate is subjected to a biaxial tension in the (x, y)-plane and the displacements of the outer boundary are fixed in z-direction. For the simulation the preferred direction was set to $\boldsymbol{a} = (1, 1, 0)^T/\sqrt{2}$ and the boundary displacements were increased in ten steps until the outer boundary was elongated up to a final displacement of 10 units on each side in horizontal and vertical direction. Thus the final dimensions of the plate in the (x, y)-plane are width/ height = $40/40$. The undeformed state and the final state are depicted in Figure 2. Furthermore, the right part of Figure 2 shows the qualitative stress distribution of the S_{11} stress components, whereby high stress-components are plotted in a light color and low stresses in a dark color. Due to the effect that the preferred direction is not aligned with the symmetry planes of the plate we observe a rotation of the principal axis of the finally more or less elliptical hole, as expected.

7. Conclusion

In this work we have proposed the formulation of polyconvex transversely isotropic free energy functions in the framework of the invariant theory. For the free energy we have assumed an additive structure, where each individual term has to fulfill the polyconvexity condition in the sense of Ball in order to guarantee the existence of minimizers of variational principles in finite elasticity. We have constructed several anisotropic polynominal invariants which allow the approximation of linearized transverse isotropic elasticity moduli in the reference configuration and fulfill the condition of a stress free reference configuration. Furthermore, we have shown that often used functions like$I_1 I_4$ and I_5 are not elliptic and hence not polyconvex, and therefore they have to be excluded in this form.

References

[1] Ball, J.M. (1977). Convexity Conditions and Existence Theorems in Non-Linear Elasticity. *Archive of Rational Mechanics and Analysis*, **63**, 337–403.

[2] Betten, J. (1987). Formulation of Anisotropic Constitutive Equations. In: *Applications of Tensor Functions in Solid Mechanics*, J.P. Boehler (edt.), CISM Course No. 292, Springer.

[3] Boehler, J.P. (1987) Introduction to the Invariant Formulation of Anisotropic Constitutive Equations. In: *Applications of Tensor Functions in Solid Mechanics*, J.P. Boehler (edt.), CISM Course No. 292, Springer-Verlag.

[4] Ciarlet, P.G. (1988). *Mathematical Elasticity, Volume 1: Three Dimensional Elasticity*, Elsevier Science Publishers B.V., North-Holland.

[5] Dacorogna, B. (1989). Direct Methods in the Calculus of Variations.*Applied Mathematical Science*, **78**, Springer-Verlag.

[6] Hartmann, S. and Neff, P. (2001). Existence Theory for a Modified Polyconvex Hyperelastic Relation of Generalized Polynomial-Type in the Case of Nearly-Incompressibility. Submitted to *Journal of the Mechanics and Physics of Solids.*

[7] Morrey, C.B. (1952). Quasi-Convexity and the Lower Semicontinuity of Multiple Integrals. *Pacific Journal Math.*, **Vol. 2**, 25–53.

[8] Raoult, A. (1986). Non-Polyconvexity of the Stored Energy Function of a St.-Venant-Kirchhoff Material. *Aplikace Matematiky*, Vol. 6, 417–419.

[9] Schröder, J. and Neff, P. (2001). Invariant Formulation of Hyperelastic Transverse Isotropy Based on Polyconvex Free Energy Functions. Submitted to *International Journal of Solids and Structures*

[10] Smith, G.F. and Rivlin, R.S. (1957). The Anisotropic Tensors. *Q. Appl. Math.*, **15**, 309–314.

[11] Smith, G.F. (1971). On Isotropic Functions of Symmetric Tensors, Skew-Symmetric Tensors and Vectors. *Int. J. Engrg. Sci.*, **19**, 899–916.

[12] Spencer, A.J.M. (1971). Theory of Invariants. In: *Continuum Physics Vol. 1*, A.C. Eringen (edt.), Academic Press, New York, 239–353.

[13] Wang, C.C. (1969). On Representations for Isotropic Functions: Part I. Isotropic Functions of Symmetric Tensors and Vectors. *Arch. Rational Mech. Anal.*, **33**, 249–267.

[14] Wang, C.C. (1971). Corrigendum to my recent papers on: Representations for Isotropic Functions. *Arch. Rational Mech. Anal.*, **43**, 392–395.

Formulation and Computation of Geometrically Nonlinear Anisotropic Inelasticity

A. Menzel, P. Steinmann

Chair of Applied Mechanics, University of Kaiserslautern

P.O. Box 3049, 67653 Kaiserslautern, Germany

amenzel@rhrk.uni-kl.de, ps@rhrk.uni-kl.de

Abstract: The main objective of this contribution is the development of a thermodynamically consistent and modular formulation of anisotropic multiplicative elasto-plasticity. Based on the framework of standard dissipative materials, we introduce as a key-ingredient additional symmetric arguments – typically structural tensors – into the relevant scalar-valued isotropic tensor functions. Then the fundamental covariance relation allows a set up directly in terms of spatial fields and a convenient implementation within any numerical setting.

Keywords: Anisotropy, structural tensors, multiplicative elasto-plasticity.

1. Introduction

Phenomenological models of inelasticity should be able to capture at least some of the dominating anisotropic effects which are exhibited by many materials. Therefore this contribution aims at a thermodynamically consistent formulation of geometrically nonlinear anisotropic multiplicative elasto-plasticity. In particular, we adopt the theory of standard dissipative materials embedded into the framework of nonlinear continuum mechanics. As a key-ingredient structural tensors are incorporated into the relevant scalar-valued isotropic tensor functions. Obviously this approach results in a reduced representation of general anisotropy, however, we are still able to monitor important anisotropy classes of a wide range of engineering materials whereby initial anisotropy as well as deformation induced anisotropy are taken into account. Concerning the algorithmic setting it is recalled that for the general anisotropic case no exponential-type integrators for the governing evolution equations are conveniently available due to the non-coaxiality of the tensorial internal variables and their flow directions. Nevertheless, standard integrators like e.g. Runge-Kutta methods can be applied and as a demonstration of the developed methodology, a simple prototype model problem is proposed and discussed within a homogeneous deformation in simple shear.

C. Miehe (ed.), IUTAM Symposium on Computational Mechanics of Solid Materials at Large Strains, 181-190.

2. Kinematics

Let the reference configuration of the body B of interest be denoted by $\mathcal{B}_0 \subset \mathbb{E}^3$, the spatial configuration by $\mathcal{B}_t \subset \mathbb{E}^3$ and the intermediate configuration of multiplicative elasto-plasticity by $\mathcal{B}_\mathrm{p}$. The corresponding tangent and dual spaces are represented by $T\mathcal{B}_0, T^*\mathcal{B}_0, T\mathcal{B}_t, T^*\mathcal{B}_t$ and $T\mathcal{B}_\mathrm{p}, T^*\mathcal{B}_\mathrm{p}$, respectively, and in addition to the co- and contra-variant metric tensors we apply mixed-variant identity tensors

$$
\begin{aligned}
&\boldsymbol{g}^\flat : T\mathcal{B}_t \to T^*\mathcal{B}_t\,, \quad \boldsymbol{g}^\sharp : T^*\mathcal{B}_t \to T\mathcal{B}_t\,, \quad \boldsymbol{g}^\natural : T\mathcal{B}_t \to T\mathcal{B}_t\,,\\
&\widehat{\boldsymbol{G}}^\flat : T\mathcal{B}_\mathrm{p} \to T^*\mathcal{B}_\mathrm{p}\,, \quad \widehat{\boldsymbol{G}}^\sharp : T^*\mathcal{B}_\mathrm{p} \to T\mathcal{B}_\mathrm{p}\,, \quad \widehat{\boldsymbol{G}}^\natural : T\mathcal{B}_\mathrm{p} \to T\mathcal{B}_\mathrm{p}\,,\\
&\boldsymbol{G}^\flat : T\mathcal{B}_0 \to T^*\mathcal{B}_0\,, \quad \boldsymbol{G}^\sharp : T^*\mathcal{B}_0 \to T\mathcal{B}_0\,, \quad \boldsymbol{G}^\natural : T\mathcal{B}_0 \to T\mathcal{B}_0\,.
\end{aligned}
\tag{1}
$$

The underlying non-singular linear tangent maps, which transform tangent vectors to curves in one configuration into tangent vectors in another configuration, are abbreviated by

$$
\begin{aligned}
&\boldsymbol{F}^\natural : T\mathcal{B}_0 \to T\mathcal{B}_t\,, \quad \boldsymbol{F}^\natural_\mathrm{p} : T\mathcal{B}_0 \to T\mathcal{B}_\mathrm{p}\,, \quad \boldsymbol{F}^\natural_\mathrm{e} : T\mathcal{B}_\mathrm{p} \to T\mathcal{B}_t\,,\\
&\boldsymbol{f}^\natural : T\mathcal{B}_t \to T\mathcal{B}_0\,, \quad \boldsymbol{f}^\natural_\mathrm{p} : T\mathcal{B}_\mathrm{p} \to T\mathcal{B}_0\,, \quad \boldsymbol{f}^\natural_\mathrm{e} : T\mathcal{B}_t \to T\mathcal{B}_\mathrm{p}\,.
\end{aligned}
\tag{2}
$$

With these metric tensors and linear mappings at hand, several kinematic tensors can be introduced, namely

$$
\begin{aligned}
&\boldsymbol{b}^\sharp = \boldsymbol{F}^\natural{}_\star\,\boldsymbol{G}^\sharp\,, \quad \boldsymbol{c}^\flat = \boldsymbol{f}^{\natural\star}\,\boldsymbol{G}^\flat\,, \quad \boldsymbol{b}^\sharp_\mathrm{e} = \boldsymbol{F}^\natural_{\mathrm{e}\star}\,\widehat{\boldsymbol{G}}^\sharp\,, \quad \boldsymbol{c}^\flat_\mathrm{e} = \boldsymbol{f}^{\natural\star}_\mathrm{e}\,\widehat{\boldsymbol{G}}^\flat\,,\\
&\widehat{\boldsymbol{b}}^\sharp_\mathrm{p} = \boldsymbol{f}^\natural_{\mathrm{p}\star}\,\boldsymbol{G}^\sharp\,, \quad \widehat{\boldsymbol{c}}^\flat_\mathrm{p} = \boldsymbol{F}^{\natural\star}_\mathrm{p}\,\boldsymbol{G}^\flat\,, \quad \widehat{\boldsymbol{B}}^\sharp_\mathrm{p} = \boldsymbol{f}^\natural_{\mathrm{e}\star}\,\boldsymbol{g}^\sharp\,, \quad \widehat{\boldsymbol{C}}^\flat_\mathrm{p} = \boldsymbol{F}^{\natural\star}_\mathrm{e}\,\boldsymbol{g}^\flat\,.\\
&\boldsymbol{B}^\sharp = \boldsymbol{f}^\natural{}_\star\,\boldsymbol{g}^\sharp\,, \quad \boldsymbol{C}^\flat = \boldsymbol{F}^{\natural\star}\,\boldsymbol{g}^\flat\,, \quad \boldsymbol{B}^\sharp_\mathrm{p} = \boldsymbol{f}^\natural_{\mathrm{p}\star}\,\widehat{\boldsymbol{G}}^\sharp\,, \quad \boldsymbol{C}^\flat_\mathrm{p} = \boldsymbol{F}^{\natural\star}_\mathrm{p}\,\widehat{\boldsymbol{G}}^\flat\,,
\end{aligned}
\tag{3}
$$

whereby the notations $[\bullet]_\star$, $[\bullet]^\star$ denote pushforward and pullback operations, respectively. For a detailed review on the kinematical framework within continuum mechanics we refer to Lodge [5], Chap. 1 and for special emphasis on elasto-plasticity to Haupt [4], Sects. 1.10 & 11.3.

3. Free Helmholtz energy density

Let the free Helmholtz energy be determined via

$$
\psi_0 = \psi_0(\boldsymbol{g}^\flat, \boldsymbol{F}^\natural_\mathrm{e}, \widehat{\boldsymbol{G}}^\sharp, \widehat{\boldsymbol{A}}^\sharp_{1,\ldots,n}, \kappa)
\tag{4}
$$

whereby $\widehat{\boldsymbol{A}}^\sharp_{1,\ldots,n}$ in $\mathcal{B}_\mathrm{p}$ represents a set of additional tensorial arguments, κ is a scalar-valued hardening variable and for notational simplicity the explicit dependence on material points has been omitted. Now, the fundamental covariance relation can be applied and a pullback operation with respect to the intermediate configuration results in

$$\psi_0 \doteq \psi_0(\boldsymbol{F}_{\rm e}^{\natural\star}\,\boldsymbol{g}^\flat, \boldsymbol{f}_{\rm e\star}^{\natural}\,\boldsymbol{F}_{\rm e}^{\natural}, \widehat{\boldsymbol{G}}^\sharp, \widehat{\boldsymbol{A}}^\sharp_{1,\dots,n}, \kappa) = \psi_0^{\rm p}(\widehat{\boldsymbol{C}}^\flat_{\rm e}, \widehat{\boldsymbol{G}}^\sharp, \widehat{\boldsymbol{A}}^\sharp_{1,\dots,n}, \kappa)\,, \tag{5}$$

since the incorporation of the mixed-variant identity $\widehat{\boldsymbol{G}}^\natural = \boldsymbol{f}_{\rm e\star}^{\natural}\,\boldsymbol{F}_{\rm e}^{\natural}$ is naturally redundant. For a general outline of the covariance principle we refer to Marsden and Hughes [7], Chap. 2 & 3 and to Menzel and Steinmann [8] and Lu and Papadopoulos [6] for special emphasis on anisotropic hyper-elasticity. Next, based on Eq. (5), further transformations in terms of $\boldsymbol{F}_{\rm e}^{\natural}$ and $\boldsymbol{F}_{\rm p}^{\natural}$ result in the spatial and material setting

$$\psi_0 \doteq \psi_0^{\rm p}(\boldsymbol{f}_{\rm e}^{\natural\star}\,\widehat{\boldsymbol{C}}^\flat_{\rm e}, \boldsymbol{F}_{\rm e\star}^{\natural}\,\widehat{\boldsymbol{G}}^\sharp, \boldsymbol{F}_{\rm e\star}^{\natural}\,\widehat{\boldsymbol{A}}^\sharp_{1,\dots,n}, \kappa) = \psi_0^{t}(\boldsymbol{g}^\flat, \boldsymbol{b}_{\rm e}^\sharp, \boldsymbol{a}^\sharp_{1,\dots,n}, \kappa) \tag{6}$$

$$\psi_0 \doteq \psi_0^{\rm p}(\boldsymbol{F}_{\rm p}^{\natural\star}\,\widehat{\boldsymbol{C}}^\flat_{\rm e}, \boldsymbol{f}_{\rm p\star}^{\natural}\,\widehat{\boldsymbol{G}}^\sharp, \boldsymbol{f}_{\rm p\star}^{\natural}\,\widehat{\boldsymbol{A}}^\sharp_{1,\dots,n}, \kappa) = \psi_0^{0}(\boldsymbol{C}^\flat, \boldsymbol{B}_{\rm p}^\sharp, \boldsymbol{A}^\sharp_{1,\dots,n}, \kappa) \tag{7}$$

by taking standard pushforward and pullback operations for the set $\widehat{\boldsymbol{A}}^\sharp_{1,\dots,n}$ into account, i.e. $\boldsymbol{a}^\sharp_{1,\dots,n} = \boldsymbol{F}_{\rm e\star}^{\natural}\,\widehat{\boldsymbol{A}}^\sharp_{1,\dots,n}$ and $\boldsymbol{A}^\sharp_{1,\dots,n} = \boldsymbol{f}_{\rm p\star}^{\natural}\,\widehat{\boldsymbol{A}}^\sharp_{1,\dots,n}$. As a typical application, we interprete these elements in the following as initially rank one structural tensors of second order which, practically speaking, characterize preferred fibre orientations and define the material symmetry group $\mathbb{G}$ of the body B in $\mathcal{B}_0$

$$\mathbb{G} = \left\{ \boldsymbol{Q}^\natural \in \mathbb{O}^3 \;\middle|\; \boldsymbol{Q}^\natural_\star\,\boldsymbol{A}^\sharp_{1,\dots,n} = \boldsymbol{A}^\sharp_{1,\dots,n} \right\}\,. \tag{8}$$

The specific choice when $\boldsymbol{F}_{\rm e}^{\natural}$ or $\boldsymbol{f}_{\rm p}^{\natural}$, respectively, boil down to the linear tangent map $\widehat{\boldsymbol{Q}}^\natural$ of an isometry leads us to the definition of an isotropic scalar-valued tensor function

$$\psi_0^{\rm p} = \widehat{\boldsymbol{Q}}^\natural_\star\,\psi_0^{\rm p} = \psi_0^{\rm p}(\widehat{\boldsymbol{Q}}^\natural_\star\,\widehat{\boldsymbol{C}}^\flat_{\rm e}, \widehat{\boldsymbol{Q}}^\natural_\star\,\widehat{\boldsymbol{G}}^\sharp, \widehat{\boldsymbol{Q}}^\natural_\star\,\widehat{\boldsymbol{A}}^\sharp_{1,\dots,n}) \quad \forall \quad \widehat{\boldsymbol{Q}}^\natural \in \mathbb{O}^3_+\,. \tag{9}$$

Recall that a superposed spatial rigid body motion represents a specific application of the general covariance principle, namely the action of a spatial isometry on Eq. (6). Moreover, note that the material symmetry group as given in Eq. (8) refers to $\mathcal{B}_0$ while the isotropic tensor function $\psi_0^{\rm p}$ has been introduced with respect to $\mathcal{B}_{\rm p}$. Based on this, the general representation theorem can be applied which enables us to compute the underlying set of invariants as summarized in Table 1[1], compare the contributions in Boehler [1] and references cited therein.

Remark 3.1 *Please recall, that the particular case where all tensorial arguments* $\boldsymbol{a}^\sharp_{1,\dots,n}$ *remain spherical during the deformation process (or even vanish) results in an isotropic setting.*

Table 1. Complete set of invariants $\psi_0^t(I_{(i),\dots,(xi)})$ for all possible choices of $i, j, k, l \in [1, n]$ but $i \neq l$.

(i)	$\boldsymbol{b}_\mathrm{e}^\sharp : \boldsymbol{g}^\flat$	(iv)	$\boldsymbol{a}_i^\sharp : \boldsymbol{g}^\flat$	$(viii)$	$\boldsymbol{a}_i^\sharp : \boldsymbol{c}_\mathrm{e}^\flat$
(ii)	$\boldsymbol{b}_\mathrm{e}^\sharp : [\boldsymbol{g}^\flat \cdot \boldsymbol{b}_\mathrm{e}^\sharp \cdot \boldsymbol{g}^\flat]$	(v)	$\boldsymbol{a}_i^\sharp : [\boldsymbol{g}^\flat \cdot \boldsymbol{b}_\mathrm{e}^\sharp \cdot \boldsymbol{g}^\flat]$	(ix)	$\boldsymbol{a}_i^\sharp : [\boldsymbol{c}_\mathrm{e}^\flat \cdot \boldsymbol{a}_j^\sharp \cdot \boldsymbol{c}_\mathrm{e}^\flat]$
(iii)	$\boldsymbol{b}_\mathrm{e}^\sharp : \left[\boldsymbol{g}^\flat \cdot [\boldsymbol{b}_\mathrm{e}^\sharp \cdot \boldsymbol{g}^\flat]^2\right]$	(vi)	$\boldsymbol{a}_i^\sharp : [\boldsymbol{g}^\flat \cdot \boldsymbol{a}_j^\sharp \cdot \boldsymbol{c}_\mathrm{e}^\flat]$	(x)	$\boldsymbol{a}_i^\sharp : \left[\boldsymbol{c}_\mathrm{e}^\flat \cdot [\boldsymbol{a}_{j,k}^\sharp \cdot \boldsymbol{c}_\mathrm{e}^\flat]^2\right]$
		(vii)	$\boldsymbol{a}_i^\sharp : [\boldsymbol{g}^\flat \cdot \boldsymbol{a}_i^\sharp \cdot \boldsymbol{g}^\flat]$	(xi)	$\boldsymbol{a}_i^\sharp : \left[\boldsymbol{c}_\mathrm{e}^\flat \cdot [\boldsymbol{a}_{i,l,l}^\sharp \cdot \boldsymbol{c}_\mathrm{e}^\flat]^3\right]$

4. Standard dissipative materials

In the sequel we adopt the framework of standard dissipative materials as advocated by Halphen and Nguyen [3] and therefore compute the material time derivative D_t of the appropriate ingredients of the free Helmholtz energy density in $\mathcal{B}_t$, namely

$$\mathrm{D}_t\, \boldsymbol{b}_\mathrm{e}^\sharp = \mathrm{L}_t\, \boldsymbol{b}_\mathrm{e}^\sharp + 2\,[\,\boldsymbol{l}^\natural \cdot \boldsymbol{b}_\mathrm{e}^\sharp\,]^\mathrm{sym} \quad \text{and} \quad \mathrm{D}_t\, \boldsymbol{a}_i^\sharp = \mathrm{L}_t\, \boldsymbol{a}_i^\sharp + 2\,[\,\boldsymbol{l}^\natural \cdot \boldsymbol{a}_i^\sharp\,]^\mathrm{sym} \tag{10}$$

with $\mathrm{L}_t\, [\bullet]^\sharp|_{\mathcal{B}_t} = \boldsymbol{F}^\natural{}_\star\, \mathrm{D}_t\, (\boldsymbol{f}^\natural{}_\star\, [\bullet]^\sharp)$ abbreviating the Lie derivative or rather the Oldroyd rate and the spatial velocity gradient $\boldsymbol{l}^\natural$ decomposes additively into

$$\boldsymbol{l}^\natural = \mathrm{D}_t\, \boldsymbol{F}^\natural \cdot \boldsymbol{f}^\natural = \mathrm{D}_t\, \boldsymbol{F}_\mathrm{e}^\natural \cdot \boldsymbol{f}_\mathrm{e}^\natural + \boldsymbol{F}_\mathrm{e}^\natural \cdot \mathrm{D}_t\, \boldsymbol{F}_\mathrm{p}^\natural \cdot \boldsymbol{f}^\natural = \boldsymbol{l}_\mathrm{e}^\natural + \boldsymbol{l}_\mathrm{p}^\natural\,. \tag{11}$$

Based on this, the pointwise and isothermal format of the dissipation inequality with respect to $\mathcal{B}_t$ consequently results in

$$\begin{aligned} \mathcal{D}_0^t &= [\boldsymbol{m}^\natural]^\mathrm{t} : \boldsymbol{l}^\natural - \mathrm{D}_t\, \psi_0^t(\boldsymbol{g}^\flat, \boldsymbol{b}_\mathrm{e}^\sharp, \boldsymbol{a}_{1,\dots,n}^\sharp, \kappa, ; \boldsymbol{X}) \\ &= \left[[\boldsymbol{m}^\natural]^\mathrm{t} - 2\,\partial_{\boldsymbol{b}_\mathrm{e}^\sharp}\, \psi_0^t \cdot \boldsymbol{b}_\mathrm{e}^\sharp - 2\, \sum_{i=1}^{n} \partial_{\boldsymbol{a}_i^\sharp}\, \psi_0^t \cdot \boldsymbol{a}_i^\sharp\right] : \boldsymbol{l}^\natural \\ &- \partial_{\boldsymbol{b}_\mathrm{e}^\sharp}\, \psi_0^t : \mathrm{L}_t\, \boldsymbol{b}_\mathrm{e}^\sharp - \sum_{i=1}^{n} \partial_{\boldsymbol{a}_i^\sharp}\, \psi_0^t : \mathrm{L}_t\, \boldsymbol{a}_i^\sharp - \partial_\kappa\, \psi_0^t\, \mathrm{D}_t\, \kappa \;\geq\; 0 \end{aligned} \tag{12}$$

whereby $[\boldsymbol{m}^\natural]^\mathrm{t} : \boldsymbol{l}^\natural$ defines the stress power and $[\boldsymbol{m}^\natural]^\mathrm{t} = \boldsymbol{g}^\flat \cdot \boldsymbol{\tau}^\sharp$ represents the spatial Mandel tensor with $\boldsymbol{\tau}^\sharp$ denoting a Kirchhoff-type stress field. Following the argumentation of rational thermodynamics, the hyper-elastic constitutive function for the mixed-variant stress tensor reads

$$[\boldsymbol{m}^\natural]^\mathrm{t} = 2\,\partial_{\boldsymbol{b}_\mathrm{e}^\sharp}\, \psi_0^t \cdot \boldsymbol{b}_\mathrm{e}^\sharp + 2\, \sum_{i=1}^{n} \partial_{\boldsymbol{a}_i^\sharp}\, \psi_0^t \cdot \boldsymbol{a}_i^\sharp\,. \tag{13}$$

which allows interpretation as the generalization of the celebrated isotropic Murnaghan-Truesdell formula (see Murnaghan [10], Truesdell and Noll [13], Eq. (85.15)) to anisotropic multiplicative elasto-plasticity. However, it is a non-trivial task to show that the contra-variant representation of Eq. (13) remains symmetric and determines the Kirchhoff stress, i.e.

$$\boldsymbol{\tau}^\sharp = \boldsymbol{g}^\sharp \cdot [\boldsymbol{m}^\natural]^{\mathrm{t}} = 2\,\boldsymbol{g}^\sharp \cdot \left[\partial_{\boldsymbol{b}_{\mathrm{e}}^\sharp}\psi_0^t \cdot \boldsymbol{b}_{\mathrm{e}}^\sharp + \sum_{i=1}^n \partial_{\boldsymbol{a}_i^\sharp}\psi_0^t \cdot \boldsymbol{a}_i^\sharp\right] \doteq 2\,\partial_{\boldsymbol{g}^\flat}\psi_0^t\,. \quad (14)$$

The corresponding proof is based on the general representation theorem of isotropic tensor functions or alternatively on the direct application of the co-variance relation and fully outlined in Menzel and Steinmann [9].

Remark 4.1 *Apparently, the reduced dissipation inequality allows representation in terms of the Mandel stress and the plastic part of the spatial velocity gradient. In this direction, recall the relation*

$$\mathrm{L}_t\,[\bullet]^\sharp = \boldsymbol{F}_{\mathrm{e}\star}^\natural\,\mathrm{L}_t^{\mathrm{p}}\,[\widehat{\bullet}]^\sharp = \boldsymbol{F}_{\mathrm{e}\star}^\natural\,\mathrm{D}_t\,[\widehat{\bullet}]^\sharp - 2\left[\boldsymbol{l}_{\mathrm{p}}^\natural \cdot [\bullet]^\sharp\right]^{\mathrm{sym}} \quad (15)$$

with $\mathrm{L}_t^{\mathrm{p}}\,[\bullet]^\sharp|_{\mathcal{B}_{\mathrm{p}}} = \boldsymbol{F}_{\mathrm{p}\star}^\natural\,\mathrm{D}_t\,(\boldsymbol{f}_{\mathrm{p}\star}^\natural\,[\bullet]^\sharp)$. *Now the specific format*

$$^{\mathrm{red}}\mathcal{D}_0^t = [\boldsymbol{m}^\natural]^{\mathrm{t}} : \boldsymbol{l}_{\mathrm{p}}^\natural - \sum_{i=1}^n \partial_{\boldsymbol{a}_i^\sharp}\psi_0^t : \left[\boldsymbol{F}_{\mathrm{e}\star}^\natural\,\mathrm{D}_t\,\widehat{\boldsymbol{A}}_i^\sharp\right] - \partial_\kappa\psi_0^t\,\mathrm{D}_t\,\kappa \geq 0 \quad (16)$$

is a straightforward exercise.

4.1. Thermodynamic forces

The computation of appropriate thermodynamic forces in $\mathcal{B}_t$ are a natural outcome, namely form Eq. (12) we obtain

$$\boldsymbol{y}^\flat = -\,\partial_{\boldsymbol{b}_{\mathrm{e}}^\sharp}\psi_0^t\,, \quad \boldsymbol{z}_i^\flat = -\,\partial_{\boldsymbol{a}_i^\sharp}\psi_0^t\,, \quad h = -\,\partial_\kappa\psi_0^t\,. \quad (17)$$

These definitions are of cardinal importance since they determine the inelastic potentials. In particular the plastic potential is defined by $\boldsymbol{y}^\flat$ which is obviously not identical to the stress measure that enters the underlying equilibrium equation.

4.2. Inelastic potentials

Let an admissible elastic domain be defined in terms of a yield function $^{\mathrm{yie}}\Phi^t$ and a hardening contribution via the additive composition

$$\mathbb{E}^t = \left\{(\boldsymbol{y}^\flat, h)\,\middle|\,{}^{\mathrm{yie}}\Phi^t = {}^{\mathrm{pla}}\Phi^t(\boldsymbol{y}^\flat, \boldsymbol{g}^\sharp, \boldsymbol{g}^\flat, \boldsymbol{a}_{n+1,\dots,m}^\sharp) + {}^{\mathrm{har}}\Phi(h) \leq 0\right\} \quad (18)$$

whereby the additional tensor series $\boldsymbol{a}^\sharp_{n+1,\dots,m}$ has been introduced and the explicit dependence on the position in space is omitted for clarity's sake. Next, by adopting a Lemaitre-type model, we account for a dissipation potential in terms of the yield function and an additional potential ${}^{\rm fib}\Phi^t$ that optionally accounts for the (fibre-) evolution of $\boldsymbol{a}^\sharp_{1,\dots,n}$

$$ {}^{\rm pot}\Phi^t = {}^{\rm yie}\Phi^t(\boldsymbol{y}^\flat, h, \boldsymbol{g}^\sharp, \boldsymbol{g}^\flat, \boldsymbol{a}^\sharp_{n+1,\dots,m}) + {}^{\rm fib}\Phi^t(\boldsymbol{z}^\flat_{1,\dots,n}, \boldsymbol{g}^\sharp, \boldsymbol{a}^\sharp_{1,\dots,n}) \,. \tag{19} $$

With these potentials at hand, the definition of appropriate evolution equations is a natural consequence

$$ \begin{aligned} \mathrm{L}_t\, \boldsymbol{b}^\sharp_{\rm e} &= \mathrm{D}_t\lambda\ \partial_{\boldsymbol{y}^\flat}{}^{\rm pot}\Phi^t &&= \mathrm{D}_t\lambda\ \partial_{\boldsymbol{y}^\flat}{}^{\rm pla}\Phi^t &&\doteq \mathrm{D}_t\lambda\, \boldsymbol{\xi}^\sharp_{y^\flat}\,, \\ \mathrm{L}_t\, \boldsymbol{a}^\sharp_i &= \mathrm{D}_t\lambda\ \partial_{\boldsymbol{z}^\flat_i}{}^{\rm pot}\Phi^t &&= \mathrm{D}_t\lambda\ \partial_{\boldsymbol{z}^\flat_i}{}^{\rm fib}\Phi^t &&\doteq \mathrm{D}_t\lambda\, \boldsymbol{\xi}^\sharp_{z^\flat_i}\,, \\ \mathrm{D}_t\, \kappa &= \mathrm{D}_t\lambda\ \partial_h\, {}^{\rm pot}\Phi^t &&= \mathrm{D}_t\lambda\ \partial_h\, {}^{\rm har}\Phi &&\doteq \mathrm{D}_t\lambda\, \xi_h\,, \end{aligned} \tag{20} $$

whereby Eqs. $(20)_{1,3}$ are obviously of associated format and Eq. $(20)_2$ is non-associated. In analogy to the free Helmholtz energy density we assume that the additional tensorial fields $\boldsymbol{a}^\sharp_{n+1,\dots,m}$ are of second order, symmetric and have rank one. Moreover, usual pull-back operations hold – like $\widehat{\boldsymbol{A}}^\sharp_{n+1,\dots,m} = \boldsymbol{f}^\natural_{\rm e\star}\, \boldsymbol{a}^\sharp_{n+1,\dots,m}$ and $\boldsymbol{A}^\sharp_{n+1,\dots,m} = \boldsymbol{f}^\natural{}_\star\, \boldsymbol{a}^\sharp_{n+1,\dots,m}$ – and the inelastic potentials are chosen as isotropic tensor functions

$$ {}^{\rm pla}\Phi^t = \boldsymbol{q}^\natural{}_\star\, {}^{\rm pla}\Phi^t\,, \qquad {}^{\rm fib}\Phi^t = \boldsymbol{q}^\natural{}_\star\, {}^{\rm fib}\Phi^t \quad \forall \quad \boldsymbol{q}^\natural \in \mathbb{O}^3_+\,. \tag{21} $$

Please note, that the symmetry group of the plastic potential ${}^{\rm pla}\Phi^t$ is defined in $\mathcal{B}_0$ via

$$ \mathbb{G} = \left\{ \boldsymbol{Q}^\natural \in \mathbb{O}^3 \,\middle|\, \boldsymbol{Q}^\natural{}_\star\, \boldsymbol{A}^\sharp_{n+1,\dots,m} = \boldsymbol{A}^\sharp_{n+1,\dots,m} \right\} . \tag{22} $$

Based on this framework, the general representation theorem of isotropic tensor functions enables us to compute the appropriate sets of invariants which define ${}^{\rm yie}\Phi^t$ and ${}^{\rm fib}\Phi^t$. Table 2[1] summarizes the complete list for the plastic potential ${}^{\rm pla}\Phi^t$ (see endnote for notational details). The outline for ${}^{\rm fib}\Phi^t$ is similar but even more extensive since we deal with the complete set $\boldsymbol{z}^\flat_{1,\dots,n}$ instead of only one thermodynamic force $\boldsymbol{y}^\flat$. Apparently, we have different options to correlate the sets $\boldsymbol{a}^\sharp_{1,\dots,n}$ and $\boldsymbol{a}^\sharp_{n+1,\dots,m}$ which enter ψ^t_0, ${}^{\rm fib}\Phi^t$ and ${}^{\rm pla}\Phi^t$, respectively. Reasonable choices are e.g. (i) $\boldsymbol{a}^\sharp_{1,\dots,n} \bigcup \boldsymbol{a}^\sharp_{n+1,\dots,m} \doteq \emptyset$ or (ii) $\boldsymbol{a}^\sharp_{1,\dots,n} \bigcup \boldsymbol{a}^\sharp_{n+1,\dots,m} \doteq \boldsymbol{a}^\sharp_{1,\dots,n}$ with $m = 2\,n$ among other possibilities.

4.3. Reduction to Isotropy

For an isotropic setting with $\boldsymbol{a}^\sharp_{1,\dots,n} \doteq \boldsymbol{0}$ the thermodynamic force $\boldsymbol{y}^\flat$ results in

$$ \boldsymbol{y}^\flat = -\tfrac{1}{2}\,\boldsymbol{g}^\flat \cdot \boldsymbol{\tau}^\sharp \cdot \boldsymbol{c}^\flat_{\rm e} = -\tfrac{1}{2}\,[\boldsymbol{m}^\natural]^{\rm t} \cdot \boldsymbol{c}^\flat_{\rm e}\,. \tag{23} $$

Table 2. Complete set of invariants ${}^{\rm yie}\Phi^t(I_{(i),\ldots,(xi)})$ for all possible choices of $i, j, k, l \in [1, n]$ but $i \neq l$.

(i)	$\boldsymbol{y}^\flat : \boldsymbol{g}^\sharp$	(iv)	$\boldsymbol{a}_i^\sharp : \boldsymbol{y}^\flat$	$(viii)$	$\boldsymbol{a}_i^\sharp : \boldsymbol{g}^\flat$
(ii)	$\boldsymbol{y}^\flat : [\boldsymbol{g}^\sharp \cdot \boldsymbol{y}^\flat \cdot \boldsymbol{g}^\sharp]$	(v)	$\boldsymbol{a}_i^\sharp : [\boldsymbol{y}^\flat \cdot \boldsymbol{a}_i^\sharp \cdot \boldsymbol{y}^\flat]$	(ix)	$\boldsymbol{a}_i^\sharp : [\boldsymbol{g}^\flat \cdot \boldsymbol{a}_j^\sharp \cdot \boldsymbol{g}^\flat]$
(iii)	$\boldsymbol{y}^\flat : \left[\boldsymbol{g}^\sharp \cdot [\boldsymbol{y}^\flat \cdot \boldsymbol{g}^\sharp]^2\right]$	(vi)	$\boldsymbol{a}_i^\sharp : [\boldsymbol{g}^\flat \cdot \boldsymbol{a}_j^\sharp \cdot \boldsymbol{y}^\flat]$	(x)	$\boldsymbol{a}_i^\sharp : \left[\boldsymbol{g}^\flat \cdot [\boldsymbol{a}_{j,k}^\sharp \cdot \boldsymbol{g}^\flat]^2\right]$
		(vii)	$\boldsymbol{a}_i^\sharp : [\boldsymbol{y}^\flat \cdot \boldsymbol{a}_i^\sharp \cdot \boldsymbol{y}^\flat]$	(xi)	$\boldsymbol{a}_i^\sharp : \left[\boldsymbol{g}^\flat \cdot [\boldsymbol{a}_{i,l,l}^\sharp \cdot \boldsymbol{g}^\flat]^3\right]$

and the representation of the spatial Mandel stress boils down to the Murnaghan-Truesdell formula $[\boldsymbol{m}^\natural]^{\rm t} = -\,2\,\boldsymbol{c}_{\rm e}^\flat \cdot \partial_{\boldsymbol{c}_{\rm e}^\flat}\,\psi_0^t = 2\,\partial_{\boldsymbol{b}_{\rm e}^\sharp}\,\psi_0^t \cdot \boldsymbol{b}_{\rm e}^\sharp$. Thereby it is clear that in the case of metal plasticity, where the elastic strains are usually small, i.e. $\boldsymbol{c}_{\rm e}^\flat \to \boldsymbol{g}^\flat$, the appropriate norms of $\boldsymbol{y}^\flat$ and $\boldsymbol{\tau}^\sharp$ are related by the factor two. Moreover, the reduced isothermal pointwise dissipation inequality without hardening contributions reads

$$\begin{aligned} {}^{\rm red}\mathcal{D}_0^t &= [\,\boldsymbol{y}^\flat \cdot \boldsymbol{b}_{\rm e}^\sharp\,] : [\,{\rm L}_t\,\boldsymbol{b}_{\rm e}^\sharp \cdot \boldsymbol{c}_{\rm e}^\flat\,] = -\tfrac{1}{2}\,[\boldsymbol{m}^\natural]^{\rm t} : [\,{\rm L}_t\,\boldsymbol{b}_{\rm e}^\sharp \cdot \boldsymbol{c}_{\rm e}^\flat\,] \\ &= \left[[\boldsymbol{m}^\natural]^{\rm t} \cdot \boldsymbol{c}_{\rm e}^\flat\right] : [\,\boldsymbol{l}_{\rm p}^\natural \cdot \boldsymbol{b}_{\rm e}^\sharp\,]^{\rm sym} = \left[[\boldsymbol{m}^\natural]^{\rm t} \cdot \boldsymbol{c}_{\rm e}^\flat\right] : [\,\boldsymbol{l}_{\rm p}^\natural \cdot \boldsymbol{b}_{\rm e}^\sharp\,] \\ &= [\boldsymbol{m}^\natural]^{\rm t} : \boldsymbol{l}_{\rm p}^\natural \geq 0 \end{aligned} \tag{24}$$

and thus an associated evolution equation takes the natural format

$$-\tfrac{1}{2}\,{\rm L}_t\,\boldsymbol{b}_{\rm e}^\sharp \cdot \boldsymbol{c}_{\rm e}^\flat = {\rm D}_t\,\lambda\,\boldsymbol{g}^\sharp \cdot \partial_{\boldsymbol{\tau}^\sharp}\,{}^{\rm pla}\Phi^t \quad \text{or} \quad \boldsymbol{l}_{\rm p}^\natural = {\rm D}_t\,\lambda\,\partial_{[\boldsymbol{m}^\natural]^{\rm t}}\,{}^{\rm pla}\Phi^t\,, \tag{25}$$

compare e.g. Simo and Miehe [11]. The applied relation $\partial_{[\boldsymbol{m}^\natural]^{\rm t}}\,{}^{\rm pla}\Phi^t = \boldsymbol{g}^\sharp \cdot \partial_{\boldsymbol{\tau}^\sharp}\,{}^{\rm pla}\Phi^t$ is a direct consequence of the Mandel-type stress tensor $[\boldsymbol{m}^\natural]^{\rm t}$ being a product of two symmetric tensors with one throughout remaining positive definite and, moreover, the spectral decomposition $[\boldsymbol{m}^\natural]^{\rm t} = \sum_{i=1}^3 \lambda_i\,{}^{\rm rig}\boldsymbol{n}_i^\flat \otimes {}^{\rm lef}\boldsymbol{n}_i^\sharp$ can be applied, see Ericksen [2], Sect. 37.

5. Simple prototype model

In view of the subsequent numerical examples we choose a typical prototype model (fully outlined in Menzel and Steinmann [9]) based on the introduction of two initial structural tensors $\boldsymbol{a}_{1,2}^\sharp$ into ψ_0^t, ${}^{\rm fib}\Phi^t$ and two further structural tensors $\boldsymbol{a}_{3,4}^\sharp$ are incorporated into ${}^{\rm pla}\Phi^t$, respectively. A standard isotropic Neo-Hooke material is enlarge to orthotropic behavior whereby the applied formulation follows the lines in Spencer [12] and additionally accounts for exponential terms as e.g. highlighted in Weiss et al. [14]. The plastic potential implies a similar structure than ψ_0^t and is defined in terms of the deviator of $\boldsymbol{y}^\flat$. Finally, the ansatz for ${}^{\text{fib}}\Phi^t$ is quadratic in $\boldsymbol{z}_{3,4}^\flat$ (i.e. determined by $[\boldsymbol{z}_{3,4}^\flat : \boldsymbol{g}^\sharp]^2$ and $[\boldsymbol{z}_{3,4}^\flat]^2 : \boldsymbol{g}^\sharp$) and the contribution ${}^{\text{har}}\Phi$ (quadratic in h) accounts for proportional, saturation-type hardening.

6. Numerical example

The homogeneous deformation in simple shear is discussed ($\boldsymbol{F} \doteq \boldsymbol{I} + \gamma\, \boldsymbol{e}_1 \otimes \boldsymbol{e}_2$ w.r.t. a Cartesian frame) and a simple Euler backward integration technique has been adopted. Now let $\boldsymbol{a}_{1,2}^\sharp$ be defined by the fibres $\boldsymbol{F}_\star\,[2\,\boldsymbol{e}_1 + \boldsymbol{e}_2]/\sqrt{5}$, $\boldsymbol{F}_\star\,[-\boldsymbol{e}_1 + 2\,\boldsymbol{e}_2 + \boldsymbol{e}_3]/\sqrt{6}$ and assume $\boldsymbol{a}_{3,4}^\sharp$ to be represented via $\boldsymbol{F}_\star\,[\boldsymbol{e}_1 + 2\,\boldsymbol{e}_2 + \boldsymbol{e}_3]/\sqrt{6}$, $\boldsymbol{F}_\star\,[2\,\boldsymbol{e}_1 - \boldsymbol{e}_2]/\sqrt{5}$ (w.r.t. a Cartesian frame), respectively. Based on these relations, Figure 1 highlights the simple shear setting for $\boldsymbol{f}^\natural{}_\star\, \boldsymbol{a}_{1,2}^\sharp \doteq \text{const}$ (case (i)). Thereby, the anisotropy measure $\delta(\boldsymbol{b}_{\text{e}}^\sharp, \boldsymbol{\tau}^\sharp) = [\boldsymbol{b}_{\text{e}}^\sharp \cdot \boldsymbol{g}^\flat \cdot \boldsymbol{\tau}^\sharp - \boldsymbol{\tau}^\sharp \cdot \boldsymbol{g}^\flat \cdot \boldsymbol{b}_{\text{e}}^\sharp]/\|\boldsymbol{b}_{\text{e}}^\sharp\|\,\|\boldsymbol{\tau}^\sharp\|$ represents a typical indicator for anisotropy and can be applied to any set of two second order tensors in order to check the underlying coaxiality. On the contrary, the numerical results monitored in Figure 2 account for an evolution of the anisotropy tensors $\boldsymbol{f}^\natural{}_\star\, \boldsymbol{a}_{1,2}^\sharp$ which obviously influences the corresponding material symmetry group and, moreover, $\boldsymbol{a}_{3,4}^\sharp \doteq \boldsymbol{a}_{1,2}^\sharp$ is assumed (case (ii)).

7. Discussion

Due to the conceptional beauty, the proposed framework is not restricted to any specific configuration but can be carried out in $\mathcal{B}_t$, $\mathcal{B}_{\text{p}}$ or $\mathcal{B}_0$ and is well suited for general finite element settings. All incorporated second order internal variables and corresponding flow directions remain symmetric and thus, as an advantage, simple integration techniques can conveniently be applied.

Notes

1. In order to keep the notation of the invariants manageable, we apply the simplification

$$[\underbrace{\boldsymbol{a}_{i,j,\ldots,k}^\sharp}_{q\times} \cdot \boldsymbol{c}_{\text{e}}^\flat]^q = \underbrace{[\boldsymbol{a}_i^\sharp \cdot \boldsymbol{c}_{\text{e}}^\flat] \cdot [\boldsymbol{a}_j^\sharp \cdot \boldsymbol{c}_{\text{e}}^\flat] \cdot \ldots \cdot [\boldsymbol{a}_k^\sharp \cdot \boldsymbol{c}_{\text{e}}^\flat]}_{q\,\times\,[\ldots]} = \left[[\boldsymbol{c}_{\text{e}}^\flat \cdot \underbrace{\boldsymbol{a}_{k,j,\ldots,i}^\sharp}_{q\times}]^q\right]^{\text{t}}.$$

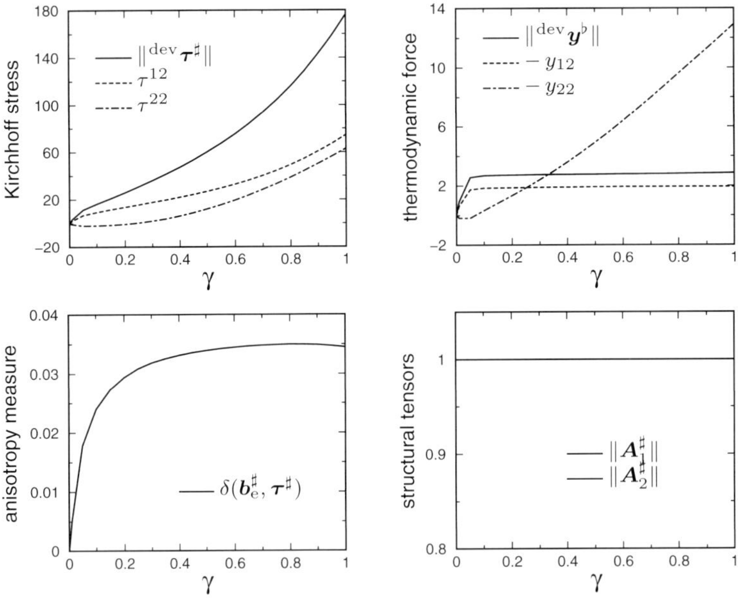

Figure 1. Kirchhoff stress $\boldsymbol{\tau}^{\sharp}$ (top left) and thermodynamic force $\boldsymbol{y}^{\flat}$ (top right); anisotropy measure $\delta(\boldsymbol{b}_{\mathrm{e}}^{\sharp}, \boldsymbol{\tau}^{\sharp})$ (bottom left) and structural tensors $\boldsymbol{A}_{1,2}^{\sharp}$ (bottom right).

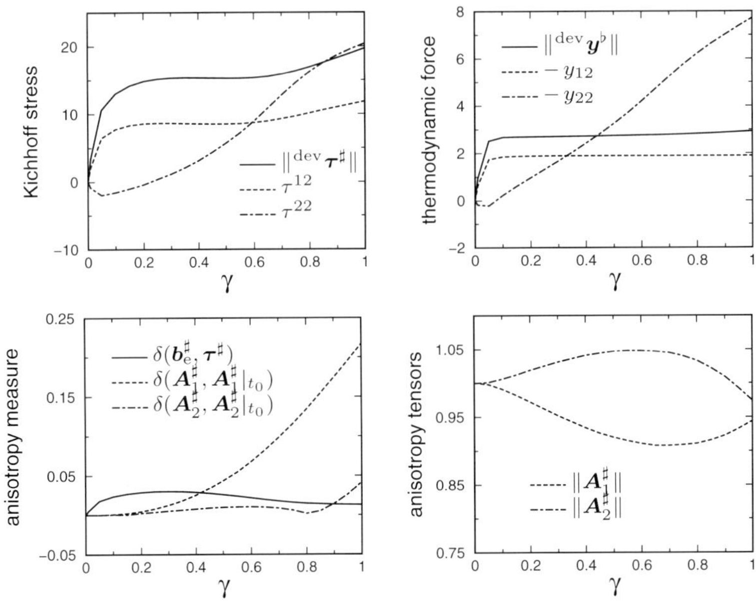

Figure 2. Kirchhoff stress $\boldsymbol{\tau}^{\sharp}$ (top left) and $\boldsymbol{y}^{\flat}$ (top right); anisotropy measure $\delta(\boldsymbol{b}_{\mathrm{e}}^{\sharp}, \boldsymbol{\tau}^{\sharp})$, $\delta(\boldsymbol{A}_{1,2}^{\sharp}, \boldsymbol{A}_{1,2}^{\sharp}|_{t_0})$ (bottom left) and anisotropy tensors $\boldsymbol{A}_{1,2}^{\sharp}$ (bottom right).

References

[1] Boehler, J. (1987). *Applications of Tensor Functions in Solid Mechanics*, **292**, CISM Courses and Lectures, Springer.

[2] Ericksen, J. (1960). Tensor Fields. In Flügge, S. (ed.), *Encyclopedia of Physics*,Springer, **III/1**, 794–858.

[3] Halphen, B. and Nguyen, Q. (1975). Sur les materiaux standards generalises. *Journal de Mecanique*, **14**, 39–62.

[4] Haupt, P. (2000). *Continuum Mechanics and Theory of Materials*, Advanced Texts in Physics, Springer.

[5] Lodge, A. (1974). *Body Tensor Fields in Continuum Mechanics (With Application to Polymer Rheology)*, Academic Press.

[6] Lu, J. and Papadopoulos, P. (2000). A covariant constitutive description of anisotropic non-linear elasticity. *Zeit. Angew. Mat. Phys.*, **51**, 204–217.

[7] Marsden J. and Hughes, T. (1994). *Mathematical Foundations of Elasticity*, Dover.

[8] Menzel, A. and Steinmann, P. (2000). On she spatial formulation of anisotropic hyperplasticity. *Technical Report*, **J00-09**, UKL/LTM.

[9] Menzel, A. and Steinmann, P. (2001). On she spatial formulation of anisotropic multiplicative elasto-plasticity. *Technical Report*, **J01-07**, UKL/LTM.

[10] Murnaghan, F. (1937). Finite deformations of an elastic solid. *Am. J. Math.*, **59**, 235–260.

[11] Simo, J. and Miehe, C. (1992). Associated coupled thermoplasticity at finite strains: Formulation, numerical analysis and implementation. *Comp. Meth. Appl. Mech. Engr.*, **98**, 41–104.

[12] Spencer, A. (1984). Constitutive theory of strongly anisotropic solids. In Spencer, A. (ed) *Continuum Theory of the Mechanics of Fibre-Reinforced Composites*, **282**, CISM Courses and Lectures, Springer.

[13] Truesdell, C. and Noll, W. (1992). *The Non-Linear Field Theories of Mechanics*, Springer, 2nd edition.

[14] Weiss, J., Maker, B. and Govindjee, S. (1996). Finite element implementation of incompressible, transversely isotropic hyperelasticity. *Comp. Meth. Appl. Mech. Engr.*, **135**, 107–128.

On the Representation of Anisotropic Viscoplasticity

P. Haupt, T. Kersten
Institute of Mechanics, University of Kassel
Mönchebergstraße 7, 34109 Kassel, Germany
haupt@ifm.maschinenbau.uni-kassel.de

Abstract: A constitutive model is developed, based on a free energy, which decomposes into an elastic and an inelastic part. The elastic part is an anisotropic function of the elastic strain and an orthogonal tensor-valued internal variable representing the orientation of the crystallographic axes. This orientation develops according to an appropriate evolution equation. The inelastic part of the free energy is assumed as a quadratic function of the resolved shear strains and further internal variables to incorporate hardening effects on the slip system level. All evolution equations are thermodynamically consistent and satisfy the principle of full invariance. While the general theory applies to all types of lattices, specific test calculations refer to cubic symmetry (fcc), small elastic strains and uniaxial tension.

Keywords: Anisotropy, viscoplasticity, crystal plasticity.

1. Introduction

One principal origin for the anisotropic behavior of real materials is the crystalline structure, which is encountered e.g. in the case of metallic bodies. Regarding a single crystal, we observe a natural anisotropy resulting from the periodic arrangement of atoms performing a lattice. From the micromechanical point of view the material properties are usually inhomogeneously distributed in a material body: actually, metallic bodies are composed of grains, i.e. the orientation of the crystallites varies from one grain to another. The idea of an isotropic material fits into this picture, if we imagine a quite large number of grains, whose crystallographic orientations have a quasi-isotropic distribution. In fact, this is the notion of a polycrystal. It is a general result of crystal plasticity, that the type of the crystal lattice remains unchanged during inelastic deformations (see e.g. Asaro [1]). Neglecting the phenomena of mechanical twinning and grain-boundary sliding, one can regard the crystallographic slip, evoked by the propagation of the dislocations through the lattice, as a main source of inelastic deformations. On the other hand, it is a basic observation that the orientation of the crystal together with the slip systems changes if inelastic deformations occur. Thus, if a polycrystal deforms, the quasi-isotropic orientation of the crystallographic axes becomes non-uniform and preferred directions appear. This type of anisotropy, induced by the mechanical process, is called texture. In the following a constitutive model of

C. Miehe (ed.), IUTAM Symposium on Computational Mechanics of Solid Materials at Large Strains, 191-200.

viscoplasticity is presented, which incorporates these basic phenomena of natural and process-induced anisotropy. The structure of the proposed theory does not depend on the particular type of crystal symmetry, although the numerical test calculations refer to a special example, namely the case of a face-centered cubic crystal, characterizing most common materials, such as Al, Cu, Ni,γ-Fe etc.

2. Strain, stress and properties of invariance

A three-dimensional model of viscoplasticity is usually based on the multiplicative decomposition of the deformation gradient into an elastic and an inelastic part:

$$\mathbf{F} = \hat{\mathbf{F}}_e \mathbf{F}_i. \tag{1}$$

This decomposition introduces an intermediate configuration. The rate of change of the intermediate configuration is the inelastic deformation rate, often called the "inelastic velocity gradient",

$$\hat{\mathbf{L}}_i = \dot{\mathbf{F}}_i \mathbf{F}_i^{-1}. \tag{2}$$

The multiplicative decomposition of the deformation gradient is not unique, because an arbitrary orthogonal transformation $\bar{\mathbf{Q}}$ can be inserted:

$$\mathbf{F} = \hat{\mathbf{F}}_e \mathbf{F}_i = \hat{\mathbf{F}}_e \bar{\mathbf{Q}}^T \bar{\mathbf{Q}} \mathbf{F}_i. \tag{3}$$

In view of the polar decomposition of $\mathbf{F}_i$ ($\mathbf{F}_i = \mathbf{R}_i \mathbf{U}_i$), $\bar{\mathbf{Q}}$ changes the rotation of the intermediate state. If the total GREEN strain tensor is transformed to operate on the intermediate configuration, it decomposes additively into a purely elastic GREEN strain and a purely inelastic ALMANSI strain:

$$\hat{\boldsymbol{\Gamma}} = \mathbf{F}_i^{T-1} \mathbf{E} \mathbf{F}_i^{-1} = \underbrace{\frac{1}{2}\left(\hat{\mathbf{F}}_e^T \hat{\mathbf{F}}_e - \mathbf{1}\right)}_{\text{elastic GREEN strain}} + \underbrace{\frac{1}{2}\left(\mathbf{1} - \mathbf{F}_i^{T-1} \mathbf{F}_i^{-1}\right)}_{\text{inelastic ALMANSI strain}} = \hat{\boldsymbol{\Gamma}}_e + \hat{\boldsymbol{\Gamma}}_i. \tag{4}$$

The GREEN strain rate $\dot{\mathbf{E}}$ is subjected to the same transformation. The result is a covariant Oldroyd derivative, calculated with $\hat{\mathbf{L}}_i$:

$$\overset{\triangle}{\hat{\boldsymbol{\Gamma}}} = \mathbf{F}_i^{T-1} \dot{\mathbf{E}} \mathbf{F}_i^{-1} = \mathbf{F}_i^{T-1} \left(\mathbf{F}_i^T \hat{\boldsymbol{\Gamma}} \mathbf{F}_i\right)^{\boldsymbol{\cdot}} \mathbf{F}_i^{-1} = \dot{\hat{\boldsymbol{\Gamma}}} + \hat{\mathbf{L}}_i^T \hat{\boldsymbol{\Gamma}} + \hat{\boldsymbol{\Gamma}} \hat{\mathbf{L}}_i. \tag{5}$$

On account of (4), the total strain rate decomposes additively as well into

$$\overset{\triangle}{\hat{\boldsymbol{\Gamma}}} = \overset{\triangle}{\hat{\boldsymbol{\Gamma}}}_e + \overset{\triangle}{\hat{\boldsymbol{\Gamma}}}_i. \tag{6}$$

The elastic part is the Oldroyd derivative of the elastic strain, while the inelastic part turns out to be the symmetric part of the inelastic deformation rate:

$$\overset{\triangle}{\hat{\boldsymbol{\Gamma}}}_e = \dot{\hat{\boldsymbol{\Gamma}}}_e + \hat{\mathbf{L}}_i^T\hat{\boldsymbol{\Gamma}}_e + \hat{\boldsymbol{\Gamma}}_e\hat{\mathbf{L}}_i, \tag{7}$$

$$\overset{\triangle}{\hat{\boldsymbol{\Gamma}}}_i = \dot{\hat{\boldsymbol{\Gamma}}}_i + \hat{\mathbf{L}}_i^T\hat{\boldsymbol{\Gamma}}_i + \hat{\boldsymbol{\Gamma}}_i\hat{\mathbf{L}}_i = \frac{1}{2}\left(\hat{\mathbf{L}}_i + \hat{\mathbf{L}}_i^T\right). \tag{8}$$

The basic stress tensor in constitutive theories is the secondPIOLA-KIRCHHOFF stress $\tilde{\mathbf{T}} = \mathbf{F}^{-1}\mathbf{S}\mathbf{F}^{T-1}$, defined on the basis of the weighted CAUCHY stress (KIRCHHOFF stress) $\mathbf{S} = (\det \mathbf{F})\,\mathbf{T}$, where $\mathbf{T}$ denotes the CAUCHY stress tensor. Transformation of $\tilde{\mathbf{T}}$ to the intermediate configuration leads to

$$\hat{\mathbf{S}} = \mathbf{F}_i\tilde{\mathbf{T}}\mathbf{F}_i^T = \hat{\mathbf{F}}_e^{-1}\mathbf{S}\hat{\mathbf{F}}_e^{T-1}. \tag{9}$$

On application of (5) in combination with (6), (7) and (8), we arrive at the following decomposition of the total stress power into elastic and inelastic parts:

$$\tilde{\mathbf{T}}\cdot\dot{\mathbf{E}} = \hat{\mathbf{S}}\cdot\overset{\triangle}{\hat{\boldsymbol{\Gamma}}} = \hat{\mathbf{S}}\cdot\dot{\hat{\boldsymbol{\Gamma}}}_e + \hat{\mathbf{P}}\cdot\hat{\mathbf{L}}_i. \tag{10}$$

In particular, the inelastic part is the product of the inelastic deformation rate $\hat{\mathbf{L}}_i$ and a stress tensor $\hat{\mathbf{P}}$ of the MANDEL-type (Mandel [7], p. 127), defined by

$$\hat{\mathbf{P}} = \left(\mathbf{1} + 2\hat{\boldsymbol{\Gamma}}_e\right)\hat{\mathbf{S}} = \hat{\mathbf{F}}_e^T\mathbf{S}\hat{\mathbf{F}}_e^{T-1}. \tag{11}$$

Due to the elastic anisotropy of the material, this stress tensor is non-symmetric. For physical reasons, a constitutive model must have certain invariance properties. These should be required with respect to a change of frame (material objectivity), defined by a time-dependent proper rotation, $t \mapsto \mathbf{Q}(t)$, and with respect to a local rotation of the intermediate configuration, $(\boldsymbol{X}, t) \mapsto \bar{\mathbf{Q}}(\boldsymbol{X}, t)$, which may be time- and space-dependent (Green and Naghdi [4], Casey and Naghdi [2]). Clearly, the deformation gradient transforms according to

$$\mathbf{F} \mapsto \mathbf{F}^* = \mathbf{Q}\mathbf{F}, \tag{12}$$

and its elastic and inelastic parts according to

$$\hat{\mathbf{F}}_e \mapsto \hat{\mathbf{F}}_e^* = \mathbf{Q}\hat{\mathbf{F}}_e\bar{\mathbf{Q}}^T \quad \text{and} \quad \mathbf{F}_i \mapsto \mathbf{F}_i^* = \bar{\mathbf{Q}}\mathbf{F}_i. \tag{13}$$

From $(13)_2$ we infer the transformation behavior of the inelastic deformation rate, i.e.

$$\hat{\mathbf{L}}_i \mapsto \hat{\mathbf{L}}_i^* = \dot{\bar{\mathbf{Q}}}\bar{\mathbf{Q}}^T + \bar{\mathbf{Q}}\hat{\mathbf{L}}_i\bar{\mathbf{Q}}^T. \tag{14}$$

Obviously, the symmetric part $\hat{\mathbf{D}}_i$ of $\hat{\mathbf{L}}_i$ is objective with respect to $\bar{\mathbf{Q}}$, while the antisymmetric part $\hat{\mathbf{W}}_i$ is not: $\hat{\mathbf{W}}_i^* = \dot{\bar{\mathbf{Q}}}\bar{\mathbf{Q}}^T + \bar{\mathbf{Q}}\hat{\mathbf{W}}_i\bar{\mathbf{Q}}^T$. To obtain the transformation behavior of the stress tensors is a straight-forward matter: We have $\mathbf{S}^* = \mathbf{Q}\mathbf{S}\mathbf{Q}^T$, $\tilde{\mathbf{T}}^* = \tilde{\mathbf{T}}$, $\hat{\mathbf{S}}^* = \bar{\mathbf{Q}}\hat{\mathbf{S}}\bar{\mathbf{Q}}^T$ and $\hat{\mathbf{P}}^* = \bar{\mathbf{Q}}\hat{\mathbf{P}}\bar{\mathbf{Q}}^T$.

3. Strain energy

The constitutive model to be set up is based on the definition of a free energy function, which is assumed to be the sum of one elastic and one inelastic part:

$$\psi = \psi_e + \psi_i \tag{15}$$

Following the ideas of TSAKMAKIS (Häusler et al. [5], Tsakmakis [11]), the strain energy should depend on the elastic deformation gradient, but also on the orientation of the crystal, incorporated by means of an orthogonal tensor-valued internal variable $\mathbf{\Phi}$ $\left(\mathbf{\Phi}\mathbf{\Phi}^T = \mathbf{1}\right)$:

$$\psi_e = \psi_e\left(\hat{\mathbf{F}}_e, \mathbf{\Phi}\right). \tag{16}$$

$\mathbf{\Phi}$ is transforms in the same way as the intermediate configuration:

$$\mathbf{\Phi} \mapsto \mathbf{\Phi}^* = \bar{\mathbf{Q}}\mathbf{\Phi}. \tag{17}$$

Then, arguments of invariance suggest ψ_e to be a function of only one tensor-valued argument, namely the elastic strain $\hat{\mathbf{\Gamma}}_e$, rotated by $\mathbf{\Phi}$ (Häusler et al. [5], Tsakmakis [11]):

$$\psi_e = \hat{\varphi}\left(\mathbf{\Phi}^T\hat{\mathbf{\Gamma}}_e\mathbf{\Phi}\right). \tag{18}$$

According to known results in the literature (Smith and Rivlin [10]), the strain energy in the context of elastic anisotropy is a (polynomial) function of a finite set of invariants, which are uniquely determined if a particular symmetry group is given. However, in this case these invariants are formed on the basis of the strain tensor $\mathbf{\Phi}^T\hat{\mathbf{\Gamma}}_e\mathbf{\Phi}$, which is the elastic strain rotated back from the intermediate configuration to the so-called isoclinic intermediate configuration:

$$\psi_e = \hat{\varphi}\left(\mathbf{\Phi}^T\hat{\mathbf{\Gamma}}_e\mathbf{\Phi}\right) = \hat{\varphi}\left(\hat{I}_1, \hat{I}_2, ...\hat{I}_N\right). \tag{19}$$

4. Entropy inequality and evolution equations

Applying the decompositions (10) and (15) and the strain energy (19), the dissipation inequality $\tilde{\mathbf{T}} \cdot \dot{\mathbf{E}} - \rho_R\dot{\psi}(t) \geq 0$ transforms to

$$\left(\hat{\mathbf{S}} - \rho_R\mathbf{\Phi}\frac{d\hat{\varphi}(\mathbf{\Phi}^T\hat{\mathbf{\Gamma}}\mathbf{\Phi}_e)}{d(\mathbf{\Phi}^T\hat{\mathbf{\Gamma}}\mathbf{\Phi}_e)}\mathbf{\Phi}^T\right) \cdot \dot{\hat{\mathbf{\Gamma}}}_e +$$

$$+ \hat{\mathbf{P}} \cdot \hat{\mathbf{L}}_i - 2\rho_R\left(\hat{\mathbf{\Gamma}}_e\mathbf{\Phi}\frac{d\hat{\varphi}(\mathbf{\Phi}^T\hat{\mathbf{\Gamma}}\mathbf{\Phi}_e)}{d(\mathbf{\Phi}^T\hat{\mathbf{\Gamma}}\mathbf{\Phi}_e)}\mathbf{\Phi}^T\right) \cdot \dot{\mathbf{\Phi}}\mathbf{\Phi}^T - \rho_R\dot{\psi}_i(t) \geq 0. \tag{20}$$

This leads to the stress relation

$$\hat{\mathbf{S}} = \rho_R \frac{\partial \hat{\varphi}(\mathbf{\Phi}^T \hat{\mathbf{\Gamma}} \mathbf{\Phi}_e)}{\partial \hat{\mathbf{\Gamma}}_e} \tag{21}$$

and with definition (11) to the residual inequality

$$\hat{\mathbf{P}} \cdot \hat{\mathbf{L}}_i - \hat{\mathbf{P}} \cdot \left(\dot{\mathbf{\Phi}} \mathbf{\Phi}^T \right) - \rho_R \dot{\psi}_i \geq 0 \,. \tag{22}$$

For metallic materials with crystalline structure, inelastic deformations are dynamic processes of simple shearing, taking place at various slip systems, which are determined from the geometry of a particular crystal. If $\dot{\kappa}_\alpha$ is the shear rate belonging to the slip system $\{\boldsymbol{N}_\alpha, \boldsymbol{S}_\alpha\}$ in the reference configuration, the superposition of all shear rates yields an evolution equation for the field $\tilde{\mathbf{F}}_i(\boldsymbol{X}, t)$, the so-called isoclinic intermediate configuration (Mandel [7], Dafalias [3]):

$$\dot{\tilde{\mathbf{F}}}_i(t) \tilde{\mathbf{F}}_i^{-1}(t) = \sum_\alpha \dot{\kappa}_\alpha \boldsymbol{S}_\alpha \otimes \boldsymbol{N}_\alpha \tag{23}$$

The intermediate configuration $\mathbf{F}_i(\boldsymbol{X}, t)$ differs from the isoclinic intermediate configuration by the rotation $\mathbf{\Phi}$; we have

$$\mathbf{F}_i = \mathbf{\Phi} \tilde{\mathbf{F}}_i \quad \text{or, in view of (1)} \quad \mathbf{F} = \hat{\mathbf{F}}_e \mathbf{\Phi} \tilde{\mathbf{F}}_i. \tag{24}$$

In particular, $\{\boldsymbol{n}_\alpha^0, \boldsymbol{s}_\alpha^0\}$ is the image of the slip system $\{\boldsymbol{N}_\alpha, \boldsymbol{S}_\alpha\}$ in the intermediate configuration, i.e. $\boldsymbol{n}_\alpha^0 = \mathbf{\Phi} \boldsymbol{N}_\alpha$ and $\boldsymbol{s}_\alpha^0 = \mathbf{\Phi} \boldsymbol{S}_\alpha$. $\{\boldsymbol{n}_\alpha, \boldsymbol{s}_\alpha\}$ is the same slip system in the current configuration, with

$$\boldsymbol{n}_\alpha = \hat{\mathbf{F}}_e^{T-1} \boldsymbol{n}_\alpha^0 = \hat{\mathbf{F}}_e^{T-1} \mathbf{\Phi} \boldsymbol{N}_\alpha \quad \text{and} \quad \boldsymbol{s}_\alpha = \hat{\mathbf{F}}_e \boldsymbol{s}_\alpha^0 = \hat{\mathbf{F}}_e \mathbf{\Phi} \boldsymbol{S}_\alpha. \tag{25}$$

From $(24)_1$ we conclude that the evolution equation (23) for the isoclinic intermediate configuration is equivalent to the following evolution equation for the intermediate configuration $\mathbf{F}_i(\boldsymbol{X}, t)$:

$$\dot{\mathbf{F}}_i(t) \mathbf{F}_i^{-1}(t) = \dot{\mathbf{\Phi}}(t) \mathbf{\Phi}^T(t) + \sum_\alpha \dot{\kappa}_\alpha \boldsymbol{s}_\alpha^0 \otimes \boldsymbol{n}_\alpha^0 \tag{26}$$

The equivalence of (23) and (26) holds for all functions $\mathbf{\Phi}(\boldsymbol{X}, t)$. Together with $(11)_2$ eq. (26) implies

$$\hat{\mathbf{P}} \cdot \left(\dot{\mathbf{F}}_i \mathbf{F}_i^{-1} - \dot{\mathbf{\Phi}} \mathbf{\Phi}^T \right) = \left(\hat{\mathbf{F}}_e^T \mathbf{S} \hat{\mathbf{F}}_e^{T-1} \right) \cdot \left(\sum_\alpha \dot{\kappa}_\alpha \boldsymbol{s}_\alpha^0 \otimes \boldsymbol{n}_\alpha^0 \right) = \sum_\alpha \tau_\alpha \dot{\kappa_\alpha}, \tag{27}$$

where $\tau_\alpha = \mathbf{s}_\alpha^0 \cdot \hat{\mathbf{P}} \mathbf{n}_\alpha^0 = \mathbf{s}_\alpha \cdot \mathbf{S} \mathbf{n}_\alpha$ are the resolved shear stresses (SCHMID-stresses, see Schmid and Boas [9]). Inserting (27) into (22), the dissipation inequality becomes

$$\sum_\alpha \tau_\alpha \dot{\kappa_\alpha} - \rho_R \dot{\psi}_i \geq 0. \tag{28}$$

For simplicity the inelastic part of the free energy is assumed as a quadratic function of all shear strains related to the slip systems:

$$\rho_R \psi_i = \sum_\alpha \frac{1}{2} c_\alpha \left(\kappa_\alpha - y_\alpha\right)^2. \tag{29}$$

The y_α are strain-like internal variables, which may be introduced to represent hardening effects. The c_α are material parameters. Differentiation of (29) and introduction of the backstresses $x_\alpha = c_\alpha \left(\kappa_\alpha - y_\alpha\right)$ transforms the dissipation inequality (28) into

$$\sum_\alpha (\tau_\alpha - x_\alpha)\dot{\kappa}_\alpha + \sum_\alpha x_\alpha (\dot{\kappa}_\alpha - \frac{1}{c_\alpha}\dot{x}_\alpha) \geq 0. \tag{30}$$

There are various ways to fulfill this inequality for all thermomechanical processes in the sense of sufficient conditions. Most simple possibilities are flow rules like (see e.g. Perzyna [8])

$$\dot{\kappa}_\alpha(t) = \frac{1}{\eta_\alpha} \left\langle \frac{|\tau_\alpha - x_\alpha| - k_\alpha}{r_\alpha} \right\rangle^m \frac{\tau_\alpha - x_\alpha}{|\tau_\alpha - x_\alpha|} \tag{31}$$

and ARMSTRONG-FREDERICK equations for the backstresses:

$$\dot{x}_\alpha(t) = c_\alpha \dot{\kappa}_\alpha(t) - b_\alpha |\dot{\kappa}_\alpha(t)| x_\alpha \tag{32}$$

The k_α are yield stresses, η_α viscosities, m, c_α, b_α and r_α are material constants.

5. Lattice spin

Since the internal variable $\mathbf{\Phi}$ is introduced to represent the orientation of the crystal, the associated rotation rate $\dot{\mathbf{\Phi}}\mathbf{\Phi}^T$ represents the spin of the crystallographic lattice. A rotation $\bar{\mathbf{Q}}$ of the intermediate configuration may be interpreted as a local change of frame, where the origin remains the same. Obviously, the lattice spin is not objective under the rotation $\bar{\mathbf{Q}}$,

$$\dot{\mathbf{\Phi}}\mathbf{\Phi}^T \mapsto \dot{\mathbf{\Phi}}^\star \mathbf{\Phi}^{T\star} = \dot{\bar{\mathbf{Q}}}\bar{\mathbf{Q}}^T + \bar{\mathbf{Q}} \left(\dot{\mathbf{\Phi}}\mathbf{\Phi}^T\right) \bar{\mathbf{Q}}^T. \tag{33}$$

However, the following can be shown: If $\bar{\mathbf{Q}}(\boldsymbol{X}, t)$ represents a change of frame, going from one relative frame to another, the rate of rotation with respect to the absolute frame is an objective tensor:

$$\dot{\mathbf{\Phi}}^\star \mathbf{\Phi}^{T\star} - \bar{\mathbf{\Omega}}^\star = \bar{\mathbf{Q}} \left(\dot{\mathbf{\Phi}}\mathbf{\Phi}^T - \bar{\mathbf{\Omega}}\right) \bar{\mathbf{Q}}^T. \tag{34}$$

This point of view can be applied in crystal viscoplasticity to set up an evolution equation for the rotation of the crystallographic lattice. As a tentative assumption, we postulate

$$\dot{\mathbf{\Phi}}\mathbf{\Phi}^T - \bar{\mathbf{\Omega}} = \beta \mathbf{\Phi} \tilde{\mathbf{W}}_i \mathbf{\Phi}^T + \gamma \hat{\mathbf{P}}_A \tag{35}$$

with $\bar{\mathbf{\Omega}} = \dot{\bar{\mathbf{Q}}}\bar{\mathbf{Q}}^T, \quad \tilde{\mathbf{W}}_i = \frac{1}{2}\sum_{\alpha} \dot{\kappa}_\alpha \left(\boldsymbol{S}_\alpha \otimes \boldsymbol{N}_\alpha - \boldsymbol{N}_\alpha \otimes \boldsymbol{S}_\alpha\right),$

$\hat{\mathbf{P}}_A = \frac{1}{2}\left(\hat{\mathbf{P}} - \hat{\mathbf{P}}^T\right)$ and the material constants β and γ. The left-hand side of (35) represents the rate of rotation with respect to the absolute frame, consisting of the relative rate $\dot{\mathbf{\Phi}}\mathbf{\Phi}^T$ plus the rotation of the relative frame, $-\bar{\mathbf{\Omega}} = -\dot{\bar{\mathbf{Q}}}\bar{\mathbf{Q}}^T$. Since both sides of (35) are objective under a transformation from one relative frame to another, the postulate of full invariance is satisfied for the constitutive equation (35) as it is for all other constitutive equations of this theory. However, it should be recognized that the evolution equation (35) does not meet the assumption of objectivity in the usual formulation, because an information about the applied frame is included in it, namely its rate of rotation $\bar{\mathbf{\Omega}}$. For practical use we set $\bar{\mathbf{\Omega}} = \mathbf{0}$, i.e. $\bar{\mathbf{Q}} = \bar{\mathbf{Q}}(\boldsymbol{X}) =$ const. in time. Then eqn. (35) specializes to

$$\dot{\mathbf{\Phi}}\mathbf{\Phi}^T = \beta \mathbf{\Phi} \tilde{\mathbf{W}}_i \mathbf{\Phi}^T + \gamma \hat{\mathbf{P}}_A. \tag{36}$$

6. Numerical simulations

The numerical simulations refer to face-centered cubic crystals. If we restrict the theory to incorporate only small elastic deformations, there are three invariants of second order. In the sense of physical linearization the strain energy would then read

$$\rho_R \psi_e = \rho_R \hat{\varphi}\left(\mathbf{\Phi}^T \hat{\mathbf{\Gamma}}_e \mathbf{\Phi}\right) = \frac{1}{2} C_1 \hat{I}_1 + C_2 \hat{I}_2 + C_3 \hat{I}_3, \tag{37}$$

with the three second-order invariants (Haupt [6], p. 372) $\hat{I}_1$, $\hat{I}_2$, $\hat{I}_3$ and no more than three elasticity constants C_1, C_2, C_3. In order to illustrate the physical meaning of the proposed constitutive theory, numerical test calculations are necessary. To this end, numerical values for the material parameters were assumed as follows: $C_1 = 170000\,$MPa, $C_2 = 124000\,$MPa, $C_3 = 76000\,$MPa, $\tau_{cr} = 150\,$MPa, $\eta = 100000\,$s, $m = 2.5$, $c_\alpha = 0\,$MPa, $b_\alpha = 0$, $\beta = 0\ (0.5)\ (0.9)$ and $\gamma = 0$. All yield limits and viscosities are set equal and all hardening effects are omitted for simplicity. The following results of numerical simulations refer to strain-controlled monotonic tension of single-crystals. Accordingly, the total velocity gradient is prescribed to

$$\mathbf{L} = (-0.0005(\mathbf{e}_x \otimes \mathbf{e}_x + \mathbf{e}_y \otimes \mathbf{e}_y) + 0.001(\mathbf{e}_z \otimes \mathbf{e}_z))\,[s^{-1}].$$

As a first test of the material model the constitutive equations are integrated to simulate the stress-strain behavior of a single crystal. Four different initial orientations of the crystal were assumed, characterized by theEULERian angles $\psi = 0^o$, $\vartheta = 10^o, 20^o, 30^o, 40^o$ and $\varphi = -30^o$. The choice of the initial orientations implies that the representation of the tension axis falls into the standard triangle of the inverse polefigure (see figure 2).

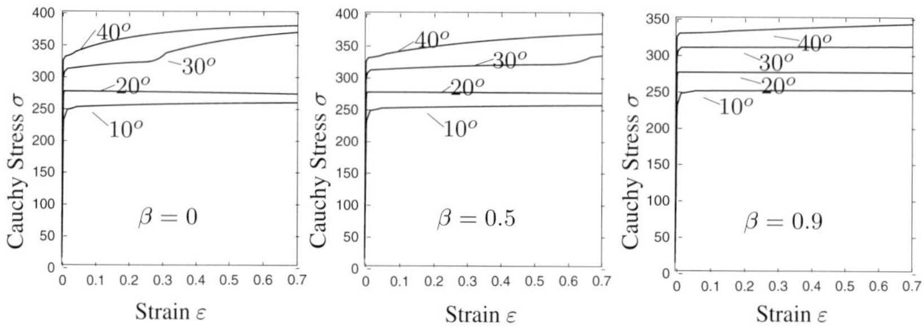

Figure 1. Uniaxial tension: stress-strain curve.

In figure 1 four stress-strain curves (axial Cauchy stress versus axial engineering strain up to 70%) are depicted, belonging to the four different initial orientations and different values for β. We notice that in case of $\beta = 0$ two different values of the saturation stress are reached: the saturation stresses for$\vartheta = 30^0$ and $\vartheta = 40^0$ seem to be equal as well as the saturation stresses for $\vartheta = 20^0$ and $\vartheta = 10^0$. This attribute disappears more and more for an increase ofβ, so that we obtain in the case of $\beta = 0.9$ four different values of saturation stress belonging to the four different initial values ofϑ. The stress-strain curves are not smooth at various points; this is particularly true for$\vartheta = 30^0$ and results from the changing activities of the different slip systems.
As shown in figure 2, the tension-axis tends either towards the [-110] direction

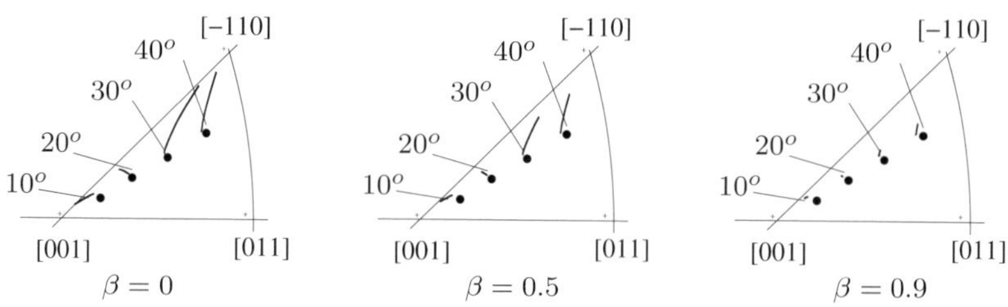

Figure 2. Inverse polefigure for $\beta = 0$, $\beta = 0.5$ and $\beta = 0.9$, uniaxial tension.

of the lattice or towards the [001] direction. This depends on the initial orientation and suggests that there might exist a separation line within the triangle, so that for any initial direction above that line, the [-110] direction is a point of attraction, whereas each initial direction below tends asymptotically to the [001] direction. Thus, it becomes clear that the two final orientations and the end of the process corresponds to no more than two different values of the saturation stress. In figure 2 the influence of the material parameter β becomes apparent. This parameter represents the possibility to accelerate or decelerate the free rotation of the crystal. In the present investigation, where β was chosen to lie within 0 and 1, the trajectories become shorter, which means that the free rotation of the crystal is decelerated. Further simulations show that acceleration occurs for negative values of β.

In order to simulate the behavior of a polycrystal a space-dependent initial distribution of the orientation was assumed, $\mathbf{\Phi}(\boldsymbol{X}, t_0) = \mathbf{\Phi}(\boldsymbol{X})$. In spite of the fact that this defines an inhomogeneous material, strain-controlled homogeneous deformation processes were analyzed on the level of the constitutive model, i.e. in the sense of the quite common TAYLOR-assumption. The results, which are omitted here, seem to be reasonable. In particular, they demonstrate again the physical meaning of the parameter β. Process-induced texture is less pronounced for positive values and becomes sharper for negative values of β.

7. Conclusion

The presented constitutive model includes the classical theory of crystal viscoplasticity as a special case, namely for $\beta = \gamma = 0$, i.e. $\dot{\mathbf{\Phi}}(\boldsymbol{X}, t) = \mathbf{0}$. The proposed theory generalizes crystal plasticity in the sense that the lattice spin is represented by means of an independent evolution equation, which may be formulated without being restricted by the dissipation inequality. On principle, the theory applies to crystals of any symmetry property, and it satisfies the postulate of full invariance. First test calculations related to face-centered cubic crystals without incorporating hardening effects show physically reasonable results, however, more detailed investigations are required in future work.

References

[1] Asaro, R.J. (1983). Crystal Plasticity. *Journal of Applied Mechanics*, **50**, 921–934.

[2] Casey, J. and Naghdi, P.M. (1980). A Remark on the Use of the Decomposition $\mathbf{F} = \mathbf{F}_e\mathbf{F}_p$ in Plasticity. *Journal of Applied Mechanics*, **47**, 672–675.

[3] Dafalias, Y.F. (1998). Plastic Spin: Necessity or Redundancy. *Int. Journal of Plasticity*, **14**, 909–931.

[4] Green, A.E. and Naghdi, P.M. (1971). Some Remarks on Elastic-Plastic Deformation at Finite Strain, *Journal of Engineering Science*, **8**, 1219–1229.

[5] Häusler, O., Tsakmakis, Ch. and Aifantis, E.C. (1999). On Anisotropic Plastic Flow. In: *IUTAM Symposium on Micro- and Macrostructural Aspects of Thermoplasticity*, O.T. Bruhns, E. Stein (eds.), Solid Mechanics and its Applications, Kluwer Academic Publishers, Dordrecht, Boston, London, **62**, 69–78.

[6] Haupt, P. (2000). *Continuum Mechanics and Theory of Materials*, Springer.

[7] Mandel, J. (1971). Plasticité Classique et Viscoplasticité. *CISM*, **397**, Udine, Springer, New York.

[8] Perzyna, P. (1966). Fundamental Problems in Viscoplasticity. *G. Kuerti (ed.), Advances in Applied Mechanics*, New York, **9**, 243–377.

[9] Schmid, E. and Boas, W. (1935). *Kristallplastizität*, Springer, Berlin.

[10] Smith, G.F. and Rivlin, R.S. (1958). The Strain Energy Function for Anisotropic Elastic Materials. *Trans. Amer. Math. Soc.*, **88**, 175–193.

[11] Tsakmakis, Ch. (2001). Description of Plastic Anisotropy Effects at Large Deformations. PartI: Restrictions Imposed by the Second Law and the Postulate of Il'iushin.; submitted for publication. *Int. Journ. of Plasticity*.

Anisotropic Elastoplastic Material Behavior in Fabric Structures

S. Reese
Department of Civil Engineering, Ruhr University Bochum
IA 01/128, 44780 Bochum, Germany
reese@nm.ruhr-uni-bochum.de

Abstract: Many composites consist of a fabric structure embedded in a matrix material. The fibres are often made of material which shows noticeable plastic deformation. The overall stiffness of the specimens is usually determined by the stiffness of these fibres, such that the correct modeling of the orthotropy of the composite is very important. In addition, the structure experiences large deformations which must be accounted for. In the present paper, suitable models for this type of materials are therefore derived in the framework of finite anisotropic plasticity. A main problem is, however, the lack of experimental data in the literature. For this reason, a computer model of the composite is set up for numerical experiments. The results serve to determine the material parameters of the developed continuum mechanical model.

Keywords: Composite, fibres, large deformation, orthotropy, finite elements, inflated structures.

1. Introduction

Anisotropic elastoplasticity has up to now mainly attracted attention in the context of crystal plasticity, where the structural vectors in the intermediate configuration are fixed and the yield condition is formulated in terms of the Schmidt stress. Refer e.g. to the classical works of Hill [9], Rice [17] and Asaro [1]. For the computational aspects see Cuitiño and Ortiz [7] and Miehe [13]. In the present work, the situation is different in the regard that the fibre orientations represent so-called material directions. The structural vectors transform like material line elements.

Anisotropic material behavior of the latter kind occurs e.g. in the context of biomechanical problems (Holzapfel et al. [10, 11], Weiss et al. [24], Bonet and Burton [6], Gasser and Holzapfel [8]). Among these papers, anisotropic elasto*plasticity* is only included in the recent works of Holzapfel and coworkers. An alternative approach to model fabric-reinforced composites, based on a rate-type formulation using the Bingham theory, has been proposed by Spencer [19]. An efficient finite element implementation, although restricted to purely*elastic* material behavior can be found in Reese et al. [15]. The present paper extends the latter model to include the important effect of plasticity.

C. Miehe (ed.), IUTAM Symposium on Computational Mechanics of Solid Materials at Large Strains, 201-210.

2. Computer model

Figure 1 shows "real" experimental results (uniaxial tension) for the two fibre directions of the fabric (polyester threads). The stiffer thread is called warp thread, the other one weft thread. The strain values are given in percent and refer to the linearized strain measure. One observes a noticeable permanent deformation, when unloading takes place. The hysteresis visible during reloading can be neglected and is therefore not shown in the picture.

The material behavior of the fibres seems to be somewhat rubber-like. This suggests a model which is usually applied in the context of rubber elasticity (see Ogden [14]). The extension to one-dimensional elastoplasticity is straightforward (see also Reese [16]).

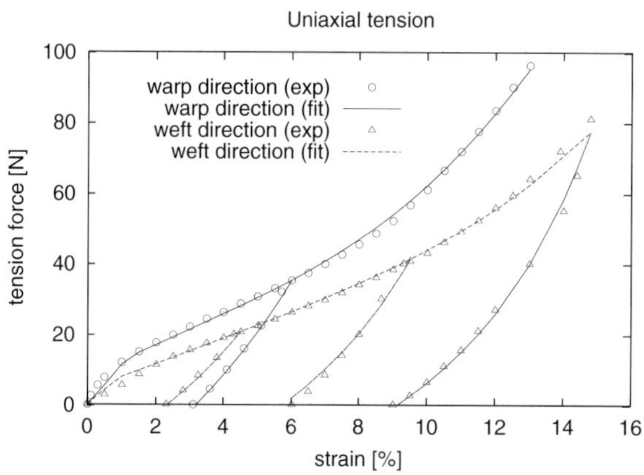

Figure 1. Experimental results and fit for uniaxial tension.

Experimental data for composites are frequently found in the form of internal reports of companies but are poorly documented in the available literature. In order to overcome this problem, we set up a "computer testing device" to obtain appropriate stress-strain data. The material parameters of the orthotropic continuum mechanical model will then be fit to these "experimental" results.

In Figure 2, the computer model for the fibre-reinforced membrane is shown. It consists of three-dimensional non-linear truss elements modeling the fibres and special low-order brick elements describing the rubber coating.

To get a representative response, ten (to be precise nine in the interior and two cut in half at the boundary) fibres are placed in each direction. With this two-phase approach, one is able to model the two constituents separately. For the rubber, the standard Neo-Hookean approach is used. The material behavior of the polyester fibres is described by means of the model discussed before.

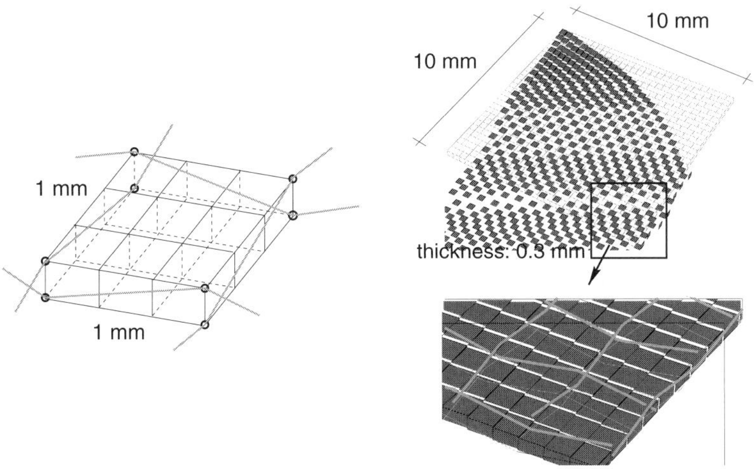

Figure 2. Computer model: discretization (left), geometry (right).

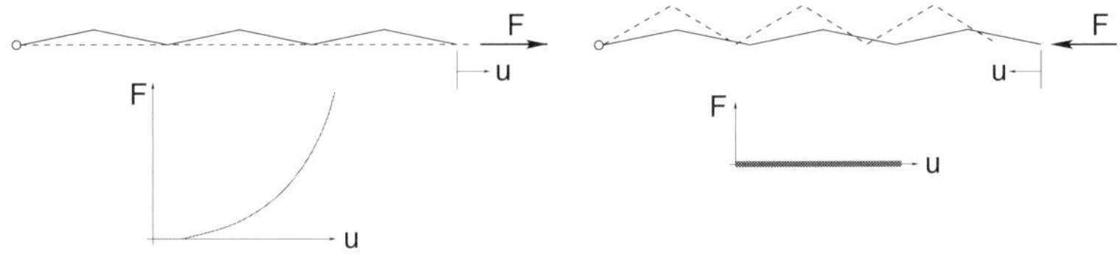

Figure 3. Fibre behavior under tension and compression.

Since the fibres are not fully extended in their initial configuration (solid line in Figure 3), one can pull them in a stress-free state up to the point, where they are stretched straight. The strain at this time is easily calculated from comparing the initial length of the fibres ($L = 10.44\,\mathrm{mm}$) with the length of the test sample (10 mm). So, the stresses measured up to a strain of about $\Delta L/L = 0.44/10 = 4.4\,\%$ are due to the influence of the rubber matrix alone and therefore very small (rubber elastic range, Figure 4). Beyond this limit, the stress in the fibres increases rapidly and the stress-strain relationship of the membrane is dominated by the behavior of the fibres (anisotropic elastic range). The yield point is indicated by the kink. The threads are permanently lengthened by the evolution of the plastic strain. Thus, on the unloading path, they return to the stress-free state already at a strain of about 10% (warp thread) or $13\,\%$ (weft thread), respectively. From this point on, the load is carried again only by the rubber coating. So, although there is a permanent deformation in the fibres, the deformation of the "computer" membrane is completely reversible. In the direction of compression, the situation is in general similar to the small strain case.

3. Continuum model

Strain energy function. We describe the deformation of a continuous body by means of the right Cauchy-Green tensor $\mathbf{C} = \mathbf{F}^T \mathbf{F}$, where $\mathbf{F}$ denotes the deformation gradient. We assume further the existence of a scalar potential $\Psi = W(\mathbf{C}, \mathbf{F}_p) + f(\boldsymbol{\varepsilon})$ (Helmholtz free energy function per reference volume), where $\mathbf{F}_p$ represents the plastic part of the deformation gradient. The vector $\boldsymbol{\varepsilon}$ contains certain internal variables which are specified later in the context of the model. The part $W(\mathbf{C}, \mathbf{F}_p)$ is usually called strain energy function. If $\mathbf{F}_p$ is considered as elastic isomorphism (Wang and Bloom [23], Bertram [2, 3], Svendsen [21]). W can be represented as a function of the "elastic" right Cauchy-Green tensor $\mathbf{C}_e = \mathbf{F}_p^{-T} \mathbf{C} \, \mathbf{F}_p^{-1} = \mathbf{F}_e^T \mathbf{F}_e$, the elastic part of the deformation gradient defined by $\mathbf{F}_e = \mathbf{F} \, \mathbf{F}_p^{-1}$. In the case of orthotropic material behavior, $W(\mathbf{C}_e)$ reduces further to an isotropic function of $\mathbf{C}_e$ and the structural tensors

$$\widetilde{\mathbf{M}}_1 = \widetilde{\mathbf{N}}_1 \otimes \widetilde{\mathbf{N}}_1, \qquad \widetilde{\mathbf{M}}_2 = \widetilde{\mathbf{N}}_2 \otimes \widetilde{\mathbf{N}}_2. \tag{1}$$

See for a more detailed discussion the theoretical works of Boehler [4, 5], Liu [12], Zhang and Rychlewski [25] and Svendsen [20, 22]. It is important to note that $\mathbf{C}_e$, $\widetilde{\mathbf{M}}_1$, and $\widetilde{\mathbf{M}}_2$ are tensors in the intermediate configuration. The vectors $\widetilde{\mathbf{N}}_i$ $(i = 1, 2)$ are given via the equation $(i = 1, 2)$

$$\widetilde{\mathbf{N}}_i = \mathbf{F}_p \mathbf{N}_i \, \frac{1}{|\mathbf{F}_p \mathbf{N}_i|} \quad \Rightarrow \quad \widetilde{\mathbf{M}}_i = \mathbf{F}_p \mathbf{M}_i \, \mathbf{F}_p^T \, \frac{1}{\mathbf{C}_p : \mathbf{M}_i} \tag{2}$$

with $\mathbf{M}_i = \mathbf{N}_i \otimes \mathbf{N}_i$, where the vectors $\mathbf{N}_i$ are oriented parallel to the fibres in the *reference* configuration (see Figure 5). The strain energy function W can be written in terms of the three invariants of $\mathbf{C}_e$ $(I_i,\ i = 1, 2, 3)$ and the first invariants of $\mathbf{C}_e \widetilde{\mathbf{M}}_1$, $\mathbf{C}_e^2 \widetilde{\mathbf{M}}_1$, $\mathbf{C}_e \widetilde{\mathbf{M}}_2$, $\mathbf{C}_e^2 \widetilde{\mathbf{M}}_2$ and $\mathbf{C}_e \widetilde{\mathbf{M}}_1 \widetilde{\mathbf{M}}_2$ $(I_j,\ j = 4, 5, 6, 7, 8)$, respectively. For simplicity, the dependence on the latter tensor, i.e. I_8, will be neglected. A possible form for the strain energy function is then the following: $W = W_{\mathsf{NH}}(I_1, I_3) + W_{\mathsf{iso}}^{+}(I_1, I_2) + W_{\mathsf{ani}}(I_4, I_5, I_6, I_7)$. For the concrete form see Reese [16].

Dissipation inequality. The invariants I_i $(i = 1, ..., 7)$ may be rewritten in the form

$$I_1 = \mathrm{tr}\, \mathbf{C}_e = \mathrm{tr}\, (\mathbf{C}\, \mathbf{C}_p^{-1}), \quad I_2 = \mathrm{tr}\, \mathbf{C}_e^2 = \mathrm{tr}\, (\mathbf{C}\, \mathbf{C}_p^{-1})^2 \tag{3}$$

$$I_3 = \mathrm{tr}\, \mathbf{C}_e^3 = \mathrm{tr}\, (\mathbf{C}\, \mathbf{C}_p^{-1})^3 \tag{4}$$

$$I_4 = \mathrm{tr}\, \mathbf{C}_e \widetilde{\mathbf{M}}_1 = \frac{\mathrm{tr}\, (\mathbf{C}\, \mathbf{M}_1)}{\mathrm{tr}\, (\mathbf{C}_p \mathbf{M}_1)}, \quad I_5 = \mathrm{tr}\, (\mathbf{C}_e^2 \widetilde{\mathbf{M}}_1) = \frac{\mathrm{tr}\, (\mathbf{C}\, \mathbf{C}_p^{-1} \mathbf{C}\, \mathbf{M}_1)}{\mathrm{tr}\, (\mathbf{C}_p \mathbf{M}_1)} \tag{5}$$

The quantities I_6 and I_7 are formulated analogously to I_4 and I_5, using $\mathbf{M}_2$. Obviously, W can be represented as a function of $\mathbf{C}$ and $\mathbf{C}_p^{-1}$. The structural tensors $\mathbf{M}_1$ and $\mathbf{M}_2$ are considered as constants. Thus, the material time derivative of W takes the form

$$\dot{W} = 2\,\frac{\partial W}{\partial \mathbf{C}} : \frac{1}{2}\,\dot{\mathbf{C}} + (2\,\mathbf{F}_p\,\frac{\partial W}{\partial \mathbf{C}_p}\,\mathbf{F}_p^T) : \mathbf{d}_p, \tag{6}$$

where $\mathbf{d}_p$ represents the symmetric part of the "plastic" velocity gradient$\mathbf{l}_p = \dot{\mathbf{F}}_p\,\mathbf{F}_p^{-1}$.

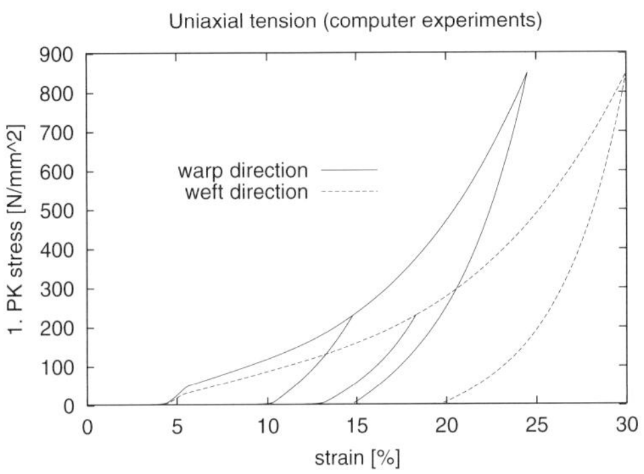

Figure 4. Computer experiments.

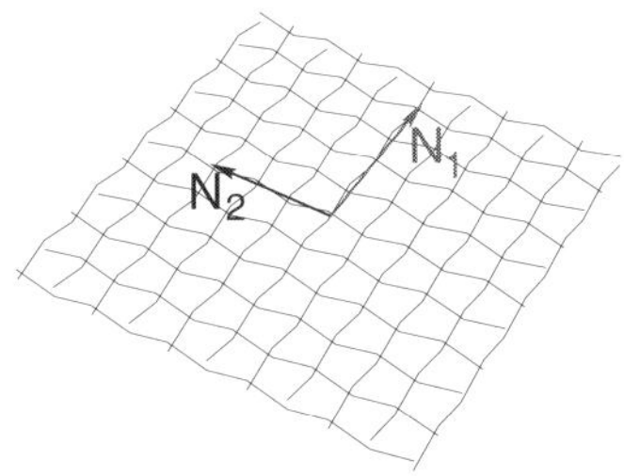

Figure 5. Roven-woven structure

Using the latter specification in the Clausius-Duhem form of the second law of thermodynamics leads with the Helmholtz free energy function$\Psi = W\,(\mathbf{C}, \mathbf{C}_p) + f\,(\varepsilon)$ to the so-called dissipation inequality

$$-\dot{\Psi} + \mathbf{S} : \frac{1}{2}\,\dot{\mathbf{C}} = (\mathbf{S} - 2\,\frac{\partial W}{\partial \mathbf{C}}) : \frac{1}{2}\,\dot{\mathbf{C}} + \boldsymbol{\Sigma} : \mathbf{d}_p + \mathbf{q}\cdot\dot{\varepsilon} \geq 0, \tag{7}$$

in which the stress tensor $\boldsymbol{\Sigma}$ and $\mathbf{q}$ are defined by

$$\boldsymbol{\Sigma} = -2\,\mathbf{F}_p\,\frac{\partial W}{\partial \mathbf{C}_p}\,\mathbf{F}_p^T \quad \text{and} \quad \mathbf{q} = -\frac{\partial f\,(\varepsilon)}{\partial \varepsilon}, \tag{8}$$

respectively. The vector ε contains internal variables describing the hardening of the material. The inequality (7) is fulfilled by $\mathbf{S} = 2\,\frac{\partial W}{\partial \mathbf{C}}$, $\boldsymbol{\Sigma} : \mathbf{d}_p \geq 0$ and $\mathbf{q}\cdot\varepsilon \geq 0$. The requirement $\boldsymbol{\Sigma} : \mathbf{d}_p \geq 0$ gives us a restriction for the choice of $\mathbf{d}_p$, i.e. the evolution of the "plastic" right stretch tensor $\mathbf{U}_p = \mathbf{R}_p^T\mathbf{F}_p$. But the "plastic" rotation tensor $\mathbf{R}_p$ and the plastic spin $\mathbf{w}_p = \text{skew}\,\mathbf{l}_p$ are undetermined.

Yield functions and flow rules. The results of the computer modelling show that the fibres in warp and weft direction hardly interact with each other. This is understandable from the fact that the warp threads are not directly linked to the weft threads (see Figure 2). The interaction is only due to the rubber matrix, the stiffness of which is very low in comparison with the one of the fibres. So, we are able to treat the composite as two one-dimensional systems which are in its initial state oriented orthogonal to each other. Keeping this in mind, it can be easily accepted that plastification begins, when the Cauchy stress in the fibre direction exceeds a certain value. If one takes additionally into account, that the material is nearly incompressible, we may use instead of the Cauchy stress also the Kirchhoff stress and express the projection of this stress on $\mathbf{n}_i = \mathbf{F} \cdot \mathbf{N}_i / |\mathbf{F} \cdot \mathbf{N}_i|$ in the form $\boldsymbol{\tau} : \mathbf{m}_i$ $(i = 1, 2)$, where $\mathbf{m}_i$ is given by $\mathbf{m}_i = \mathbf{n}_i \otimes \mathbf{n}_i$ and the Kirchhoff stress tensor by $\boldsymbol{\tau} = \mathbf{F}\,\mathbf{S}\,\mathbf{F}^T$. In order to fulfill the dissipation inequality, it is important to clarify how the latter quantity is related to the stress tensor $\boldsymbol{\Sigma}$. After some steps, one obtains the result

$$\boldsymbol{\tau} : \mathbf{m}_i = \boldsymbol{\Sigma} : \mathrm{sym}\,(\widetilde{\mathbf{M}}_j \cdot \mathbf{C}_e)\,\frac{1}{I_{2\,i+2}}, \tag{9}$$

which is valid, if one of the statements

(a) the fibres remain orthogonal to each other ($\widetilde{\mathbf{N}}_1 \cdot \widetilde{\mathbf{N}}_2 = \mathbf{n}_1 \cdot \mathbf{n}_2 = 0$)
(b) the elastic part of the deformation is small ($\mathbf{b}_e \approx \mathbf{1}$)

holds. The results obtained with the computer model hardly deviate from (a) or (b) such that the yield functions ($i = 1, 2$)

$$\Phi_i = \left| \boldsymbol{\Sigma} : \mathrm{sym}\,(\widetilde{\mathbf{M}}_i \cdot \mathbf{C}_e)\,\frac{1}{I_{2\,i+2}} \right| - (\sigma_{Y\,i} + H_{\mathsf{lin}\,i}\,\xi_i + H_{\mathsf{nl}\,i}\,\xi_i^{\chi_i}) \tag{10}$$

can be considered to be a physically reasonable choice. The coefficients of $\mathbf{q}$ are here given with $q_i = H_{\mathsf{lin}\,i}\,\xi_i + H_{\mathsf{nl}\,i}\,\xi_i^{\chi_i}$ $(i = 1, 2)$. The evolution equations are chosen as

$$\mathbf{d}_p = \sum_{i=1}^{2} \dot{\gamma}_i\,\frac{\partial \Phi_i}{\partial \boldsymbol{\Sigma}}, \qquad \dot{\xi}_i = \dot{\gamma}_i, \tag{11}$$

where the vector $\boldsymbol{\varepsilon}$ introduced in (7) is specified with $\boldsymbol{\varepsilon}^T = \{\xi_1, \xi_2\}$. Above that, the Kuhn-Tucker conditions $\dot{\gamma}_i \geq 0$, $\Phi_i \leq 0$, $\dot{\gamma}_i\,\Phi_i = 0$ $(i = 1, 2)$ have to be fulfilled. Obviously, (11) fulfills the residual dissipation inequality (7). Although the expression $\boldsymbol{\tau} : \mathbf{m}_i$ cannot be exactly transferred into an expression including $\boldsymbol{\Sigma}$, it would be certainly still preferable from the physical point of view to work with the yield criteria ($i = 1, 2$)

$$\bar{\Phi}_i = |\,\boldsymbol{\tau} : \mathbf{m}_i\,| - (\sigma_{Y\,i} + H_{\mathsf{lin}\,i}\,\xi_i + H_{\mathsf{nl}\,i}\,\xi_i^{\chi_i}). \tag{12}$$

This can be achieved by exploiting $(11)_1$ to yield

$$\begin{aligned} \mathbf{F}_e \cdot \mathbf{d}_p \cdot \mathbf{F}_e^T &= \sum_{i=1}^{2} \dot{\gamma}_i \, \mathrm{sign}\,(\boldsymbol{\tau} : \mathbf{m}_i) \;\; \mathrm{sym}\,(\mathbf{m}_i \cdot \mathbf{b}_e) \\ &= \sum_{i=1}^{2} \dot{\gamma}_i \, \mathrm{sym} \left(\frac{\partial \bar{\Phi}_i}{\partial \boldsymbol{\tau}} \cdot \mathbf{b}_e \right) \end{aligned} \tag{13}$$

It is obvious that in the case of $\bar{\Phi}_i$ being an isotropic function of $\boldsymbol{\tau}$ and $\boldsymbol{\tau}$ being coaxial with $\mathbf{b}_e$, we obtain the classical evolution equation for finite isotropic elastoplasticity written in terms of the Oldroyd derivative of $\mathbf{b}_e$, the tensor $\mathcal{L}_v \, \mathbf{b}_e = -2 \, \mathbf{F}_e \, \mathbf{d}_p \, \mathbf{F}_e^T$ (see Simo and Miehe [18]). It can be easily shown that also (13) fulfills the dissipation inequality. The implementation of (13) into a finite element code is discussed in Reese [16].

4. Parameter identification

The material model presented in Section 3 is subsequently fit to the results of the computer experiment already discussed in Section 2. The fit is at first restricted to uniaxial and biaxial tension (see Figure 6). Obviously, the correlation between the "experimental" results and the results obtained with the continuum mechanical model is very good. The part W_{iso}^{+} in the strain energy function can be neglected. W_{NH} has the classical Neo-Hooke form. For W_{ani}, one obtains

$$W_{\mathsf{ani}} = K_1^{\mathsf{ani}\,1} \, (I_4 - c_4)^{\beta_1} + K_1^{\mathsf{ani}\,2} \, (I_6 - c_6)^{\gamma_1}, \tag{14}$$

where the parameters $K_1^{\mathsf{ani}\,i}$ $(i = 1, 2)$ are only unequal to zero, if $I_{2\,i+2} \geq c_{2\,i+2}$ holds (tension in the fibre). Otherwise, the fibre is not stretched straight and therefore does not carry any load. See fore more details the work of Reese [16]. The material parameters used to fit the experimental results in Figure 6 are given in what follows:

$$\begin{aligned} K_1^{\mathsf{ani}\,1} &= 1000 \text{ N/mm}^2 & K_1^{\mathsf{ani}\,2} &= 1000 \text{ N/mm}^2 \\ \sigma_{Y\,1} &= 50 \text{ N/mm}^2 & \sigma_{Y\,2} &= 25 \text{ N/mm}^2 \\ H_{\mathsf{lin}\,1} &= 2480 \text{ N/mm}^2 & H_{\mathsf{lin}\,2} &= 2150 \text{ N/mm}^2 \\ H_{\mathsf{nl}\,1} &= 65 \cdot 10^6 \text{ N/mm}^2 & H_{\mathsf{nl}\,2} &= 11 \cdot 10^6 \text{ N/mm}^2 \\ \beta_1 &= 2.3 & \gamma_1 &= 2.3 \\ \chi_1 &= 4.69 & \chi_2 &= 4.69 \end{aligned}$$

Note that the small strain stiffness is practically only determined by the Neo-Hooke part of the model, because the other parts of the stress are not active in this strain range. The Neo-Hooke parameters can therefore be chosen equal to the values in the matrix of the computer model ($\mu = 1.4\,\text{N/mm}^2$,

$\Lambda = 300\,\mathrm{N/mm^2}$). It should be emphasized that the model is fit to the three experiments *simultaneously* (uniaxial tension in weft direction not shown in Figure 6). At last, it is certainly important to test the model for the case that the fibres rotate. For this purpose, we look at the shearing of the computer test piece (Figure 7a). The boundary conditions in the third coordinate direction are chosen such that a plane stress state is obtained. For the results shown in Figure 7b, the material parameters have not been fit again. The agreement is very satisfactory.

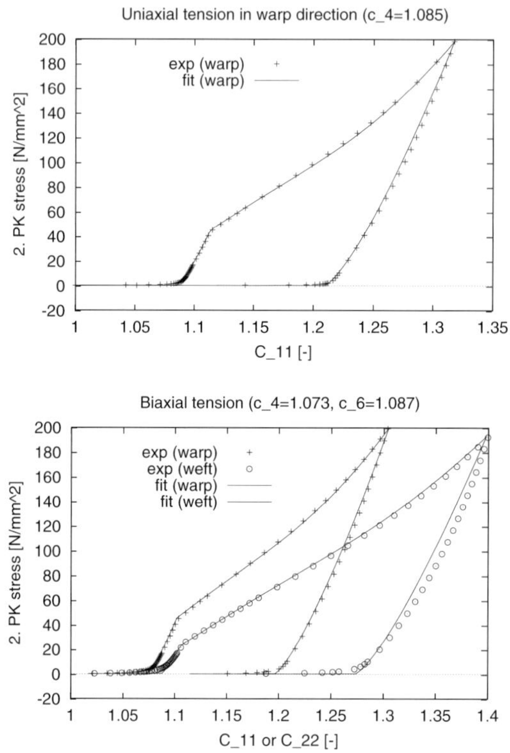

Figure 6. Experimental results and fit for uniaxial and biaxial tension.

5. Conclusions

The main goal of this paper has been to model the anisotropically elastoplastic material behavior of fibre-reinforced membranes. Due to the lack of enough experimental data, a computer model has been set up.

As an additional benefit, one gains a deeper understanding of what happens inside the material. It is crucial to consider plastic deformations in order to obtain a realistic description of the material behavior. Interestingly, one can show

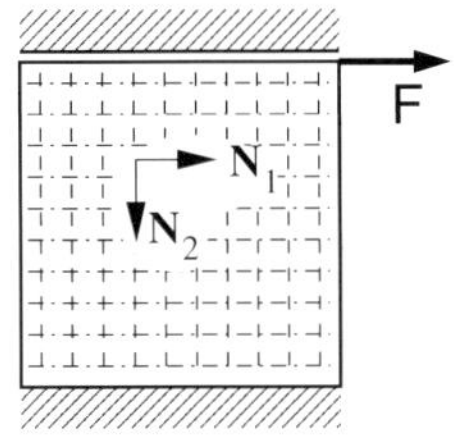

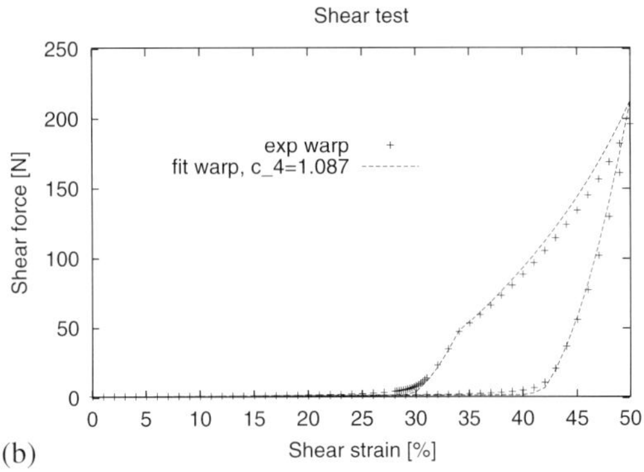

Figure 7. (a) Shear test. (b) Experimental results and fit for shear test.

that based on the assumption that the structural vectors transform like material line elements, the plastic deformation is represented by means of a symmetric tensor-valued internal variable. Thus, the plastic spin remains undetermined. This is a surprising result in the context of anisotropic material behavior.

Acknowledgments

The author wishes to thank Bob Svendsen (University of Dortmund, Germany) for useful discussions on important aspects of this work.

References

[1] Asaro, R. (1983). Micromechanics of crystals and polycrystals. *Advances in Applied Mechanics*, **23**, 1–115.

[2] Bertram, A. (1992). Description of finite inelastic deformations. In: *Proceedings of MECAMAT '92, Multiaxial plasticity*, Benallal A., Billardon R., Marquis D. [eds.], 821–835.

[3] Bertram, A. (1998). An alternative approach to finite plasticity based on material isomorphisms. *International Journal of Plasticity*, **52**, 353–374.

[4] Boehler, J.P. (1977). On irreducible representations for isotropic functions. *Zeitschrift für Angewandte Mathematik und Mechanik*, **57**, 323–327.

[5] Boehler, J.P. (1979). A simple derivation of representations for non-polynomial constitutive equations in some cases of anisotropy. *Zeitschrift für Angewandte Mathematik und Mechanik*, **59**, 157–167.

[6] Bonet, J. and Burton, A.J. (1998). A simple orthotropic, transversely isotropic hyperelastic constitutive equation for large strain computations. *Computer Methods in Applied Mechanics and Engineering*, **162**, 151–164.

[7] Cuitiño, A.M. and Ortiz, M. (1992). Computational modeling of single crystals. *Modelling Simul. Mat. Sci. Engng*, **1**, 225–263.

[8] Gasser, T.C. and Holzapfel, G.A. (2001). A rate-independent elastoplastic constitutive model for fiber-reinforced composites at finite strains: Continuum basis, algorithmic formulation and finite element implementation. In preparation.

[9] Hill, R. (1966). Generalized constitutive relations for incremental deformation of metal crystals by multislip. *Journal of the Mechanics and Physics of Solids*, **14**, 95–102.

[10] Holzapfel, G.A., Eberlein, R., Wriggers, P. and Weizsäcker, H.W. (1996). A new axisymmetrical membrane element for anisotropic, finite strain analysis of arteries. *Communications in Numerical Methods in Engineering*, **12**, 507–517.

[11] Holzapfel, G.A., Schulze-Bauer, C.A.J. and Stadler, M. (2000). Mechanics of angioplasty: wall, balloon and stent. In: *Mechanics in Biology*, Casey J. and Bao G. [eds.], AMD-Vol. 242/BED-Vol. 46, 141-156, The American Society of Mechanical Engineers (ASME), New York, USA.

[12] Liu, S. (1982). On representations of anisotropic invariants. *International Journal of Engineering Science*, **20**, 1099–1109.

[13] Miehe, C. (1996). Exponential map algorithm for stress updates in anisotropic multiplicative elastoplasticity for single crystals. *International Journal for Numerical Methods in Engineering*, **39**, 3367–3390.

[14] Ogden, R.W. (1984). *Nonlinear Elastic Deformations*, Ellis Horwood, Chichester.

[15] Reese, S., Raible, T. and Wriggers P. (2001). Finite element modeling of orthotropic material behavior in pneumatic membranes. *International Journal of Solids and Structures.* To be published 2001.

[16] Reese, S. (2001). Meso-macro modelling of fibre-reinforced composites exhibiting elastoplastic material behavior. In: *Proceedings of the Second European Conference on Computational Mechanics*, Cracow, Poland.

[17] Rice, J.R. (1971). Inelastic constitutive relations for solids: an internal-variable theory and its application to metal plasticity. *Journal of the Mechanics and Physics of Solids*, **19**, 433–455.

[18] Simo, J.C. and Miehe, C. (1992). Associative coupled thermoplasticity at finite strains: formulation, numerical analysis and implementation. *Computer Methods in Applied Mechanics and Engineering*, **98**, 41–104.

[19] Spencer, A.J.M. (2001). A theory of viscoplasticity for fabric-reinforced composites. *Journal of the Mechanics and Physics of Solids*, **49**, 2667–2687.

[20] Svendsen, B. (1994). On the representation of constitutive relations using structure tensors. *International Journal of Engineering Science*, **32**, 1889–1892.

[21] Svendsen, B. (1998). A thermodynamic formulation of finite-deformation elastoplasticity with hardening based on the concept of material isomorphism. *International Journal of Plasticity*, **14**, 473–488.

[22] Svendsen, B. (2001). On the modelling of anisotropic elastic and inelastic material behavior at large deformation. *International Journal of Solids and Structures.* To be published 2001.

[23] Wang, C.C. and Bloom, J. (1974). Material uniformity and inhomogeneity in anelastic bodies. *Archive for Rational Mechanics and Analysis*, **53**, 246.

[24] Weiss, J.A., Maker, B.N. and Govindjee, S. (1996). Finite element implementation of incompressible, transversely isotropic hyperelasticity. *Computer Methods in Applied Mechanics and Engineering*, **135**, 107–128.

[25] Zhang, Q.-S. and Rychlewski, J.M. (1990). Structural tensors for anisotropic solids. *Archives of Mechanics*, **42**, 267–277.

Necking Phenomena of a Fiber-Reinforced Bar modeled by Multisurface Plasticity

T.C. Gasser, G.A. Holzapfel
Institute for Structural Analysis, Computational Biomechanics, Graz University of Technology
Schiesstattgasse 14b, 8010 Graz, Austria
gh@biomech.tu-graz.ac.at

Abstract: A rate-independent elastoplastic constitutive model for fiber-reinforced composites at finite strains is presented. The model is based on multisurface plasticity and its finite element implementation uses the classical elastic predictor/plastic corrector method. We present a technique for computing the incremental slip parameters satisfying the incremental Karush-Kuhn-Tucker loading/unloading conditions *a priori*. This technique circumvents the need for a search of the active slip system. Particularized constitutive equations are used to illustrate the peculiar anisotropic necking phenomena of a fiber-reinforced bar at finite strains.

Keywords: Anisotropy, reinforced composite, multisurface plasticity, necking.

1. Introduction

This paper focuses on the elastoplastic modeling of fiber-reinforced composites and demonstrates the necking phenomena of a geometrical imperfect fiber-reinforced bar under extension. Wide applicability of the constitutive framework presented here may be found in many engineering applications and in biomechanics of soft tissues (see, for example, Holzapfel et al. [7], Holzapfel [4]).

In order to demonstrate the plastic mechanism of the model we consider a bar (with length 1) reinforced in the axial direction by one family of fibers. The structural arrangement of the fibers is characterized by a unit direction vector $\mathbf{a}_0^1$, as illustrated in Figure 1. The specimen is now stretched in the axial direction until plastic deformation occurs. After subsequent (elastic) unloading a plastic stretch λ^{p} remains and this determines the (plastically deformed) intermediate configuration. It is assumed that due to loading the fibers of the composite slip relative to each other so that plastic deformation in the matrix of the composite evolves. The fibers, however, always behave elastically. This mechanism is shown in Figure 1. Figure 1a illustrates the undeformed (reference) configuration of the bar, while Figure 1b shows the intermediate configuration of the bar after plastic deformation. Note that the length of the fibers do not change between these configurations.

C. Miehe (ed.), IUTAM Symposium on Computational Mechanics of Solid Materials at Large Strains, 211-220.

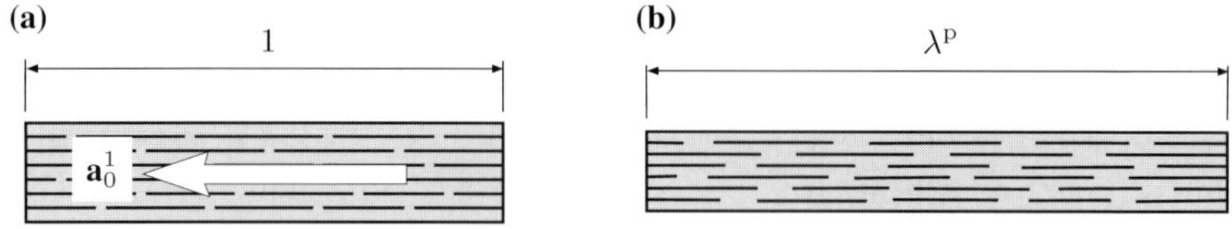

Figure 1. Fiber-reinforced composite with one family of fibers, characterized by the unit direction vector $\mathbf{a}_0^1$: (a) before plastic deformation (length 1), and (b) after plastic deformation (plastic stretch λ^{p}). Plastic deformation is accumulated in the matrix of the composite by assuming that the fibers of the composite slip relative to each other.

The mechanism demonstrated above may be generalized to n families of fibers characterized by the set $\{\mathbf{a}_0^\alpha; \alpha = 1, \ldots, n\}$ of unit direction vectors. For modeling the anisotropic *elastic* response of the material we use the theory of fiber-reinforced composites (see Spencer [17], Holzapfel and Gasser [5]). The *plastic* response of the composite is described by the concept of rate-independent multisurface plasticity. For that, a convex, but non-smooth, yield surface in stress space is introduced. Multisurface plasticity is well-established in engineering science (see, for example, Asaro [1], Simo [13], Miehe [10]).

The irreversible nature of the plastic deformation is enforced by the Karush-Kuhn-Tucker loading/unloading conditions. For the finite element implementation of the constitutive model we employ the classical elastic predictor/plastic corrector method, which is based on a backward-Euler discretization of the evolution equation.

2. Multiplicative multisurface plasticity

In this section the continuum mechanical framework is briefly outlined. We derive governing constitutive equations suitable for describing the response of strongly anisotropic materials in the elastic and plastic regions. In order to split the deformation gradient $\mathbf{F} = \mathbf{F}^{\mathrm{e}}\mathbf{F}^{\mathrm{p}}$ into elastic $\mathbf{F}^{\mathrm{e}}$ and plastic $\mathbf{F}^{\mathrm{p}}$ parts we introduce a locally defined macro-stress-free intermediate configuration which is standard in computational plasticity (Simo and Hughes [15]).

2.1. Anisotropic elastic response

The elastic response of the fiber-reinforced material is defined by keeping the plastic deformation fixed, i.e. the intermediate configuration is fixed in space. It is assumed that the Helmholz free-energy Ψ may be decomposed into macroscopic and microscopic counterparts, i.e $\Psi(\mathbf{F}^{\mathrm{e}}, \mathbf{a}_0^1, \ldots, \mathbf{a}_0^n, A) = \Psi_{\mathrm{macro}}(\mathbf{F}^{\mathrm{e}}, \mathbf{a}_0^1, \ldots, \mathbf{a}_0^n) + \Psi_{\mathrm{micro}}(A)$. Herein A denotes an internal (scalar) variable which accounts for irreversible changes in the microstructure. The anisotropic nature of Ψ is reflected by the set $\{\mathbf{a}_0^\alpha; \alpha = 1, \ldots, n\}$ of unit direction vectors characterizing the structure of the material.

Furthermore, we assume an additive decomposition of the macroscopic free energy into two parts, namely $\Psi_{\mathrm{macro}}(\mathbf{F}^{\mathrm{e}}, \mathbf{a}_0^1, .., \mathbf{a}_0^n) = U(J^{\mathrm{e}}) + \overline{\Psi}(\overline{\mathbf{F}}^{\mathrm{e}}, \mathbf{a}_0^1, \ldots, \mathbf{a}_0^n)$, where $J^{\mathrm{e}} = \det \mathbf{F}^{\mathrm{e}}$ and $\overline{\mathbf{F}}^{\mathrm{e}} = J^{\mathrm{e}\,-1/3}\mathbf{F}^{\mathrm{e}}$ denote the elastic volume ratio and the modified elastic deformation gradient, respectively. Since we assume isochoric deformations, U serves as an augmented function to enforce the internal kinematic constraint in an approximate manner.

In order to particularize the elastic response with a suitable number of material parameters we choose the form

$$\overline{\Psi}(\overline{\mathbf{F}}^{\mathrm{e}}, \mathbf{a}_0^1, \ldots, \mathbf{a}_0^n) = \overline{\Psi}_{\mathrm{m}}(\bar{I}_1) + \sum_{\alpha \in \mathcal{B}} \overline{\Psi}_{\mathrm{f}}^{\alpha}(\bar{I}_4^{\alpha}), \tag{1}$$

where $\bar{I}_1 = \mathrm{tr}\overline{\mathbf{C}}^{\mathrm{e}}$ is the first invariant of the modified elastic right Cauchy-Green tensor $\overline{\mathbf{C}}^{\mathrm{e}} = \overline{\mathbf{F}}^{\mathrm{e}\,\mathrm{T}}\overline{\mathbf{F}}^{\mathrm{e}}$, and $\bar{I}_4^\alpha$ are additional invariants of $\overline{\mathbf{C}}^{\mathrm{e}}$ and the structural tensors $\mathbf{a}_0^\alpha \otimes \mathbf{a}_0^\alpha$, given by $\bar{I}_4^\alpha = \overline{\mathbf{C}}^{\mathrm{e}} : \mathbf{a}_0^\alpha \otimes \mathbf{a}_0^\alpha$, $\alpha = 1, \ldots, n$. Furthermore, we assume that the fibers do not carry any compressive load. This is expressed by the set $\mathcal{B} = \{\alpha \in 1, \ldots, n \,|\, \bar{I}_4^\alpha > 1\}$ of active (stretched) fibers.

In (1) we have used the assumption that the material is composed of an isotropic matrix material, characterized by $\overline{\Psi}_{\mathrm{m}}$, and the fiber-reinforcement determined through the functions $\overline{\Psi}_{\mathrm{f}}^{\alpha}$ (Holzapfel et al. [6]).

2.2. Anisotropic inelastic response

To derive equations suitable for determining the plastic part of the deformation we follow the approach described in Simo and Hughes [15]. We introduce an elastic domain, which is bounded by a convex, but non-smooth, yield surface $\partial\mathbb{E}$, defined as $\partial\mathbb{E} = \{\mathrm{dev}\boldsymbol{\Sigma}, \mathbf{a}_0^1, \ldots, \mathbf{a}_0^n, B | \hat{\Phi}^\alpha = 0, \alpha = 1, \ldots, n\}$. Here we have introduced the (non-symmetric) Mandel stress tensor $\boldsymbol{\Sigma}$ and the micro-stress $B = \partial\Psi_{\mathrm{micro}}/\partial A$.

In the definition of $\partial\mathbb{E}$ we have used n-independent (nonredundant) yield criterion functions $\hat{\Phi}^\alpha$, $\alpha = 1, \ldots, n$. Following Miehe [10] we use the additive decomposition

$$\hat{\Phi}^\alpha = \Phi_{\mathrm{macro}}^\alpha(\mathrm{dev}\boldsymbol{\Sigma}, \mathbf{a}_0^1, \mathbf{a}_0^2, \ldots, \mathbf{a}_0^n) - \Phi_{\mathrm{micro}}(B) \tag{2}$$

of $\hat{\Phi}^\alpha$ into macro-stress dependent functions $\Phi_{\mathrm{macro}}^\alpha$ and a micro-stress dependent function Φ_{micro}.

A generalization of the plastic mechanism to n slip systems is outlined in Section 1. The slip systems are characterized by the same set $\{\mathbf{a}_0^\alpha; \alpha = 1, \ldots, n\}$ of unit direction vectors that also determine the anisotropic elastic response of the material. Similarly to the description used in crystal plasticity (see, e.g., Kahn and Huang [8], p. 357) a generalized flow rule reads

$$\mathbf{L}^{\mathrm{p}} = \sum_{\alpha=1}^{n} \lambda^{\alpha} \mathrm{dev}(\mathbf{a}_0^{\alpha} \otimes \mathbf{a}_0^{\alpha}), \tag{3}$$

where $\mathbf{L}^{\mathrm{p}}$ is the plastic part of the rate tensor $\mathbf{L}$ (work-conjugate to the Mandel stress tensor $\boldsymbol{\Sigma}$), and $\mathbf{L}$ is obtained by a consistent transformation of the spatial velocity gradient $\mathbf{l} = \dot{\mathbf{F}}\mathbf{F}^{-1}$ to the intermediate configuration, i.e. $\mathbf{L} = \mathbf{F}^{\mathrm{e}\,-1}\mathbf{l}\mathbf{F}^{\mathrm{e}}$ (Miehe [9]).

In eq. (3) $\lambda^{\alpha} \geq 0$ are $n \geq 1$ consistency parameters (slip rates) and in view of the irreversibility of plastic flow they are assumed to obey the n (Karush-Kuhn-Tucker) loading/unloading conditions, i.e. $\lambda^{\alpha} \geq 0$, $\hat{\Phi}^{\alpha} \leq 0$, $\lambda^{\alpha}\hat{\Phi}^{\alpha} = 0$ as well as the consistency conditions $\lambda^{\alpha}\dot{\hat{\Phi}}^{\alpha} = 0$ (Simo and Hughes [15]). In order to complete the formulation, it is assumed that the evolution of the hardening variable A is determined by the summation $\dot{A} = \sum_{\alpha=1}^{n} \lambda^{\alpha}$ of the consistency parameters λ^{α} over all slip systems.

3. Integration of the evolution equations

In this section we derive the algebraic expressions required for an efficient use of the constitutive model within the finite element method. The approach is based on a *backward Euler* integration of the rate equations. In order to determine the incremental slip (plastic) parameters we present a technique which satisfies the incremental Karush-Kuhn-Tucker loading/unloading conditions *a priori*. This approach has advantages over standard approaches (Miehe [10]) and omits a search of the active slip system (known as active set search).

3.1. Update equations based on backward-Euler integration

In a finite element implementation of the model, the governing rate equations (3) and the evolution of the hardening variable A have to be integrated numerically over a certain time step $\Delta t = t - t_n$ that is defined between the previous time t_n and the current time t. Based on an unconditionally stable and first-order accurate backward-Euler integration scheme the flow rule (3) reads, after some algebraic manipulations,

$$\overline{\mathbf{F}}^{\mathrm{e}} = \overline{\mathbf{F}}^{\mathrm{e\,tr}} - \sum_{\alpha=1}^{n} \gamma^{\alpha} \left(\mathbf{a}^{\alpha\,\mathrm{tr}} \otimes \mathbf{a}_0^{\alpha} - \frac{1}{3}\overline{\mathbf{F}}^{\mathrm{e\,tr}} \right), \tag{4}$$

where the notation $(\bullet)^{\mathrm{tr}}$ indicates *elastic trial (predictor)* state values that are achieved by freezing the plastic deformation (for further details see, for example, Simo and Hughes [15], or Gasser and Holzapfel [3]). In particular, the elastic trial values of the modified elastic deformation gradient $\overline{\mathbf{F}}^{\mathrm{e}}$ and the Eulerian vectors $\mathbf{a}^{\alpha} = \overline{\mathbf{F}}^{\mathrm{e}}\mathbf{a}_0^{\alpha}$ read $\overline{\mathbf{F}}^{\mathrm{e\,tr}} = \overline{\mathbf{F}}\,\overline{\mathbf{F}}_n^{\mathrm{p}\,-1}$ and $\mathbf{a}^{\alpha\,\mathrm{tr}} = \overline{\mathbf{F}}^{\mathrm{e\,tr}}\mathbf{a}_0^{\alpha}$, respectively. In the update equation (4) for the modified elastic deformation gradient, $\gamma^{\alpha} := \lambda^{\alpha}\Delta t$ denote the *incremental* slip (plastic) parameters on each slip system α. They have to obey the incremental versions of the loading/unloading conditions

$$\gamma^{\alpha} \geq 0, \qquad \hat{\Phi}^{\alpha} \leq 0, \qquad \gamma^{\alpha}\hat{\Phi}^{\alpha} = 0, \tag{5}$$

and the consistency conditions $\gamma^{\alpha}\dot{\hat{\Phi}}^{\alpha} = 0$. Discretization of the evolution equations $\dot{A} = \sum_{\alpha=1}^{n}\lambda^{\alpha}$ for the hardening variable A leads to the update formula

$$A = A^{\mathrm{tr}} + \sum_{\alpha=1}^{n}\gamma^{\alpha}, \tag{6}$$

where the elastic trial value A^{tr} denotes the hardening variable at the previous time t_n, i.e. $A^{\mathrm{tr}} = A_n$.

3.2. Computation of the incremental slip parameters

In order to work with the update equations (4), (6) one has to determine the incremental slip parameters γ^{α}. The standard approach (direct method) aims to satisfy the yield criteria $\hat{\Phi}^{\alpha} = 0$ in conjunction with a search strategy of the active set in order to meet the incremental loading/unloading conditions (5) (see Miehe [10]).

We present here a different approach (indirect method), where we focus on the equalities

$$\Theta^{\alpha} = \sqrt{(\hat{\Phi}^{\alpha})^2 + (\gamma^{\alpha})^2} + \hat{\Phi}^{\alpha} - \gamma^{\alpha} = 0, \qquad \alpha = 1, \ldots, n, \tag{7}$$

equivalent to the Karush-Kuhn-Tucker equalities and inequalities (5) (see Bertsekas [2] and Schmidt-Baldassari [12]). Note that eq. (7) includes all slip systems, i.e. active *and* inactive, so that an active set search is omitted. A numerical solution of (7) uses Newton's method and leads to recurrence formulas for the incremental slip parameters γ^{α}. Thus,

$$\gamma^{\alpha} \Leftarrow \gamma^{\alpha} - \sum_{\beta=1}^{n} H^{\alpha\beta\,-1}\hat{\Theta}^{\beta}, \qquad H^{\alpha\beta} = \frac{\partial\Theta^{\alpha}}{\partial\gamma^{\beta}} \quad \text{for} \quad \alpha, \beta = 1, \ldots, n, \tag{8}$$

where the closed form solution for the $n \times n$ matrix $H^{\alpha\beta}$ is provided in Gasser and Holzapfel [3]. A detailed derivation of the algorithmic expressions required for the direct and indirect methods is also given in Gasser and Holzapfel [3].

For the case of linearly dependent (or redundant) constraints the matrix $[H^{\alpha\beta}]$ is singular, whereas it is required to be invertible in eq. $(8)_1$. Hence, γ^{α} are not uniquely defined. For a solution procedure to circumvent the problem of linearly dependent constraints the reader is referred to, for example, Miehe [10] and Miehe and Schröder [11].

4. Example: 3D necking problem

In this section we particularize the continuum mechanical framework and consider the elastoplastic extension of a geometrically imperfect, fiber-reinforced bar on the basis of a moderate scale finite element simulation. It is assumed that the bar is reinforced by two ($n = 2$) families of fibers characterized by $\mathbf{a}_0^1$ and $\mathbf{a}_0^2$.

4.1. PARTICULARIZATION OF THE CONTINUUM MECHANICAL FRAMEWORK

We particularize the derived continuum mechanical framework and start with the elastic part (1) of the model. We use a standard *neo-Hookean* model for the isotropic matrix material, i.e. $\overline{\Psi}_{\rm m}(\bar{I}_1) = \mu/2(\bar{I}_1 - 3)$ and an exponential function for describing the fiber-reinforcement, i.e. $\overline{\Psi}_{\rm f}^{\alpha}(\bar{I}_4^{\alpha}) = (0.5k_1/k_2)$ $\{\exp[k_2(\bar{I}_4^{\alpha} - 1)^2] - 1\}$, $\alpha = 1, 2$. Herein μ and k_1, k_2 are material parameters associated with the matrix material and the fiber-reinforcement, respectively. For a motivation of these functions see Holzapfel et al. [6].

In order to quantify the inelastic part of the model we particularize the yield criterion functions (2). For the macro-stress dependent functions $\Phi_{\rm macro}^{\alpha}$ we omit coupling effects between the different slip systems and introduce the new functions τ^{α}. Furthermore, we assume that τ^{α} is defined by the projection of the deviatoric Mandel stress tensor onto the particular slip system $\mathbf{a}_0^{\alpha}$, i.e. $\Phi_{\rm macro}^{\alpha}(\mathrm{dev}\boldsymbol{\Sigma}, \mathbf{a}_0^1, \mathbf{a}_0^2, \ldots, \mathbf{a}_0^n) = \tau^{\alpha}(\mathrm{dev}\boldsymbol{\Sigma}, \mathbf{a}_0^{\alpha}) = \mathrm{dev}\boldsymbol{\Sigma} : \mathbf{a}_0^{\alpha} \otimes \mathbf{a}_0^{\alpha}$.

The particularization of the micro-stress dependent function $\Phi_{\rm micro}$ is based on isotropic Taylor hardening, decomposed additively into an initial yield stress τ_0 and a micro-stress B. For simplicity we assume linear hardening so that $B = hA$, with h denoting the linear hardening parameter. Hence, for the yield criterion functions we find finally that $\hat{\Phi}^{\alpha} = \tau^{\alpha} - (\tau_0 + hA)$, $\alpha = 1, \ldots, n$.

Based on these particularizations it can be shown that the plastic dissipation $\mathcal{D}^{\rm p} = \boldsymbol{\Sigma} : \mathbf{L}^{\rm p} - B\dot{A} = \tau_0 \dot{A} \geq 0$ of the constitutive model is always non-negative (Gasser and Holzapfel [3]).

4.2. NUMERICAL EXAMPLE

A bar reinforced by two families of fibers is investigated within the elastoplastic domain. Two independent slip systems arise; they are associated with the two fiber families. The symmetric geometry and loading conditions allow us to model an eighth of the bar (cross-section $a \times b$, length l and fiber angle α), as

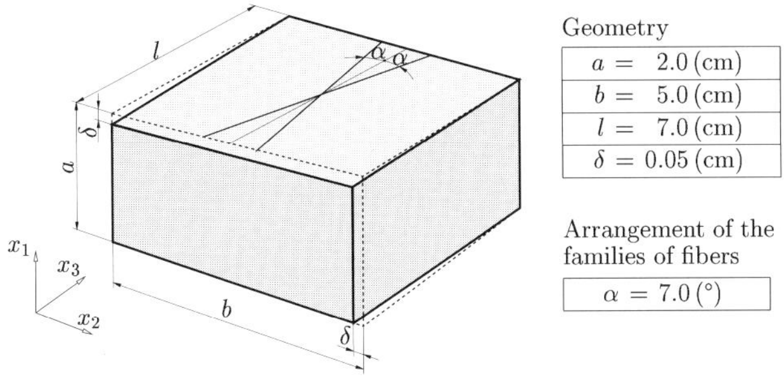

Figure 2. Referential geometry and structural arrangement of two families of fibers of one eighth of a bar. A (small) perturbation of the rectangular geometry is characterized by the value δ.

depicted in Figure 2, with appropriate boundary conditions. In view of material stability investigations we introduce a (small) perturbation δ for initializing a necking zone during elongation of the bar.

The elastic material parameters (μ, k_1, k_2) and the plastic material parameters (τ_0, h) used for the numerical analysis are summarized in Table 1. In order to investigate the evolution of the plastic deformation during extension, the distribution of the hardening variable A of the (whole) bar at different states of extension is shown in Figure 3. The three-dimensional computation illustrated uses $QM1/E12$-elements (Simo et al. [14]) for one eighth of the bar, while Figure 3(a) shows the reference configuration.

Table 1. Material parameters for the necking problem.

Elastic				*Plastic*			
μ	=	30.0	(kPa)	τ_0	=	70.0	(kPa)
k_1	=	4.0	(kPa)	h	=	100.0	(kPa)
k_2	=	2.3	(—)				

Figure 3(b) shows the hardening variable A at a displacement of $u = 2.55\,(\mathrm{cm})$. At this state slight plastic deformation is present. It is initiated in the middle of the bar by the geometrical imperfection δ introduced. In Figure 3(c), at the (final) displacement of $u = 6.5\,(\mathrm{cm})$, a very well pronounced necking zone can be seen. The figure shows localization of plastic strains in terms of the hardening variable A. A further increase of stretch leads to a softening response with loss of ellipticity of the governing equations (Simo et al. [16]). This phenomenon is accompanied by a strain-localization instability and mesh-dependent solutions occur (Zienkiewicz and Taylor [18]). Furthermore, in Figure 3(c), we see that plastic deformation in the necking zone accumulates on two sides of the bar. Plastic deformation is not distributed uniformly across the cross-section. Although we used the $QM1/E12$-element the deformed configuration, as illustrated in Figure 3(c), shows hourglass instabilities.

5. Conclusion

A rate-independent elastoplastic constitutive model for fiber-reinforced composites was presented. The constitutive framework is based on multisurface plasticity and suitable for describing the mechanical response of composites in the finite elastic and plastic strain domains.

The model is based on the multiplicative decomposition of the deformation gradient into elastic and plastic parts. The isochoric elastic response of the composite is determined by the Helmholtz free-energy functions $\overline{\Psi}_{\mathrm{m}}$ and $\overline{\Psi}_{\mathrm{f}}^{\alpha}$ associated with the isotropic matrix material and the fiber-reinforcement, respectively. The isochoric inelastic response of the composite is governed by a flow rule in terms of $n \geq 1$ consistency parameters obeying the (Karush-Kuhn-Tucker) loading/unloading conditions. A plastic corrector scheme has been derived on the basis of an Eulerian vector update. The governing rate equations are integrated numerically and the incremental slip parameters are determined using an indirect method satisfying the loading/unloading conditions *a priori*.

Particularized governing equations have been used for showing the applicability of the constitutive model to a moderate scale finite element simulation. In particular, the anisotropic necking phenomena of a fiber-reinforced bar was studied in some detail. The analysis has shown pronounced localization of plastic deformation at both sides of the bar leading to softening response and loss of ellipticity of the governing equations.

Acknowledgments

Financial support for this research was provided by the *Austrian Science Foundation* under *START-Award Y74-TEC*.

References

[1] Asaro, R.J. (1983). Crystal plasticity. *J. Appl. Mech.*, **50**, 921–934.

[2] Bertsekas, D.P. (1996). *Constrained Optimization and Lagrange Multiplier Methods*, Athena Scientific, Belmont.

[3] Gasser, T.C. and Holzapfel, G.A. (2001). A rate-independent elastoplastic constitutive model for fiber-reinforced composites at finite strains: Continuum basis, algorithmic formulation and finite element implementation. Submitted.

[4] Holzapfel, G.A. (2001). Biomechanics of soft tissue. In Lemaitre, J., editor, *The Handbook of Materials Behavior Models. Volume III, Multiphysics Behaviors, Chapter 10, Composite Media, Biomaterials*, Boston. Academic Press, 1049–1063.

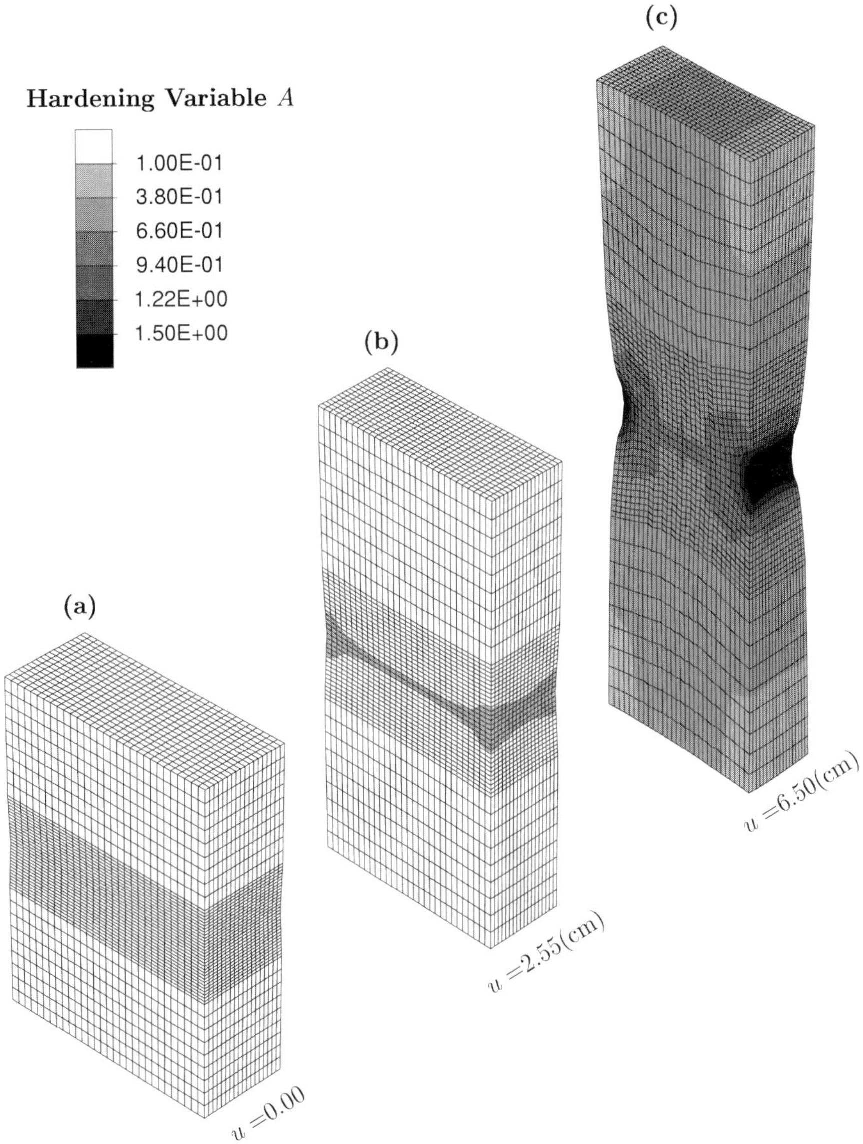

Figure 3. Evolution of the plastic deformation in terms of the hardening variable A during extension. (a) Reference configuration, (b) exceeding elastic limit and evolution of plastic deformation, (c) pronounced localization of plastic deformation accumulated at two sides of the bar.

[5] Holzapfel, G.A. and Gasser, T.C. (2001). A viscoelastic model for fiber-reinforced composites at finite strains: Continuum basis, computational aspects and applications. *Comput. Meth. Appl. Mech. Engr.*, **190**, 4379–4403.

[6] Holzapfel, G.A., Gasser, T.C. and Ogden, R.W. (2000). A new constitutive framework for arterial wall mechanics and a comparative study of material models. *J. Elasticity*, **61**, 1–48.

[7] Holzapfel, G.A., Schulze-Bauer, C.A.J. and Stadler, M. (2000). Mechanics of angioplasty: Wall, balloon and stent. In Casey, J. and Bao, G., editors, *Mechanics in Biology*, New York, The American Society of Mechanical Engineers (ASME). AMD-Vol. 242/BED-Vol. 46, 141–156.

[8] Kahn, A.S. and Huang, S. (1995). *Continuum Theory of Plasticity*. John Wiley & Sons, New York.

[9] Miehe, C. (1994). On the representation of prandtl-reuss tensors within the framework of multiplicative plasticity. *Int. J. Plasticity*, **10**, 609–621.

[10] Miehe, C. (1996). Multisurface thermoplasticity for single crystals at large strains in terms of Eulerian vector updates. *Int. J. Solids and Structures*, **33**, 3103–3130.

[11] Miehe, C. and Schröder, J. (2001). A comparative study of stress update algorithms for rate-independent and rate-dependent crystal plasticity. *Int. J. Numer. Meth. Engr.*, **50**, 273–298.

[12] Schmidt-Baldassari, M. (2001). Numerical concepts for rate-independent single crystal plasticity. submitted.

[13] Simo, J.C. (1987). A J_2-flow theory exhibiting a corner-like effect and suitable for large-scale computation. *Comput. Meth. Appl. Mech. Engr.*, **62**, 169–194.

[14] Simo, J.C., Armero, F. and Taylor, R.L. (1993). Improved versions of assumed enhanced strain tri–linear elements for 3d finite deformation problems. *Comput. Meth. Appl. Mech. Engr.*, **110**, 359–386.

[15] Simo, J.C. and Hughes, T.J.R. (1998). *Computational Inelasticity*, Springer-Verlag, New York.

[16] Simo, J.C., Oliver, J. and Amero, F. (1993). An analysis of strong discontinuities induced by strain softening in rate-independent inelastic solids. *Int. J. Numer. Meth. Engr.*, **12**, 277–296.

[17] Spencer, A.J.M. (1984). Constitutive theory for strongly anisotropic solids. In Spencer, A.J.M., editor, *Continuum Theory of the Mechanics of Fibre-Reinforced Composites*, Springer-Verlag, Wien, 1–32. CISM Courses and Lectures No. 282, International Centre for Mechanical Sciences.

[18] Zienkiewicz, O.C. and Taylor, R.L. (2000). *The Finite Element Method. Solid Mechanics*, **2**, Butterworth Heinemann, Oxford, 5th edition.

Material Growth in Solid-Like Materials

G.A. Maugin
Laboratoire de Modélisation en Mécanique, UMR 7607, Université Pierre et Marie Curie
Case 162, 8 rue du Capitaine Scott, 75015 Paris, France
gam@ccr.jussieu.fr

S. Imatani
Departement of Energy Conversion Science, Kyoto University
Sakyo-ku, Kyoto 606-8501, Japan
imatani@energy.kyoto-u.ac.jp

Abstract: A thermodynamically admissible model of volumetric growth is presented which exploits the notion of material transplant or local structural rearrangement issued from the Epstein-Maugin theory of material inhomogeneities. The driving force appears to be the Mandel stress (a part of the Eshelby stress tensor). Anisotropy of growth is characterized by a vector field slaved to the principal directions of that tensor. The model is very much like one of viscoelasticity in finite strains. It is applicable to self-organization or adaptation. The numerical solution of specific problems is based on a finite-element formulation obtained with reference to the total Lagrangian approach. The validity of the model is thus assessed in terms of circumferential (monotomic) growth/resorption behavior, stress induction in a ring, and the dynamical effect (repeated alternate loading) on the material growth in a cantilever beam. The model proves to possess a sufficiently rich potential for a further comprehensive description of growth/adaptation phenomena.

Keywords: Finite strain, material growth, nonlinear viscoelasticity, structural rearrangements, anisotropy, numerics.

1. Introduction

This contribution proposes a constitutive modeling of volumetric growth and simple related applications on the basis of a thermodynamically admissible finite-strain model devised by Epstein and Maugin [7]. It must be reminded that adaptation (re-organization or re-modeling) as well as volumetric growth are of particular importance when dealing with biomaterials such as bones, tissues, etc. To name a few, such works as those of Fung et al. [8], Cowin et al. [3], Rodriguez et al. [20], Hart et al. [10], Kuehn and Hauger [12] and the review of Taber [22] are particularly enlightening on this subject matter. In the present modeling which borrows ideas from the general theory of material uniformity and inhomogeneity (cf. Maugin [15, 17], Epstein and Maugin [6]) framed on the material manifold, the multiplicative decomposition of the deformation gradient plays a fundamental role. Accordingly, basic concepts that are well

C. Miehe (ed.), IUTAM Symposium on Computational Mechanics of Solid Materials at Large Strains, 221-234.

recognized in finite-strain plasticity can readily be applied to the present context, not only in the mechanical/mathematical modeling, but also in the numerical methodology. In particular, it is shown that the Mandel stress is the most likely driving force for the remodeling/growth behavior, whereas the finite-element formulation can be based on the total-Lagrangian approach (cf. Simo [21]). While the theoretical frame yields a kind of nonlinear viscoelastic behavior, the numerics help to provide elementary illustrations of monotonic growth, adaptation, and dynamical effects. It must be emphasized that the approach is purely "volumetric" without diffusion. The reason for this is to be found in the work of Epstein and Maugin [7] where these authors have shown that a consistent description of growth with diffusion would require a more complex geometrical background.This is left for further work, if feasible at all.

2. The notion of local structural rearrangement

In order to make ideas clear let us consider *the case of quasi-statics in the absence of body force* with an elastic energy density given by $W = \bar{W}(\mathbf{F};\mathbf{X})$ per unit reference volume, where $\mathbf{F} := \partial\chi/\partial\mathbf{X}$ is the deformation gradient, $\boldsymbol{x} = \chi(\boldsymbol{X},t)$ is the deformation mapping from the reference configuration K_R to the actual configuration K_t and $\boldsymbol{X}$ denotes the material point. Then the basic field equation – bulk equilibrium – in the Piola formalism reduces to

$$\mathrm{div}_R \boldsymbol{T} = \boldsymbol{0}\ , \quad \boldsymbol{T} = \partial\bar{W}/\partial\boldsymbol{F} \tag{1}$$

Following Epstein and Maugin [4], we consider (thought experiment) the case where the material inhomogeneity can be artificially removed*at each* material point $\boldsymbol{X}$, by effecting a change of reference configuration. This is done at each material point, and the reference change is therefore *local* and generally *not integrable* over the whole body. Such a change is called a*local structural rearrangement.* Let $\mathbf{K}(\mathbf{X})$ this reference change, which brings the neighborhood of $\boldsymbol{X}$ into the so-called crystal reference (this is made modulo the material symmetry; cf. Epstein and Maugin [5], Epstein and Maugin [7]), so that accounting for the accompanying volume change $J_K = \det\boldsymbol{K}$, it combines multiplicatively to the right with $\boldsymbol{F}$ and, for energies, we can write

$$\bar{W}(\boldsymbol{F};\boldsymbol{X}) = J_K^{-1}\hat{W}(\boldsymbol{F}\boldsymbol{K}(\boldsymbol{X})) = \tilde{W}(\boldsymbol{F},\boldsymbol{K}). \tag{2}$$

Obviously we can compute the partial derivatives of the last mentioned function $\tilde{W}$, obtaining thus, as easily checked,

$$\boldsymbol{T} = \frac{\partial\tilde{W}}{\partial\boldsymbol{F}} = \frac{\partial\bar{W}}{\partial\boldsymbol{F}}\ , \quad \tilde{\boldsymbol{b}} = -\frac{\partial\tilde{W}}{\partial\boldsymbol{K}} = -(\boldsymbol{T}.\boldsymbol{F} - W\mathbf{1}_R).\boldsymbol{K}^{-T}. \tag{3}$$

Accordingly,

$$\boldsymbol{b} \equiv \tilde{\mathbf{b}}.\mathbf{K}^T = -\frac{\partial\tilde{W}}{\partial\boldsymbol{K}}.\boldsymbol{K}^T \equiv W\mathbf{1}_R - \boldsymbol{T}.\boldsymbol{F}. \tag{4}$$

This provides an elegant definition of the quasi-static *Eshelby stress* $\boldsymbol{b}$ (referred to as the energy-momentum or Maxwell stress by Eshelby) via the notion of local structural rearrangement. Assuming that we just know (1) at all regular points $\boldsymbol{X}$, $J_F := \det \boldsymbol{F} > 0$, we can then compute the material divergence of $\boldsymbol{b}$ resulting in (*expl* means the material gradient at fixed field $\boldsymbol{F}$)

$$\mathrm{div}_R \boldsymbol{b} + \boldsymbol{f}^{inh} = \mathbf{0}, \boldsymbol{f}^{inh} := \partial \bar{W} / \partial \boldsymbol{X} \Big|_{expl} \tag{5}$$

where $\boldsymbol{f}^{inh}$ is the so-called "material force of inhomogeneity" (cf. Maugin [15]). But we can also express this material co-vector through the operation

$$\left.\frac{\partial W}{\partial \boldsymbol{X}}\right|_{expl} = (\nabla_R \boldsymbol{K}) : \frac{\partial \tilde{W}(\boldsymbol{F}, \boldsymbol{K})}{\partial \boldsymbol{K}} = -(\nabla_R \boldsymbol{K}) : (\boldsymbol{b}.\boldsymbol{K}^{-T}) = \boldsymbol{b} : \boldsymbol{\Gamma}, \tag{6}$$

where $\boldsymbol{\Gamma}$ is the (geometrical) *connection* based on the non-integrable mapping $\boldsymbol{K}$; that is, in components (to avoid any misunderstanding):

$$\Gamma^{A}_{B.K} := -(\boldsymbol{K}^{-1})^{\alpha}_{.B} K^{A}_{.\alpha,K}. \tag{7}$$

Therefore, eqn. (5) also reads (Epstein and Maugin [4])

$$\mathrm{div}_R \boldsymbol{b} = \boldsymbol{b} : \boldsymbol{\Gamma}. \tag{8}$$

In some geometrical theories of *continuous distributions of dislocations*, the connection $\boldsymbol{\Gamma}$ is directly related to the *density of dislocations* (cf. Maugin [15]). Accordingly, we can say that in such "continuously dislocated" elastic bodies, dislocations create a material force density which is responsible for the *non-divergence-free nature of the Eshelby stress tensor.* Dislocations, which originally are discrete defects, act thus as a materially *distributed inhomogeneity force* in agreement with equation (8). We do not pursue further here this geometrical approach to continuously distributed defects (see more on differential geometry, the notions of material uniformity and homogeneity, the role of material symmetry groups, crystallographic basis, transplants, G-structure, and G-covariance in Epstein and Maugin [6]).

3. The case of dissipative media with internal variables

In modern continuum mechanics, we account for a variety of microscopic phenomena responsible for macroscopic dissipation through the notion of *internal variables of state* (Maugin [14, 18]). These variables have for essential property to be uncontrollable directly by external stimuli, so that they expend power only in the bulk. We note collectively α these variables whose choice and tensorial nature depend on the physical acuteness of the theoretician helped by the experimentalist who should uncover those most representative variables (e.g., density of dislocation, plastic strain, work-hardening variables, etc). Such a thermodynamical framework is particularly powerful in the field of study of *anelastic behaviors* of solid-like materials. Remarkably enough, compared to what was given in the preceding paragraph, the only initial change is the introduction of α and of the absolute temperature θ in the list of functional arguments of the free energy $W = \bar{W}(\boldsymbol{F}, \theta, \alpha; \boldsymbol{X})$ for an anisotropic, possibly anelastically inhomogeneous material in finite strains, whose basic behavior is elastic, but it may present combined anelasticity. Classically, $\bar{W}$ will depend on $\boldsymbol{F}$ only through another quantity such as an element of a multiplicative decomposition while α itself may contain another element of this decomposition (case of finite-strain elasto-plasticity and elasto-viscoplasticity). Just the same as in Maugin [17], but introducing the entropy density S and the new thermodynamical force A associated with α, we have the following equations of state, definitions, and limit condition :

$$W = W(\boldsymbol{F}, \theta, \alpha; \boldsymbol{X}), \quad \boldsymbol{T} = \frac{\partial W}{\partial \boldsymbol{F}}, \quad S = -\frac{\partial W}{\partial \Theta}, \quad A = -\frac{\partial W}{\partial \alpha}, \tag{9}$$

and

$$\begin{gathered} L^{th} = K - W; \quad K = \frac{1}{2}\rho_0(\boldsymbol{X})\boldsymbol{v}^2, \quad \boldsymbol{b} = -(L\boldsymbol{1}_R + \boldsymbol{T}.\boldsymbol{F}), \\ \boldsymbol{f}^{inh} = \left.\frac{\partial L^{th}}{\partial \boldsymbol{X}}\right|_{expl}, \quad \boldsymbol{f}^{th} = S\nabla_R\theta, \quad \boldsymbol{f}^{int} = A.(\nabla_R\alpha)^T, \end{gathered} \tag{10}$$

$$\boldsymbol{Q}(\boldsymbol{F}, \theta, \alpha; \nabla_R\theta; \boldsymbol{X}) \to \boldsymbol{0} \quad \text{as} \quad \nabla_R\theta \to \boldsymbol{0}. \tag{11}$$

The classical dynamical equations of the thermo-elasticity of conductors are left formally unchanged. But, obviously, there exists now an *intrinsic dissipation* so that the heat equation is generalized to

$$\theta\frac{\partial S}{\partial t} + \nabla_R.\boldsymbol{Q} = \boldsymbol{\Phi}^{int} \equiv A\left.\frac{\partial \alpha}{\partial t}\right|_{\boldsymbol{X}}. \tag{12}$$

In the present case (dynamics without external body force), the mass conservation equation and the balance of linear (physical) momentum read

$$\left.\frac{\partial \rho_0}{\partial t}\right|_{\boldsymbol{X}} = 0 \tag{13}$$

and

$$\left.\frac{\partial \boldsymbol{p}}{\partial t}\right|_{X} - \text{div}_R\boldsymbol{T} = \boldsymbol{0}, \quad \boldsymbol{p} = \rho_0(\boldsymbol{X})\boldsymbol{v}, \quad \boldsymbol{v} = \partial\chi/\left.\partial t\right|_{\boldsymbol{X}}. \tag{14}$$

The balance of canonical momentum (equation in material space) is accordingly modified. Indeed, performing on eqn.$(14)_1$ the pull-back to the reference configuration by applying $\boldsymbol{F}$ to the right and making some rearrangements, we arrive at the following *inhomogeneous* balance of canonical momentum on the material manifold :

$$\left.\frac{\partial P}{\partial t}\right|_{\boldsymbol{X}} - (\mathrm{div}_R \boldsymbol{b} + \boldsymbol{f}^{inh}) = \boldsymbol{f}^{th} + \boldsymbol{f}^{int}, \quad \boldsymbol{V} = -\boldsymbol{F}^{-1}.\boldsymbol{v}, \tag{15}$$

where $P = -\rho_0 \boldsymbol{v}.\boldsymbol{F} = \rho_0 \boldsymbol{C}.\boldsymbol{V}$ is the canonical momentum, and $\rho_0(\boldsymbol{X})$ is the reference density. Clearly, insofar as this equation is concerned, internal variables produce a source term – a material force – that resembles the thermal material force $\boldsymbol{f}^{th}$ in both general expression and effects! In particular, we may say that the microscopic irreversibility for which α accounts, manifests itself as a *quasi-inhomogeneity* in this equation, just the same as thermal effects. We also note that the new additional terms in (12) and (15) exhibit a consistency, one involving the time derivative, the other the space derivative, of the internal variable. This seems more symmetric than for temperature effects. But this is only apparent because (12) can be rewritten as (cf. Maugin [19])

$$\left.\frac{\partial (S\theta)}{\partial t}\right|_{\boldsymbol{X}} + \nabla_R.\boldsymbol{Q} = \boldsymbol{\Phi}^{th} + \boldsymbol{\Phi}^{int}, \quad \boldsymbol{\Phi}^{th} := S\frac{\partial \theta}{\partial t}. \tag{16}$$

The similarity between variables α and θ is thus enhanced; but while the former is governed by the *heat equation*, the former has to be governed by a pure *evolution equation* subjected to the second law of thermodynamics (non-negative dissipation).

The identification of the driving force of anelasticity is central to a thermomechanical approach to this type of irreversible behavior. Anelastic behavior is a macroscopic manifestation of microstructural material rearrangements. The corresponding time evolution in fact is the evolution of the distribution of microstructural material inhomogeneities, e.g., dislocation patterns. Prime importance is thus attributed to the *inhomogeneity mapping* $\boldsymbol{K}$ of Section 2 and its time evolution. The inhomogeneity finite strain and the inhomogeneity velocity gradient can be defined by (Epstein and Maugin [5], Cleja-Tigoiu and Maugin [2])

$$\boldsymbol{C}_{\boldsymbol{K}} := \boldsymbol{K}^{-T}.\boldsymbol{K}^{-1}, \quad \boldsymbol{C}_{\boldsymbol{K}}^{-1} = \boldsymbol{K}.\boldsymbol{K}^{T}, \quad \boldsymbol{L}_{\boldsymbol{K}} = \dot{\boldsymbol{K}}.\boldsymbol{K}^{-1}, \tag{17}$$

where a superimposed dot denotes partial time differentiation. The local configuration reached after the $\boldsymbol{K}$ mapping at $\boldsymbol{X}$ is called the *reference crystal* C_R (Epstein and Maugin [4]). But if $\boldsymbol{K}$ is taken to represent anelastic effects, then by simple composition we see that $\boldsymbol{F}.\boldsymbol{K}(\boldsymbol{X}) = \boldsymbol{F}_e$ is the *elastic deformation gradient.* But $\boldsymbol{K}$ being generally non-integrable, this is also the case of $\boldsymbol{F}_e$. Accordingly we can write the *Bilby-Bullough-Stroh-Kröner-Lee multiplicative decomposition* $\boldsymbol{F} = \boldsymbol{F}_e.\boldsymbol{F}_p$, $\boldsymbol{F}_p \equiv \boldsymbol{K}^{-1}$, where $\boldsymbol{F}_p$ may be referred to as the *plastic* (in general, *anelastic*) deformation gradient, in fact *not* a true gradient. Usually, unless some porosity effects are predominant, the plastic part of the decomposition of $\boldsymbol{F}$ is viewed as incompressible. It is of interest to notice that the so-called Mandel stress tensor $\boldsymbol{M}$ which plays a crucial role in many theories of finite-strain anelasticity (Lubliner [13]) is intimately related to the Eshelby stress as it is given by

$$\boldsymbol{M} = \boldsymbol{T}.\boldsymbol{F} = \boldsymbol{C}.\boldsymbol{S} = -(L\boldsymbol{1}_R + \boldsymbol{b}), \tag{18}$$

where $\boldsymbol{S}$ is Piola thermodynamic stress tensor. A modeling of anelasticity may be based on the above-presented notion, cf. Maugin [16], Epstein and Maugin [5], Cleja-Tigoiu and Maugin [2].

Another totally different interesting case of local structural rearrangement is that of *volumetric growth* to which the remainder of this work is devoted.

4. Volumetric growth as a structural rearrangement

Basing on a recent work by Epstein and Maugin [7], we view volumetric growth as a process that happens to a uniform body, which we describe in terms of a fixed reference configuration. As time goes on, more material of the *same* kind is "squished-in" smoothly into the body, but in such a way that material points preserve their identity. In other words, the process of growth can be seen as an evolution of material-point neighborhoods in a fixed reference configuration, an evolution that finds its expression in *temporal changes of density* and concomitant distortions of material-point neighborhoods. While preserving the uniform material properties, the growth process will in general cause the developments of *inhomogeneities*, responsible for *residual stresses* in the body. Other types of growth, such as inclusions and surface growth (accretion), are beyond the scope of this study. Accretive growth is dealt with by Gurtin [9] and the school of Arutyunyan et al. [1]. In contrast to previous sections (cf. eqn. (13)), here mass density, even in the reference configuration, can evolve in time. Accordingly, discarding diffusive effects, the global mass balance equation reads locally on the material manifold

$$\left.\frac{\partial \rho_0}{\partial t}\right|_{\boldsymbol{X}} = \Pi. \tag{19}$$

Epstein and Maugin [7] have shown that, for this theory, the first Piola-Kirchhoff stress $\boldsymbol{T}$, and the Eshelby stress tensor $\boldsymbol{b}$ are left formally unchanged while, a *material growth force* $\boldsymbol{f}^{gr} := \Pi\boldsymbol{C}.\boldsymbol{V}$ appears in such a way that the balance equation of canonical momentum reads

$$\left.\frac{\partial P}{\partial t}\right|_{\boldsymbol{X}} - (\mathrm{div}_R\boldsymbol{b} + \boldsymbol{f}^{inh}) = \boldsymbol{f}^{th} + \boldsymbol{f}^{int} + \boldsymbol{f}^{gr} \tag{20}$$

The Clausius-Duhem inequality is also left formally unchanged compared to the case of thermoelasticity, i.e.,

$$-(\dot{W} + S\theta) + \mathrm{tr}\left(\boldsymbol{T}(\nabla_R \boldsymbol{v})^T\right) - \theta^{-1}\boldsymbol{Q}.\nabla_R\theta \geq 0. \tag{21}$$

Now we introduce the inhomogeneity mapping $\boldsymbol{K}$ called *transplant* in this growth theory with physiological applications. Let $J_K = \det\boldsymbol{K}$, and ρ_c the truly invariant mass density in the crystal reference so that

$$\rho_0 = J_K^{-1}\rho_C. \tag{22}$$

By time differentiation this shows that *growth* (increase in density) corresponds to $\partial(\det\boldsymbol{K})/\partial t < 0$ and *resorption* (decrease in density) corresponds to the opposite condition. With a simplified functional dependence of W taken as $W = \bar{W}(\boldsymbol{F}, \theta, \boldsymbol{X}, t)$ we can write $W = J_K^{-1}W_c(\boldsymbol{F}.\boldsymbol{K}(\boldsymbol{X}, t), \theta)$. We also introduce the following two transplant velocity gradients:

$$\boldsymbol{L_K} = \dot{\boldsymbol{K}}.\boldsymbol{K}^{-1}, \quad \bar{\boldsymbol{L}}_{\boldsymbol{K}} = \boldsymbol{K}^{-1}.\dot{\boldsymbol{K}} = -\dot{\boldsymbol{K}}^{-1}.\boldsymbol{K}. \tag{23}$$

On account of (22) we have

$$\left.\frac{\partial\rho_0}{\partial t}\right|_{\boldsymbol{X}} = -\rho_0 \,\mathrm{tr}\boldsymbol{L_K} = \Pi. \tag{24}$$

This shows that there is no need of an evolution equation for the volumetric source, once we know the evolution of $\boldsymbol{K}$. This must be ruled by the second law of thermodynamics. As a matter of fact, $\boldsymbol{T}$ and S receiving their classical form as state laws, the reduced dissipation reads as follows:

$$\Phi = \mathrm{tr}(\boldsymbol{M}.\boldsymbol{L_K}) - \theta^{-1}\boldsymbol{Q}.\nabla_R\theta \geq 0, \tag{25}$$

where $\boldsymbol{M}$ is the Mandel stress defined in (18). The system of constitutive equations must now be closed by a heat conduction law (e.g., Fourier's) and the evolution equation for the transplant $\boldsymbol{K}$. The latter can be envisaged as the general relationship between various arguments:

$$\varphi(\boldsymbol{K}, \dot{\boldsymbol{K}}, \boldsymbol{M} \text{ or } \boldsymbol{b}, \boldsymbol{F}, \dot{\boldsymbol{F}}) = 0, \tag{26}$$

reflecting thus some kind of *nonlinear viscoelasticity*. Using invariance arguments it can be shown (Epstein and Maugin [7]), that this can be reduced to the following relationship:

$$\varphi(\bar{\boldsymbol{L}}_{\boldsymbol{K}}, \boldsymbol{b}_c, \boldsymbol{C}_e, \dot{\boldsymbol{C}}_e) = 0, \tag{27}$$

where $\boldsymbol{C}_e$ and $\dot{\boldsymbol{C}}_e$ come from $\boldsymbol{F}_e = \boldsymbol{F}.\boldsymbol{K}$ – the elastic deformation gradient, $\bar{\boldsymbol{L}}_{\boldsymbol{K}}$ comes from an invariant combination of $\boldsymbol{K}$ and $\dot{\boldsymbol{K}}$, and $\boldsymbol{b}_c$ is given by $\boldsymbol{b}_c =$

$J_K \boldsymbol{K}^{-1}.\boldsymbol{b}.\boldsymbol{K}$. As a simple example of isotropic behavior for the symmetric part $\bar{\boldsymbol{L}}_S$ of $\bar{\boldsymbol{L}}_{\boldsymbol{K}}$ we have the evolution equation

$$\bar{\boldsymbol{L}}_S = \varphi_0 \boldsymbol{1} + \varphi_1 \boldsymbol{b}_c + \varphi_2 \boldsymbol{b}_c^2, \tag{28}$$

where the coefficients are functions of the fundamental invariants of $\boldsymbol{b}_c$. Such a law is akin to *nonlinear viscoelasticity.* In the more realistic modeling of growth by Imatani and Maugin [11], anisotropic effects are taken into account (see below).

5. Constitutive modeling for anisotropic growth

Let $\boldsymbol{\sigma}$ the Cauchy stress and set

$$\boldsymbol{C}^* = \boldsymbol{F}_e^T.\boldsymbol{F}_e, \quad \boldsymbol{S}^* = J_{\boldsymbol{F}_e} \boldsymbol{F}_e^{-1} \boldsymbol{\sigma} \boldsymbol{F}_e^{-T}. \tag{29}$$

These are conjugated strain and stress on the material manifold in the *local reference crystal* (i.e., elastically released configuration). The corresponding Mandel stress $\boldsymbol{M}^*$ is given by (cf. (18)) $\boldsymbol{M}^* = \boldsymbol{S}^*.\boldsymbol{C}^*$. For a hyperelastic material $\boldsymbol{S}^* = 2\partial W_c / \partial \boldsymbol{C}^*$, where $W_c = \hat{W}$ is the strain energy function per unit volume in the reference crystal. A possible expression for W_c is

$$W_c = C_1(I_1^* I_3^{*-1/3} - 3) + C_2(I_2^* I_3^{*-2/3} - 3) + \frac{1}{2} C_3 (I_3^{*1/2} - 1)^2, \tag{30}$$

where the I_α^* are the invariants of $\boldsymbol{C}^*$ and the C_α are material constants. This results in an $\boldsymbol{S}^*$ of the form

$$\boldsymbol{S}^* = \phi_0 \boldsymbol{1} + \phi_1 \boldsymbol{C}^* + \phi_2 \boldsymbol{C}^{*2}, \tag{31}$$

with coefficients depending on the I_α. The evolution equation for the transplant tensor is taken in a similar form (compare to (28)) but in terms of the Mandel stress:

$$\bar{\boldsymbol{L}}_S = \psi_0 \boldsymbol{1} + \psi_1 \boldsymbol{M}^* + \psi_2 \boldsymbol{M}^{*2} = \boldsymbol{G}^*(\boldsymbol{M}^*). \tag{32}$$

For example, $\psi_0 < 0$, $\psi_1 = \psi_2 = 0$, yields growth. But *anisotropy* of growth is an unavoidable feature. Let $\boldsymbol{w}$ ($\boldsymbol{w}^*$ in the reference crystal) be the unit vector that characterizes anisotropy. This influences the strain-stress and we do need to specify an *evolution* equation for $\boldsymbol{w}$. Introducing the associated skewsymmetric tensor $\boldsymbol{W}^*$ such that $\boldsymbol{W}\boldsymbol{a} = \boldsymbol{a} \times \boldsymbol{w}, \forall \boldsymbol{a}$, and using representation theorems we can now propose a generalization of (32) as (see Imatani and Maugin [11] for details and the full expression)

$$\bar{\boldsymbol{L}}_S = \boldsymbol{G}_W^*(\boldsymbol{M}^*, \boldsymbol{W}^*). \tag{33}$$

A useful simple expression is given by

$$\begin{aligned}\bar{\boldsymbol{L}}_S &= \psi_0 \boldsymbol{1} + \psi_1 \boldsymbol{M}^* + \psi_3 \boldsymbol{N}^* + \frac{1}{2}\psi_7(\boldsymbol{M}^*\boldsymbol{N}^* + \boldsymbol{N}^*\boldsymbol{M}^*) \\ &= \boldsymbol{G}^*(\boldsymbol{M}^*, \boldsymbol{N}^*),\end{aligned} \tag{34}$$

where $\boldsymbol{N}^*$ is such that $\boldsymbol{W}^{*2} = \boldsymbol{N}^* - \boldsymbol{1}$. The advantage of using $\boldsymbol{N}$ instead of $\boldsymbol{W}$ is that the nonzero component in the out-of-plane direction disappears in this form and that is useful to clarify the characteristics of the vector field $\boldsymbol{w}$ especially in the plane-deformation problem. Accordingly, the functions ψ_0 and ψ_1 are related to isotropic growth/remodeling while ψ_3 and ψ_7 relate to the anisotropic growth. As to the time development of anisotropy it is quite natural to slave it to the evolution of the transplant tensor, hence to consider a time evolution in the following objective form

$$\dot{\boldsymbol{w}}^* = (\boldsymbol{w}^*.(\boldsymbol{G}^*\boldsymbol{w}^*))\boldsymbol{w}^* - \boldsymbol{G}^*.\boldsymbol{w}^*, \tag{35}$$

or using the Mandel stress

$$\dot{\boldsymbol{w}}^* = \boldsymbol{M}^*.\boldsymbol{w}^* - (\boldsymbol{w}^*(\boldsymbol{M}^*.\boldsymbol{w}^*))\boldsymbol{w}^*. \tag{36}$$

Therefore, the vector field $\boldsymbol{w}$ “floats” in the material, and its rate will vanish when the direction of $\boldsymbol{w}$ coincides with one of the principal directions of $\boldsymbol{M}^*$. Hence material anisotropy evolves in such a way that the particular direction of growth is specified by the current state of the Mandel stress. This model may be applicable to self-organization or adaptation.

6. Numerical examples

6.1. Methodology

The numerics are based on an *incremental* formulation in the reference configuration while the constitutive model was set in the local reference crystal. Mechanical variables need to be updated in a step-by-step procedure. This numerical formulation is documented in Imatani and Maugin [11] for two-dimensional problems. The starting point obviously is the principle of virtual work in the reference configuration :

$$\int_{\Omega_0} \rho_0 \dot{\boldsymbol{v}}.\delta\boldsymbol{u}\, dV + \int_{\Omega_0} \boldsymbol{S} : \delta\boldsymbol{E}\, dV = \int_{\Gamma_S} \bar{\boldsymbol{t}}.\delta\boldsymbol{u}\, dS, \tag{37}$$

where $\boldsymbol{E}$ is the Green-Lagrange strain. For *quasi-statics* we write this as

$$\int_{\Omega_0} \boldsymbol{S}^{\tau} : \Delta(\delta\boldsymbol{E})\, dV + \int_{\Omega_0} \Delta\boldsymbol{S} : \delta\boldsymbol{E}^{\tau}\, dV = \int_{\Gamma_S} \Delta\bar{\boldsymbol{t}}.\delta\boldsymbol{u}\, dS, \tag{38}$$

where Δ denotes the variation from time τ to time $\tau+\Delta\tau$, while in *full* dynamics this takes the form

$$\int_{\Omega_0} \rho_0 \dot{\boldsymbol{v}}^{\tau+\Delta\tau} . \delta \boldsymbol{u} \, dV + \int_{\Omega_0} \boldsymbol{S}^\tau : \delta \boldsymbol{E}^\tau \, dV + \int_{\Omega_0} \boldsymbol{S}^\tau : \Delta(\delta \boldsymbol{E}) \, dV \\ + \int_{\Omega_0} \Delta \boldsymbol{S} : \delta \boldsymbol{E}^\tau \, dV = \int_{\Gamma_S} \bar{\boldsymbol{t}}^{\tau+\Delta\tau} . \delta \boldsymbol{u} \, dS, \tag{39}$$

where we implement Newmark's method according to which quantities are evaluated at the next time $\tau + \Delta\tau$ and then the results are unconditionally stable. The finite-element scheme is given in Imatani and Maugin [11].

6.2. Examples and Results

Basically three different processes are illustrated. The first one is *monotonic* growth and *adaptation.* Figure 1 shows monotonic growth without any constraint where a quarter part (fragment) of a ring is analyzed under surface-force free conditions. Here ψ_1 and ψ_2 are set equal to zero in order to exemplify *isotropic growth.* A circumferential growth is revealed under the conditions ($\psi_0 = 0.0, \psi_3 = -0.1$) and anisotropy vector $\boldsymbol{w}$ also lying circumferentially. Under the action of constant inner pressure, the mean radius varies as indicated in Figure 2a for different constitutive assumptions regarding the coefficients. The case b-11r shows the *adaptative behavior* (path B is the purely hyperelastic case without growth). The (Cauchy) stress distribution in a growing material in the form of a ring is illustrated in Figure 2b where the nodal displacement at the inner surface is prescribed, moving radially during one second. The pure hyperelastic case corresponds to path B. For other cases, competition between growth in the radial and circumferential directions creates completely different distributions.

Finally, cyclic and dynamic effects in a cantilever beam are examined (Figures 3–5). A triangular pattern A (Figure 3) is applied in quasi-statics and combined wave forms such as in B or C are applied in the fully dynamical analysis. The input is composed of normal forces at the upper edge of the side walls. The anisotropic vector field $\boldsymbol{w}$ is vertically embedded in the material specimen resulting in some resistance to the bending stress. Figure 4 shows the trajectories of the neutral (central) axis of the beam. The dynamical result exhibits an asymmetry in time due to inertial effects that become important as the eigenfrequency of the beam is of the same order as the time scale of the forcing. Then the damping due to growth takes place after release of the applied force. The last Figure 5 shows the volumetric growth ($\det \boldsymbol{K}^{-1}$) during the course of cyclic loading. The quasi-static analysis clearly overestimates the growth.

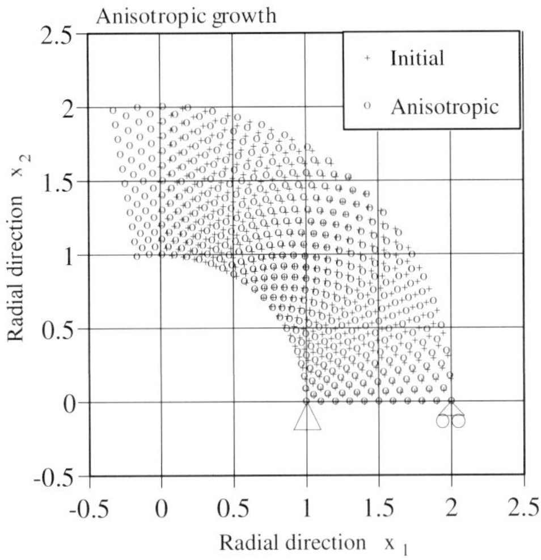

Figure 1. Anisotropic growth of a fragment of ring.

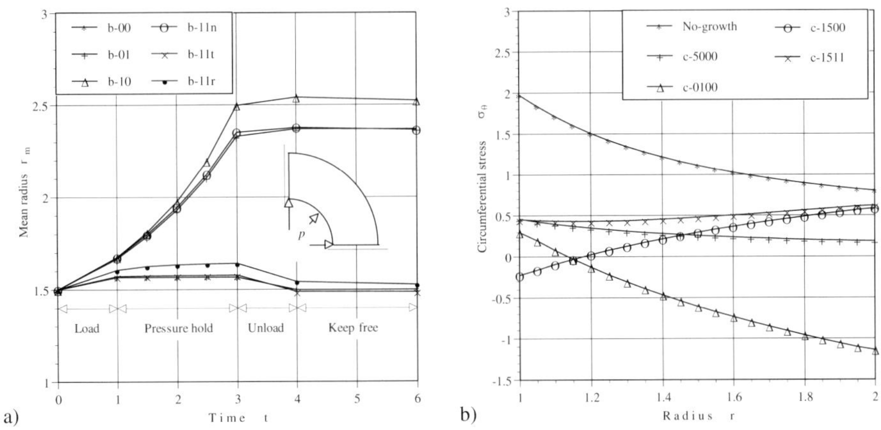

Figure 2. a) Variation of mean radius in the ring under pressure followed by unload. The path b-00 means the result by the parameters to be zero corresponding to pure hyperelasticity. Recovery effect appears in the path b-11r where the anisotropic vector is randomly distributed in the domain. b) Circumferential stress (Chauchy stress) distribution under various patterns of material parameters.

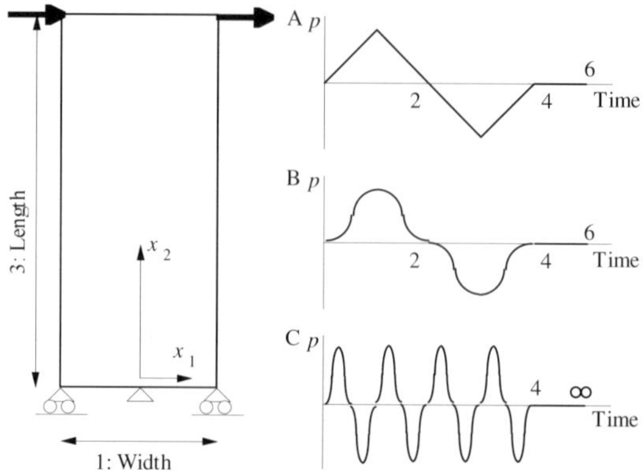

Figure 3. Schematic illustration of the boundary conditions for the cantilever beam problem. The surface force is imposed at the edge of the side wall as a normal force. The wave A is used in the static analysis while B and C are used in the dynamic analysis.

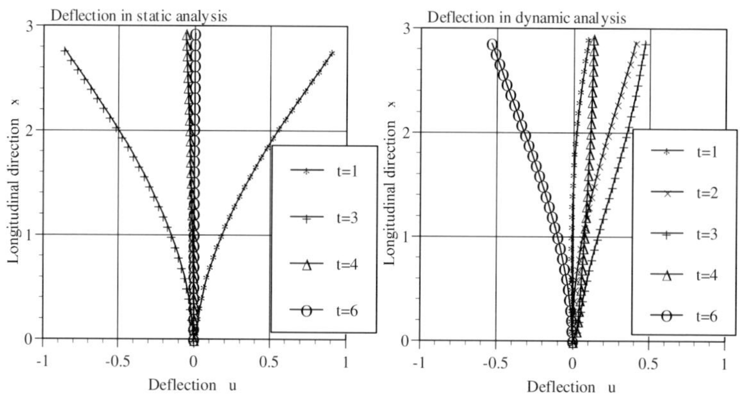

Figure 4. Deflection of cantilever under static load (left) and dynamic load (right).

7. Conclusion

The above-given theory and examples are sufficient to exhibit a rather good capture of the main features of volumetric growth. Here practically each material parameter uniquely corresponds to a characteristic phenomenon of isotropic/anisotropic growth. The model describes satisfactorily a variety of growth behaviors and it possesses a potential capability to describe adaptation in which the material changes its properties (in this model anisotropy) in connection with the Mandel stress. But, on the one hand, the model does not provide for the accompanying diffusive effects and, on the other hand, it leaves out the physico-chemical origin of growth. Much more work is needed to include these facts and properties.

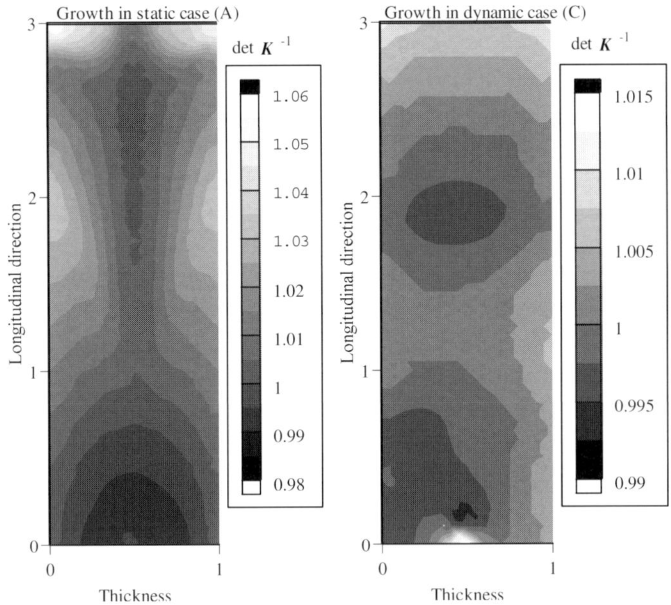

Figure 5. Growth behavior in the course of cyclic loads.

Acknowledgments

This work was performed within the European Network TMR 98-0229 "Phase Transitions in Solids". S.I. benefited from a French Government Fellowship during his stay at LMM UPMC & CNRS, Paris in 1999-2000.

References

[1] Arutyunyan, N.Kh., Drozdov, A.D. and Naumov, V.E. (1985). *Mechanics of Growing Viscoelastic-plastic Solids*, (in Russian), Nauka, Moscow.

[2] Cleja-Tigoiu, S. and Maugin, G.A. (2000). Eshelby's Stress Tensors in Finite Elastoplasticity. *Acta Mechanica*, **139**, 231–249.

[3] Cowin, S.C., Sadegh, A.M. and Luo, G.M. (1992). An Evolutionary Wolff's Law for Trabecular Architecture. *ASME J. Biomech. Engng.*, **114**, 129–136.

[4] Epstein, M. and Maugin, G.A. (1990). The Energy-momentum Tensor and Material Uniformity in Finite Elasticity. *Acta Mechanica*, **83**, 127–133.

[5] Epstein, M. and Maugin, G.A. (1995). On the Geometrical Structure of Anelasticity. *Acta Mechanica*, **115**, 119–131.

[6] Epstein, M. and Maugin, G.A. (1997) *Notions of Material Uniformity and Homogeneity (ICTAM'96, Kyoto)*, T. Tatsumi et al. (eds.), Elsevier, Amsterdam, 201–215.

[7] Epstein, M. and Maugin, G.A. (2000). Thermomechanics of Volumetric Growth in Uniform Bodies. *Int. J. Plasticity*, **16**, 951–978.

[8] Fung, Y.C, Liu, S.Q. and Zhou, J.B. (1993). Remodeling the Constitutive Equation While a Blood Vessel Remodels itself under Stress. *ASME J. Biomech. Engng.*, **115**, 453–459.

[9] Gurtin, M.E. (1993). The Dynamics of Solid-solid Phase Transitions.I. Coherent Interfaces. *Arch. Rat. Mech. Anal.*, **123**, 303–335.

[10] Hart, R.T., Davy, D.T. and Heiple, K.G. (1984). A Computational Method for Stress Analysis of Adaptive Elastic Materials with a View Toward Applications in Strain-induced Bone Remodeling. *ASME J. Biomech. Engng.*, **106**, 342–350.

[11] Imatani, S. and Maugin, G.A. (2001). A Constitutive Model for Growing Materials and its Applications to Finite-element Analysis. *Trans. ASME J. Appl. Mech.*, submitted.

[12] Kuehn, S. and Hauger, W. (2000). A Theory of Adaptive Growth of Biological Materials. *Arch. Appl. Mech.*, **70**, 183–192.

[13] Lubliner, J. (1990). *Plasticity Theory*, McMillan, New York.

[14] Maugin, G.A. (1992). *Thermomechanics of Plasticity and Fracture*, Cambridge University Press, U.K.

[15] Maugin, G.A. (1993). *Material Inhomogeneities in Elasticity*, Chapman and Hall, London.

[16] Maugin, G.A. (1994). Eshelby Stress in Elastoplasticity and Fracture. *Int. J. Plasticity*, **10**, 393–408.

[17] Maugin, G.A. (1995). Material Forces: Concepts and Applications. *ASME Appl. Mech. Rev.*, **48**, 213–245.

[18] Maugin, G.A. (1999). Thermomechanics of Nonlinear Irreversible Behaviors. *World Scientific*, Singapore and River Edge, N.J.

[19] Maugin, G.A. (2000). On the Universality of the Thermomechanics Forces Driving Singular Sets. *Arch. Appl. Mech.*, **70**, 31–45.

[20] Rodriguez, E.K., Hoger, A. and McCullogh, A.D. (1994). Stress-dependent Finite Growth in Soft Elastic Tissues. *J. Biomechanics*, **27**, 455–467.

[21] Simo, J.C. (1988). A Framework for Finite Strain Elastoplasticity Based on Maximum Dissipation and the Multiplicative Decomposition, Parts 1 and 2. *Comp. Methods Appl. Mech. Eng.*, **66**, 199–219, **68**, 1–31.

[22] Taber, L.A. (1995). Biomechanics of Growth, Remodeling and Morphogenesis. *ASME Appl. Mech. Rev.*, **48**, 487–545.

Theoretical and Computational Simulation of Viscoelastic Polymeric Foams at Finite Strains

W. Ehlers, B. Markert
Institute of Applied Mechanics (CE), University of Stuttgart
Pfaffenwaldring 7, 70569 Stuttgart, Germany
ehlers@mechbau.uni-stuttgart.de, markert@mechbau.uni-stuttgart.de

Abstract: Viscoelastic polymer foams (soft foams) as well as synthetic sponges are of great interest in various fields of engineering. As a result of the foaming process, these materials exhibit an open or closed cellular micro structure leading to a macroscopic porosity of more than 90 %. Following this, extremely large deformations can occur giving these materials their outstanding mechanical characteristics.

The goal of this paper is to present an efficient macroscopic continuum mechanical model based on the Theory of Porous Media (TPM) in order to carry out numerical investigations at suitable means of computational costs. In particular, the model accounts for the complex cellular structure including the intrinsic viscoelastic behavior of the polymeric skeleton which is saturated by an independent compressible pore-gas constituent.

Keywords: Finite viscoelasticity, porous media, polymeric foams, mixed finite elements.

1. Introduction

Foamed polymers (soft foams) are subject to large viscoelastic deformations (mostly compression) during their practical application, e. g. as bumpers, packaging or seat cushions. They consist of a multiphasic micro structure built by a combination of open and/or closed fluid-filled cells. In particular, each cell of the micro structure undergoes complex deformation mechanisms under external loads resulting in the characteristic highly non-linear stress-strain behavior of the macroscopic foam (Gibson and Ashby [8]).

With new methods, e. g. the computer tomography (CT), it is possible to obtain detailed information of the inner cellular structure. However, even by increased computer power, the numerical simulation of real applications on the micro scale using discrete cell models is not reasonable until now. Furthermore, the description of the moving pore-fluid and, as a consequence, the frictional drag forces on the cell faces is problematic. A further possibility to analyze the response of foams to stress is to combine the beam theory with scaling laws, i. e. to approximate an average cell by a simple beam model and determine the effective material parameters via homogenization, see e. g. Gibson and Ashby [8]. This leads to a good description concerning honeycombs and open-cell foams

C. Miehe (ed.), IUTAM Symposium on Computational Mechanics of Solid Materials at Large Strains, 235-244.

but the model reaches its limits in describing closed-cell foams and, in general, in describing the pore-fluid motion properly. Thus, statistical, multiphasic continuum mechanical models are necessary to successfully represent the behavior of foam materials under external loads.

In the present contribution, the macroscopic description of fluid-saturated polymer foams is considered in the framework of the well-founded Theory of Porous Media (TPM), cf. Ehlers [5]. Following this, we proceed from a biphasic description of polymeric foams consisting of a materially incompressible, cellular polymer matrix saturated by a compressible viscous pore-gas. The assumption of a materially incompressible solid skeleton is used based on the fact that the material compressibility of the polymer material itself is much smaller than the bulk compressibility of the entire foam. Therefore, there exists a so-called point of compaction, which is defined as that deformation state, where all pores are closed and no further volume reduction is possible. This concept of the compaction point implies the characteristic property of structural densification of solid foams. The energy absorbing behavior of the cellular polymer matrix is described by an appropriate *Ogden*-type finite viscoelasticity law which can be descriptively deduced from rheological and thermodynamical considerations. Concerning the description of finite deformation viscoelasticity of non-porous solids, a lot of work can be found in the literature, see e. g. Reese and Govindjee [10] and citations therein. The solid-fluid interaction is taken into account by the viscous drag force, where a deformation-dependent permeability is considered, since experiments have shown that it plays an important role on the transient compressive response of open-cell foams at higher deformation rates. For simplicity, thermal effects as well as mass exchanges between the two constituents are excluded. The arising pure mechanical, finite viscoelastic two-phase model is capable to describe the complex material response of open- and closed-cell polymeric foams under mechanical loading, where the biphasic formulation is essential for the consideration of the basic physical properties.

The numerical treatment of the presented model is carried out by use of the finite element method (FEM), where the semi-discretization of the governing field equations in the spatial domain leads to a system of differential-algebraic equations (DAE) in the time domain. The numerical integration of this semi-discrete system can be efficiently performed by suitable time integration schemes, e. g. *Runge-Kutta* methods, where an embedded error estimator and a time step control can be added (Ellsiepen [7]). To set an example, the model is adapted to a highly porous, open-cell polyurethane (PU) foam, where the parameter adjustment is principally based on the sequential quadratic programming (SQP) method (Spellucci [11]). Finally, numerical simulations of uniaxial compression experiments with compressions up to $86\,\%$ show the applicability of the presented macroscopic approach.

2. Theoretical framework

In the framework of the Theory of Porous Media (TPM), a fluid-saturated cellular foam can be treated as an immiscible mixture of constituents φ^α ($\alpha = S$: cellular solid skeleton; $\alpha = F$: pore-fluid) which are assumed to be in a state of ideal disarrangement. Following this, the prescription of a real or a virtual averaging process over a representative elementary volume (REV) leads to a model φ of superimposed and interacting continua, i. e. the homogenized or "smeared" model. In the arising biphasic macro model, the local structure of the mixture is represented by scalar variables, the volume fractions $n^\alpha = \mathrm{d}v^\alpha/\mathrm{d}v$, which are defined as the local ratios of the constituent volumes v^α with respect to the bulk volume v. Thus, assuming fully saturated conditions, i. e. avoiding any vacant space, the saturation condition yields $n^S + n^F = 1$.

The model under consideration incorporates three independent fields, the solid displacement $\mathbf{u}_S$, the seepage velocity $\mathbf{w}_F$, and the effective fluid pressure p. The corresponding three equations for quasi-static considerations can be obtained from the kinematics, the balance relations, and the constitutive equations discussed in detail by Ehlers [5]:

- Mixture balance of momentum:

$$\operatorname{div}(\mathbf{T}_E^S - p\,\mathbf{I}) + (n^S \rho^{SR} + n^F \rho^{FR})\,\mathbf{b} = \mathbf{0}\,, \tag{1}$$

- Fluid balance of mass:

$$n^F (\rho^{FR})'_S + \rho^{FR} \operatorname{div}(\mathbf{u}_S)'_S + \operatorname{div}(n^F \rho^{FR}\,\mathbf{w}_F) = 0\,, \tag{2}$$

- Generalized *Darcy*'s filter law (isotropic):

$$\frac{K^S}{\mu^{FR}}\,(\operatorname{grad} p - \rho^{FR}\,\mathbf{b}) + n^F\,\mathbf{w}_F = \mathbf{0}\,. \tag{3}$$

Herein, $\mathbf{T}_E^S$ is the solid Cauchy extra (effective) stress, $\mathbf{I}$ is the identity tensor, and $\mathbf{b}$ is the body force density. Furthermore, μ^{FR} is the effective pore-fluid viscosity, K^S is the intrinsic permeability, and $(\,\cdot\,)'_S$ denotes the material time derivative of $(\,\cdot\,)$ corresponding to the individual motion of φ^S. Concerning gas-filled foams, the pore-fluid is compressible, i. e. the effective (material) fluid density is a function of the pore-fluid pressure, whereas the cellular solid skeleton is supposed to be materially incompressible, i. e. $\rho^{SR} = \text{const.}$ Following this, the pore-gas is assumed to be governed by the ideal gas law

$$\rho^{FR}(p) = \frac{M_m^F}{R\,\theta}\,p \qquad \text{or} \qquad p(\rho^{FR}) = \frac{R\,\theta}{M_m^F}\,\rho^{FR}\,, \tag{4}$$

where R is the universal gas constant, θ the absolute *Kelvin*'s temperature, and M_m^F the molecular mass of φ^F.

In general, the intrinsic permeability in *Darcy*'s law (3) depends on the deformation state, i. e. the value of K^S decreases when the pore volume decreases. Thus, it is assumed that the permeability depends on the porosity n^F via

$$K^S(n^F) = K_0^S \left(\frac{n^F}{n_{0S}^F} \right)^\kappa , \tag{5}$$

where K_0^S is the initial permeability and the exponent κ governs the deformation dependency. Therefore, K^S governs the property of an open-cell, i. e. permeable $(K^S > 0)$, or a closed-cell, i. e. impermeable $(K^S \to 0)$, foam.

3. Finite viscoelasticity law

To describe the intrinsic dissipative phenomena of the cellular polymer skeleton, an adequate viscoelasticity law is required including the property of structural densification towards the point of compaction. The fundamental approach is based on the rheological structure of the generalized *Maxwell* model, where for the geometrically linear case, the discrete relaxation spectrum algorithm can be included straight forward into porous media theories (Ehlers and Markert [6]).

In the framework of a finite deformation theory of porous materials with elastic and inelastic (here: viscous) behavior, it is convenient to proceed from a multiplicative split of the solid deformation gradient

$$\mathbf{F}_S = (\mathbf{F}_{Se})_n \, (\mathbf{F}_{Si})_n \, , \quad n = 1, ..., N \tag{6}$$

into elastic parts $(\mathbf{F}_{Se})_n$ and inelastic parts $(\mathbf{F}_{Si})_n$, where N denotes the total number of considered *Maxwell* elements. Furthermore, from thermodynamics with internal state variables (Coleman and Gurtin [4]) and rheological considerations, one obtains the ansatz of a decomposed solid extra stress

$$\boldsymbol{\tau}_E^S[\mathbf{F}_S, (\mathbf{F}_{Se})_n] = \boldsymbol{\tau}_{EQ}^S[\mathbf{F}_S] + \boldsymbol{\tau}_{NEQ}^S[(\mathbf{F}_{Se})_n] \, , \quad n = 1, ..., N \tag{7}$$

into equilibrium parts (Index *EQ*) describing the basic elasticity and non-equilibrium parts (Index *NEQ*) vanishing in the thermodynamic equilibrium. Note that the *Kirchhoff* extra stress is a weighted *Cauchy* stress $\boldsymbol{\tau}_E^S = J_S \, \mathbf{T}_E^S$, where $J_S = \det \mathbf{F}_S$ is the solid *Jacobi*an.

In general, the stress-strain behavior of the cellular polymer matrix is very complex and highly non-linear concerning both the basic elasticity and the viscoelastic overstress. Therefore, assuming isotropic material behavior, we proceed from an *Ogden*-type material formulation (cf. Ogden [9]) which is extended towards an inelastic porous media application. In particular, volumetric extension terms are developed which describe the finite volume change including the concept of the compaction point. Thus, the equilibrium part of the solid extra stress can be obtained by

$$
\begin{aligned}
\boldsymbol{\tau}^S_{EQ} = \mu^S_0 \sum_{j=1}^{3} \sum_{k=1}^{K_0} \mu^*_k \, (\lambda_j^{\alpha_k/2} - 1) \, \mathbf{N}_j + \\
+ \frac{\Lambda^S_0}{\gamma_0 - 1 + \frac{\gamma_0 + 1}{(1 - n^S_{0S})^2}} \left(J_S^{\gamma_0} - \frac{J_S \, (1 - n^S_{0S})^{\gamma_0}}{(J_S - n^S_{0S})^{\gamma_0 + 1}} + \frac{J_S \, n^S_{0S}}{1 - n^S_{0S}} \right) \mathbf{I} \, ,
\end{aligned}
\tag{8}
$$

whereas the non-equilibrium part is computed from

$$
\begin{aligned}
\boldsymbol{\tau}^S_{NEQ} = \sum_{n=1}^{N} \Bigg(\mu^S_n \sum_{j=1}^{3} \sum_{k=1}^{K_n} (\mu^*_k)_n \left[(\lambda_{ej})_n^{(\alpha_k)_n/2} - 1 \right] \\
+ \frac{\Lambda^S_n}{\gamma_n - 1 + \frac{\gamma_n + 1}{[1 - (n^S_i)_n]^2}} \Bigg[(J_{Se})_n^{\gamma_n} - \\
- \frac{(J_{Se})_n \, [1 - (n^S_i)_n]^{\gamma_n}}{[(J_{Se})_n - (n^S_i)_n]^{\gamma_n + 1}} - \frac{(J_{Se})_n \, (n^S_i)_n}{1 - (n^S_i)_n} \Bigg] \mathbf{I} \Bigg) \, .
\end{aligned}
\tag{9}
$$

In (8) and (9), λ_j are the eigenvalues of the *Cauchy-Green* deformation tensors $\mathbf{C}_S = \mathbf{F}_S^T \mathbf{F}_S$ or $\mathbf{B}_S = \mathbf{F}_S \mathbf{F}_S^T$, respectively, $(\lambda_{ej})_n$ are the eigenvalues of the elastic *Cauchy-Green* deformation tensors $(\mathbf{C}_{Se})_n = (\mathbf{F}^T_{Se})_n \, (\mathbf{F}_{Se})_n$ or $(\mathbf{B}_{Se})_n = (\mathbf{F}_{Se})_n \, (\mathbf{F}^T_{Se})_n$, respectively, and $\mathbf{N}_j = \partial \lambda_j / \partial \mathbf{B}_S$ and $(\mathbf{N}_{ej})_n = \partial (\lambda_{ej})_n / \partial (\mathbf{B}_{Se})_n$ denote the eigentensors corresponding to the eigenvalues λ_j and $(\lambda_{ej})_n$. Furthermore, μ^S_0 and μ^S_n are the first macroscopic *Lamé* constants and μ^*_k, α_k $(k = 1, \dots, K_0)$ and $(\mu^*_k)_n$, $(\alpha_k)_n$ $(k = 1, \dots, K_n)$ are the dimensionless *Ogden* parameters, where the number of required *Ogden* terms K_0 and K_n depends on the complexity of the material behavior under study. Moreover, in the volumetric extension terms, Λ^S_0 and Λ^S_n are the second *Lamé* constants, γ_0 and γ_n are parameters that influence the volumetric non-linearity, $(J_{Se})_n = \det(\mathbf{F}_{Se})_n$ are the determinants of the elastic parts of the deformation gradient, n^S_{0S} is the initial solidity, and $(n^S_i)_n = n^S_{0S} \det(\mathbf{F}_{Si})_n^{-1}$ are the inelastic solid volume fractions with respect to the intermediate configuration. To guarantee the downward compatibility to *Hooke*an laws and to maintain stability, cf. Ball [1], the introduced parameters must hold the conditions

$$
\begin{aligned}
\sum_{k=1}^{K_0} \mu^*_k \, \alpha_k = 2 \wedge \{ \mu^*_k > 0, \alpha_k > 1 \vee \mu^*_k < 0, \alpha_k < -1 \} \, , \\
\sum_{k=1}^{K_n} (\mu^*_k)_n \, (\alpha_k)_n = 2 \wedge \{ (\mu^*_k)_n > 0, (\alpha_k)_n > 1 \vee (\mu^*_k)_n < 0, (\alpha_k)_n < -1 \} \, , \\
\gamma_0 \geq 1 \, , \quad \gamma_n \geq 1 \, .
\end{aligned}
\tag{10}
$$

The inelastic strains as internal state variables are obtained from linear evolution equations formulated with respect to the intermediate configuration, since this configuration acts as the actual configuration of the viscous deformation:

$$\begin{aligned}
(\hat{\mathbf{D}}_{Si})_n &= \overset{4}{\hat{\mathbf{D}}}{}^{-1}_n \, (\hat{\boldsymbol{\tau}}^S_{NEQ})_n \,, \\
\overset{4}{\hat{\mathbf{D}}}{}^{-1}_n &= \frac{1}{2\eta^S_n} \, (\mathbf{I} \otimes \mathbf{I})^{\overset{23}{T}} - \frac{\zeta^S_n}{2\eta^S_n \, (2\eta^S_n + 3\zeta^S_n)} \, (\mathbf{I} \otimes \mathbf{I}) \,.
\end{aligned} \tag{11}$$

Therein, $(\hat{\mathbf{D}}_{Si})_n$ are the inelastic solid deformation rates, $\overset{4}{\hat{\mathbf{D}}}{}^{-1}_n$ are the positive definite, isotropic, fourth order viscous compliances, where η^S_n and ζ^S_n are the macroscopic viscosity parameters, and $(\hat{\boldsymbol{\tau}}^S_{NEQ})_n = (\mathbf{F}_{Se})^{-1}_n \, (\boldsymbol{\tau}^S_{NEQ})_n \, (\mathbf{F}_{Se})^{T-1}_n$ are the respective non-equilibrium stress tensors. Furthermore, the transposition $(\,\cdot\,)^{\overset{ik}{T}}$ indicates an exchange of the i-th and k-th basis systems, and $(\,\hat{\cdot}\,)$ indicates the belonging to the intermediate configuration.

4. Numerical treatment

For the numerical treatment within the FEM, weak forms of the governing field equations (1) and (2) are required. Therefore, after eliminating the seepage velocity by use of the *Darcy* filter law (3), the balance relations weighted by independent test functions and integrated over the spatial domain Ω with surface $\partial\Omega$ result in the respective weak formulations:

$$\begin{aligned}
\mathcal{G}_{\text{MM}}(\mathbf{u}_S, \delta\mathbf{u}_S, p) \equiv & \int_\Omega \operatorname{grad} \delta\mathbf{u}_S \cdot (\mathbf{T}^S_E - p\,\mathbf{I}) \, \mathrm{d}v - \\
& - \int_\Omega \delta\mathbf{u}_S \cdot (n^S \rho^{SR} + n^F \rho^{FR}) \, \mathbf{b} \, \mathrm{d}v - \int_{\partial\Omega} \delta\mathbf{u}_S \cdot \bar{\mathbf{t}} \, \mathrm{d}a = 0 \,,
\end{aligned} \tag{12}$$

$$\begin{aligned}
\mathcal{G}_{\text{MF}}(\mathbf{u}_S, p, \delta p) \equiv & \int_\Omega \delta p \left[n^F \, (\rho^{FR})'_S + \rho^{FR} \operatorname{div} \, (\mathbf{u}_S)'_S \right] \mathrm{d}v + \\
+ & \int_\Omega \operatorname{grad} \delta p \cdot \left[\frac{K^S}{\mu^{FR}} \, \rho^{FR} \left(\operatorname{grad} p - \rho^{FR} \, \mathbf{b} \right) \right] \mathrm{d}v + \int_{\partial\Omega} \delta p \, \bar{q} \, \mathrm{d}a = 0 \,.
\end{aligned} \tag{13}$$

Herein, $\delta\mathbf{u}_S$ and δp are the test functions corresponding to the solid displacement $\mathbf{u}_S$ and the pore pressure p, $\bar{\mathbf{t}} = (\mathbf{T}^S_E - p\,\mathbf{I})\,\mathbf{n}$ is the external load vector, $\bar{q} = n^F \rho^{FR} \, \mathbf{w}_F \cdot \mathbf{n}$ denotes the filter mass flow of the fluid draining through the surface $\partial\Omega$, and $\mathbf{n}$ is the outward oriented unit surface normal.

Proceeding from (12) and (13), the spatial semi-discretization is carried out by the finite element method. Applying the *Bubnov-Galerkin* procedure, the unknown fields $\mathbf{u}_S$ and p and the corresponding test functions $\delta\mathbf{u}_S$ and δp are approximated by the same shape functions. Concerning the treatment of coupled solid-fluid problems within the FEM, mixed element formulations are used for the spatial discretization. In mixed formulations, the inf-sup or LBB condition is a crucial criterion for the stability of the numerical solution (Brezzi and Fortin [3]). One possible choice of a stable discretization is to use quadratic shape functions for the solid displacement $\mathbf{u}_S$ and linear shape functions for the pressure p which is known as the *Taylor-Hood* element. In the three-dimensional case, the quadratic-linear approximation leads to an enormous number of mid nodes in finer meshes which causes a large number of degrees of freedom increasing the computational costs rapidly. To overcome this problem, one has to reduce the order of the displacement interpolation. Using an equal interpolation element with a linear-linear approximation yields an unpredictable, mesh dependent behavior due to strange instabilities in the pressure field [3].

However, to get a stable pressure continuous element while keeping the linear approximation for the displacement, one has to add bubble functions to the displacement field. This results in the so-called MINI element, where the additional bubbles enlarge the trial space for the displacement which guarantees stability (Braess [2]). Since the bubble functions are restricted to a single finite element, the barycentric bubble nodes can be removed via static condensation. Therefore, for the numerical treatment of large coupled solid-fluid problems in the sense of the TPM approach, the MINI element is the best choice concerning quality and computational costs in the framework of the finite element method. For illustration (Figure 1), the mixed element formulations are shown in a triangle and can directly be transferred to tetrahedrons.

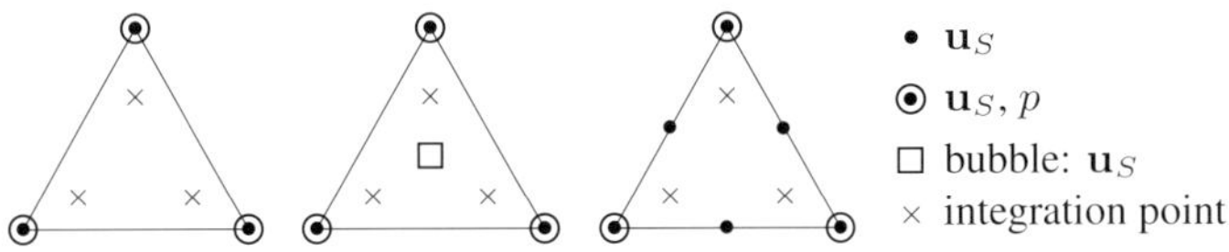

Figure 1. Mixed element formulations in a triangle: linear-linear approximation T1P1 (left), MINI element (middle), *Taylor-Hood* element T2P1 (right).

For the time discretization, the degrees of freedom $\mathbf{u}_S$ and p at all FEM nodes are collected in a vector $\boldsymbol{u}$ of external variables, whereas the internal (history) variables $(\mathbf{C}_{Si})_n$, i. e. the inelastic deformation tensors, at all integration points are collected in a vector $\boldsymbol{q}$. Then, with the vector of unknowns $\boldsymbol{y} := (\boldsymbol{u}, \boldsymbol{q})^T$, the semi-discrete initial-value problem in time with $t \geq t_0$ and $\boldsymbol{y}(t_0) = \boldsymbol{y}_0$ reads

$$\boldsymbol{F}[\boldsymbol{y}, (\boldsymbol{y})'_S, t] \equiv \begin{bmatrix} \boldsymbol{F}_1[t, \boldsymbol{u}, (\boldsymbol{u})'_S, \boldsymbol{q}] \\ \boldsymbol{F}_2[t, \boldsymbol{q}, (\boldsymbol{q})'_S, \boldsymbol{u}] \end{bmatrix} \equiv \begin{bmatrix} \boldsymbol{M}\,(\boldsymbol{u})'_S + \boldsymbol{k}[\boldsymbol{u}, \boldsymbol{q}] - \boldsymbol{f} \\ (\boldsymbol{q})'_S - \boldsymbol{g}[\boldsymbol{q}, \boldsymbol{u}] \end{bmatrix} \overset{!}{=} \mathbf{0}\,. \quad (14)$$

Therein, $\boldsymbol{F}_1$ represents the field equations, where $\boldsymbol{M}$ is the generalized mass matrix, $\boldsymbol{k}$ is the generalized stiffness vector, $\boldsymbol{f}$ is the generalized external force vector, and $\boldsymbol{F}_2$ represents the evolution equations of the viscoelastic model. Due to the quasi-static formulation, the mixture momentum balance (12) delivers no contribution to the generalized mass matrix which is related to $(\mathbf{u}_S)'_S$, i. e. $\boldsymbol{M}$ does not possess the full rank. Thus, (14) becomes an index one system of differential-algebraic equations (DAE) of first order in time which must be handled by a suitable time integration method. In order to integrate large FEM systems with internal variables, one-step methods with embedded error estimators, e. g. singly diagonally implicit *Runge-Kutta* methods (SDIRK), provide a suitable means at moderate storage and computational costs, see Ellsiepen [7].

Within the finite element discretization, the evolution equations (11) for the inelastic deformation tensors $(\mathbf{C}_{Si})_n$ are computed only at the integration points of the numerical quadrature. Therefore, the evolution equations are rewritten with respect to the reference configuration leading to

$$\begin{aligned} &[(\mathbf{C}_{Si})_n]'_S - \frac{1}{\eta_n^S}\,(\mathbf{C}_{Si})_n\,(\mathbf{S}^S_{NEQ})_n\,(\mathbf{C}_{Si})_n + \\ &+ \frac{\zeta_n^S}{\eta_n^S\,(2\eta_n^S + 3\zeta_n^S)}\,[(\mathbf{S}^S_{NEQ})_n \cdot (\mathbf{C}_{Si})_n]\,(\mathbf{C}_{Si})_n = \mathbf{0}\,, \end{aligned} \tag{15}$$

where $\mathbf{S}^S_{NEQ} = \mathbf{F}_S^{-1}\,\boldsymbol{\tau}^S_{NEQ}\,\mathbf{F}_S^{T-1}$ is the second *Piola-Kirchhoff* solid stress tensor. The ordinary differential equations (15) represent $\boldsymbol{F}_2$ of (14) and are treated with the same time integration method as is used for the global FEM system.

5. Application

In the present contribution, the model is applied to a highly porous, viscoelastic, open-cell polyurethane (PU) foam (bulk density $48\,\mathrm{kg/m^3} \rightarrow$ porosity $96\,\%$) which finds its application, e. g. as seat cushions, in the automotive industry.

To get a manageable optimization problem, it is useful to separate the basic elasticity from the inelastic properties. Therefore, uniaxial compression experiments with holding times were performed in our lab to allow the complete relaxation of the non-equilibrium stresses at several deformation states in order to obtain the purely elastic response of the material. In particular, cubic PU foam specimens of size $70 \times 70 \times 70\,\mathrm{mm}^3$ have been cemented on aluminum platens at the top side and at the bottom side and have been loaded and unloaded displacement driven with a constant deformation rate of $v(t) = 7\,\mathrm{mm/s}$ and a maximal compression of $86\,\%$, whereas the lateral walls were unconfined (Figure 2). During the loading and unloading cycles, several holding times of $30\,\mathrm{min}$ and $120\,\mathrm{min}$ were incorporated while recording the responding force. The experimental results are shown in the stress-strain diagram in Figure 3. It can easily be seen that even holding times of $120\,\mathrm{min}$ are too short, since the overstresses are not completely relaxed. Therefore, the mid points of the relaxed

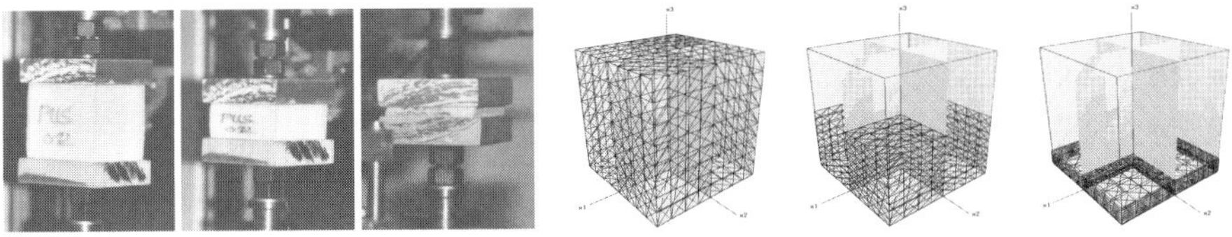

Figure 2. Uniaxial compression experiment on PU foam and FE simulation with PANDAS.

stresses between the loading and unloading cycles were taken as input for the numerical adjustment of the *Ogden* law (8) describing the equilibrium stress. In particular, the optimization algorithm DONLP2 (© Spellucci) is applied which is based on a variation of the SQP method and the gradient projection technique (Spellucci [11]). Running a numerical simulation of the uniaxial experiment with PANDAS shows the perfect fit of the basic elasticity to the experimental data (Figure 3).

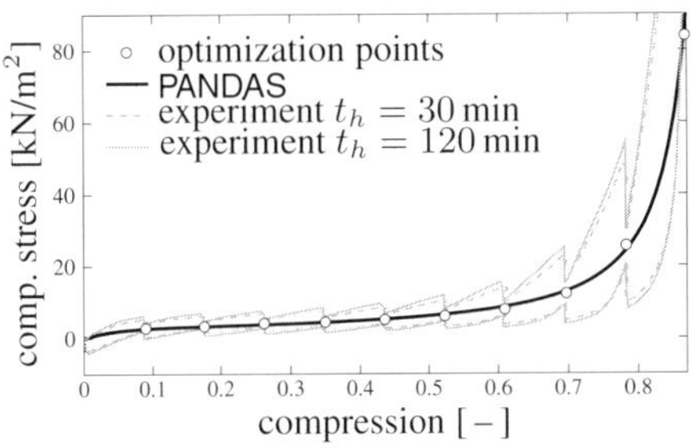

Figure 3. Experimental results and computed equilibrium stress (PANDAS).

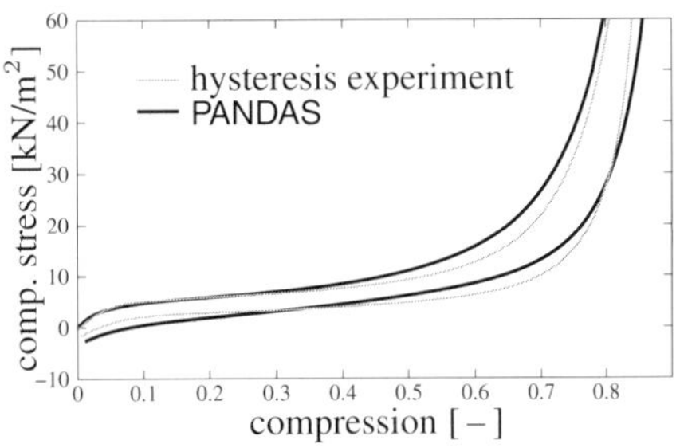

Figure 4. Experimental result and computed total stress (PANDAS).

After the adjustment of the basic elasticity model, the parameters of the viscous material law are identified from hysteresis experiments without holding times. For simplicity, only one *Maxwell* element is used for the description. Again, the respective parameters are optimized with the optimization tool DONLP2 as was also used for the basic elasticity. The result of the parameter adjustment is shown in Figure 4. As can be seen, the *Ogden* model is able to describe the complex time-dependent behavior of the considered PU foam.

6. Conclusions

Proceeding from the Theory of Porous Media (TPM), the macroscopic model equations for the description of cellular polymer foams have been derived. Therefore, an appropriate finite strain viscoelasticity formulation which implies the property of structural densification of the cellular polymer skeleton has been developed. Furthermore, the model accounts for an independent compressible pore-gas constituent and a deformation-dependent permeability, since this is an important property of open-cell polymer foams at large deformations.

In the framework of the finite element method (FEM), a weak form of the governing model equations has been presented. Proceeding from quasi-static situations, the application of the *Bubnov-Galerkin* procedure leads to a system of differential-algebraic equations in the time domain which can efficiently be solved by one-step time integration methods using embedded error estimators for the time-adaptive scheme. Moreover, different mixed element formulations have been presented and discussed.

To show its suitability, the presented biphasic viscoelastic model has been applied for the description of a highly porous, open-cell polyurethane (PU) foam which finds its application, e. g. as seat cushions, in the automotive industry. Furthermore, the model has been adapted to experimental data by use of the sequential quadratic programming (SQP) method. Finally, the numerical simulation of large deformation uniaxial compression experiments shows the applicability of the presented model, where the biphasic formulation allows the consideration of the in fact transient, however strong influence of the independent gas-phase at higher deformation rates, e. g. in impact or crash situations.

References

[1] Ball, J.M. (1977). Convexity conditions and existence theorems in nonlinear elasticity. *Arch. Rational Mech. Anal.*, **63**, 337–403.

[2] Braess, D. (1997). *Finite Elemente*, Springer-Verlag, Berlin.

[3] Brezzi, F. and Fortin, M. (1991). *Mixed and Hybrid Finite Element Methods*, Springer-Verlag, New York.

[4] Coleman, B.D. and Gurtin, M.E. (1967). Thermodynamics with internal state variables. *J. Chem. Phys.*, **47**, 597–613.

[5] Ehlers, W. (2002). Foundations of multiphasic and porous materials. In: *Porous Media: Theory, Experiments and Numerical Applications*, W. Ehlers, J. Bluhm (eds.), Springer-Verlag, Berlin, 3–86.

[6] Ehlers, W. and Markert, B. (2000). On the viscoelastic behaviour of fluid-saturated porous materials. *Granular Matter*, **2**, 153–161.

[7] Ellsiepen, P. (1999). *Zeit- und ortsadaptive Verfahren angewandt auf Mehrphasenprobleme poröser Medien*, Dissertation, Bericht Nr. II-3 aus dem Institut für Mechanik (Bauwesen), Universität Stuttgart.

[8] Gibson, L.J. and Ashby, F. (1997). *Cellular Solids, Structure and Properties, 2nd ed.*, University Press, Cambridge.

[9] Ogden, R.W. (1984). *Nonlinear elastic deformations*, Ellis Horwood, Chichester.

[10] Reese, S. and Govindjee, S. (1998). A theory of finite viscoelasticity and numerical aspects. *Int. J. Solids Structures*, **35**, 3455–3482.

[11] Spellucci, P. (1998). A new technique for inconsistent QP problems in the SQP method. *Math. Meth. OR*, **47**, 335–400.

Multiscale Analyses and Homogenization Methods

Analysis of Inhomogeneous Materials at Large Strains using Fast Fourier Transforms

N. Lahellec, J.C. Michel, H. Moulinec, P. Suquet
Laboratoire de Mécanique et d'Acoustique / C.N.R.S.
31 Chemin Joseph Aiguier, 13402 Marseille Cedex 20, France
suquet@lma.cnrs-mrs.fr

Abstract: A computational technique based on fast Fourier transforms has been recently developed to analyze the response of inhomogeneous materials with complex microstructure, both at the global level (effective properties) and local level (stress and strain fields). Initially proposed for elastic constituents and subsequently extended to elastoplastic constituents in the approximation of infinitesimal strains, the method is extended here to large strains, using either a Lagrangian formulation in the reference configuration, or an Eulerian formulation in the deformed configuration. In the latter case, the algorithm inspired by "particle-in-cell" method, makes use of two grids, a computational grid and a material grid. Applications to hyperelastic materials reinforced by stiff inclusions (Lagrangian formulation) and to the motion of a rigid particle in a viscous fluid (Eulerian formulation) are given.

Keywords: Composite materials, large strains, fast Fourier transforms, particle-in-cell methods.

1. Introduction

Every material is naturally inhomogeneous at a small enough scale. The aim of *homogenization* is to replace a microscopically inhomogeneous material by a macroscopically homogeneous medium, the properties of which are the *effective properties* of the initial inhomogeneous material. However, nonlinear phenomenas such as fatigue or rupture are often triggered by local stress or strain concentrations at small scale. Therefore, besides effective properties, it is also essential to investigate how local fields are influenced by, and influence in turn, the microstructure of a material.

Precise determination or accurate estimation of local fields may involve considerable computational efforts. Direct numerical simulations of composites with periodic microstructures have been performed by the Finite Element Method (FEM) since the early seventies, but despite numerous achievements, the application of this method to three-dimensional problems is still limited by the complexity of meshes. The need for investigating materials with complex microstructure without meshing provided a first motivation for developing a new computational method, specifically devised for microstructures, which could make direct use of images of microstructures, as delivered by Scanning Electron

C. Miehe (ed.), IUTAM Symposium on Computational Mechanics of Solid Materials at Large Strains, 247-258.

Microscopy or Computed Tomography (see Moulinec and Suquet [6], Moulinec and Suquet [7] and Michel et al. [4]).

A second motivation for developing a method capable of simulating easily the response of different microstructures stems from the recent theoretical models of the effective properties of nonlinear composites (see the review article Ponte Castañeda and Suquet [9] and the references herein). Direct comparison of these models with experimental data is often difficult, in that many different phenomena may occur simultaneously in the nonlinear regime (plasticity but also damage, interphase debonding, grain boundary sliding...). A comparison with numerical simulations, although only a step towards a complete validation, permits to "activate" all nonlinear mechanisms independently.

2. Solving a periodic elasticity problem with the FFT

As a reminder for readers who are not familiar with the FFT method, the case of linear elasticity at small strains is discussed first. The microstructure of the composite material is described by a *representative volume element* (r.v.e.) V comprised of N homogeneous phases. The N individual constituents are *linear elastic*, with stiffness $\boldsymbol{c}(\boldsymbol{x})$ and perfectly bonded across their interfaces. The volume element is subjected to an average strain $\boldsymbol{E}$ and the *local problem* to be solved for the local stress and strain fields reads as

$$\boldsymbol{\sigma}(\boldsymbol{x}) = \boldsymbol{c}(\boldsymbol{x}) : \varepsilon(\boldsymbol{u}(\boldsymbol{x})), \ \text{div}\,(\boldsymbol{\sigma}(\boldsymbol{x})) = \boldsymbol{0}, \quad \langle \varepsilon \rangle = \boldsymbol{E}, \tag{1}$$

where $\langle . \rangle$ denotes the spatial average over V. The boundary conditions applied on ∂V should reproduce as closely as possible the *in situ* state of the r.v.e. This state is very seldomly known and periodicity conditions are assumed to close the problem: the local strain field is decomposed into its average and a periodic fluctuation $\varepsilon(\boldsymbol{u}(\boldsymbol{x})) = \boldsymbol{E} + \varepsilon(\boldsymbol{u}^*(\boldsymbol{x}))$ where $\boldsymbol{u}^*$ is periodic (notation $\boldsymbol{u}^*\#$). The tensions $\boldsymbol{\sigma}.\boldsymbol{n}$ take opposite values on opposite sides of the r.v.e. (notation $\boldsymbol{\sigma}.\boldsymbol{n} - \#$). The reader is referred to Suquet [13] for more details about periodicity conditions.

The local problem (1) can be re-written by introducing a homogeneous reference material with elastic stiffness $\boldsymbol{c}^0$:

$$\boldsymbol{\sigma}(\boldsymbol{x}) = \boldsymbol{c}^0 : \varepsilon(\boldsymbol{u}^*(\boldsymbol{x})) \ + \ \boldsymbol{\tau}(\boldsymbol{x}), \ \text{div}\,(\boldsymbol{\sigma}(\boldsymbol{x})) = \boldsymbol{0}, \quad \boldsymbol{u}^*\ \#, \ \boldsymbol{\sigma}.\text{n}\ -\#, \tag{2}$$

where the polarization field $\boldsymbol{\tau}(\boldsymbol{x})$ is given by:

$$\boldsymbol{\tau}(\boldsymbol{x}) = \delta\boldsymbol{c}(\boldsymbol{x}) : (\varepsilon(\boldsymbol{u}^*(\boldsymbol{x})) + \boldsymbol{E}) + \boldsymbol{c}^0 : \boldsymbol{E}, \quad \delta\boldsymbol{c}(\boldsymbol{x}) \ = \ \boldsymbol{c}(\boldsymbol{x}) - \boldsymbol{c}^0. \tag{3}$$

The solution of problem (2) can be expressed by means of the *periodic Green's operator* associated with the elasticity tensor $\boldsymbol{c}^0$ and reads, in real space and

Fourier space respectively:

$$\varepsilon(\boldsymbol{u}^*(\boldsymbol{x})) = -\boldsymbol{\Gamma}^0 * \boldsymbol{\tau}(\boldsymbol{x}), \quad \hat{\varepsilon}^*(\boldsymbol{\xi}) = -\hat{\boldsymbol{\Gamma}}^0(\boldsymbol{\xi}) : \hat{\boldsymbol{\tau}}(\boldsymbol{\xi}) \;\; \forall \boldsymbol{\xi} \neq \boldsymbol{0}, \quad \hat{\varepsilon}^*(\boldsymbol{0}) = \boldsymbol{0}.$$

After substituting back the expression (3) of $\boldsymbol{\tau}$ in this relation, the initial local problem (1) reduces to the *periodic Lippmann-Schwinger integral equation* for $\varepsilon(\boldsymbol{u})$ which reads, in real space:

$$\varepsilon(\boldsymbol{u}(\boldsymbol{x})) = -\boldsymbol{\Gamma}^0 * (\boldsymbol{\delta c}(\boldsymbol{x}) : \varepsilon(\boldsymbol{u}(\boldsymbol{x}))) + \boldsymbol{E}. \tag{4}$$

The operator $\boldsymbol{\Gamma}^0$ is explicitely known in Fourier space *for arbitrary anisotropy* of the reference medium:

$$\hat{\Gamma}^0_{ijkh} = N^0_{ik}\xi_j\xi_h|_{(ijkh)}, \quad K^0_{ik} = c^0_{ijkh}\xi_j\xi_h, \quad \boldsymbol{N}^0 = \left(\boldsymbol{K}^0\right)^{-1}, \tag{5}$$

where the symbol $_{(ijkh)}$ denotes symmetrization with respect to the four indices $(ijkh)$ resulting in minor and major symmetries for $\hat{\boldsymbol{\Gamma}}^0$. Specific expressions for $\boldsymbol{\Gamma}^0$ can be found in Mura [8] for different classes of anisotropy of the reference medium.

The integral equation (4) is solved by a fixed-point method:

$$\varepsilon(\boldsymbol{u}^{\mathrm{i}+1}) = -\boldsymbol{\Gamma}^0 * \left(\boldsymbol{\delta c} : \varepsilon(\boldsymbol{u}^{\mathrm{i}})\right) + \boldsymbol{E}. \tag{6}$$

This algorithm can be further simplified by noting that $\boldsymbol{\Gamma}^0 * (\boldsymbol{c}^0 : \varepsilon) = \varepsilon(\boldsymbol{u}^*)$ and the iterative method reads

$$\varepsilon(\boldsymbol{u}^{\mathrm{i}+1}) = \varepsilon(\boldsymbol{u}^{\mathrm{i}}) - \boldsymbol{\Gamma}^0 * \boldsymbol{\sigma}^{\mathrm{i}}, \quad \boldsymbol{\sigma}^{\mathrm{i}}(\boldsymbol{x}) = \boldsymbol{c}(\boldsymbol{x}) : \varepsilon(\boldsymbol{u}^{\mathrm{i}}(\boldsymbol{x})). \tag{7}$$

The $\mathrm{i}+1$-th iterate of the numerical algorithm typically reads:

$$\left.\begin{array}{ll} & \varepsilon^{\mathrm{i}} \text{ and } \boldsymbol{\sigma}^{\mathrm{i}} \text{ being known} \\ a) & \hat{\boldsymbol{\sigma}}^{\mathrm{i}} = \mathrm{FT}(\boldsymbol{\sigma}^{\mathrm{i}}), \\ b) & \text{Convergence test}, \\ c) & \hat{\varepsilon}^{\mathrm{i}+1}(\boldsymbol{\xi}) = \hat{\varepsilon}^{\mathrm{i}}(\boldsymbol{\xi}) - \hat{\boldsymbol{\Gamma}}^0(\boldsymbol{\xi}) : \hat{\boldsymbol{\sigma}}^{\mathrm{i}}(\boldsymbol{\xi}) \; \forall \boldsymbol{\xi} \neq \boldsymbol{0} \text{ and } \hat{\varepsilon}^{\mathrm{i}+1}(\boldsymbol{0}) = \boldsymbol{E}, \\ d) & \varepsilon^{\mathrm{i}+1} = \mathrm{FT}^{-1}(\hat{\varepsilon}^{\mathrm{i}+1}) \\ e) & \boldsymbol{\sigma}^{\mathrm{i}+1}(\boldsymbol{x}) = \boldsymbol{c}(\boldsymbol{x}) : \varepsilon^{\mathrm{i}+1}(\boldsymbol{x}), \quad \forall \boldsymbol{x} \in V. \end{array}\right\} \tag{8}$$

FT stands for the Fourier transform. Convergence is reached when $\boldsymbol{\sigma}^{\mathrm{i}+1}$ is in equilibrium.

In practice the microstructure is given in the form of an image, consisting of pixels (or voxels in dimension three) of a given size. The spatial resolution of the image is the number of pixels along each coordinate axis. The image sampling in real space generates a corresponding sampling in Fourier space. The algorithm (8) is implemented in discretized form, the FT being replaced by the Fast Fourier Transform (FFT) attached to the above sampling in real and Fourier space (Moulinec and Suquet [7]).

The FFT method can be used in most problems where the FEM is used, provided that periodicity conditions can be adopted for the problem at hand. Extensions of the method to plasticity with or without hardening, to viscoplasticity, to phase transformation were given in Moulinec and Suquet [7], Michel et al. [5] and Dreyer and Müller [1] in the context of infinitesimal strains. The next sections are concerned with the extension of the method to large strains.

3. Hyperelastic materials

Consider now the case where the deformations applied to the composite material are large. The shape of the r.v.e. V evolves with time and classically two configurations play a particular role, the initial configuration V_0 and the current configuration $V(t)$. When the individual constituents are hyperelastic, a Lagrangian formulation can be adopted and the local problem can be posed on the initial configuration V_0 only. A particle which was initially at point $\boldsymbol{X}$ moves to a location $\boldsymbol{x} = \boldsymbol{x}(\boldsymbol{X}, t) = \boldsymbol{X} + \boldsymbol{u}(\boldsymbol{X}, t)$ at time t. The deformation gradient and the first Piola-Kirchhoff stress tensor $\boldsymbol{\pi}$ are related by:

$$\boldsymbol{f} = \boldsymbol{I} + \nabla_{\boldsymbol{X}} \boldsymbol{u}(\boldsymbol{X}, t), \quad \boldsymbol{\pi}(\boldsymbol{X}) = \frac{\partial w}{\partial \boldsymbol{f}}(\boldsymbol{X}, \boldsymbol{f}), \tag{9}$$

where $w(\boldsymbol{X}, .)$ is the strain energy at point $\boldsymbol{X}$. The average gradient $\boldsymbol{F}$ of the transformation is applied incrementally along a prescribed path in the loading space. This path is parametrized by a scalar variable t. The local problem to be solved reads:

$$\left.\begin{array}{c} \boldsymbol{\pi}(\boldsymbol{X}) = \dfrac{\partial w}{\partial \boldsymbol{f}}(\boldsymbol{X}, \boldsymbol{f}), \quad \mathrm{div}_{\boldsymbol{X}} \boldsymbol{\pi} = \boldsymbol{0} \text{ in } V_0, \\[2ex] \boldsymbol{f}(\boldsymbol{X}) = \boldsymbol{F} + \nabla_{\boldsymbol{X}} \boldsymbol{u}^*, \quad \boldsymbol{u}^* \, \#, \ \boldsymbol{\pi}.\boldsymbol{N} \ - \# \text{ on } \partial V_0 \end{array}\right\} \tag{10}$$

where $\boldsymbol{N}$ denotes the outer normal unit vector to ∂V_0.

The local problem (10) is similar to (2). It is posed on a *fixed* configuration which allows us to use the FFT method with fixed grids in real space and Fourier space. The constitutive relations are nonlinear and the whole gradient $\boldsymbol{f}$ (and not only its symmetric part) is taken into account in (10). The problem (10) is solved step-by-step in time, with a Newton-Raphson algorithm at each time step.

3.1. NEWTON-RAPHSON ALGORITHM

$\boldsymbol{F}_{t+\Delta t}$ is imposed at time $t+\Delta t$. The principal unknown is the gradient $\nabla \boldsymbol{u}^*_{t+\Delta t}$ of the periodic fluctuation of the displacement. This gradient is determined by imposing that $\boldsymbol{\pi}_{t+\Delta t}$ is in equilibrium.
Let $\boldsymbol{\pi}^i$, $\boldsymbol{f}^i$ and $(\boldsymbol{u}^*)^i$ denote the iterates approximating $\boldsymbol{\pi}_{t+\Delta t}$, $\boldsymbol{f}_{t+\Delta t}$ and $\boldsymbol{u}^*_{t+\Delta t}$.

<u>Iterate i</u> : $\nabla(\boldsymbol{u}^*)^{i-1}$ being known:

(1) Compute $\boldsymbol{f}^{i-1}$ and $\boldsymbol{\pi}^{i-1}$

$$\boldsymbol{f}^{i-1}(\boldsymbol{X}) = \boldsymbol{F}_{t+\Delta t} + \nabla(\boldsymbol{u}^*)^{i-1}(\boldsymbol{X}),\ \boldsymbol{\pi}^{i-1}(\boldsymbol{X}) = \frac{\partial w}{\partial \boldsymbol{f}}(\boldsymbol{X}, \boldsymbol{f}^{i-1}(\boldsymbol{X})). \quad (11)$$

Check if $\boldsymbol{\pi}^{i-1}$ is in equilibrium. If not, then

(2) Solve the linear tangent problem for the periodic field $\nabla(\delta \boldsymbol{u}^*)$

$$\langle \frac{\partial^2 w}{\partial \boldsymbol{f}^2}(\boldsymbol{f}^{i-1}) : \nabla(\delta \boldsymbol{u}^*) : \nabla \boldsymbol{v}\rangle = -\langle \boldsymbol{\pi}^{i-1} : \nabla \boldsymbol{v}\rangle \quad \forall \boldsymbol{v}\ \#. \qquad (12)$$

(3) Update $\nabla(\boldsymbol{u}^*)^i$: $\nabla(\boldsymbol{u}^*)^i(\boldsymbol{X}) = \nabla(\boldsymbol{u}^*)^{i-1}(\boldsymbol{X}) + \nabla(\delta\boldsymbol{u}^*)(\boldsymbol{X})$.

The criterion which serves to check equilibrium reads:

$$\max_{\boldsymbol{\xi}} |\boldsymbol{\xi}.\widehat{\boldsymbol{\pi}}^i(\boldsymbol{\xi})| \ \leq\ \varepsilon|\widehat{\boldsymbol{\pi}}^i(\boldsymbol{0})|, \quad \text{where typically } \varepsilon = 10^{-4}.$$

The problem (12) is solved iteratively using the FFT method described in section 2. More specifically an internal loop is performed to find $\nabla(\delta\boldsymbol{u}^*)(\boldsymbol{X})$

$$\nabla(\delta \boldsymbol{u}^*)^k = \nabla(\delta \boldsymbol{u}^*)^{k-1} - \tilde{\boldsymbol{\Gamma}}^0 * \left(\boldsymbol{L}^{i-1} : \nabla(\delta\boldsymbol{u}^*)^{k-1} + \boldsymbol{\pi}^{i-1}\right), \qquad (13)$$

where $\boldsymbol{L}^{i-1}(\boldsymbol{X}) = \dfrac{\partial^2 w}{\partial \boldsymbol{f}^2}(\boldsymbol{X}, \boldsymbol{f}^{i-1}(\boldsymbol{X}))$. $\tilde{\boldsymbol{\Gamma}}^0$ is defined in Fourier space as

$$\hat{\tilde{\Gamma}}^0_{ijkh} = N^0_{ik}\xi_j\xi_h|_{(ik)(jh)},$$

where the symbol $_{(ik)(jh)}$ denotes symmetrization with respect to the indices i, k and j, h only.

3.2. EXAMPLE: FIBER-REINFORCED ELASTOMER

The above algorithm has been applied to model the deformation of an elastomeric matrix reinforced by stiff fibers. In its initial configuration the unit

Figure 1. Hyperelastic matrix reinforced by 64 circular fibers. (a) Initial configuration. (b)–(f) Successive deformed states for different values of the macroscopic logarithmic strain $E_1 = \ln F_{11}$.

cell was a square containing 64 circular identical impenetrable fibers, with 25% volume fraction, arranged randomly in the unit cell. The matrix and the fibers were compressible Mooney-Rivlin materials with strain energy:

$$w(\boldsymbol{f}) = \frac{\mu}{2}(i_1 - 3) - \mu \ln j + \frac{\kappa}{2}(j-1)^2, \; i_1 = \mathrm{tr}({}^{\mathsf{T}}\boldsymbol{f}.\boldsymbol{f}), \; j = \det \boldsymbol{f}. \quad (14)$$

The fibers were 10 times stiffer than the matrix. 10 different configurations were tested. The macroscopic deformation was a biaxial isochoric deformation $\boldsymbol{F} = \exp(E_1)\, \boldsymbol{e}_1 \otimes \boldsymbol{e}_1 + \exp(-E_1)\, \boldsymbol{e}_2 \otimes \boldsymbol{e}_2 + \boldsymbol{e}_3 \otimes \boldsymbol{e}_3$, where the logarithmic strain in the first direction E_1 was negative (contraction in direction 1, extension in direction 2).

The deformed states of a typical configuration are shown in Figure 1. The stress-strain response of all 10 different configurations are shown in Figure 2. They are seen to be very close to one another. Also shown in Figure 2 is the prediction of the second-order estimate (SOE) of Ponte Castañeda and Tiberio [10] implemented with the Hashin-Shtrikman estimate (lower bound) for circular fibers in a matrix. As can be seen from this figure, the agreement between the theory and the numerical calculations is excellent.

4. Eulerian formulation

4.1. Local problem

We now turn to the case where the constitutive relations of the constituents can be expressed by means of quantities defined on the current configuration only, the Cauchy stress $\boldsymbol{\sigma}$ and the Eulerian strain-rate $\boldsymbol{d}$ for instance. This is the case

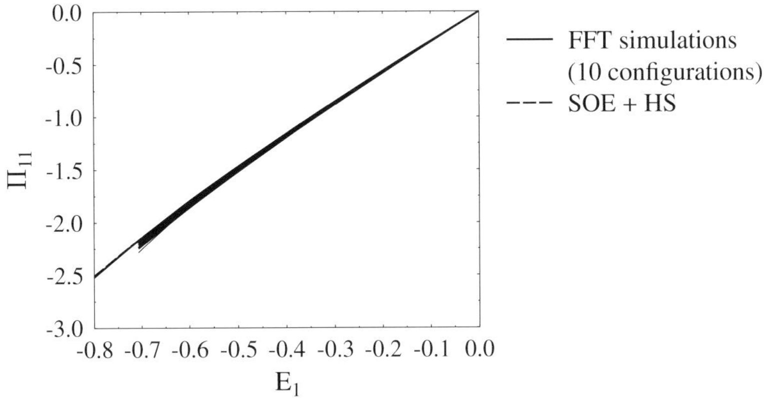

Figure 2. Macroscopic stress-strain curves. First Piola-Kirchhoff stress as a function of the logarithmic strain in direction 1. 10 different configurations. Comparison between the FFT simulations and the second order estimate (SOE) with the lower Hashin-Shtrikman estimate.

for Newtonian fluids for which the constitutive relations read

$$\boldsymbol{\sigma}(\boldsymbol{x}) = \boldsymbol{L}(\boldsymbol{x}) : \boldsymbol{d}(\boldsymbol{x}), \tag{15}$$

and more generally for viscous materials when elasticity effects are neglected (the case of power-law viscous materials is considered in Lebensohn [3]). For simplicity we restrict our attention to Newtonian fluids as described by (15).

The geometry of the r.v.e. evolves with time in two ways. First the unit vectors $\boldsymbol{Y}_i$ defining the unit cell are convected by the macroscopic velocity gradient:

$$\dot{\boldsymbol{Y}}_i(t) = \langle \text{grad } \boldsymbol{v}(t) \rangle . \boldsymbol{Y}_i(t). \tag{16}$$

Second, the position of the phases within the r.v.e. changes with time, the material particles being convected by the microscopic velocity field:

$$\dot{\boldsymbol{x}}(t) = \boldsymbol{v}(\boldsymbol{x}, t) \tag{17}$$

The local problem to be solved in the Eulerian configuration consists of the constitutive relations, together with the equilibrium and compatibility equations. At each time t the local problem on the current configuration can be solved using the FFT method described in section 2. It remains to see how the unit cell changes with time.

4.2. A MULTI-PARTICLE METHOD

The two grids. The proposed method falls in the category of multi-particle methods and follows rather closely the Particle-In-Cell (PIC) method proposed by Sulsky et al. [11, 12], with a difference stemming from the updating of the computational grid (the method proposed by Lebensohn [3] is also similar in spirit). The underlying idea of the PIC method is to consider two separate grids:

- *The computational grid* is used for applying the FFT method described in section 2. It is a regular grid but it does not have to be rectangular (Fourier transforms can be defined on noncubic lattices).
- *The material grid* is attached to the material particles. It does not have to be structured or regular and can be seen as a collection of particles rather than a structured mesh. It is only used to apply the constitutive relations (which are material relations). In the FFT method it is used in real space only. In most methods the material grid is finer than the computational one (by a factor of 4 to 9).

Each grid carries its own set of unknowns. In other words,

- the computational grid carries unknowns labeled with lowerscript c (as in "computations"): $\boldsymbol{x}_c^{(i,j)} = (i-1)\boldsymbol{Y}_1 + (j-1)\boldsymbol{Y}_2$ denote the nodes of the computational grid, $\boldsymbol{\sigma}_c$, $\boldsymbol{v}_c$ denote the stress and velocity fields at these nodes.
- the material grid carries unknowns labeled with lowerscript p (as in "particles"): $\boldsymbol{x}_p$ denote the location of the particles, $\boldsymbol{\sigma}_p$, $\boldsymbol{v}_p$ denote the stress and velocity field at the particles.

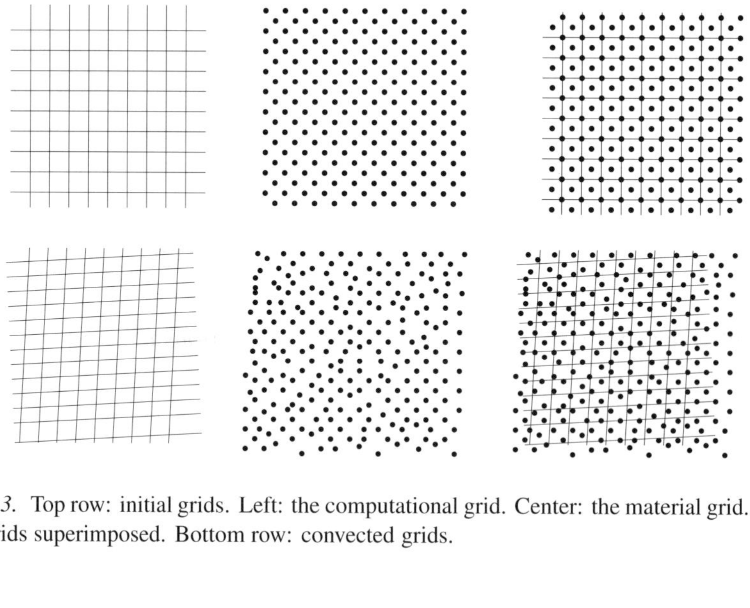

Figure 3. Top row: initial grids. Left: the computational grid. Center: the material grid. Right: both grids superimposed. Bottom row: convected grids.

4.3. UPDATING THE GRIDS

In the initial PIC method (see Sulsky et al. [11, 12]), the computational grid is fixed. However in the problem under consideration here the computational grid is associated with the unit cell which generates the whole microstructure by periodicity and cannot remain fixed. The unit vectors $\boldsymbol{Y}_i$ defining the periodic lattice are material vectors and follow the *macroscopic* velocity field according to (16). The computational grid is updated accordingly. The material grid is updated using the *microscopic* velocity field according to (17).

The grids are updated using an explicit scheme. At time t_n, the velocity field $\boldsymbol{v}_c(t_n)$ is computed on the computational grid and transferred (see below) onto the material grid to get a field $\boldsymbol{v}_p(t_n)$. Then:

$$\left.\begin{array}{r}\boldsymbol{Y}_i(t_{n+1}) = \boldsymbol{Y}_i(t_n) + \Delta t \langle \text{grad } \boldsymbol{v}_c(t_n)\rangle.\boldsymbol{Y}_i(t),\\ \boldsymbol{x}_c(t_{n+1}) = (i-1)\boldsymbol{Y}_1(t_{n+1}) + (j-1)\boldsymbol{Y}_2(t_{n+1}),\\ \boldsymbol{x}_p(t_{n+1}) = \boldsymbol{x}_p(t_n) + \Delta t\ \boldsymbol{v}_p(\boldsymbol{x}_p(t_n)).\end{array}\right\} \quad (18)$$

Transfer operators. Transfer operators are needed in order to transfer information from the computational grid to the material grid and *vice-versa*. Devising consistent transfer operators is certainly a crucial part of the algorithm.

Consider a computational grid with an initially rectangular pattern. In the initial configuration the grid nodes have coordinates $((i-1)\Delta x_0, (j-1)\Delta y_0)$. In the current configuration the grid nodes have coordinates $(x^c_{i,j}, y^c_{i,j})$. Let us introduce a family of shape functions $N^{(i,j)}, i = 1, ..., I, j = 1, ..., J$ (as is classical in the FEM) on the computational grid, where I and J denote the number of sampling points on the horizontal and vertical axis (for the initial grid) respectively. Given the regular geometry of the computational grid a natural choice for the $N^{(i,j)}$ are the interpolation functions of the 4 point quadrilateral element in the FEM (these functions were used in the example presented in section 4.4). Then, given the values $f_c^{(i,j)} = f(\boldsymbol{x}_c(i,j))$ of any function f at the computational nodes, the interpolated field at any other point $\boldsymbol{x}$ reads as

$$f(\boldsymbol{x}) = \sum_{(i,j)=(1,1)}^{I,J} N^{(i,j)}(\boldsymbol{x}) f_c^{(i,j)}. \quad (19)$$

These relations apply in particular at the particle positions $\boldsymbol{x}_p$.

The inverse transfer from the material grid to the computational grid is performed using the same interpolation functions. Each particle is assigned a mass m_p and the mass of a computational node is defined as

$$m(\boldsymbol{x}_c^{(i,j)}) = \sum_{p=1}^{P} m_p N^{(i,j)}(\boldsymbol{x}_p) \quad (20)$$

Then the values f_c of f at any computational node $\boldsymbol{x}_c$ is given by:

$$f_c(\boldsymbol{x}_c^{(i,j)}) = \frac{1}{m(\boldsymbol{x}_c^{(i,j)})} \sum_{p=1}^{P} m_p N^{(i,j)}(\boldsymbol{x}_p) f_p(\boldsymbol{x}_p) \tag{21}$$

4.4. A TEST EXAMPLE

To check the accuracy of the updating scheme, the problem of a rigid particle rotating in a shear flow of a Newtonian fluid with uniform and constant viscosity has been investigated. The fluid is subjected to a macroscopic flow with strain-rate $\boldsymbol{D}$ and rotation-rate $\boldsymbol{\Omega}$. Inertia effects are neglected. The solution (rotation of the particle) for a spheroidal particle with infinitesimal volume fraction$c^{(p)}$ in an incompressible Newtonian fluid goes back to Jeffery [2]. A generalization of Jeffery's result, also applying to compressible fluids and viscous particles, can also be derived by means of Eshelby's result for ellipsoidal inclusions. This derivation will not be given here. For a rigid particle the evolution equation for a unit normal vector $\boldsymbol{u}$ attached to the particle is

$$\frac{d\boldsymbol{u}}{dt} = \boldsymbol{\omega}^I.\boldsymbol{u}, \quad \boldsymbol{\omega}^I = \boldsymbol{\Omega} - \boldsymbol{\Pi} : \boldsymbol{S}^{-1} : \boldsymbol{D}, \tag{22}$$

where $\boldsymbol{S}$ and $\boldsymbol{\Pi}$ are the two Eshelby tensors of the particle in the matrix giving respectively the deformation-rate in the inclusion and the rotation-rate of the inclusion.

This general three-dimensional solution can be specialized to dimension 2. The result is

$$\frac{d\boldsymbol{u}}{dt} = \boldsymbol{\Omega}.\boldsymbol{u} + B\left(\boldsymbol{D}.\boldsymbol{u} - (\boldsymbol{u}.\boldsymbol{D}.\boldsymbol{u})\boldsymbol{u}\right), \tag{23}$$

with $B = (r^2-1)/\left(r^2+1+2r\frac{1-2\nu}{2(1-\nu)}\right)$, where r is the aspect-ratio of the particle and ν is the Poisson ratio of the linearly viscous matrix.

In a shear flow,

$$x_1(t) = X_1 + \gamma(t)X_2, \quad x_2(t) = X_2, \tag{24}$$

(23) gives the the rotation of the unit normal vector$\boldsymbol{u} = \sin\varphi \boldsymbol{e}_1 + \cos\varphi \boldsymbol{e}_2$ in the form:

$$\frac{d\varphi}{dt} = \frac{\dot{\gamma}}{2}\left[1 + B(\cos^2\varphi - \sin^2\varphi)\right], \tag{25}$$

which can be integrated into

$$\varphi(t) = \varphi(0) + \text{Arctg}\left(R\,\text{tg}\left(\frac{\gamma(t)}{R+1/R}\right)\right), \quad \text{with } R = \sqrt{\frac{1+B}{1-B}}. \tag{26}$$

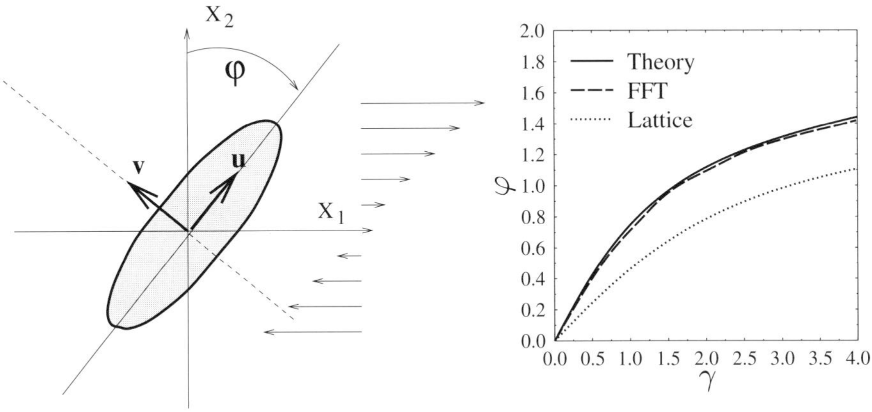

Figure 4. Particle rotating in a shear flow.

A numerical simulation of the rotation of a particle in a shear flow has been conducted using the method described in section 4.2 with the following data

$$r = 3, \quad \nu = 0.499, \quad c^{(p)} = 0.3\%. \tag{27}$$

The comparison between the analytical and the numerical results is good as shown in Figure 4b. Also shown in this figure is the lattice rotation. As can be seen the particle rotates faster than the lattice.

5. Concluding remarks

Two extensions to finite strains of the numerical method based on Fast Fourier Transforms (see Moulinec and Suquet [6, 7]) have been proposed and test examples have been analyzed. A few numerical issues deserve attention in future work:

- In the Lagrangian approach presented in section 3, the rate of convergence of the iterative FFT algorithm applied to the tangent problem (13) can be rather poor. This is probably due to the fact that the tangent operator is only strongly elliptic (and not very strongly elliptic) and highly contrasted. An accelerated scheme would be very helpful.

- In the Eulerian approach presented in section 4, consistent transfer operators between the computational grid and the material grid are to be found.

Acknowledgments

The authors greatly benefited from discussions with R. Lebensohn, D. Sulsky and N. Triantafyllidis during the completion of this research. Part of it was carried out while Pierre Suquet was a Visiting Professor at Caltech. Wonderful interactions with K. Bhattacharya and M. Ortiz, financial support from Caltech and from the *Délégation Générale à l'Armement* are gratefully acknowledged.

References

[1] Dreyer, W. and Müller, W.H. (2000). A study of the coarsening in Pb/Sn solders. *Int. J. Solids Structures*, **37**, 3841–3871.

[2] Jeffery, G.B. (1922). The motion of ellipsoidal particles immersed in a viscous fluid. *Proc. Royal Soc. London A*, **102**, 161–179.

[3] Lebensohn, R. (2001). N-site modelling of a 3d viscoplastic polycrystal using Fast Fourier Transforms. *Acta Materiala*, **49**, 2723–2737.

[4] Michel, J.C. Moulinec, H. and Suquet, P. (1999). Effective properties of composite materials with periodic microstructure: a computational approach. *Comp. Meth. Appl. Mech. Engng.*, **172**, 109–143.

[5] Michel, J.C., Moulinec, H., and Suquet, P. (2001). A computational method for linear and nonlinear composites with arbitrary phase contrast. *Int. J. Numer. Meth. Engng*, **52**, 139–160.

[6] Moulinec, H. and Suquet, P. (1994). A fast numerical method for computing the linear and nonlinear properties of composites. *C. R. Acad. Sc. Paris II*, **318**, 1417–1423.

[7] Moulinec, H. and Suquet, P. (1998). A numerical method for computing the overall response of nonlinear composites with complex microstructure. *Comp. Meth. Appl. Mech. Engng.*, **157**, 69–94.

[8] Mura, T. (1978). *Micromechanics of Defects in Solids*, Martinus Nijhoff Publishers, Dordrecht.

[9] Ponte Castañeda, P. and Suquet, P. (1997). Nonlinear composites. In: *Advances in Applied Mechanics*, E. Van der Giessen and T.Y. Wu (eds.), Academic Press, New York, **34**, 171–302.

[10] Ponte Castañeda, P. and Tiberio, E. (2000). A second-order homogenization method in finite elasticity and applications to black-filled elastomers. *J. Mech. Phys. Solids*, **48**, 1389–1411.

[11] Sulsky, D., Chen, Z. and Schreyer, H.L. (1994). A particle method for history-dependent materials. *Comp. Methods Appl. Mech. Engrg*, **118**, 179–196.

[12] Sulsky, D., Zhou, S.J. and Schreyer, H.L. (1995). Application of a particle-in-cell method to solid mechanics. *Computer Physics Communications*, **87**, 236–252.

[13] Suquet, P. (1987). Elements of Homogenization for Inelastic Solid Mechanics. In: *Homogenization Techniques for Composite Media*, E. Sanchez-Palencia and A. Zaoui (eds.), *Lecture Notes in Physics*, New York, Springer Verlag, **272**, 193–278.

Multiscale Characterization of Deformation Behavior of Particulate-Reinforced Metal-Matrix Composite

Y. Tomita
Graduate School of Science and Technology, Kobe University
1-1 Rokkodai-cho Nada-ku Kobe, Hyogo 657-8501, Japan
tomita@mech.kobe-u.ac.jp

Y. Higa
Graduate School of Engineering, Osaka University
2-1 Yamada-oka Suita, Osaka 565-0871, Japan
higa@ams.eng.osaka-u.ac.jp

Abstract: In order to estimate the characteristic feature of deformation behavior of particulate-reinforced metal-matrix composites with a length scale, the strain gradient plasticity theories and a homogenization method are used. The effects of the volume fraction, size and distribution pattern of the reinforcement particles and of the direction of the applied force on the macroscopic mechanical properties of the composites are examined. It has been clarified that the deformation resistance of the composite increases with the decrease of the particle size under a constant volume fraction of the reinforcement and strongly depends on the distribution pattern of reinforcement particles and the direction of applied force. The main mechanism of the increase in deformation resistance in the plastic range is the high strain gradient which appears in the matrix material and increases with a reduction in the distance between the particles.

Keywords: Multiscale modeling, homogenization method, particulate-reinforced metal-matrix composite, strain gradient theory.

1. Introduction

It is essential to understand the mechanism of plastic flow and the associated phenomena of patterning and localization of plastic deformation into shear bands in order to model and estimate the mechanical properties of the materials that depend on the characteristic length scale. Studies employing the generalized constitutive equation with a length scale that accounts for the microstructural interaction and evolution of microscopic texture are based on pioneering works (Aifantis [1, 2] and Fleck [3]). In Aifantis [1, 2], the derived constitutive equations contain the higher order terms of the strain gradient, whereas, in Fleck [3], the constitutive equation contains the first gradient of strain, which is derived from the geometrically necessary dislocation. The related review and details of the origin of the strain-gradient-dependent nature of the deformation behaviors have been extensively discussed (Zbib and

C. Miehe (ed.), IUTAM Symposium on Computational Mechanics of Solid Materials at Large Strains, 259-268.

Aifantis [14]). Subsequently, the length-scale dependence of the localized zone and internal structures has been characterized using the constitutive equations that were developed based on the strain gradient plasticity theories (Zbib and Aifantis [13], Vardoulakis and Aifantis [12]). Furthermore, it has been clarified that regularization with strain gradient plasticity theories substantially reduces the mesh dependence of the results obtained by finite element methods (Tomita and Nakao [9], Tomita [8]), which is well accepted in the field of computational mechanics. Thus, the strain gradient theory seems to provide a direct means of modeling the size effect of plastic flow within a continuum mechanics approach. In this paper, the homogenization method for materials obeying strain gradient plasticity theory and its finite element formulation are given. Then, the finite element method is applied to estimate the mechanical characteristics of the metal-matrix composites under plane strain conditions. The effects of the volume fraction, size and distribution pattern of the reinforcement particles on the macroscopic mechanical properties of the composite under external force applied in different directions with respect to the distribution pattern of reinforcements are discussed.

2. Constitutive equation and homogenization method

To derive the constitutive equation, an updated Lagrangian formulation is employed. For elasto-plastic solids, the constitutive equation for the elastic part of the strain rate $\boldsymbol{D}^e$ is assumed to be expressed by Hooke's law (Tomita [7]):

$$\overset{\triangledown}{S}_{ij} = \overset{\triangledown}{\sigma}_{ij} + \sigma_{ij} D_{kk} = C^e_{ijkl} D^e_{kl}, \tag{1}$$

$$C^e_{ijkl} = 2G \left\{ \frac{1}{2} \left(\delta_{ik}\delta_{jl} + \delta_{il}\delta_{jk} \right) + \frac{\nu}{1-2\nu} \delta_{ij}\delta_{kl} \right\},$$

where G is the elastic shear modulus, ν is Poisson's ratio and $\overset{\triangledown}{\boldsymbol{S}}$ and $\overset{\triangledown}{\boldsymbol{\sigma}}$ are the Jaumann rates of Kirchhoff stress $\boldsymbol{S}$ and Cauchy stress $\boldsymbol{\sigma}$ in the current state, respectively. Next, a constitutive equation for the plastic strain rate $\boldsymbol{D}^p$ with strain gradient dependence is indicated, with the help of the Mises-type flow rule, as

$$D^p_{ij} = \frac{3\dot{\overline{\varepsilon}}^p}{2\overline{\sigma}} \sigma'_{ij}, \quad \overline{\sigma} = \left(\frac{3}{2} \sigma'_{ij} \sigma'_{ij} \right)^{1/2}, \quad \dot{\overline{\varepsilon}}^p = \left(\frac{2}{3} D^p_{ij} D^p_{ij} \right)^{1/2}, \tag{2}$$

where $\boldsymbol{\sigma}'$ is the deviatoric part of Cauchy stress, $\overline{\sigma}$ is the equivalent stress and $\dot{\overline{\varepsilon}}^p$ is the equivalent plastic strain rate. The strain-gradient-dependent flow stress $\overline{\sigma}$ can be expressed similarly as indicated by Zbib and Aifantis [13] as

$$\overline{\sigma} = \overline{\overline{\sigma}} \left(\overline{\varepsilon}^p \right) - c \overline{\varepsilon}^p_{,ii}, \tag{3}$$

where $\overline{\overline{\sigma}}\,(\overline{\varepsilon}^p)$ is the conventional work hardening function and the second term indicates the strain-gradient-dependent terms. c is the coefficient describing the degree of dependence on the strain gradient and will be given in Section 3. Zbib and Aifantis [13] further discussed more elaborate constitutive equations. Combining Eqs. (1)–(3), we derive the elastoplastic constitutive equation

$$\dot{S}_{ij} = L^e_{ijkl} D_{kl} - P'_{ij}, \quad P'_{ij} = L_{ijkl} \frac{3\dot{\overline{\varepsilon}}^p}{2\overline{\sigma}} \sigma'_{kl}\,. \tag{4}$$

The equivalent strain rate $\dot{\overline{\varepsilon}}^p$ can be treated as an independent variable; as a result, the variational principle becomes very simple (Muhlhaus and Aifantis [6], Tomita and Fujimoto [10]) and we can derive a mixed-type finite element equation with independent variables of velocity $\boldsymbol{v}$ and equivalent plastic strain rate $\dot{\overline{\varepsilon}}^p$, which are expressed using a shape function with conventional C^0 continuity. On the other hand a conventional finite element equation can be derived with independent variable of velocity $\boldsymbol{v}$ and constitutive equation (4), where the strain gradient is estimated using the strain distribution in the previous calculation step.

Figure 1 indicates the plastic strain rate distributions in the compressed plane strain blocks obeying constitutive equation (4) evaluated by the mixed-type finite element method and the conventional finite element method. The latter provides highly accurate results compared to the former, therefore, we formulate the present boundary value problems based on the virtual work principle with an independent variable of velocity $\boldsymbol{v}$, which provides the simplified homogenization method.

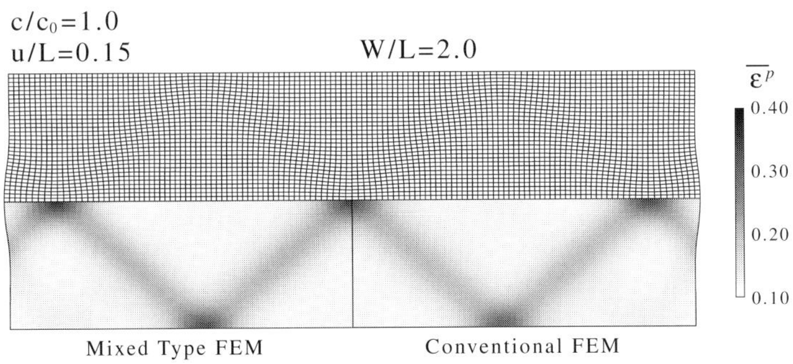

Figure 1. Plastic strain rate distributions in plane strain blocks under compression. Results due to mixed type FEM and conventional FEM.

In order to estimate the macroscopic deformation behavior of a particulate reinforced composite material with periodically distributed reinforcement, the homogenization method [4] is employed. Consider a two-dimensional problem with domain Ω and boundary S subjected to surface force on S_t and prescribed velocity on S_u. The body is formed by the spatial repetition of a base cell made of different materials. Assuming that the base cell is very small, of order η compared with the dimensions of the whole body, the global coordinate is $\boldsymbol{x}$ for the whole body and the local coordinate is $\boldsymbol{y}$ for the single base cell, then $\boldsymbol{y}=\boldsymbol{x}/\eta$. Similar to the assumption used in the case of linear elastic materials (Guedes and Kikuchi [4]), the velocity and velocity gradients are presumed to be expressed as an asymptotic expansion with respect to the parameter η. Substituting these into the virtual work principle and rearranging the same order in η, we can arrive at the homogenized expression of the virtual work principle.

$$\int_\Omega \left[L^H_{ijkl}\dot{E}^0_{kl}(\boldsymbol{v}) - P^H_{ij} + \sigma^H_{ij} + \tau^H_{ijkl}\frac{\partial v^0_k}{\partial x_l}\right]\frac{\partial \delta v_i}{\partial x_j}d\Omega = \int_{S_t}\dot{P}_i\delta v_i dS \tag{5}$$

$$\left.\begin{aligned}
L^H_{ijkl} &= \frac{1}{|Y|}\int_Y \left[L_{ijkl} - L_{ijpq}\frac{1}{2}\left(\frac{\partial \chi^{kl}_p}{\partial y_q} + \frac{\partial \chi^{kl}_q}{\partial y_p}\right)\right]dY\ ,\\
P^H_{ij} &= \frac{1}{|Y|}\int_Y \left[P'_{ij} - L_{ijkl}\frac{1}{2}\left(\frac{\partial \phi_k}{\partial y_l} + \frac{\partial \phi_l}{\partial y_k}\right)\right]dY\ ,\\
\sigma^H_{ij} &= \frac{1}{|Y|}\int_Y \sigma_{mj}\frac{\partial \phi_i}{\partial y_m}dY\ ,\ \tau^H_{ijkl} = \frac{1}{|Y|}\int_Y\left(\sigma_{lj}\delta_{ki} - \sigma_{mj}\frac{\partial \chi^{kl}_i}{\partial y_m}\right)dY
\end{aligned}\right\} \tag{6}$$

$$\dot{\varepsilon}^0_{ij}(\boldsymbol{x},\boldsymbol{y}) = \dot{E}^0_{ij}(\boldsymbol{v}) - \frac{1}{2}\left(\frac{\partial \chi^{kl}_i}{\partial y_j} + \frac{\partial \chi^{kl}_j}{\partial y_i}\right)\dot{E}^0_{kl}(\boldsymbol{v}) + \frac{1}{2}\left(\frac{\partial \phi_i}{\partial y_j} + \frac{\partial \phi_j}{\partial y_i}\right) \tag{7}$$

$$\dot{S}^0_{ij}(\boldsymbol{x},\boldsymbol{y}) = L_{ijkl}\dot{\varepsilon}^0_{kl}(\boldsymbol{x},\boldsymbol{y}) - P'_{ij} \tag{8}$$

Equations (5), (6) and (7), (8) are the governing equations for the macroscopic and microscopic scales, respectively. The notation $\dot{\boldsymbol{E}}^0$ indicates the macroscopic strain rate and $\boldsymbol{\chi}^{kl}$ and $\boldsymbol{\phi}$ are characteristic functions, defined in the unit cell, which satisfy the Y-periodicity and which are determined by the following equations.

$$\begin{aligned}
\int_Y \left[L_{ijpq}\frac{1}{2}\left(\frac{\partial \chi^{kl}_p}{\partial y_q} + \frac{\partial \chi^{kl}_q}{\partial y_p}\right) + \sigma_{qj}\delta_{pi}\frac{\partial \chi^{kl}_p}{\partial y_q}\right]\frac{\partial \delta v_i}{\partial y_j}dY \\
= \int_Y (L_{ijkl} + \sigma_{lj}\delta_{ki})\frac{\partial \delta v_i}{\partial y_j}dY
\end{aligned} \tag{9}$$

$$\boldsymbol{\chi}^{kl} \cdots Y\text{–periodic}$$

$$\int_Y \left[L_{ijkl}\frac{1}{2}\left(\frac{\partial \phi_k}{\partial y_l} + \frac{\partial \phi_l}{\partial y_k}\right) + \sigma_{mj}\frac{\partial \phi_i}{\partial y_m}\right]\frac{\partial \delta v_i}{\partial y_j}dY = \int_Y P'_{ij}\frac{\partial \delta v_i}{\partial y_j}dY \tag{10}$$

$$\boldsymbol{\phi} \cdots Y\text{–periodic}$$

Thus, characteristic functions χ^{kl} and ϕ for the unit cell depend solely on the material characteristics and configuration of the microstructure of the unit cell, which are, in turn, obtained without interacting stress and strain on the macroscale. On the other hand, the macroscopic equilibrium given by Eq. (5) can be solved independently because the macroscopic characteristic functions indicated in Eq. (6) are identified through Eqs. (9) and (10).

3. Computational model

Figure 2 shows the particulate reinforced composite material under external force. The material is assumed to have the periodically distributed microstructure indicated in the unit cell which consists of the matrix phase and reinforcement particles. Here, the material characteristics of the matrix are assumed to obey the strain gradient dependent constitutive equation (4) and the reinforcement particle behaves as an isotropic elastic material. Each quadrilateral corresponds to the cross-triangular elements. As indicated in Fig. 2, in the present investigation, the reinforcement material is assumed to be square with side length d which characterizes the particle size. Furthermore, in order to investigate the effect of the loading direction with respect to the principal direction of the unit cell, namely, the y_2 coordinate, the angle θ is introduced and is varied from 0 to 45 degrees. For the purpose of comparison, two different particle distribution patterns, I and II, shown in Fig. 2c, are also introduced. The effects of the direction θ on the macroscopic deformation behavior of the composites are investigated through computational simulation. The cohesive interface condition between the matrix and reinforcement is assumed to be preserved throughout the deformation processes. The boundary conditions for the macroscopic scale are, as indicated in Fig. 2a, such that the upper and bottom surfaces are shear free with a constant displacement constraint, whereas the right and left surfaces are assumed to be stress free.

Since the present investigation focuses on the effect of the volume fraction, size and distribution pattern of particles coupled with the strain gradient dependence of the matrix material on the characteristics of particle reinforced composites, a very simple constitutive equation for the plastic deformation of the matrix material is assumed, $\bar{\bar{\sigma}} = \sigma_y \left(1 + \varepsilon/\varepsilon_0\right)^n \left(1 + \dot{\varepsilon}/\dot{\varepsilon}_0\right)^m$, with material constants $\sigma_y = 275\,\mathrm{MPa}$, $\varepsilon_0 = 0.002$, $\dot{\varepsilon}_0 = 0.002\,/\mathrm{s}$, $n = 0.25$ and $m = 0.024$, elasticity modulus $E = 69.0\,\mathrm{GPa}$ and Poisson's ratio $\nu = 0.333$. The coefficient of the strain gradient was set to be $c_0 = 360\,\sigma_y l^2$, $l = W/16$, which could suppress the effect of the size dependence in the estimation of the second derivative of the strain (Tomita and Fujimoto [10]). The reinforcement particles are assumed to be elastic with elasticity modulus $E = 450$ GPa and Poisson's ratio $\nu = 0.170$.

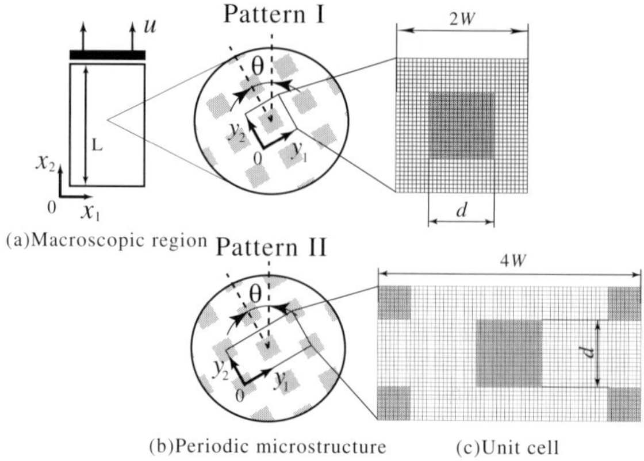

Figure 2. Computational model of particle reinforced composite material.

4. Characteristic feature of metal-matrix composite

Figure 3 indicates the maximum principal stress for different angles θ with different volume fractions of reinforcement particles ($V_f = 25.0$ and 44.4%), different particle sizes ($d = 12.0$ and $24.0\,\mu\mathrm{m}$), and particle distribution pattern I at three different deformation stages ($u/L = 0.01$, 0.03 and 0.05). The effect of strain gradients on the macroscopic stress becomes considerable as the deformation proceeds and the particle size decreases. The deformation resistance increases as the particle size decreases under a constant volume fraction of reinforcement particles, which is consistent with experimentally observed evidence (Kamat et al. [5], Ushigome et al. [11]). This tendency is intensified as the volume fraction of reinforcement increases. Furthermore, the directional dependence of the effect of the strain gradients on the resistance of deformation is notable but diminishes as the angle increases.

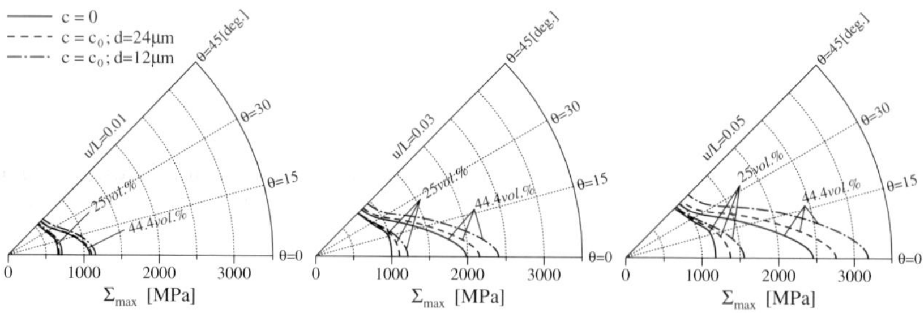

Figure 3. Macroscopic maximum principle stress $\Sigma_{\max}$ for different directions θ defined in Fig. 2, volume fraction $V_f = 25$, 44.4% and particle size d at $u/L = 0.01$, 0.03 and 0.05.

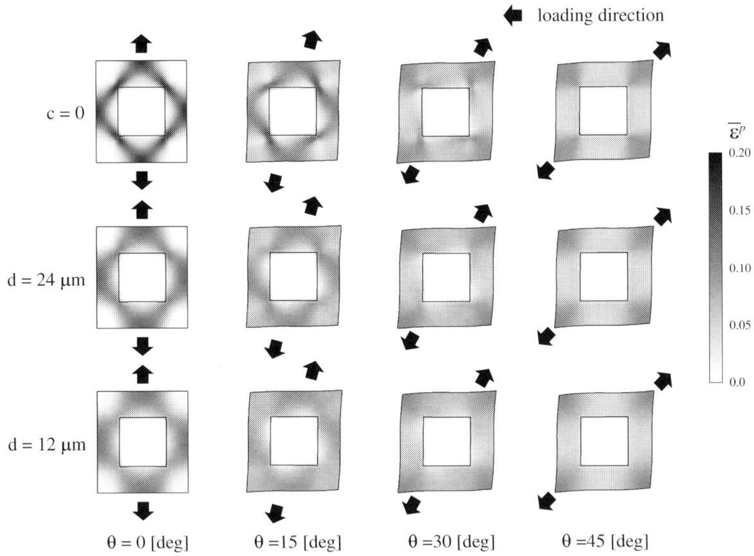

Figure 4. Equivalent plastic strain distribution for the directions $\theta = 0$ to $45°$, $d = 12$, $24\,\mu\mathrm{m}$ for strain gradient independent, $c = 0$, and dependent, $c = c_0$ materials are shown, respectively.

Next, we will further discuss these characteristic points based on the information concerning microscopic deformation. Figure 4 indicates the equivalent plastic strain distribution for the directions 0 to 45 degrees. Two different particle sizes for strain gradient independent and dependent materials are shown. The difference in the stiffness between the matrix and particles is marked so that the soft matrix area, which is an area of high strain gradient as indicated in the equivalent plastic strain distribution, absorbs the deformation applied. The effect of the direction angle on the strain distribution is considerable. That is, the high strain regions emanating from the corners of the reinforcement particles connect and form networklike structures for the case of 0 degrees. Meanwhile, for the case of 45 degrees, the strain is distributed and its concentration is not great, so that the maximum strain level is substantially lower than that for the case of $\theta = 0°$. Figure 5 shows the strain gradient distribution for two different particle sizes under different directional tension. The strain gradient increases as the size of the particle decreases, which results in a higher strength of the composite, as has been seen in Fig. 3. The effect of the direction of tension on the macroscopic resistance is also represented by the relative magnitude of the gradient term and decreases as the angle increases.

Finally, we will discuss the effect of the distribution patterns of the reinforced particles, as shown in Fig. 3c, on the deformation behavior of composites. The volume fraction is fixed at $V_f = 25.0\%$ and the size of the reinforcements is $d = 24.0\,\mu\mathrm{m}$. Figure 6 shows the average stress and strain relations for$\theta = 0°$

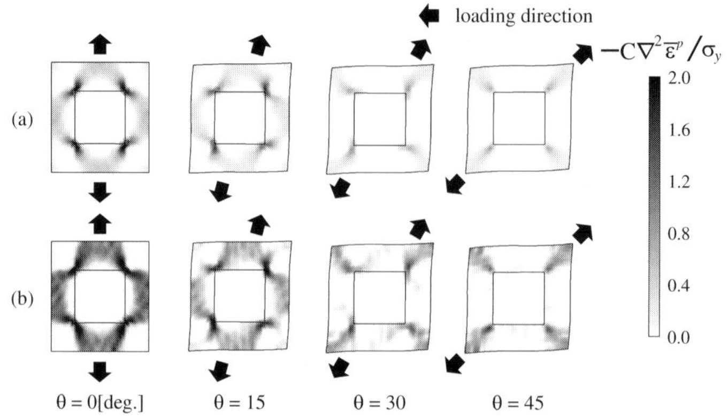

Figure 5. Distribution of positive strain gradient term ($-c\nabla^2\bar{\varepsilon}^p/\sigma_y \geq 0.0$) for particulate-reinforced composite materials with different particles size (a) $d = 24\,\mu\mathrm{m}$ and (b) $d = 12\,\mu\mathrm{m}$ for four different angles θ at $u/L = 0.05$.

and 30° with different distribution patterns of reinforcement particles. Considerable differences in average stress and strain relations for different distribution patterns and different θ are seen. Pattern I (square array) of reinforcement particles induces a tendency of the deformation resistance to increase more than that for Pattern II (hexagonal array) when $\theta = 0°$. On the other hand, for the direction $\theta = 30°$, stress-strain relations are insensitive to the change of the distribution pattern and the strain gradient term, which can be explained in terms of the distribution of plastic equivalent strain and the strain gradient term shown in Fig. 7. That is, the size of the bandlike regions such as the high strain regions emanating from the particle and matrix boundary, which is strongly related to the particle distribution, substantially affects the difference observed in the resistance of the composites in Fig. 6. This may explain why the resistance for the deformation of the composite with Pattern II is slightly higher than for that with Pattern I, as opposed to the result in the case of $\theta = 0°$.

5. Conclusion

In this investigation, to reproduce the experimentally observed evidence for particulate-reinforced composites which exhibit particle-size-dependent macroscopic characteristics, a simplified computational model employing the strain gradient constitutive equation for plastic deformation of the matrix material and a homogenization method were developed.

Using the proposed model of a particle-reinforced composite under plane strain conditions, the effects of the volume fraction, size and distribution pattern of reinforced particles and of the direction of particle distribution with respect to applied stress on the macroscopic characteristic features of the composites

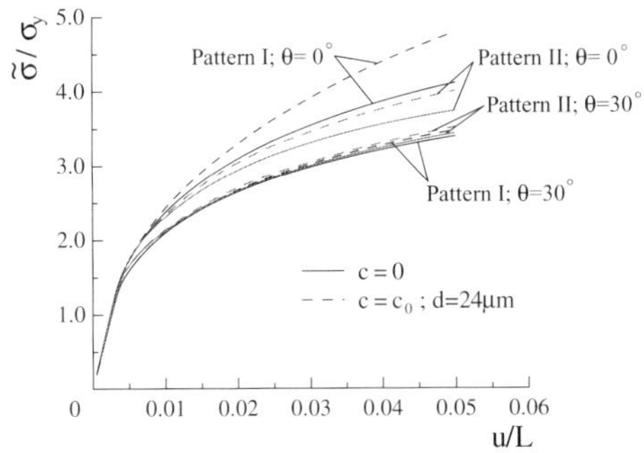

Figure 6. Average stress and strain relations for composites with different directions θ and distribution patterns for strain gradient independent and dependent materials.

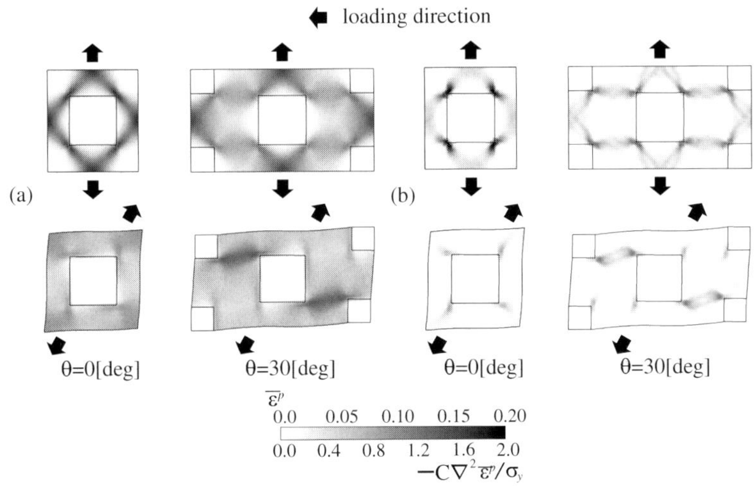

Figure 7. Distribution of (a) equivalent strain $\overline{\varepsilon}^p$ and (b) positive strain gradient term $(-c\nabla^2\overline{\varepsilon}^p/\sigma_y \geq 0.0)$ with different reinforcement particle distributions Pattern I and II for different direction angles θ at $u/L = 0.05$.

were clarified. The results indicated that the present simulation strategy has the capability to predict the experimentally observed particle-size-dependent macroscopic characteristic features of the composites. The main mechanism of the increase in the deformation resistance in the plastic range is the high strain gradient of the matrix material, which increases with decreasing distance between particles.

In order to directly compare the results predicted by computational simulation with those obtained experimentally, we must identify the material parameters for the matrix and the reinforcement particle, as well as the coefficient which determines the effect of the strain gradient term on the flow stress of the matrix material with a high strain gradient.

Acknowledgments

Financial support from the Ministry of Education, Culture, Sports, Science and Technology of Japan is gratefully acknowledged.

References

[1] Aifantis, E.C. (1984). On the microstructural original of certain inelastic models. *Trans. ASME, Eng. Mat. Tech.*, **106**, 326–330.

[2] Aifantis, E.C. (1987). The physics of plastic deformation. *Int. J. Plasticity*, **3**, 211–247.

[3] Fleck, N.A., Muller, G.M., Ashby, M.F. and Hutchinson, J.W. (1994). Strain gradient plasticity: Theory and experiment. *Acta Metall. Mater.*, **42**, 475–487.

[4] Guedes, J.M. and Kikuchi, N. (1990). Preprocessing and postprocessing for materials based on the homogenization method with adaptive finite element methods. *Comput. Meth. Appl. Mech. Engng.*, **83**, 143–198.

[5] Kamat, S.V., Hirth, J.P. and Mehrabian, R. (1989). Mechanical properties of particulate reinforced aluminium-matrix composites. *Acta Metall. Mater.*, **37**, 2395–2402.

[6] Muhlhaus, H.B. and Aifantis, E.C. (1991). A variational principle for gradient plasticity. *Int. J. Solids Struct.*, **28**, 845–858.

[7] Tomita, Y. (1990). *Computational Elasto-Plasticity*, Tokyo, YOKENDO, (in Japanese).

[8] Tomita, Y. (1994). Simulation of plastic instabilities in solid mechanics. *Trans. ASME, Appl. Mech. Rev.*, **47**, 171–205.

[9] Tomita, Y. and Nakao, T. (1992). Shear localization in thermo-elasto-viscoplastic plane strain blocks. *IUTAM Symposium, Finite Inelastic Deformations, Theory and Applications*, D. Besdo and E. Stein (eds.), Hannover, Germany, 179–188.

[10] Tomita, Y. and Fujimoto, T. (1995). Plane strain flow localization in tension blocks obeying strain-gradient-dependent constitutive equation. *Mater. Sci. Res. Int.*, **1**, 254–259.

[11] Ushigome, S., Yamamoto, K. and Fukuda, K. (1989). Mechanical properties of SiC particle reinforced aluminum alloy composite. *Tetsu-to-Hagané*, **75**, 1549–1554, (in Japanese).

[12] Vardoulakis, V. and Aifantis, E.C. (1991). A gradient flow theory of plasticity for granular materials. *Acta Mechanica*, **87**, 197–217.

[13] Zbib, H.M. and Aifantis, E.C. (1988). On the localization and postlocalization behavior of plastic deformation, Part I, II, III. *Res. Mechanica*, **23**, 261–277, 279–291, 293–305.

[14] Zbib, H.M. and Aifantis, E.C. (1989). A gradient-dependent plasticity theory of plasticity. *Trans. ASME, Appl. Mech. Rev.*, **42**, 295–304.

Homogenization-Based Predictions for Texture Evolution in Halite

P. Gilormini

Laboratoire de Mécanique et Technologie, ENS de Cachan-CNRS-Université de Paris 6
94235 Cachan, France
Pierre.Gilormini@lmt.ens-cachan.fr

Y. Liu, P. Ponte Castañeda

Department of Mechanical Engineering and Applied Mechanics, University of Pennsylvania
220 S. 33rd. Street, Philadelphia, PA 19104-6315, U.S.A
yiliu2@seas.upenn.edu, ponte@seas.upenn.edu

Abstract: The "variational" homogenization method developed by deBotton and Ponte Castañeda [2] is used here to predict texture development in halite polycrystals at room and high temperatures accounting for hardening and grain shape changes. The new predictions are compared with those of the Taylor and "tangent" model of Molinari et al. [5] for uniaxial tension and compression. The predictions of the "variational" model are found to be intermediate between the Taylor and "tangent" predictions, although not too different from either, as a consequence of the relatively high isotropy of the halite single crystal grains.

Keywords: Polycrystals, plasticity, texture evolution, halite.

1. Introduction

Halite (NaCl) is a polycrystalline aggregate of ionic single crystals with cubic structure. In this study, the crystals are assumed to deform by crystallographic slip only, thus neglecting recrystallization and other potential deformation mechanisms. For halite, three families of slip systems are usually activated; they are the types A (6 $\{110\}\langle 110\rangle$ systems), B (6 $\{100\}\langle 110\rangle$ systems) and C (12 $\{111\}\langle 110\rangle$ systems; the same as in fcc crystals) families. A power-law relation is assumed to hold between the resolved shear stress and the slip rate on a given system, which is a reasonable approximation for halite at least at high temperature according to Carter and Heard [1] and Heard [3]. For simplicity, a value of $n = 7$ will be assumed for all slip families in the high-temperature simulations, which provides a reasonably good fit with the results of Heard [3] at 400°C . On the other hand, an exponent of $n = 10$ is used for the simulations at room temperature, which was close to the value of 10.5 reported in Carter and Heard [1] for the A systems (no values were given for the B or C systems). In addition, a hardening model incorporating latent hardening and saturation

C. Miehe (ed.), IUTAM Symposium on Computational Mechanics of Solid Materials at Large Strains, 269-278.

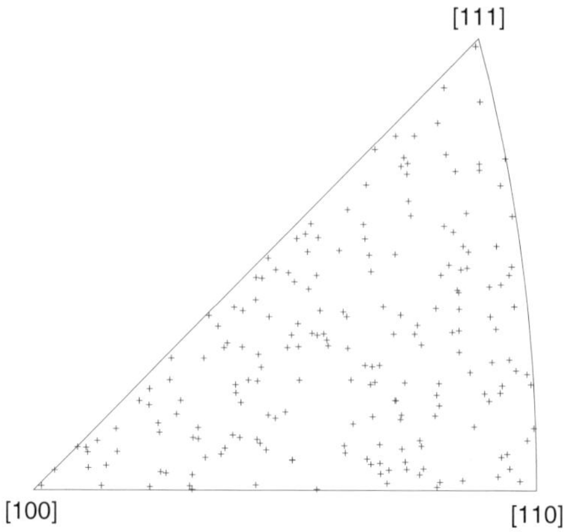

Figure 1. Inverse pole figure of the 200-grain initial random texture.

values was used, as described in more detail later, to fit the experimental data for halite single crystals at both room and high temperature.

An initially random texture of 200 grains was used (Figure 1) for all simulations, having verified that use of other initially random distributions of 200 orientations had little effect on the results. In this work, we will make use of the "variational" homogenization method of deBotton and Ponte Castañeda [2] and Nebozhyn et al. [6] to estimate the evolution of crystallographic texture in halite polycrystals, following the detailed proposal of Ponte Castañeda [9] to account for microstructure evolution. The results will be compared with the predictions of the Taylor estimate and "tangent" model of Molinari et al. [5].

2. Formulation

The halite polycrystals are assumed to occupy region in space Ω and can be approximated as N-phase composites, where each "phase" is defined as the subregion $\Omega^{(r)}$ ($r = 1, \cdots, N$) occupied by all grains of a given orientation, as measured by a rotation tensor $\mathbf{Q}^{(r)}$ relative to a "reference" crystal with known orientation. Each grain is assumed to undergo viscous deformation on a set of K ($k = 1, \cdots, K$) preferred crystallographic slip systems, which are characterized by the second-order tensors

$$\boldsymbol{\mu}_{(k)}^{(r)} = \frac{1}{2}\left(\mathbf{n}_{(k)}^{(r)} \otimes \mathbf{m}_{(k)}^{(r)} + \mathbf{m}_{(k)}^{(r)} \otimes \mathbf{n}_{(k)}^{(r)}\right) , \tag{1}$$

where $\mathbf{n}_{(k)}^{(r)}$ and $\mathbf{m}_{(k)}^{(r)}$ are the unit vectors normal to the slip plane and along the slip direction in the kth system.

Local behavior. For a given Cauchy stress $\boldsymbol{\sigma}$, acting on the kth slip system with orientation r, we introduce the resolved shear stresses $\tau_{(k)}^{(r)} = \boldsymbol{\sigma} \cdot \boldsymbol{\mu}_{(k)}^{(r)}$, so that the response of the grains with orientation r may be written in the form

$$\varepsilon = \frac{\partial u^{(r)}}{\partial \boldsymbol{\sigma}}(\boldsymbol{\sigma}), \quad u^{(r)}(\boldsymbol{\sigma}) = \sum_{k=1}^{K} \phi_{(k)}(\tau_{(k)}^{(r)}), \tag{2}$$

where ε is the Eulerian strain rate and $u^{(r)}$ is the stress potential for the crystals with orientation r. In this last relation, the functions $\phi_{(k)}$, characterizing the slip potentials of the reference crystal, are assumed to be convex. Here, they will be chosen to be of the power-law type:

$$\phi_{(k)}(\tau) = \frac{\dot{\gamma}_0 \, \tau_k}{n+1} \left(\frac{|\tau|}{\tau_k} \right)^{n+1}, \tag{3}$$

where $0 \leq m = 1/n \leq 1$, $\tau_k > 0$ and $\dot{\gamma}_0 = 1$/sec are the strain-rate sensitivity, reference stress of the kth slip system and reference strain rate, respectively. Then, the slip strain rate $\dot{\gamma}$ in system k is given by $\dot{\gamma} = \dot{\gamma}_0(|\tau|/\tau_k)^n \mathrm{sign}(\tau)$.

Effective behavior. Assuming that the size of of the typical grain is small relative to the size of the specimen under consideration, one defines volume averages $\langle . \rangle$ and $\langle . \rangle^{(r)}$ over Ω and $\Omega^{(r)}$, respectively, so that the *effective response* for the polycrystal may be written in the form

$$\overline{\varepsilon} = \frac{\partial \tilde{U}}{\partial \overline{\boldsymbol{\sigma}}}(\overline{\boldsymbol{\sigma}}), \quad \tilde{U}(\overline{\boldsymbol{\sigma}}) = \min_{\boldsymbol{\tau} \in \mathcal{S}(\overline{\boldsymbol{\sigma}})} \sum_{r=1}^{N} c^{(r)} \, \langle u^{(r)}(\boldsymbol{\tau}) \rangle^{(r)}, \tag{4}$$

where $\tilde{U}$ is the *effective potential* for the polycrystal, $\overline{\boldsymbol{\sigma}} =< \boldsymbol{\sigma} >$ and $\overline{\varepsilon} =< \varepsilon >$ are the *average stress and strain rate*, the $c^{(r)} = \langle \chi^{(r)} \rangle$ characterize the texture of the polycrystal, and $\mathcal{S}$ is set of statically admissible stresses: $\mathcal{S}(\overline{\boldsymbol{\sigma}}) = \{\boldsymbol{\tau}, \text{ div } \boldsymbol{\tau} = 0 \text{ in } \Omega, \ \boldsymbol{\tau}\mathbf{n} = \overline{\boldsymbol{\sigma}}\mathbf{n} \text{ on } \partial\Omega\}$.

Variational homogenization method. Following an analogous development by Ponte Castañeda [8] for N-phase composites with isotropic phases, deBotton and Ponte Castañeda [2] derived an alternative variational form for the effective potential (4) of the nonlinear polycrystal defined by (2), for some given set of slip potentials $\phi_{(k)}$. This variational representation involves a "linear comparison polycrystal" with phase potentials defined by

$$u_0^{(r)}(\boldsymbol{\sigma}) = \frac{1}{2} \boldsymbol{\sigma} \cdot \mathbf{M}^{(r)} \boldsymbol{\sigma}, \quad \mathbf{M}^{(r)} = 2 \sum_{k=1}^{K} \alpha_{(k)}^{(r)} \, \boldsymbol{\mu}_{(k)}^{(r)} \otimes \boldsymbol{\mu}_{(k)}^{(r)}. \tag{5}$$

Under appropriate hypotheses, a general bound for $\tilde{U}$ may be then obtained:

$$\tilde{U}(\overline{\boldsymbol{\sigma}}) \geq \max_{\substack{\alpha_{(k)}^{(r)}>0, \\ r=1,\ldots,N, \\ k=1,\ldots,K}} \left\{ \frac{1}{2}\overline{\boldsymbol{\sigma}} \cdot \tilde{\mathbf{M}}(\boldsymbol{\alpha})\overline{\boldsymbol{\sigma}} - \sum_{r=1}^{N}\sum_{k=1}^{K} c^{(r)} V_{(k)}\left(\alpha_{(k)}^{(r)}\right) \right\}, \tag{6}$$

where

$$V_{(k)}(\alpha) = \max_{\tau} \left\{ \alpha |\tau|^2 - \phi_{(k)}(\tau) \right\} \tag{7}$$

serves to characterize the nonlinearity of the material behavior. Here also, $\boldsymbol{\alpha}$ denotes the full set of positive slip compliances $\alpha_{(k)}^{(r)}$ and $\tilde{\mathbf{M}}$ is the effective compliance tensor of the linear comparison polycrystal with grain compliances $\mathbf{M}^{(r)}$, as defined by (5) in terms of the slip compliances $\boldsymbol{\alpha}$. As suggested by Nebozhyn et al. [6], use will be made of the standard self-consistent approximation to estimate the effective compliance tensor $\tilde{\mathbf{M}}$ of the linear comparison polycrystal.

If the maximum in (6) is attained at the stationarity condition, it follows that the effective constitutive relation for the polycrystal may be approximated as

$$\overline{\boldsymbol{\varepsilon}} = \tilde{\mathbf{M}}\left(\hat{\boldsymbol{\alpha}}; s^{(\alpha)}, \tau_k\right)\overline{\boldsymbol{\sigma}} \tag{8}$$

where $\hat{\boldsymbol{\alpha}}$ now denotes the set of optimal $\alpha_{(k)}^{(r)}$ from (6). Here, the variables $s^{(\alpha)}$ and τ_k denote the instantaneous values of the microstructural variables characterizing the "state" of the crystallographic and morphological textures of the polycrystal, and of the reference slip stresses, respectively. In general, they evolve in time as a consequence of the finite deformation imposed on the polycrystal.

Microstructure evolution. For the "granular" microstructures of the viscoplastic polycrystals, it will be assumed that the distribution of the grains is controlled by the mean flow. In other words, the "shape" and "orientation" of the grains, characterizing the *morphological texture*, are assumed to be controlled by the average strain rate $\overline{\boldsymbol{\varepsilon}}$ and spin $\overline{\boldsymbol{\omega}}$ in the composite. This is required to preserve the integrity of the composite – for example, if every grain was to deform with the average strain and spin tensor in the grains, the composite would likely disintegrate, as each grain would in general be expected to have a different strain and spin field. In addition, it is also necessary to keep track of the orientation of the material within each grain, characterizing the *crystallographic texture.* The evolution of this orientation distribution function will be assumed here to be controlled by the average spin $\overline{\boldsymbol{\omega}}^{(r)}$ of the material within the grains, as determined by the self-consistent estimates of the previous section for nonlinearly viscous polycrystals. Although a continuum distribution of orientations could be considered, for simplicity, only a discrete set of N orientations, corresponding to N different "phases," will be considered here. For more details, refer to Ponte Castañeda [9].

Slip systems hardening law. A reasonably good fit with the uniaxial extension data of Carter and Heard [1] for halite single crystals could be obtained with a hardening law incorporating latent hardening, as well as flow stress saturation levels. For each temperature, the sum of the squared differences with the 12 experimental points was minimized with respect to the 10 parameters of the model. Because of the nonuniqueness, and because the lowest minimum was not necessarily reasonable, a set of parameters has been selected that both gave good agreement with the measurements and used physically meaningful values (especially for latent hardening). This hardening law is described by the following relations:

$$\dot{\tau}_k = \sum_{l=1}^{24} H_{kl} \left|\dot{\gamma}_l\right| \tag{9}$$

where

$$H_{kl} = h_k^0 \left(\frac{\tau_k^{sat} - \tau_k}{\tau_k^{sat} - \tau_k^0} \right) Q_{kl} \quad \text{(no sum over } k\text{)} \tag{10}$$

with $k = 1, \ldots, 6$ for the A family, etc. The Q_{kl} values define the effect of slip rate $\dot{\gamma}_l$ of system l on the hardening of reference stress τ_k on system k. Here, a simple latent hardening model is used:

$$Q_{kl} = \begin{cases} 1 & \text{if systems } k \text{ and } l \text{ are coplanar} \\ q & \text{otherwise} \end{cases} \tag{11}$$

This law leads to different hardening rates on the various families, and also on the various systems in each family. The hardening rate H_{kl} equals either h_k^0 (self-hardening) or qh_k^0 (latent hardening) initially (when $\tau_k = \tau_k^0$), and it decreases to 0 when τ_k increases and tends to τ_k^{sat}, which corresponds to a stabilized value of τ_k. This hardening law differs from one used, for instance, by Kalidindi et al. [4], where H_{lk} replaces H_{kl} (the function between h_k and Q_{kl} was also changed to a Voce law in the present work). Such a modification of the subscripts is already mentioned in Paquin et al. [7], but without comment. The reason for using H_{kl} rather than H_{lk} is for keeping the physical significance of the saturation value: when $\tau_k = \tau_k^{sat}$, H_{kl} is zero for all l, and consequently $\dot{\tau}_k = 0$ in multislip conditions, whereas $\dot{\tau}_k$ can be nonzero if H_{lk} is used in (10), which nullifies a column rather than a row of the $[H]$ matrix.

Although results are not shown for brevity, it was found that the values given in Table 1 for the 10 parameters involved in the hardening law gave good agreement with the experimental data. The fit was better for high temperature than for room temperature. The τ_A^0, τ_B^0 and τ_C^0 values shown in Table 1 indicate that the single crystals will be almost isotropic at high temperature, as discussed above, at least initially. By contrast, some anisotropy is expected at room temperature.

Table 1. Halite single crystal parameters for a hardening law of the Voce type on each family of slip systems (h and τ values in MPa).

Parameter	25°C	400°C
n	10	7
q	1.5	1.14
h_A^0	170	13.0
h_B^0	740	42.2
h_C^0	45	21.5
τ_A^0	8.7	5.5
τ_B^0	20.7	6.0
τ_C^0	45.0	5.5
τ_A^{sat}	62	8.0
τ_B^{sat}	110	12.2
τ_C^{sat}	250	9.0

3. Textures predictions at 100% strain

In this section, the above-described "variational" model is used to generate predictions for the crystallographic texture evolution of halite polycrystals. Following Wenk et al. [11], both uniaxial tension and compression are considered, and the two cases discussed above (i.e., room temperature and 400°C) in the context of hardening are implemented. Comparisons are given with the Taylor model, as well as the "tangent" model. In this last connection, it should be noted that the "tangent" predictions shown here use the hypothesis that all the grains evolve with the same aspect ratio – in contrast with what was done by Wenk et al., but in agreement with what is done in our 'variational" model.

Figures 2 and 3 show the inverse pole figures obtained in tension and compression, respectively. The general trend is a strong similitude between the results given by the "variational" and Taylor models, and some differences with the "tangent" model. For tension at room temperature, the "tangent" model predicts a sharp texture at [100] and [111], whereas the "variational" model gives slightly lower values (but higher than the Taylor model). At high temperature (the more "isotropic" case), the three models give even closer results, with the above trends still present but much less pronounced. For compression (Figure 3), where textures are less sharp after a 100% strain, the "tangent" model predicts a texture component at [100], but the other models do not. The results of the "variational" and Taylor models are quite similar globally, but exhibit lower values than those predicted by the "tangent" model.

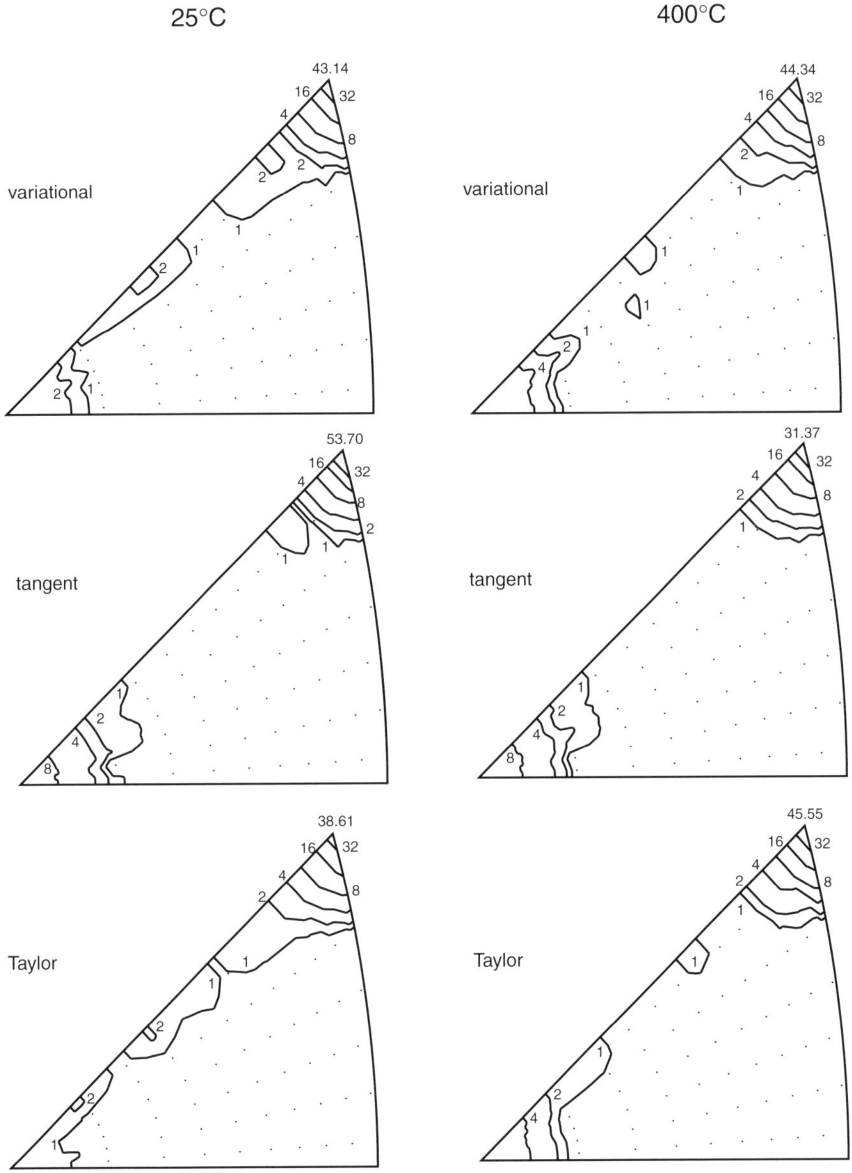

Figure 2. Inverse pole figures obtained after an axial strain of 100% applied under uniaxial tension at low (25°C) and high (400°C) temperatures. Comparison between the variational self-consistent, tangent self-consistent and Taylor models. Dotted areas are below the average level of 1.

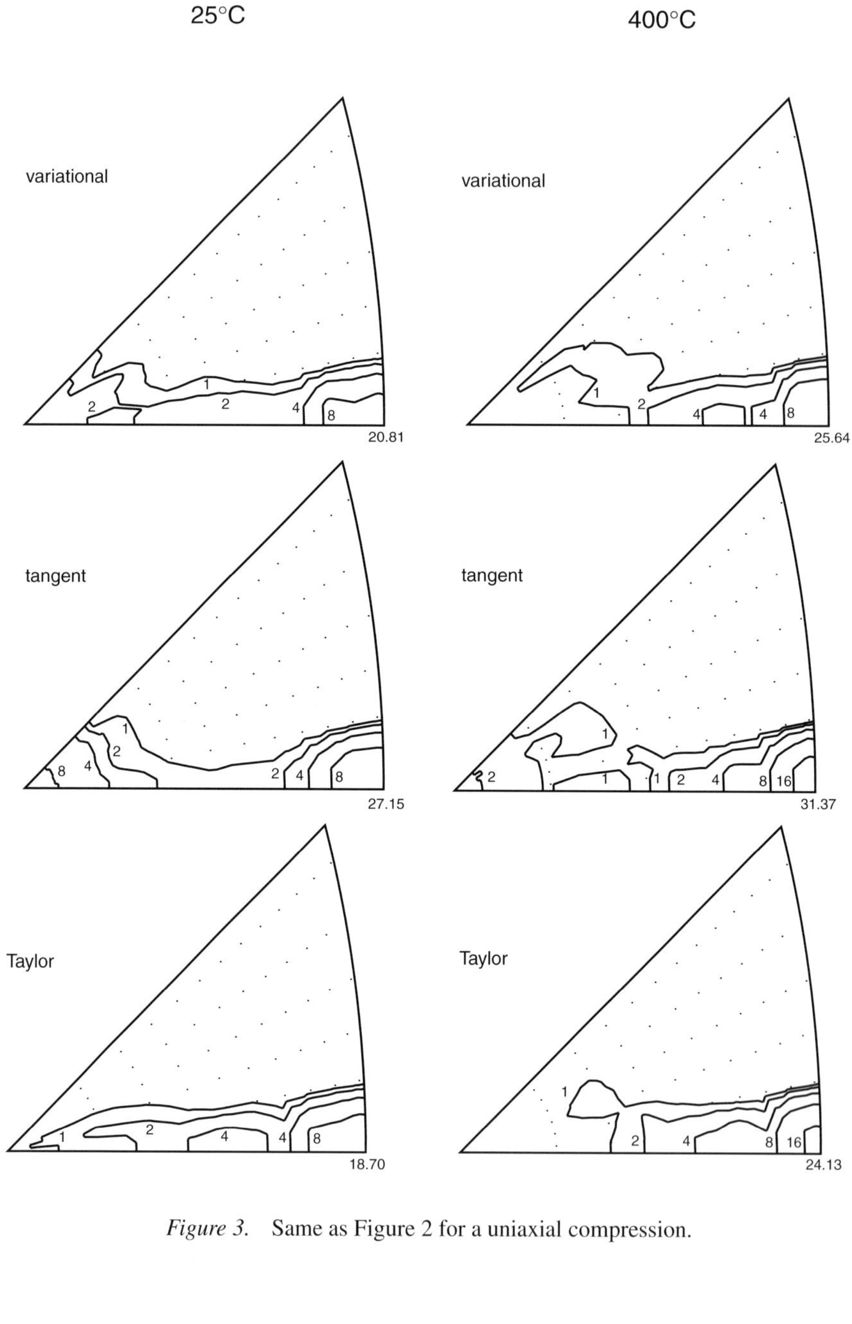

Figure 3. Same as Figure 2 for a uniaxial compression.

These figures can be compared with Figures 7 and 8 of Wenk et al. [11]. Since $n = 9$ was used in that work, comparisons with our room-temperature simulations ($n = 10$) are more appropriate, but it must also be kept in mind that the hardening laws used are different. The agreement is qualitatively good, although the values obtained by Wenk et al. [11] (with hardening) at $[100]$ in both tension and compression with the tangent model appear to be higher.

Another important consideration in comparing the "variational" and "tangent" self-consistent models is the computation time required. Using the same computer, and minimizing the outputs of both programs, it has been found that the variational procedure requires about 6.5 times the computation time of the tangent model. This is still a question of a few tens of minutes on modern computers and is perfectly affordable. On the other hand, this also represents a very significant improvement with respect to earlier implementations (Nebozhyn et al. [6]) of the "variational" method, which used a much slower minimization procedure and would have made the texture evolution computations infeasible. Of course, the Taylor model requires much shorter computation times (tens of seconds) and is rather close to the "variational" model in the present context.

4. Conclusions

This paper was concerned with a first implementation of the "variational" homogenization method of deBotton and Ponte Castañeda [2] to predict texture development following the program proposed by Ponte Castañeda [9] for halite polycrystals. As already anticipated by the work of Nebozhyn et al. [6], the predictions of the "variational" method are intermediate between those of the Taylor and "tangent" methods. However, unfortunately, the experimental data that was used here corresponds to a situation where the relevant cubic single crystals are highly isotropic, as expected from the fact that the flow stresses of the various slip families are quite close to each other, even at room temperature (refer to Table 1). Because of this the relative differences in the predictions were rather small. We are presently applying the procedure to more highly anisotropic HCP polycrystals, where preliminary results show significant differences in the predictions of the various models. These will be reported elsewhere.

Acknowledgments

YL and PPC acknowledge the support of the NSF through grants CMS-99-72234 and INT-97-26521. We are grateful to B. Bacroix for supplying a program to draw texture contour plots and to O. Castelnau for providing a program to compute the "tangent" estimates.

References

[1] Carter, N.L. and Heard, H.C. (1970). *Am. J. Sci.*, **269**, 193.

[2] deBotton, G. and Ponte Castañeda, P. (1995). *Proc. R. Soc. Lond. A*, **448**, 121.

[3] Heard, H.C. (1972). *Am. Geoph. U. Monogr.*, **16**, 191.

[4] Kalidindi, S.R., Bronkhorst, C.A. and Anand, L. (1992). *J. Mech. Phys. Solids*, **40**, 537.

[5] Molinari, A., Canova, G.R. and Ahzi, S. (1987). *Acta Metall. Mater.*, **35**, 2983.

[6] Nebozhyn, M.V., Gilormini, P. and Ponte Castañeda, P. (2001). *J. Mech. Phys. Solids*, **49**, 313.

[7] Paquin, A., Berbenni, S., Favier, V., Lemoine, X. and Berveiller, M. (2001). *Int. J. Plasticity*, **17**, 1267.

[8] Ponte Castañeda, P. (1991). *J. Mech. Phys. Solids*, **39**, 45.

[9] Ponte Castañeda, P. (1998). In: *Continuum Models and Discrete Systems, Proceedings of the 9th International Symposium*, Inan E. and Markov K. Z. [eds.], World Scientific Publishing Co., 228.

[10] Skrotzki, W. and Welch, P. (1983). *Tectonophysics*, **99**, 47.

[11] Wenk, H.R., Canova, G., Molinari, A. and Mecking, H. (1989). *Acta Metall.*, **37**, 2017.

On the Influence of Texture Model Types on Simulations of Sheet Metal Forming Processes

D. Besdo
Institut für Mechanik, Universität Hannover
Appelstraße 11, 30167 Hannover, Germany
besdo@ifm.uni-hannover.de

Abstract: Simulations of sheet metal forming processes including anisotropy and spring back need very sophisticated material laws. Phenomenological denotations are often not precise enough. Taylor-models or reference volume elements (RVE) are much better but need ideas how to compose the RVE. To both types of models ideas and examples are presented.

Keywords: Anisotropy, phenomenological laws, texture development, reference volume elements.

1. Introduction: model types and anisotropy

1.1. PHENOMENOLOGICAL LAWS

Phenomenological laws with isotropic or kinematical hardening and with a Hill-anisotropy consist of three parts:

- a *"yield criterion"*, expressed in terms of Cauchy-stresses $\underline{\underline{\sigma}}$, a yield stress Y, and internal variables $\underline{\underline{a}}$ and $\underline{\underline{\underline{\underline{k}}}}$, e.g.:

$$f = \left(\underline{\underline{\sigma}}^D - \underline{\underline{a}}\right) \cdot \cdot \underline{\underline{\underline{\underline{k}}}} \cdot \cdot \left(\underline{\underline{\sigma}}^D - \underline{\underline{a}}\right) - \frac{2}{3} Y^2 \leq 0 \quad , \tag{1}$$

- a *"flow rule"*, combining the stresses with strain rates $\underline{\underline{\lambda}}$ with the state of stresses, normally by the v.Mises-potential law:

$$\underline{\underline{\lambda}} = \xi \frac{\partial f}{\partial \underline{\underline{\sigma}}} \tag{2}$$

- and *"evolutional rules"* for the variables Y, $\underline{\underline{a}}$, and $\underline{\underline{\underline{\underline{k}}}}$.

Difficulties arise in connection with the evolutional rules for $\underline{\underline{a}}$:
The well-known Prager's hardening rule in its original version

$$\dot{\underline{\underline{a}}} = c \underline{\underline{\lambda}} \ ; \ \underline{\underline{\lambda}} = sym\left(grad\, \underline{v}\right) = \frac{1}{2}\left[\nabla \otimes \underline{v} + \left(\nabla \otimes \underline{v}\right)^T\right] \tag{3}$$

C. Miehe (ed.), IUTAM Symposium on Computational Mechanics of Solid Materials at Large Strains, 279-290.

or improved by an Armstrong-Frederick-term

$$\dot{\underline{\underline{a}}} = c\underline{\underline{\lambda}} - b\overline{\lambda}\,\underline{\underline{a}}\,, \quad \overline{\lambda} \stackrel{!}{=} \left[\frac{2}{3}\underline{\underline{\lambda}}\cdot\cdot\underline{\underline{\lambda}}\right]^{1/2} \tag{4}$$

is applied very often. But the type of rates is uncertain.

The Zaremba-Jaumann rate yields stress oscillations in case of simple shear (e.g. Lehmann [5]). Concepts to avoid these problems are:

- the use of Oldroyd-rates (Lehmann), but this is very questionable,
- the introduction of a plastic spin

$$\underline{\underline{\Omega}} = \xi\left[\underline{\underline{\lambda}}\cdot\underline{\underline{a}} - \underline{\underline{a}}\cdot\underline{\underline{\lambda}}\right], \tag{5}$$

- the use of Cosserat models.

The advantage of phenomenological laws is their simplicity and speed whereas their numerous parameters have no direct physical meaning.

1.2. TEXTURE REGARDING MODELS

Distributions of lattice orientations are regarded by Taylor-Lin- and Sachs-models. Both do not consider the crystallites size and shape. Their basic assumptions, that strains (Taylor-Lin) or stresses (Sachs) are constants, give them the character of upper or lower bounds, respectively.

Also theories regarding so-called "representative volume elements" (RVE) are in use. They recognize the influence of the crystallite sizes as well as shapes. In reality these shapes can be very irregular. A special version of such a RVE will be discussed here in detail.

2. Phenomenological laws

Isotropic hardening can be described by one variable yield stress$Y$, which is assumed to be a function of the temperature ϑ, of the equivalent strain rate $\overline{\lambda}$, and of its time-integral, the equivalent strain$\overline{\varphi}$. But kinematical hardening with its evolution needs more effort.

2.1. KINEMATICAL HARDENING

The problems concerning kinematical hardening are often overcome by use of Oldroyd-rates in eqs. (3), (4). But this is very questionable. There are four kinds of such rates for a tensor of second order. They are denoted by a circle on top, whereas the simple material rate is denoted by a triangle. The specific type of the chosen Oldroyd-rate may be classified by additional blocks at the right side. The connection between the Oldroyd- and the simple material rates can be presented with the velocity gradient

$$\underline{\underline{G}} \stackrel{!}{\equiv} (\nabla \otimes \underline{v})^T \quad : \tag{6}$$

$$\alpha) \quad \dot{\underline{\underline{a}}} \Longrightarrow \overset{0}{\underline{\underline{a}}}{}^{\square\square} \equiv \overset{0}{a}{}^{ij} \, \underline{g}_i \otimes \underline{g}_j \Longrightarrow \quad \overset{\triangle}{\underline{\underline{a}}} = \dot{\underline{\underline{a}}} + \underline{\underline{G}} \cdot \underline{\underline{a}} + \underline{\underline{a}} \cdot \underline{\underline{G}}^T \tag{7}$$

$$\beta) \quad \dot{\underline{\underline{a}}} \Longrightarrow \overset{0}{\underline{\underline{a}}}{}_{\square\square} \equiv \overset{0}{a}{}_{ij} \, \underline{g}^i \otimes \underline{g}^j \Longrightarrow \quad \overset{\triangle}{\underline{\underline{a}}} = \dot{\underline{\underline{a}}} - \underline{\underline{G}}^T \cdot \underline{\underline{a}} - \underline{\underline{a}} \cdot \underline{\underline{G}} \tag{8}$$

$$\gamma) \quad \dot{\underline{\underline{a}}} \Longrightarrow \overset{0}{\underline{\underline{a}}}{}^{\square}{}_{\square} \equiv \overset{0}{a}{}^{i}{}_{j} \, \underline{g}_i \otimes \underline{g}^j \Longrightarrow \quad \overset{\triangle}{\underline{\underline{a}}} = \dot{\underline{\underline{a}}} + \underline{\underline{G}} \cdot \underline{\underline{a}} - \underline{\underline{a}} \cdot \underline{\underline{G}} \tag{9}$$

$$\delta) \quad \dot{\underline{\underline{a}}} \Longrightarrow \overset{0}{\underline{\underline{a}}}{}_{\square}{}^{\square} \equiv \overset{0}{a}{}_{i}{}^{j} \, \underline{g}^i \otimes \underline{g}_j \Longrightarrow \quad \overset{\triangle}{\underline{\underline{a}}} = \dot{\underline{\underline{a}}} - \underline{\underline{G}}^T \cdot \underline{\underline{a}} + \underline{\underline{a}} \cdot \underline{\underline{G}}^T \tag{10}$$

The rates of the cases $\alpha)$ and $\beta)$ do not yield a deviatoric $\overset{\triangle}{\underline{\underline{a}}}$ for a deviatoric $\dot{\underline{\underline{a}}}$, whereas the other cases yield unsymmetric tensors $\overset{\triangle}{\underline{\underline{a}}}$. Appropriate corrections yield anew stress oscillations in case of simple shear.

It is well-known that the Armstrong-Frederick-terms reduce the oscillations. If simultaneous isotropic hardening occurs no longer curious effects can be observed really. Additionally, the application of the concept of plastic spin is helpful for both kinds of tensors.

Very general laws in strain-space-representation proposed in Besdo [2, 3], are able to describe very large strains without curious effects. They can, but they need not contain a modified version of plastic spin: The plastic rotation is no longer assumed to depend on the backstresses directly. It stands for a lattice rotation and not for stresses of any kind. It depends on an auxiliary tensor $\underline{\underline{h}}$, which itself has an own evolutional law. The stresses occurring in yield criteria and in flow-rules, are replaced by elastic parts of strains observed from an intermediate configuration. This is also valid for backstresses being replaced by an arbitrary number of additive backstrains ${}^{\alpha}\underline{\underline{u}}$ with different evolutions and for the yield strain y which is connected with Y by

$$y = \frac{Y}{(2G)} \tag{11}$$

These material laws can also include a variable Hill-anisotropy. The law of this kind belonging to the family of Huber-Levy-Mises-Hencky-laws, consists of the following elements:

- *Yield criterion*, basicly in the intermediate configuration $\widehat{\mathcal{K}}$:

$$\hat{g} = \left(\underline{\underline{\gamma}}^{eD} - \underline{\underline{\hat{u}}}\right) \cdot\cdot \underline{\underline{\underline{\underline{\hat{k}}}}} \cdot\cdot \left(\underline{\underline{\gamma}}^{eD} - \underline{\underline{\hat{u}}}\right) - \frac{2}{3}y^2 \le 0 \tag{12}$$

with

$$\underline{\underline{\gamma}}^{eD} = \underline{\underline{\gamma}}^{e} - \frac{1}{3}\underline{\underline{1}}\ \underline{\underline{1}} \cdot\cdot \underline{\underline{\gamma}}^{e},\ \underline{\underline{\gamma}}^{e} = \frac{1}{2}\left(\underline{\underline{F}}^{eT} \cdot \underline{\underline{F}}^{e} - 1\right),\ \underline{\underline{\hat{u}}} = \sum_{\alpha} {}^{\alpha}\underline{\underline{\hat{u}}}\,. \tag{13}$$

The pull-back rules for $\underline{\underline{\gamma}}^{e}$, ${}^{\alpha}\underline{\underline{\hat{u}}}$, and $\underline{\underline{\underline{\underline{\hat{k}}}}}$ are:

$$\underline{\underline{\gamma}}^{e} = \underline{\underline{\widehat{F}}}^{-T} \cdot \underline{\underline{\gamma}}'' \cdot \underline{\underline{\widehat{F}}}^{-1} \quad \text{with} \quad \underline{\underline{\gamma}}'' \equiv \frac{1}{2}\left(\underline{\underline{C}} - \underline{\underline{\widehat{C}}}\right) \tag{14}$$

$${}^{\alpha}\underline{\underline{\hat{u}}} = \underline{\underline{\widehat{F}}}^{-T} \cdot {}^{\alpha}\underline{\underline{U}} \cdot \underline{\underline{\widehat{F}}}^{-1} \quad \text{with} \quad {}^{\alpha}\underline{\underline{U}} \equiv {}^{\alpha}\underline{\underline{\tilde{u}}} \tag{15}$$

$$\begin{aligned}\underline{\underline{\underline{\underline{\hat{k}}}}} = \underline{\underline{\widehat{F}}} \overset{1,4,5,6}{\underset{2\text{-}3}{\otimes}} \left\{ \underline{\underline{\widehat{F}}} \overset{3,1,5,6}{\underset{2\text{-}4}{\otimes}} \left[\underline{\underline{\widehat{F}}} \overset{3,4,1,6}{\underset{2\text{-}5}{\otimes}} \left(\underline{\underline{\widehat{F}}} \overset{3,4,5,1}{\underset{2\text{-}6}{\otimes}} \underline{\underline{\underline{\underline{K}}}} \right) \right] \right\} \equiv \\ \equiv \underline{\underline{\widehat{F}}} * \underline{\underline{\underline{\underline{K}}}} \quad \text{with} \quad \underline{\underline{\underline{\underline{K}}}} \equiv \underline{\underline{\underline{\underline{\tilde{k}}}}}\,.\end{aligned} \tag{16}$$

Their application yields with $\underline{\underline{\widehat{B}}} \equiv \underline{\underline{\widehat{C}}}^{-1}$:

$$\tilde{g} = \frac{1}{4}\,\underline{\underline{\Gamma}} \cdot\cdot \underline{\underline{\underline{\underline{K}}}} \cdot\cdot \underline{\underline{\Gamma}} - \frac{2}{3}y^2 \le 0 \tag{17}$$

$$\underline{\underline{\Gamma}} \equiv \underline{\underline{C}} - \frac{1}{3}\underline{\underline{\widehat{C}}}\left(\underline{\underline{C}} \cdot\cdot \underline{\underline{\widehat{B}}}\right) - 2\underline{\underline{U}} \quad \left(\underline{\underline{U}} = \sum_{\alpha} {}^{\alpha}\underline{\underline{U}}\right) \tag{18}$$

- *The elastic potential* creating the stresses:

$$\psi = \hat{\psi}\left(\underline{\underline{\gamma}}^{e}, {}^{\alpha}\underline{\underline{\hat{u}}}\right) \stackrel{!}{=} \tilde{\psi}\left(\underline{\underline{C}}, {}^{\alpha}\underline{\underline{U}}, \underline{\underline{\widehat{B}}}\right) \tag{19}$$

In case of small elastic strains it may have a quadratic form:

$$\hat{\psi} = G\left\{\underline{\underline{\gamma}}^{e} \cdot\cdot \underline{\underline{\gamma}}^{e} + \frac{\nu}{1-2\nu}\left(\underline{\underline{\gamma}}^{e} \cdot\cdot \underline{\underline{1}}\right)^2\right\} + \sum_{\alpha} {}^{\alpha}\underline{\underline{\hat{u}}} \cdot\cdot {}^{\alpha}\underline{\underline{\hat{u}}} \tag{20}$$

Its pull-back version reads:

$$\begin{aligned}\tilde{\psi} &= \frac{G}{4}\left\{\underline{\underline{C}} \cdot\cdot \left(\underline{\underline{\widehat{B}}} \cdot \underline{\underline{C}} \cdot \underline{\underline{\widehat{B}}}\right) - 2\underline{\underline{C}} \cdot\cdot \underline{\underline{\widehat{B}}} + 3 + \right. \\ &+ \left. \frac{\nu}{1-2\nu}\left(\underline{\underline{C}} \cdot\cdot \underline{\underline{\widehat{B}}} - 3\right)^2\right\} + \sum_{\alpha} {}^{\alpha}\underline{\underline{U}} \cdot\cdot \left(\underline{\underline{\widehat{B}}} \cdot {}^{\alpha}\underline{\underline{U}} \cdot \underline{\underline{\widehat{B}}}\right).\end{aligned} \tag{21}$$

It creates 2.Piola-Kirchhoff-stresses by

$$\underline{\underline{\widetilde{T}}} = 2\frac{\partial \widetilde{\psi}}{\partial \underline{\underline{C}}} \tag{22}$$

- *Flow rule.* The flow rule can be derived from Ilyushin's postulate in the pull-back version. Its complete form is then:

$$\frac{\partial}{\partial \underline{\underline{C}}}\left(\dot{\widetilde{\psi}} - \frac{\partial \widetilde{\psi}}{\partial \underline{\underline{C}}} \cdot\cdot \dot{\underline{\underline{C}}}\right) = -\chi^* \frac{\partial \widetilde{\psi}}{\partial \underline{\underline{C}}} \quad \text{with} \quad \chi^* \geq 0\,. \tag{23}$$

  Neglecting terms in the amount of $||\underline{\underline{\gamma}}^e||$ with respect to 1, a simplified version can be found:

$$\dot{\underline{\underline{\hat{B}}}} = -\chi \frac{\partial \widetilde{\psi}}{\partial \underline{\underline{C}}} \quad \text{with} \quad \chi \geq 0\,. \tag{24}$$

  Its pull-forward version is

$$\underline{\underline{\widehat{\lambda}}} = \frac{1}{2}\chi \frac{\partial \widehat{g}}{\partial \underline{\underline{\gamma}}^e}\,. \tag{25}$$

- *Evolutional rules.* The often somehow questionable rules of evolution should contain productive terms like the original Prager's rule does, but also recovery-terms which may be of Armstrong-Frederick-type. If the considered quantity cannot vanish completely a minimal value must be provided. This holds for y and $\underline{\underline{\underline{K}}}$.

  Finally, in evolutional laws of Besdo [5] the rate is a mixture (parameter ξ between 0 and 1) of the Oldroyd-rates and a plastic spin is provided by a second factor η.

  Auxiliary quantities are tensors $\underline{\underline{W}}$ and $\underline{\underline{V}}$:

$$\underline{\underline{W}} = \left[\xi\,\dot{\underline{\underline{\widehat{C}}}} \pm \frac{1}{2}\left(\underline{\underline{H}} \cdot \underline{\underline{\widehat{B}}} \cdot \dot{\underline{\underline{\widehat{C}}}} - \dot{\underline{\underline{\widehat{C}}}} \cdot \underline{\underline{\widehat{B}}} \cdot \underline{\underline{H}}\right)\right] \cdot \underline{\underline{\widehat{B}}}\,. \tag{26}$$

$$\underline{\underline{V}} = \left[(1-\xi)\,\dot{\underline{\underline{\widehat{B}}}} \pm \frac{1}{2}\left(\underline{\underline{\widehat{B}}} \cdot \underline{\underline{H}} \cdot \dot{\underline{\underline{\widehat{B}}}} - \dot{\underline{\underline{\widehat{B}}}} \cdot \underline{\underline{H}} \cdot \underline{\underline{\widehat{B}}}\right)\right] \cdot \underline{\underline{\widehat{C}}}\,. \tag{27}$$

  where $\underline{\underline{H}}$ represents anyhow the texture development:

$$\dot{\underline{\underline{H}}} = \frac{1}{2} a_H \dot{\underline{\underline{\widehat{C}}}} - \left(b_H \overline{\widehat{\lambda}} - b_H^*\right) \underline{\underline{H}} + \underline{\underline{W}} \cdot \underline{\underline{H}} + \underline{\underline{H}} \cdot \underline{\underline{W}}^T \tag{28}$$

with

$$\overline{\overline{\lambda}} = \sqrt{\frac{1}{6}\left(\dot{\hat{\underline{\underline{B}}}} \cdot \hat{\underline{\underline{C}}} \cdot \dot{\hat{\underline{\underline{B}}}}\right) \cdot \cdot \hat{\underline{\underline{C}}}} \,. \tag{29}$$

The evolutional rules are:

$$\dot{y} = a_y \overline{\overline{\lambda}} - \left(b_y \overline{\overline{\lambda}} + b_y^*\right) \cdot (y - y_0) \,, \tag{30}$$

$$^{\alpha}\dot{\underline{\underline{U}}} = \frac{1}{2} a_\alpha \dot{\hat{\underline{\underline{C}}}} - \left(b_\alpha \overline{\overline{\lambda}} + b_\alpha^*\right) {}^{\alpha}\underline{\underline{U}} + \underline{\underline{W}} \cdot {}^{\alpha}\underline{\underline{U}} + {}^{\alpha}\underline{\underline{U}} \cdot \underline{\underline{W}}^T \,, \tag{31}$$

$$\dot{\underline{\underline{\underline{\underline{K}}}}} = -\frac{a_K}{4\overline{\overline{\lambda}}} \dot{\hat{\underline{\underline{B}}}} \otimes \dot{\hat{\underline{\underline{B}}}} - \left(b_K \overline{\overline{\lambda}} + b_K^*\right) \cdot \left(\underline{\underline{\underline{\underline{K}}}} - \underline{\underline{\underline{\underline{K}}}}_0\right)$$
$$+ \underline{\underline{V}} \overset{1,4,5,6}{\underset{2\text{-}3}{\otimes}} \underline{\underline{\underline{\underline{K}}}} + \underline{\underline{V}} \overset{3,1,5,6}{\underset{2\text{-}4}{\otimes}} \underline{\underline{\underline{\underline{K}}}} + \underline{\underline{V}} \overset{3,4,1,6}{\underset{2\text{-}5}{\otimes}} \underline{\underline{\underline{\underline{K}}}} + \underline{\underline{V}} \overset{3,4,5,1}{\underset{2\text{-}6}{\otimes}} \underline{\underline{\underline{\underline{K}}}} \,. \tag{32}$$

2.2. Possibilities offered by these laws

The possibilities offered by these laws are demonstrated by two examples: The first one shows (Fig. 1) that a pronounced yield limit during the first loading step and simultaneously an early re-plastification can be described by this type of laws if the parameters are chosen adequately.

In a second example a rolling process was simulated approximately (Fig. 2). Here the Cauchy-stresses σ_{11} and σ_{33} are plotted versus the Green-tensor coordinate γ_{11} which is in the program called G_{11}. Additionally, the three yield loci of the principal-stresses are plotted, too.

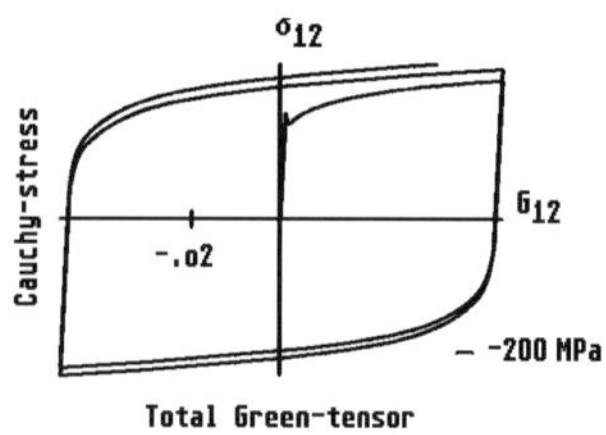

Ay,By,Dy:	1.0000	2000.000	0.0000
Ak,Bk,Dk:	0.6500	1.300	0.0000
Ah,Bh,Dh:	0.7000	1.000	0.0000
U(1):	0.2000	1000.000	0.0000
U(2):	0.0300	100.000	0.0000
U(3):	0.0050	10.000	0.0000
Y0 =	0.0010	Ya0 =	0.0020
Xi =	0.5000		

Figure 1. Numerical results of a cyclic shear loading including a pronounced yield limit and early re-plastification using Besdo's phenomenological law of 1991.

3. Application of the Taylor-Lin-Theory

The Taylor- or, in case of elastoplasticity, the Lin-theory can be applied if the behavior of a polycrystallite is to be simulated. Rotations are those of the crystal lattices. Every point of a grain or at least every grain in total has its own rotation. This can never be recognized in detail. The only disadvantage of the

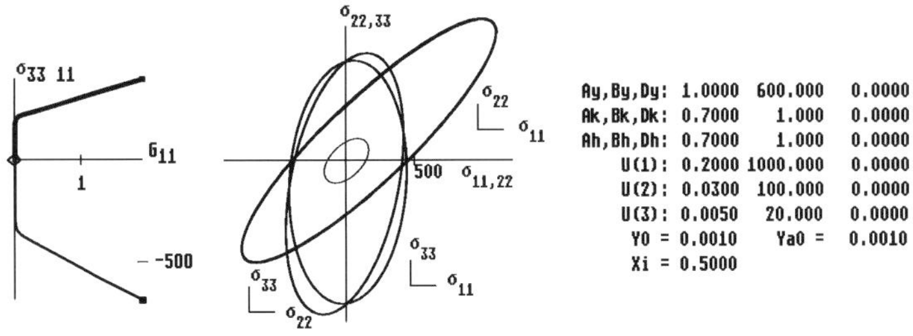

Figure 2. Numerical results a rolling process using Besdo's phenomenological law of 1991.

Taylor-Lin-theory (similar to Sachs' theory) is that the velocity gradient of all the crystallites is the same. So it gets the character of a upper-bound theory. Nevertheless, also the elastic-plastic transition can be modeled (Besdo and Wellerdick-Wojtasik [4]). Results being comparable with the results of Fig. 2, are presented in Fig. 3. In each of the cases the experimental procedure to find the yield loci by a $\Delta\varepsilon$-measure was simulated, too, with two numbers of $\Delta\varepsilon$, also two numbers of crystallites were used once following Müller [6] or the other applying a purely random procedure (see next chapter).

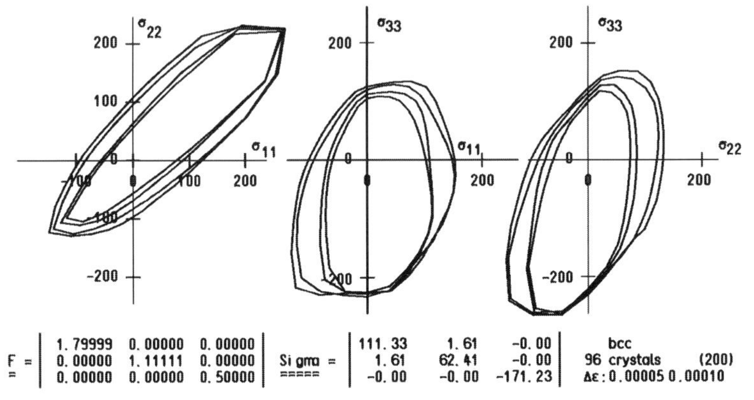

Figure 3. Yield loci as results of simulations, applications of the Taylor-Lin-theory.

4. A representative volume model

Theories with phenomenological laws do usually not take into account more than one rotation. Hence, they cannot recognize what happens really: Each of the grains has its own lattice rotation. The rotation which is looked upon by a Cosserat-model or by the ideas behind the plastic spin concept can optimally be their mean value. The one or the other lattice rotates also after very large deformations.

This effect is better described by Taylor-Lin- or also by Sachs-models. But the very strong restrictions of their assumptions make them also not natural enough. The next stage is the introduction of a representative volume element. Often the model gets a non-physical, but purely numerical anisotropy due to the shape of the single grains or to a non-appropriately chosen distribution of the starting orientations.

4.1. BASIC IDEAS

The basic ideas which should be considered in detail when a representative volume element (RVE) is constructed, are:

- *Isotropy is not "natural".*
  The isotropy which might be regarded in the virgin state, results from the ideal mixture of the orientations.
- *Regular crystallite structures are not natural, too.*
  Starting from any crystallization center crystallites grow very irregularly until other crystallites stop their growth. It may be allowed to assume that boundaries between grains are piecewise plane. A fundamental mistake which is often made, is a subdivision by orthogonal cuts.
- *Crystal lattices are responsible for material symmetries.*
  The mostly cubic or hexagonal structure of the lattices dominates the material behavior. This is valid in the elastic domain as well as in the plastic one. Whereas the linear-elastic behavior yields an over-all-isotropy for a relatively small number of properly chosen orientations already, this cannot be stated for the plastic part.
- *The ODF (Orientation distribution function) of a virgin state is responsible the over-all isotropy in this state.*
  If RVEs are to be used the virgin distribution of the discretized orientations must be looked upon carefully.
- *Pictures of crystals are nothing but plane cuts.*
  By this reason it is impossible to construct the spatial form of crystallite assemblies from figures due to microscopies.

4.2. CONSTRUCTION OF THE RVE

A method proposed already in the literature, but modified slightly here, may be described as a stepwise construction of a grain assembly:

1. step: A cubic domain which at the end will be the RVE is first "filled" by n starting points for the mathematical growth of the crystallites. In order to avoid irregular structures, new points being too near to already existing ones can be with-drawn. A "distance-factor" was introduced.

2. step: The cubic domain is looked upon as a RVE. Hence, its neighbors must have the same structure. Here this means: The 26 neighboring cubes are filled with n starting points in an identical manner. For a plane problem only 8 neighbors exist.

3. step: The grains are treated as growing from the centers with constant speed in every direction until it gets into conflict with another grain. As a result, plane boundaries come out. The limit can be due to another point belonging to the RVE, but also one belonging to its 26 (plane case: 8 neighbors). The total volume of the n crystallites at the end of this step is exactly that of the RVE, but the surface of this body is very irregular (see for a plane case Fig. 4).

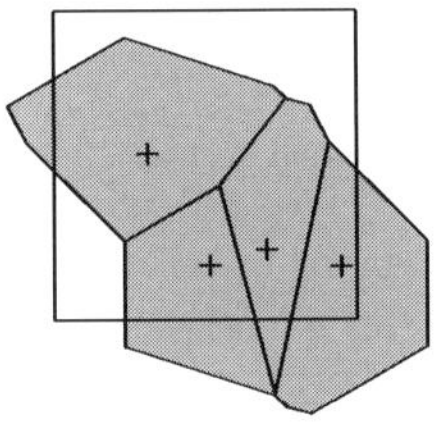

Figure 4. Plane "RVE" with 4 crystallites which started their "growth" inside the RVE.

4. step: For computational applications the representative volume must be constructed under the assumption that the neighboring RVE are subdivided in identical manner. In the case of the plane problem of Fig. 4, in total four neighboring RVEs are needed to fill the original RVE completely. At the end cuts are necessary to create the final RVE. This can be seen in Fig. 5 for the plane case with 4 starting points and in Fig. 6 for the spatial case consisting 384 grains. In Fig. 5, those grains with starting-points inside the "RVE" are visualized by double boundary lines.

4.3. DISTRIBUTION OF ORIENTATIONS

The adequate discretization of the lattice orientations in the virgin state is most important for future computations. In the present case the method uses arbitrary. The four quaternions $q_\alpha \rightarrow q_0, q_1, q_2, q_3$ which are not really independent, since they have to satisfy:

$$\alpha \equiv q_0^2 + q_1^2 + q_2^2 + q_3^2 = 1\,. \tag{33}$$

Random numbers in a domain of $-1 \leq q_\alpha \leq +1$ can be chosen, but finally they have to be projected to the shell of the hypersphere satisfying eq. (33). To

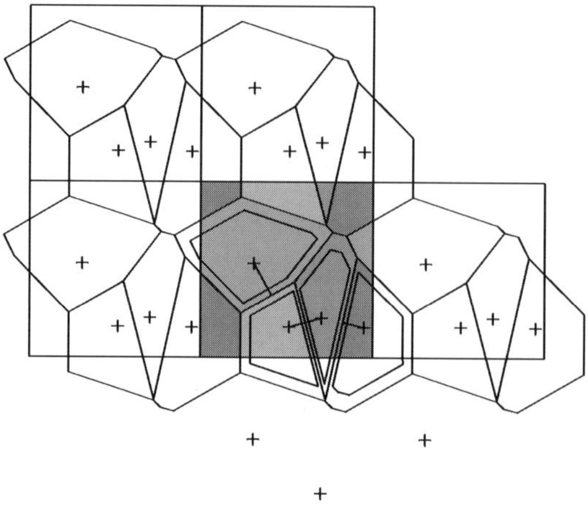

Figure 5. Plane "RVE": Final version.

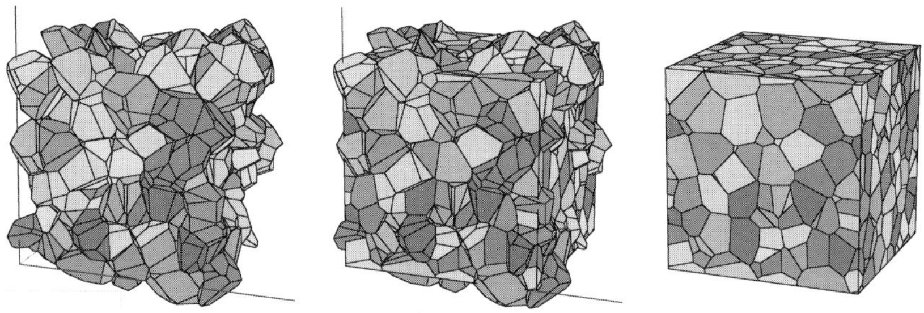

Figure 6. Construction of a spatial RVE with 384 crystallites (17778 corners, 31020 tetraeders).

avoid every anisotropy combinations of numbers which satisfy $\alpha > 1$ must be with-drawn. The transformation matrix $[e_{Ki}]$ is determined by the q_α by

$$[e_{Ki}] = \begin{bmatrix} 2\,(q_0^2 + q_1^2) - 1 & 2\,(q_1\,q_2 + q_0\,q_3) & 2\,(q_1\,q_3 - q_0\,q_2) \\ 2\,(q_1\,q_2 - q_0\,q_3) & 2\,(q_0^2 + q_2^2) - 1 & 2\,(q_2\,q_3 + q_0\,q_2) \\ 2\,(q_1\,q_2 + q_0\,q_3) & 2\,(q_2\,q_3 - q_0\,q_2) & 2\,(q_0^2 + q_3^2) - 1 \end{bmatrix} \tag{34}$$

4.4. Numerical application and results

Every grain or – if it has been cut – grain part is subdivided into tetraeder elements using boundary plane corners, boundary plane centers and (sub-)centers. The spline functions are quadratic. Since the number of variables is very high, an iterative solving algorithm is chosen. Periodic boundary conditions are applied where exact periodicity is assumed, superimposed by homogeneous strains. The linear elastic response on stresses $\sigma_{xy} = 1$ was tested. For correct isotropy the results should satisfy

$$\varepsilon_{xy} = \frac{1+\nu}{E}\sigma_{xy} : \quad \varepsilon_{xx} = \frac{1}{E}\sigma_{xx} ; \quad \varepsilon_{yy} = \varepsilon_{zz} = \frac{\nu}{E}\sigma_{xx} ; \tag{35}$$

$$\varepsilon_{xx} - \varepsilon_{yy} = \varepsilon_{xy} ; \quad \varepsilon_{yy} = \varepsilon_{zz} , \quad \text{and} \quad \varepsilon_{xz} = \varepsilon_{yz} = 0 . \tag{36}$$

The results of the computations are presented in Table 1 which shows for each of the examples:

The number n of the crystallites within the RVE, the distance-factor f_d, the strains ε_{xx}, ε_{yy}, and ε_{zz}, the mean-value $\overline{\varepsilon}_q$ of ε_{yy} and ε_{zz}, $\Delta\varepsilon_q$ being the difference of ε_{yy} and ε_{zz} which should vanish, the difference $\varepsilon_{xx} - \overline{\varepsilon}_q$ which should be equal to ε_{xy}, the strain ε_{xy} itself, the strains ε_{yz} and ε_{zx} which should vanish.

In Table 1, four results for 1000 crystallites with distance factors 0.03 and 1.0 and for 100 and 10 crystallites with factor 1 are presented first. They are computed for an artificial very strongly anisotropic medium.

Additionally three results for 200, 20, and 3 crystallites (factor 1) and α-Fe are shown. No excellent isotropic results are reached. The best ones belong to 1000 or 200 crystallites, respectively. Nevertheless, the differences between $\varepsilon_{xx} - \overline{\varepsilon}_q$ and ε_{xy} are relatively high also in these cases.

Table 1. Average strains of linear elastic simulations for $\sigma_{xx} = \sigma_{xy}$ (every value ε is multiplied by 10^7).

n	f_d	ε_{xx}	ε_{yy}	ε_{zz}	$\overline{\varepsilon}_q$	$\Delta\varepsilon_q$	$\varepsilon_{xx} - \overline{\varepsilon}_q$	ε_{xy}	ε_{yz}	ε_{zx}
1000	0.03	1659	-710	-671	-690	20	2349	2279	14	59
1000	1.00	1688	-746	-665	-710	36	2398	2318	30	20
100	1.00	1817	-794	-745	-770	25	2578	2553	-52	-76
10	1.00	917	-390	-248	-319	71	1236	2672	86	690
200	1.00	447	-118	-129	-128	10	675	587	1	2
20	1.00	516	-167	-149	-158	9	674	572	25	31
3	1.00	622	-215	-207	-211	4	833	621	86	690

These tests were necessary to ensure that the discretization of the virgin state might be an isotropic one.

The last figure shows overstretched results for pure tensile stresses in x-direction in the range of less than 1% of plastic tension. The amount of the difference between ε_{yy} and ε_{zz} is very high. But only 50 crystallites are taken into account. Also no large strains are demonstrated because their computation needs much effort.

The numbers at the right side of this picture are: NK crystallites, NP points, NL lines, NT tetraeders, and NF degrees of freedom.

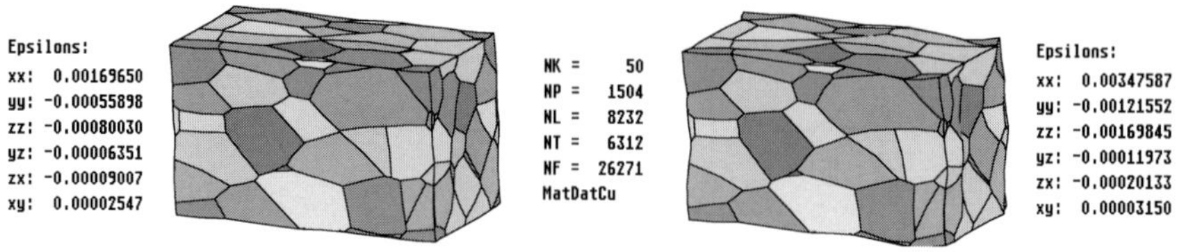

Figure 7. Results in the plastic range for 50 crystallites, deformations overestimated.

5. Conclusions

- Very quick and variable descriptions of the anisotropic behavior of polycrystals can be reached using phenomenological laws – especially in strain-space denotations. Their only disadvantage is that interpretations of the parameters are almost impossible. Hence, lots of experiments and expensive fitting procedures are necessary.
- Better agreement with the physical background is given if Taylor-Lin- or Sachs-models are applied. But even then the shape of the grains is not considered.
- Best results are possible, but with much effort, if a representative volume element (RVE) is used. During its construction a purely numerical anisotropy can occur and must be avoided. Mistakes can happen by unsatisfactory shapes of the grains (do not use cubes!) and by non-appropriate discretizations of the virgin orientation distribution. A model avoiding both of these effects is presented.

References

[1] Besdo, D. (1971). Zur anisotropen Verfestigung anfangs isotroper starrplastischer Medien. *Z. Angew. Math. Mech.*, **51**, T97–T98.

[2] Besdo, D. (1991). Constitutive laws for metal forming purposes in stress- and in strain-space representation. *VDI-Berichte*, **894**, 1–15.

[3] Besdo, D. (1998) A general material law of plasticity and its numerical application. *Steel research*, **69**, 188–193.

[4] Besdo, D. and Wellerdick-Wojtasik, N. (2001). The influence of large torsional prestrains on the texture development and yield surface of polycrystals. In: *Deutsche Forschungsgemeinschaft (DFG): Plasticity of Metals: Experiments, Models, Computation*. Wiley-VCH, 131–148.

[5] Lehmann, Th. (1972). Einige Bemerkungen zu einer allgemeinen Klasse von Stoffgesetzen für große elasto-plastische Formänderungen. *Ing.-Archiv*, **41**, 297–310.

[6] Müller, M. (1993). Plastische Anisotropie polykristalliner Materialien als Folge der Texturentwicklung. *VDI-Fortschr.-Ber.*, Reihe 18, Bd. 137, VDI-Verlag, Düsseldorf.

A Growth Law for Hooke's Tensor

T. Böhlke, A. Bertram
Institute of Mechanics, Otto-von-Guericke-University of Magdeburg
Postfach 4120, 39016 Magdeburg, Germany
boehlke@mb.uni-magdeburg.de, bertram@mb.uni-magdeburg.de

Abstract: The main purpose of the work is to formulate a phenomenological model for the evolution of the elastic properties of fcc polycrystals due to an evolving texture. The phenomenological model allows the determination of the current elasticity tensor as a functional of the deformation process. The transition from an elastically isotropic initial state to a path-dependent final anisotropic state is analyzed for polycrystalline copper. The predictions of the model are compared with micro-macro simulations based on a Taylor type model.

Keywords: Anisotropic material, constitutive behavior, polycrystalline material.

1. Introduction

The evolution of the elastic anisotropy of textured polycrystalline materials is of interest for rather different applications. Examples are the spring-back analysis for metal forming processes and the investigation of wave propagation phenomena in geological materials. The deformation induced mechanical anisotropy of polycrystalline metals is caused by crystallographic or morphological textures, i.e. non-uniform orientation distributions of crystals, grain boundaries, or grain edges (Bunge [10], Kocks et al. [12]). In the present paper we consider the anisotropy of the elastic behavior which is assumed to be affected only by the crystallographic texture development.

The main purpose of the work is to formulate a phenomenological model for the evolution of the elastic properties of aggregates of cubic crystals due to an evolving texture at room temperature. As a special case, deformations typical for metal forming operations are investigated which can be described by a symmetric velocity gradient. Texture simulations with initially isotropic polycrystals indicate that symmetric velocity gradients do not induce rotations of the macroscopic symmetry axes. As a result the plastic spin vanishes and the flow rule is symmetric (Böhlke [2]). This fact implies a much simpler set of constitutive equations, which shall be considered here. The material under consideration is copper. Hardening and damage phenomena are neglected presently for the sake of simplicity.

In the present paper the evolution equation for the texture-dependent anisotropic part of the effective elasticity tensor is extended compared to the model pre-

C. Miehe (ed.), IUTAM Symposium on Computational Mechanics of Solid Materials at Large Strains, 291-300.

sented in Böhlke and Bertram [5]. The evolution equation now consists of two parts. The first term depends only on inelastic rate of deformation and represents the driving term for the evolving anisotropy. The second term governs the saturation of anisotropy and is linear in the anisotropic portion of the effective elasticity tensor. In Böhlke and Bertram [5] the saturation of the elastic anisotropy has been modeled by a scalar function. The main advantage of the present approach is that asymptotic values for the elasticity tensors can be analytically determined for axisymmetric deformations (Böhlke and Bertram [7]), and furthermore, that axial effects observed in torsion experiments can be described by the presented model (Böhlke et al. [6]).

Notation. Throughout the text a direct tensor notation is preferred. A linear mapping of a 2nd-order tensor is written as $\mathbf{A} = \mathbb{C}[\mathbf{B}]$. The scalar product, the dyadic product, and the Frobenius norm are denoted by $\mathbf{A} \cdot \mathbf{B}$, $\mathbf{A} \otimes \mathbf{B}$, and $\|\mathbf{A}\| = (\mathbf{A} \cdot \mathbf{A})^{1/2}$, respectively. Irreducible, i.e., completely symmetric and traceless tensors of any order are designated by a prime, e.g., $\mathbf{A}'$ and $\mathbb{C}'$. $\mathbf{I}$ and $\mathbb{I}$ denote the 2nd-order identity and the identity on symmetric 2nd-order tensors, respectively. The Rayleigh product of a 2nd-order tensor $\mathbf{A}$ and a 4th-order tensor $\mathbb{C} = C_{ijkl}\mathbf{e}_i \otimes \mathbf{e}_j \otimes \mathbf{e}_k \otimes \mathbf{e}_l$ is defined by

$$\mathbf{A} \star \mathbb{C} = C_{ijkl}(\mathbf{A}\mathbf{e}_i) \otimes (\mathbf{A}\mathbf{e}_j) \otimes (\mathbf{A}\mathbf{e}_k) \otimes (\mathbf{A}\mathbf{e}_l), \tag{1}$$

where $\{\mathbf{e}_i\}$ is an arbitrary but fixed orthonormal basis. A tilde indicates that a quantity is formulated with respect to the undistorted configuration.

2. Description of the model

2.1. Elastic law

The stress-strain relation applied here is equivalent to the standard formulation of the theory of multiplicative elasto-plasticity based on a linear elastic law of the St.Venant-Kirchhoff type. The Kirchhoff stress tensor $\boldsymbol{\tau}$ is given as a linear map of the elastic Almansi strain tensor

$$\boldsymbol{\tau} = \mathbb{C}_e[\mathbf{E}_e^A], \qquad \mathbf{E}_e^A = \frac{1}{2}(\mathbf{I} - \mathbf{F}_e^{-\mathsf{T}}\mathbf{F}_e^{-1}), \qquad \mathbf{F}_e = \mathbf{F}\mathbf{F}_p^{-1} \tag{2}$$

(see, e.g., Böhlke and Bertram [5]). The Kirchhoff stress $\boldsymbol{\tau}$ is defined by the Cauchy stress $\boldsymbol{\sigma}$ and the determinant J of $\mathbf{F}$ through $\boldsymbol{\tau} = J\boldsymbol{\sigma}$. The tensor $\mathbf{F}_e$ represents the elastic portion of the deformation gradient $\mathbf{F} = \mathbf{F}_e\mathbf{F}_p$. The multiplicative decomposition of the deformation gradient into an elastic and an inelastic portion can be consistently derived by the concept of isomorphic elastic ranges (Bertram [1]). The stiffness operator $\mathbb{C}_e$ is given by the Rayleigh product of $\mathbf{F}_e$ and the reference stiffness tensor $\tilde{\mathbb{C}}$ with respect to the undistorted configuration

$$\mathbb{C}_e = \mathbf{F}_e \star \tilde{\mathbb{C}} = \tilde{C}_{ijkl}(\mathbf{F}_e\mathbf{e}_i) \otimes (\mathbf{F}_e\mathbf{e}_j) \otimes (\mathbf{F}_e\mathbf{e}_k) \otimes (\mathbf{F}_e\mathbf{e}_l). \quad (3)$$

Within the standard formulation of the theory of elasto-plasticity, i.e. without induced elastic anisotropy, $\tilde{\mathbb{C}}$ is assumed to be constant. The tensor $\mathbb{C}_e$ is generally non-constant even in the case of $\tilde{\mathbb{C}}$ being constant and isotropic.

Volume averages of stiffnesses or compliances with cubic symmetry generally allow for the following unique decomposition into three different parts, as e.g., in terms of stiffnesses

$$\tilde{\mathbb{C}} = 3K\mathbb{P}_1^I + 2G\mathbb{P}_2^I + \zeta\tilde{\mathbb{A}}', \qquad \mathbb{P}_1^I = \tfrac{1}{3}\mathbf{I} \otimes \mathbf{I}, \qquad \mathbb{P}_2^I = \mathbb{I} - \mathbb{P}_1^I. \quad (4)$$

This result holds, if the arithmetic, the harmonic mean, or the geometric mean of stiffness tensors is determined (Böhlke and Bertram [4, 5]). K, G, and ζ depend on the type of averaging whereas $\tilde{\mathbb{A}}'$ does not. The two 4th-order projectors $\mathbb{P}_1^I$ and $\mathbb{P}_2^I$ represent the isotropic part of the elastic law, which is independent of the crystallographic texture in the aggregate. K is the bulk modulus and G is the shear modulus. The tensor $\tilde{\mathbb{A}}'$ represents the purely anisotropic part of the stiffness tensor $\tilde{\mathbb{C}}$. For a polycrystal the norm $\|\tilde{\mathbb{A}}'\|$ is in the range $[0, 1)$. If there is no crystallographic texture in the aggregate, then $\tilde{\mathbb{A}}' = \mathbb{O}$ holds and $\tilde{\mathbb{C}}$ is isotropic. For a single crystal the norm $\|\tilde{\mathbb{A}}'\|$ is equal to one. Hence the scalar $\|\tilde{\mathbb{A}}'\|$ quantifies the total amount of elastic anisotropy. Furthermore $\tilde{\mathbb{A}}'$ is symmetric and traceless with respect to all pairs of indices. Because of the its irreducibility, $\tilde{\mathbb{A}}'$ contains only nine independent components, and this property leads to a rigorous simplification of the evolution equation of $\tilde{\mathbb{C}}$.

2.2. FLOW RULE

As mentioned before, we consider only symmetric velocity gradients $\mathbf{L}$. The texture simulations of coaxial deformation paths indicate that there are no rotations of the symmetry axes of the effective stiffness tensor (Böhlke [2]). The mean rotation of the symmetry axes is given by the orthogonal part $\mathbf{R}_e$ of the polar decomposition of $\mathbf{F}_e = \mathbf{V}_e\mathbf{R}_e$. Hence, $\mathbf{R}_e \approx \mathbf{I}$. Furthermore, in the case of small elastic strains, $\mathbf{V}_e \approx \mathbf{I}$ holds. Based on the general identity for the velocity gradient

$$\mathbf{L} = \mathbf{D} + \mathbf{W} = \dot{\mathbf{F}}_e\mathbf{F}_e^{-1} + \mathbf{F}_e\dot{\mathbf{F}}_p\mathbf{F}_p^{-1}\mathbf{F}_e^{-1}, \qquad \dot{\mathbf{F}}_p\mathbf{F}_p^{-1} = \tilde{\mathbf{D}}_p + \tilde{\mathbf{W}}_p, \quad (5)$$

one concludes with $\mathbf{W} = \mathbf{0}$ and $\mathbf{F}_e \approx \mathbf{I}$ that the plastic spin $\mathbf{W}_p \approx \mathbf{0}$ vanishes. Therefore, the flow rule can be assumed to be symmetric, if symmetric velocity gradients are considered and the symmetry axes are stationary.

The viscoplastic flow is assumed to be driven by an overstress and the direction of the flow is specified by the anisotropic portion of the effective elasticity tensor

$$\dot{\mathbf{F}}_p \mathbf{F}_p^{-1} = \frac{\dot{\gamma}_0}{\|\tilde{\mathbf{S}}'\|_a} \left\langle \frac{\sqrt{\frac{3}{2}}\|\tilde{\mathbf{S}}'\|_a - \sigma_F}{\sigma_D} \right\rangle^{\frac{1}{m}} \tilde{\mathbb{N}}(\eta, \tilde{\mathbb{A}}')[\tilde{\mathbf{S}}'], \tag{6}$$

$$\tilde{\mathbb{N}}(\eta, \tilde{\mathbb{A}}') = \mathbb{P}_2^I + \eta \tilde{\mathbb{A}}', \qquad \|\tilde{\mathbf{S}}'\|_a = \sqrt{\tilde{\mathbf{S}}' \cdot \tilde{\mathbb{N}}(\eta, \tilde{\mathbb{A}}')[\tilde{\mathbf{S}}']}. \tag{7}$$

The MacCauley bracket $\langle x \rangle$ is defined by $(x + |\, x\,|)/2$. The flow rule is rate-dependent and this rate-dependence is specified by the scalar m. The scalars σ_F and σ_D represent a critical flow stress and a drag stress, respectively. The scalar $\dot{\gamma}_0$ is a referential inelastic rate. With $\|\tilde{\mathbf{S}}'\|_a$ we denote an anisotropic norm of the deviator $\tilde{\mathbf{S}}'$. The anisotropic part of the norm is governed by the scalar η and the 4th-order tensor $\tilde{\mathbb{A}}'$, respectively. This implies that the anisotropy of the plastic flow is affected by the elastic anisotropy. This coupling is substantiated by experiments and theoretical considerations (Man [13, 14]). Keeping in mind that $\mathbb{P}_2^I$ is the identity on symmetric and traceless 2nd-order tensors, then it is not hard to see, that for the elastically isotropic case a rate-dependent J_2-theory is obtained.

2.3. Growth law for Hooke's tensor

For the evolution of $\tilde{\mathbb{A}}'$ the following ansatz is applied

$$\dot{\tilde{\mathbb{A}}}' = \|\tilde{\mathbf{D}}_p'\| \left(\tilde{\mathbb{G}}'(\tilde{\mathbf{N}}_p') - d(III_p)\tilde{\mathbb{A}}' \right), \qquad III_p = \det(\tilde{\mathbf{N}}_p'). \tag{8}$$

The driving term $\tilde{\mathbb{G}}'(\tilde{\mathbf{N}}_p')$ depends only on the direction $\tilde{\mathbf{N}}_p'$ of the inelastic rate of deformation $\tilde{\mathbf{D}}_p'$, which is the symmetric part of $\dot{\mathbf{F}}_p \mathbf{F}_p^{-1}$. The ansatz (8) takes into account that for constant strain rates the crystallographic texture tends to saturate for large deformations. The saturation behavior is controlled by the term $d(III_p)\tilde{\mathbb{A}}'$. The function d depends on the only non-constant principal invariant of $\tilde{\mathbf{N}}_p'$, i.e., its determinant.

The general form of the function $\tilde{\mathbb{G}}'$ can be determined by the theory of isotropic tensor functions (Böhlke and Bertram [5]). Due to the constraint that, similar to $\tilde{\mathbb{A}}'$, the function $\tilde{\mathbb{G}}'$ has to be symmetric and traceless with respect to all pairs of indices, the function $\tilde{\mathbb{G}}'$ is given by three 4th-order tensor generators $\tilde{\mathbb{G}}_\alpha'$ $(\alpha = 1, 2, 3)$ and corresponding scalar functions G_α

$$\tilde{\mathbb{G}}'(\tilde{\mathbf{N}}_p') = G_1(III_p)\tilde{\mathbb{G}}_1'(\tilde{\mathbf{N}}_p') + G_2(III_p)\tilde{\mathbb{G}}_2'(\tilde{\mathbf{N}}_p') + G_3(III_p)\tilde{\mathbb{G}}_3'(\tilde{\mathbf{N}}_p'). \tag{9}$$

The tensor generators are given by

$$\begin{aligned} \mathbb{G}_1'(\mathbf{A}) &= \langle \mathbf{A} \otimes \mathbf{A} \rangle - \tfrac{2}{7}\left(\mathrm{tr}(\mathbf{A})\langle \mathbf{A} \otimes \mathbf{I}\rangle + 2\langle \mathbf{A}^2 \otimes \mathbf{I}\rangle\right) \\ &+ \tfrac{1}{35}\left(\mathrm{tr}(\mathbf{A})^2 + 2\mathrm{tr}(\mathbf{A}^2)\right)\langle \mathbf{I} \otimes \mathbf{I}\rangle, \end{aligned} \tag{10}$$

$$\mathbb{G}_2'(\mathbf{A}) = \mathbb{G}_1'(\mathbf{A}^2), \tag{11}$$

$$\begin{aligned} \mathbb{G}_3'(\mathbf{A}) &= \langle \mathbf{A}^2 \otimes \mathbf{A} \rangle - \tfrac{2}{7}\left(\mathrm{tr}(\mathbf{A}^2)\langle \mathbf{A} \otimes \mathbf{I} \rangle + \mathrm{tr}(\mathbf{A})\langle \mathbf{A}^2 \otimes \mathbf{I} \rangle \right. \\ &+ \left. 4\langle \mathbf{A}^3 \otimes \mathbf{I} \rangle\right) + \tfrac{2}{35}\left(\mathrm{tr}(\mathbf{A})\mathrm{tr}(\mathbf{A}^2) + 2\mathrm{tr}(\mathbf{A}^3)\right)\langle \mathbf{I} \otimes \mathbf{I} \rangle. \end{aligned} \tag{12}$$

The bracket formula $\langle\cdot\rangle$ is defined by $(\mathbf{A}, \mathbf{B} \in Sym)$

$$\langle A_{ij} B_{kl} \rangle = \frac{1}{6}\left(A_{ij}B_{kl} + A_{ik}B_{jl} + A_{il}B_{kj} + B_{ij}A_{kl} + B_{ik}A_{jl} + B_{il}A_{kj}\right). \tag{13}$$

3. Numerical results

3.1. TEXTURE SIMULATION

In this section we compare the predictions of the phenomenological model with texture simulations based on the Taylor-Lin model, which treats compatibility constraints as paramount. In contrast to Taylor's original rigid-plastic approach, the Taylor-Lin model allows for elastic deformations of the crystal lattices. Bronkhorst et al. [9] have shown that the Taylor-Lin model is a reasonably good first order approximation for the texture formation in single-phase materials. This statement is made with respect to both the stress-strain response as well as the texture formation.

Table 1. Material functions on the macroscale (I).

Elastic law	$\boldsymbol{\tau}_e = \mathbb{C}_e[\mathbf{E}_e^A] \qquad \mathbb{C}_e = \mathbf{F}_e \star \tilde{\mathbb{C}}$	
Stiffness tensor	$\tilde{\mathbb{C}} = 3K\mathbb{P}_1^I + 2G\mathbb{P}_2^I + \zeta\tilde{\mathbb{A}}'$	
Properties	$\tilde{A}'_{ijkl} = \tilde{A}'_{jikl} = \tilde{A}'_{klij} = \tilde{A}'_{kjil} = \ldots$	$\tilde{A}'_{iikl} = 0$
Flow rule	$\tilde{\mathbf{D}}'_p = \mathrm{sym}(\dot{\mathbf{F}}_p\mathbf{F}_p^{-1}) = \frac{\dot{\gamma}_0}{\Vert\tilde{\mathbf{S}}'\Vert_a}\langle\frac{\sqrt{\frac{3}{2}}\Vert\tilde{\mathbf{S}}'\Vert_a - \sigma_F}{\sigma_D}\rangle^{\frac{1}{m}}\tilde{\mathbb{N}}[\tilde{\mathbf{S}}']$	
Inelast. compl.	$\tilde{\mathbb{N}} = \mathbb{P}_2^I + \eta\tilde{\mathbb{A}}'$	
Anisotr. norm	$\Vert\tilde{\mathbf{S}}'\Vert_a = \sqrt{\tilde{\mathbf{S}}' \cdot \tilde{\mathbb{N}}[\tilde{\mathbf{S}}']}$	
Growth law	$\dot{\tilde{\mathbb{A}}}' = \Vert\tilde{\mathbf{D}}'_p\Vert\left(\mathbb{G}'(\tilde{\mathbf{N}}'_p) - d(III_p)\tilde{\mathbb{A}}'\right)$	
	$\tilde{\mathbf{N}}'_p = \tilde{\mathbf{D}}'_p/\Vert\tilde{\mathbf{D}}'_p\Vert \qquad III_p = \det(\tilde{\mathbf{N}}'_p)$	
	$\mathbb{G}'(\tilde{\mathbf{N}}'_p) = \sum_{i=1}^3 G_i(III_p)\mathbb{G}'_i(\tilde{\mathbf{N}}'_p)$	

The elastic law has the same form as (2) with the only difference that the corresponding $\tilde{\mathbb{C}}$ of the single crystals is constant and of cubic symmetry. The following flow rule has been applied on the microscale

$$\dot{\mathbf{F}}_p\mathbf{F}_p^{-1} = \sum_{\alpha=1}^{12}\dot{\gamma}_\alpha\tilde{\mathbf{d}}_\alpha \otimes \tilde{\mathbf{n}}^\alpha, \qquad \dot{\gamma}_\alpha = \dot{\gamma}_0 \,\mathrm{sign}\,(\tau_\alpha)\left\langle\frac{|\tau_\alpha| - \tau^C}{\tau^D}\right\rangle^{\frac{1}{m}}, \tag{14}$$

where $\tilde{\mathbf{d}}_\alpha$ and $\tilde{\mathbf{n}}^\alpha$ are the octahedral slip system directions and normals, respectively. The variable τ^C is the critical resolved shear stress. τ^D is the drag stress. The material under consideration is OFHC copper. The elastic constants are given by $\tilde{C}_{1111} = 168$ GPa, $\tilde{C}_{1122} = 121$ GPa, $\tilde{C}_{1212} = 75$GPa (Kocks et al. [12]). The reference shear rate is taken as $\dot{\gamma}_0 = 1/1000\ \mathrm{s}^{-1}$ and the strain rate sensitivity parameter as $m = 0.012$. The critical resolved shear stress is $\tau^C = 12$ MPa, whereas the drag stress is equal to $\tau^D = 4$ MPa. The texture simulations discussed in subsequent paragraphs have been performed with 1000 filtered random orientations (Böhlke and Bertram [3]).

The stretching tensor, which is assumed to be traceless, can generally be decomposed as

$$\mathbf{D} = \|\mathbf{D}\|\mathbf{Q}_D\mathbf{N}_D\mathbf{Q}_D^{\mathsf{T}}, \tag{15}$$

where $\|\mathbf{D}\|$ is the magnitude of $\mathbf{D}$ and $\mathbf{Q}_D$ is an orthogonal tensor that rotates an arbitrary but fixed orthonormal reference basis $\{\mathbf{e}_i\}$ onto the eigenvectors of $\mathbf{D}$. These constraints, i.e., that $\mathbf{N}_D$ is traceless and has a magnitude equal to one, are identically fulfilled by the following parameterization of $\mathbf{N}_D$

$$\mathbf{N}_D = \sum_{\alpha=1}^{3} n_\alpha \mathbf{e}_\alpha \otimes \mathbf{e}_\alpha, \quad n_{1,3} = -\frac{\sqrt{6}}{6}\xi \pm \frac{\sqrt{2}}{2}\sqrt{1-\xi^2}, \quad n_2 = \frac{\sqrt{6}}{3}\xi, \tag{16}$$

$\xi \in [-1/2, +1/2]$. The ξ-values -0.5, 0, and $+0.5$ belong to the uniaxial tension, plane strain compression, and simple compression, respectively. For the subsequent considerations the amount of deformation is quantified by the v.Mises equivalent strain $\phi(t) = \sqrt{2/3}\int_0^t \|\mathbf{D}(\tilde{t})\|\, \mathrm{d}\tilde{t}$.

3.2. IDENTIFICATION OF MATERIAL FUNCTIONS AND PARAMETERS OF THE PHENOMENOLOGICAL MODEL

The phenomenological model is specified by the following set of material parameters and functions

$$K, G, \sigma_F, \sigma_D, \dot{\gamma}_0, m, \eta, G_1(III_p), G_2(III_p), G_3(III_p), d(III_p). \tag{17}$$

K and G are the bulk and shear modulus of the polycrystal without crystallographic texture. The flow stress σ_F and the drag stress σ_D can be identified by tensile tests, whereas the strain-rate sensitivity parameter m can be determined from tensile tests with different strain rates. It can be shown that the scalar parameter η governs the Lankford coefficient in the sheet plane of rolled materials (Böhlke [2]).

A difficult problem is the determination of the functions $G_i(III_p)$, which specify the rate of change of $\tilde{\mathbb{A}}'$ in terms of the direction of the inelastic strain rate. Here, texture simulations indicate the way. Inspection of simulations

Table 2. Material functions on the macroscale (II).

$G_1(III_p)$	$G_2(III_p)$	$G_4(III_p)$
$\frac{\sqrt{2}}{14}\left(1+\frac{497\sqrt{6}}{66}III_p\right)G(III_p)$	$\frac{17\sqrt{2}}{7}G(III_p)$	$\left(1-\left(\frac{629\sqrt{3}}{483}-4\right)\frac{3\sqrt{6}}{2}III_p\right)G(III_p)$

with polycrystalline copper shows that certain components of $\tilde{\mathbb{A}}'$ are approximately constant for certain deformation paths. In the case of a plane strain

Table 3. Material functions on the macroscale (III).

	$\xi=-0.50$	$\xi=-0.25$	$\xi=0.0$	$\xi=0.25$	$\xi=0.50$
$G(III_p)$	0.25	0.50	0.38	0.26	2.50
$d(III_p)$	2.35	2.05	0.44	0.04	2.00

compression deformation ($\xi = 0$) with $\mathbf{e}_1$ the rolling and $\mathbf{e}_3$ the normal direction, the following relations hold: $\tilde{A}'_{e2233} \approx const.$, $\tilde{A}'_{e3333} \approx const.$ In the case of a rate of deformation with $\xi = 0.25$ (see equation (16)), the following relations hold: $\tilde{A}'_{e1111} \approx const.$, $\tilde{A}'_{e3333} \approx const.$ Approximating the direction of the inelastic rate of deformation $\tilde{\mathbf{N}}'_p$ by the direction of the stretching tensor $\mathbf{N}'$, and assuming $G_1 = (m_1\xi + n_1)G_2$ and $G_3 = (m_2\xi + n_2)G_2$, the parameters m_1, m_2, n_1, and n_1 can be determined by the aforementioned constraints (Böhlke [2]). The solution is given in Table 2 where ξ has been

Table 4. Material parameters on the macroscale.

K	G	ζ	σ_D	$\dot{\gamma}_0$	m	η
136.93 GPa	54.56 GPa	114.14 GPa	46.5 MPa	0.001 s^{-1}	0.0125	0.999

eliminated by the relation $\xi(III_p) \approx -3\sqrt{6}III_p/2$. This ansatz approximates the exact relation $\xi(III_p) = -\cos\left(\arccos(3\sqrt{6}III_p)/3 + \pi/3\right)$, which is only weakly non-linear in the range $\xi \in [-1/2, 1/2]$. The linear approximation guarantees $\xi(\sqrt{6}/18) = -1/2$, $\xi(0) = 0$, and $\xi(-\sqrt{6}/18) = 1/2$, and, $III_p \in [-\sqrt{6}/18, \sqrt{6}/18]$. The two remaining scalar functions $G(III_p)$ and $d(III_p)$ can be estimated by texture simulations. The function $G(III_p)$ governs the initial rate of change of $\tilde{\mathbb{A}}'$, whereas $d(III_p)$ affects the saturation behavior. The thermodynamical consistency of the presented model can be shown (Böhlke [2]). The material functions and parameters which have been used for the macro simulation are summarized in Tables 1–4.

3.3. DISCUSSION OF COAXIAL DEFORMATION PATHS

Uniaxial tension ($\xi = -0.50$). The uniaxial tension test is the most reliable test for the determination of plastic properties. It is the preferred test for the determination of the yield strength and the initial strain hardening behavior. The disadvantage of the test is the formation of shear banding instabilities at large strains. Tension is an axisymmetric deformation. For initially non-textured aggregates, such a deformation induces an axisymmetric texture. Most of the grains are aligned such that either $\{111\}$ or $\{100\}$ planes become perpendicular to the loading axis.

In the case of uniaxial tension, the components of $\mathbb{A}'_e = \tilde{\mathbf{F}} \star \tilde{\mathbb{A}}'$ are approximately constant for equivalent strain values ϕ larger than 2. The maximum amount of anisotropy is given by $\|\mathbb{A}'_e\| \approx 0.2$. In Figure 1 the predictions of the phenomenological model and the Taylor-Lin model for the Eulerian stiffness $\mathbb{C}_e$ and different deformations are compared. In the case of the Taylor-Lin model, the stiffness $\mathbb{C}_e = \mathbf{F}_e \star \tilde{\mathbb{C}}$ is determined as a volume average of the Eulerian stiffness tensors of the single crystals forming the aggregate. In Figures 1 a,b,c, Young's modulus in different planes is plotted for the tension mode ($\mathbf{e}_1$ axial direction).

Plane strain compression ($\xi = 0.0$). The channel-die test is important for measuring the material response in monotonic deformations. Generally this test is better suited for obtaining plastic flow data rather than yield data. The textures produced by this deformation are generally more reliable than those determined from ill-controlled rolling experiments. Usually the copper orientation $\{112\}\langle 111\rangle$ and the brass orientation $\{011\}\langle 211\rangle$ are regarded as the dominating elements of so called "pure metal" and "alloy" textures. The experimental data show a peak near the $\{112\}\langle 111\rangle$ copper orientation. The prediction of the Taylor-Lin model shows a shift of this component towards the $\{4\ 4\ 11\}\langle 11\ 11\ 8\rangle$ Taylor orientation. Generally, the Taylor type models are not able to predict the $\{011\}\langle 211\rangle$ brass orientation. But the latter is not very pronounced in the case of pure copper.

The numerical results indicate that there is no saturation of the amount of elastic anisotropy even for equivalent strains of $\phi \approx 5$. The amount of anisotropy is then given by $\|\mathbb{A}'_e\| \approx 0.46$. This is the largest value for all $\xi \in [-0.5, +0.5]$. Therefore, the plane strain compression induces the strongest elastic anisotropy compared with all other rate of deformation tensors $\mathbf{D}$.

In Figures 1 d,e,f, Young's modulus in different planes is plotted for the plane strain compression mode ($\mathbf{e}_1$ rolling direction, $\mathbf{e}_3$ normal direction). The thickness reduction of 90 % corresponds to an equivalent strain of $\phi \approx 2.7$. In Figure 1 g,h,i, the same data are shown for the deformation with $\xi = 0.12$, which is in between plane-strain-compression and uniaxial compression.

Uniaxial compression ($\xi = +0.50$). This type of a deformation has generally two advantages compared with the uniaxial tension. First, the specimen shape provides more surface for texture measurements. Second, the specimen can be subjected to much higher strains. The main difficulty is the minimization of friction. Experimental textures indicate that in the simple bulk mode there is one dominant component of texture. The grains of the polycrystal are aligned such that $\{110\}$ planes become perpendicular to the loading axis. In Figures 1 j,k,l, Young's modulus in different planes is plotted (e_3 axial direction). It is seen that the phenomenological model is capable to predict the texture induced anisotropy as it is given by the Taylor-Lin model.

4. Summary

In this paper, a phenomenological model has been developed, which describes the evolution of the elastic properties of aggregates of cubic crystallites due to a crystallographic texture development. The transition from an elastically isotropic initial state to a path-dependent final anisotropic state is discussed for the case of polycrystalline copper. The comparison of the predictions to Taylor type texture simulations indicates, that the phenomenological model captures the gross features of the evolving elastic anisotropy.

References

[1] Bertram, A. (1992). Description of finite inelastic deformations. In: *Proceedings of MECAMAT'92, International Seminar on Multiaxial Plasticity*, Benallal, A., Billardon, R. and Marquis, D. (eds.), 821–835.

[2] Böhlke, T. (2001). *Crystallographic Texture Evolution and Elastic Anisotropy: Simulation, Modeling and Applications*, Shaker Verlag, Dissertation, Fakultät für Maschinenbau, Otto-von-Guericke-Universität Magdeburg.

[3] Böhlke, T. and Bertram, A. (1998). Simulation of texture development and induced anisotropy of polycrystals. In: *Proceedings of ICES'98, Modeling and Simulation Based Engineering*, Atluri, S. and O'Donoghue, P. (eds.), 1390–1395.

[4] Böhlke, T. and Bertram, A. (2000). A minimum problem defining effective isotropic elastic properties. *Z. angew. Math. Mech.*, **80(S2)**, 419–420.

[5] Böhlke, T. and Bertram, A. (2000). The evolution of Hooke's law due to texture development in polycrystals. *Int. J. Sol. Struct.*, **38(52)**, 9437–9459.

[6] Böhlke, T., Bertram, A. and Krempl, E. (2002). Modeling nonlinear effects in free-end torsion. To appear in: *Int. J. Plast.*

[7] Böhlke, T. and Bertram, A. (2001). Asymptotic values of elastic anisotropy in uniaxial tension and compression. *Comp. Mat. Sc.* Submitted.

[8] Böhlke, T. and Brüggemann, C. (2001). Graphical representation of the generalized Hooke's law. *Technische Mechanik*, **21(2)**, 145–158.

[9] Bronkhorst, C., Kalidindi, S. and Anand, L. (1992). Polycrystalline plasticity and the evolution of crystallographic texture in fcc metals. *R. Soc. Lond.*, **A341**, 443–477.

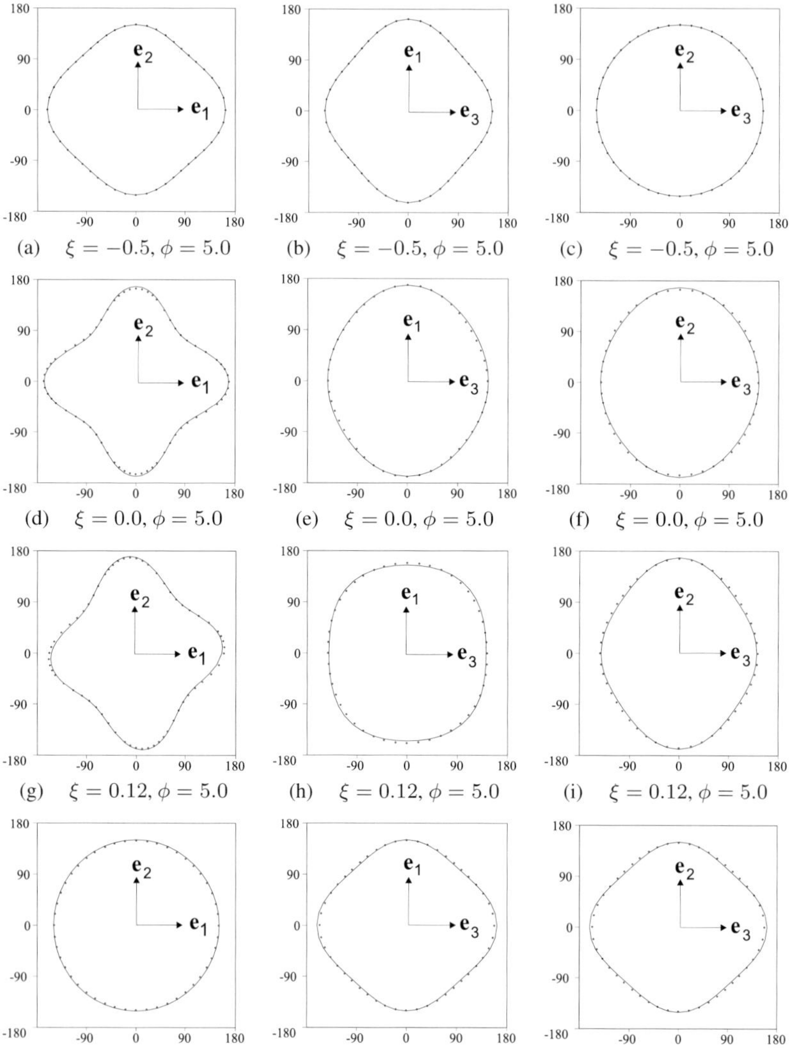

(a) $\xi = -0.5, \phi = 5.0$ (b) $\xi = -0.5, \phi = 5.0$ (c) $\xi = -0.5, \phi = 5.0$

(d) $\xi = 0.0, \phi = 5.0$ (e) $\xi = 0.0, \phi = 5.0$ (f) $\xi = 0.0, \phi = 5.0$

(g) $\xi = 0.12, \phi = 5.0$ (h) $\xi = 0.12, \phi = 5.0$ (i) $\xi = 0.12, \phi = 5.0$

(j) $\xi = 0.5, \phi = 5.0$ (k) $\xi = 0.5, \phi = 5.0$ (l) $\xi = 0.5, \phi = 5.0$

Figure 1. Young's modulus $E(\mathbf{n})$ (GPa) for different deformation modes: (–) phenomenological model, (**.**) Taylor-Lin model.

[10] Bunge, H.-J. (1993). *Texture Analysis in Material Science*, Göttingen.

[11] Federov, F. (1968). *Theory of Elastic Waves in Crystals*, Plenum Press, New York.

[12] Kocks, U., Tome, C. and Wenk, H. (1998). *Texture and Anisotropy: Preferred Orientations in Polycrystals and Their Effect on Materials Properties*, Cambridge

[13] Man, C.-S. (1995). On the correlation of elastic and plastic anisotropy in sheet metals. *J. Elast.*, **39**, 165–173.

[14] Man, C.-S. (1998). On the constitutive equations of some weakly-textured materials. *Arch. Rat. Mech. Anal.*, **143**, 77–103.

An "Affine" Micromechanical Approach for the Prediction of the Elastoplastic Behavior of Polycrystals at Finite Strain

F. Auslender, M. Bornert, A. Zaoui
Laboratoire de Mécanique des Solides, CNRS-Ecole Polytechnique
91128 Palaiseau, France
zaoui@lms.polytechnique.fr

T. Hoc
Laboratoire de Mécanique des Sols, Structures et Matériaux, Ecole Centrale de Paris
Grande Voie des Vignes, 92295 Chatenay-Malabry, France

R. Masson
EDF Research and Development
Les Renardieres, 77818 Moret sur Loing, France

Abstract: The so-called "affine" formulation, already proposed at small strain, has been proved to improve significantly on former treatments such as Hill's incremental one. This paper reports on an extension of this treatment at finite strain. A linearization procedure, consistent with the one used for small strains, is proposed. This extension has to solve general problems related to the application of classical homogenization schemes to finite strain situations, as well as new ones which are specific to the affine formulation. Preliminary results concerning the tensile stress-strain response of FCC polycrystals are favorably compared with those derived from Hill's approach.

Keywords: Polycrystals, elastoplasticity, finite strain, homogenization.

1. Introduction

Until a very recent past, Hill's incremental method was unanimously considered the most appropriate micromechanical treatment to predict the elastoplastic behavior of polycrystals. Recently, Gilormini [2] proved that this approach, applied to viscoplastic two-phase materials, can lead to an upper bound violation. This shows that the incremental treatment must be somewhat "softened" in order to yield more realistic predictions. This conclusion has motivated the conception of a new approach, the so-called "affine formulation", proposed by Masson et al. [9]. Initially defined, in the context of infinitesimal strain, for nonlinear elasticity or viscosity, it has been extended to history-dependent constitutive behavior such as rate-independent or rate-dependent elastoplasticity.

C. Miehe (ed.), IUTAM Symposium on Computational Mechanics of Solid Materials at Large Strains, 301-310.

This paper aims at extending the affine elastoplastic approach to finite strain. In the first section, we briefly recall the small strain affine formulation for nonlinear elasticity and elastoplasticity. The second section is devoted to the finite strain extension: we first focus on the choice of the appropriate (lagrangian, eulerian...) description framework; we then address the question of the single crystal constitutive equations and the specific problems of the affine approach when finite deformations are considered and when the self-consistent scheme is adopted. Preliminary predictions of the overall response of polycrystals, compared with those derived from Hill's approach, are reported in the last section.

Here, a polycrystal is considered as an aggregate of perfectly bound and randomly distributed crystals which only differ by their lattice orientation: all the grains which have the same crystallographic orientation define a "phase" of this multiphase material. Plastic deformation is only due to glide on slip systems, the system α being defined by the unit normal to its slip plane $\boldsymbol{n}_\alpha$ and its unit slip direction $\boldsymbol{m}_\alpha$.

2. Affine formulation for small strains

2.1. Nonlinear elasticity or viscoplasticity

In this section, the crystal constitutive behavior is derived from a convex stress energy function $u(\boldsymbol{\sigma})$ (a similar approach could be developed from a strain energy function). The strain response at any point $\boldsymbol{x}$ in the considered representative volume element V is independent of the loading path:

$$\boldsymbol{\varepsilon}(\boldsymbol{x}) = \frac{\partial u}{\partial \sigma}(\boldsymbol{x}, \boldsymbol{\sigma}(\boldsymbol{x})). \tag{1}$$

According to the affine approach and using the local tangent compliance $\boldsymbol{M}(\boldsymbol{x})$, the nonlinear constitutive equations (1) are linearized with respect to the current stress value at each point $\boldsymbol{x}$. The linearization takes place once and for all without any incremental process:

$$\begin{aligned} \boldsymbol{\varepsilon}(\boldsymbol{x}) &= \boldsymbol{M}(\boldsymbol{x}) : \boldsymbol{\sigma}(\boldsymbol{x}) + \boldsymbol{\eta}(\boldsymbol{x})\,, \\ \boldsymbol{M}(\boldsymbol{x}) &= \frac{\partial^2 u}{\partial \boldsymbol{\sigma}^2}(\boldsymbol{\sigma}(\boldsymbol{x}), \boldsymbol{x})\,, \\ \boldsymbol{\eta}(x) &= \frac{\partial^2 u}{\partial \boldsymbol{\sigma}^2}(\boldsymbol{\sigma}(\boldsymbol{x}), \boldsymbol{x}) - \boldsymbol{M}(\boldsymbol{x}) : \boldsymbol{\sigma}(\boldsymbol{x})\,. \end{aligned} \tag{2}$$

These linearized equations may be considered as associated to a linear thermoelastic composite constituted of an infinite number of phases (the notation $\boldsymbol{C} : \boldsymbol{D}$ means: $\boldsymbol{C}_{ijkl}\boldsymbol{D}_{kl}$, where summation is done over repeated indices). To deal with such a composite, one has, as usual, to adopt the simplification of piecewise constant compliances and prestrains per phase. Accordingly, for each phase, we assign their value at some reference stress σ_s to the tangent compliances and prestrains. In this paper, the reference stresses are the mean value of the stresses over the concerned phase. By this way we obtain a thermo-elastic linear composite with a finite number of phases. This composite represents the linear comparison medium (LCM) as developed by Ponte Castañeda [12]. The homogenization problem related to this composite can be solved by classical linear homogenization techniques for thermoelastic media, such as the self-consistent scheme when one deals with polycrystals.

Various discussions and results (Masson et al. [9]) have proved the viscoplastic affine formulation to offer an improved alternative to previous treatments such as the Hill-Hutchinson secant treatment (Hutchinson [5]) or the "tangent" model (Molinari et al. [10], Lebensohn and Tomé [7]). It differs from the second-order procedure (SOP) (Ponte Castañeda [13]) by making use of the direct stress-strain relations (1) instead of deriving the stresses from the overall potential: consequently, on the one hand, the SOP takes better into account the local intraphase heterogeneity and then gives better predictions for nonlinear elasticity or viscosity; on the other hand, the affine method, unlike the SOP, does not lead to any duality gap and can be extended to any constitutive behavior which, such as rate-dependent or rate-independent elastoplasticity, does not derive from a single potential.

2.2. Rate-independent elastoplasticity

In this section, rate-independent elastoplastic polycrystals are considered in the framework of small strains. The local strain rate is decomposed into its elastic and plastic parts:

$$\dot{\varepsilon} = \dot{\varepsilon}^e + \dot{\varepsilon}^p = \boldsymbol{K}^{-1} : \dot{\boldsymbol{\sigma}} + \sum_{\alpha} \dot{\gamma}^{\alpha} \boldsymbol{R}_{\alpha}\,, \tag{3}$$

with $\boldsymbol{K}$ the elastic moduli and $\dot{\gamma}^{\alpha}$ the non-negative plastic slip rate on system α with the orientation tensor $\boldsymbol{R}_{\alpha} = (\boldsymbol{n}_{\alpha} \otimes \boldsymbol{m}_{\alpha})^S$ ($\boldsymbol{A}^S$ denotes the symmetric part of the second-order tensor $\boldsymbol{A}$). Associate plasticity theory is used with the Schmid law, with the critical resolved shear stress τ_c^{α} on a slip system α. Such a system is a potentially active system (P.A.S.) if $\boldsymbol{R}_{\alpha} : \boldsymbol{\sigma} = \tau_{\alpha}^{c}$. It is active (A.S.) if, in addition, $\boldsymbol{R}_{\alpha} : \dot{\boldsymbol{\sigma}} = \dot{\tau}_{\alpha}^{c}$. We adopt the following hardening law:

$$\dot{\tau}_{\alpha}^{c} = \sum_{\beta} H_{\alpha\beta} \dot{\gamma}^{\beta}\,, \tag{4}$$

where the hardening matrix $H_{\alpha\beta}$ is assumed to be constant (we could also deal with a non constant matrix through an appropriate linearization procedure, but the associated technical developments are not essential here). Thus, the local elastoplastic constitutive equations read:

$$\dot{\boldsymbol{\sigma}} = \boldsymbol{L} : \dot{\boldsymbol{\varepsilon}} \,,$$
$$\boldsymbol{L} = \boldsymbol{K} : (\boldsymbol{I} - \sum_{\alpha,\beta} D^{\alpha\beta} \boldsymbol{R}_\alpha \otimes \boldsymbol{R}_\beta : \boldsymbol{K}) \,, \tag{5}$$

where $D^{\alpha\beta} = (H_{\alpha\beta} + \boldsymbol{R}_\alpha : \boldsymbol{K} : \boldsymbol{R}_\beta)^{-1}$ and the summation on (α, β) is limited to active systems.

The overall stress response at time t now depends on the whole previous loading path. Thus, a step by step procedure is needed (Masson et al. [9]). The steps can be defined as separating two consecutive "events": an event occurs at time t_n when some new system becomes active or inactive in any phase of the polycrystal. The step by step procedure (which does not lead to an incremental approach!) can be described as follows: from the stress values obtained at the end of the former step, we determine, at the beginning of the current step ($t = t_n$), which systems are potentially active. Among all these P.A.S., we select one branch, i.e., a given combination of A.S.. According to (5), this choice helps to compute the tangent moduli in each phase, up to the next event. Meanwhile, according to (5), these moduli, say $\boldsymbol{L}_s$, are constant. The local constitutive equations are then linearized in the usual affine manner:

$$\boldsymbol{\sigma}_s = \boldsymbol{L}_s : \boldsymbol{\varepsilon} + \boldsymbol{\theta}_s \,, \qquad \boldsymbol{\theta}_s = \boldsymbol{\sigma}_s - \boldsymbol{L}_s : \boldsymbol{\varepsilon}_s \,, \tag{6}$$

where $\boldsymbol{\varepsilon}_s$ and $\boldsymbol{\sigma}_s$ are the strain and stress averages in phase s obtained at the end of the previous step. From the known moduli and prestresses of the linear comparison composite, we derive the macroscopic elastoplastic tangent modulus $\tilde{\boldsymbol{L}}$ and the macroscopic prestress $\tilde{\boldsymbol{\theta}}$, which characterize the macroscopic behavior during the current step. We can also calculate the concentration tensors and then the local variables during the current step and especially at the end of the step which is defined by the occurrence of the next nearest event. From these values, the consistency of the chosen branch can be checked, i.e., the selected A.S. must be still active and the initially non active P.A.S. still inactive at the end of the step. If this consistency is verified, we go on to the next step. If not, we go back to the beginning of the step and we select another branch until consistency conditions are satisfied. Resuming the same procedure at each new increment up to t provides the expected macroscopic response of the polycrystal at time t.

3. The affine formulation at finite strain

3.1. Choice of the general description framework

In view of applications to metal forming, which involve very large deformations, an extension of the affine formulation at finite strain is needed. We have first to specify the general description framework. For several reasons listed below, we have adopted a mixed description instead of the classical eulerian or lagrangian ones (Iwakuma and Nemat-Nasser [6], Lipinski et al. [8], Harren [3]). This description makes special use of the deformation gradient tensor defined both on the initial and the current configurations. The small strains constitutive equation $\dot{\boldsymbol{\sigma}} = \boldsymbol{L} : \dot{\boldsymbol{\varepsilon}}$ of the single crystal now reads $\dot{\boldsymbol{n}} = \boldsymbol{N} : \dot{\boldsymbol{F}}$ where $\boldsymbol{n}$ is Hill's nominal stress tensor, $\boldsymbol{F} = \partial \boldsymbol{x} / \partial \boldsymbol{X}$ is the deformation gradient tensor and $\boldsymbol{x}$ and $\boldsymbol{X}$ are the position vectors on the current and the reference configuration respectively. The main advantage of this formulation relies on the fact that the macroscopic values of the nominal stress tensor and of the deformation gradient are the average over the initial configuration, which is known unambiguously, of their microscopic counterparts. Moreover, unlike for the Cauchy or the Kirchhoff stress tensors, this property also holds for their time derivatives (Hill [4]):

$$\bar{\boldsymbol{n}} = \langle \boldsymbol{n} \rangle_{\Omega_0}, \quad \bar{\boldsymbol{F}} = \langle \boldsymbol{F} \rangle_{\Omega_0}, \quad \dot{\bar{\boldsymbol{n}}} = \langle \dot{\boldsymbol{n}} \rangle_{\Omega_0}, \quad \dot{\bar{\boldsymbol{F}}} = \langle \dot{\boldsymbol{F}} \rangle_{\Omega_0}. \tag{7}$$

In the affine approach, we have rather to keep the reference configuration constant all along the loading path because we do not know the current configuration at the beginning of a new step. The main implication of this choice concerns the implicit description of the microstructure evolution. Though the derivations are performed on a fixed geometry (e.g., the constraint tensor, considered in the following, is calculated on a fixed ellipsoid), the evolution of the phase distribution is taken into account through that of the local tangent moduli which depend on the per phase average of the total and the elastic deformation (see below).

3.2. Single crystal elastoplastic constitutive equations

From the elastic-plastic decomposition $\boldsymbol{F} = \boldsymbol{F}^e.\boldsymbol{F}^p$, the plastic kinematic relation $\dot{\boldsymbol{F}}^p.(\boldsymbol{F}^p)^{-1} = \dot{\gamma}^\alpha \boldsymbol{m}_\alpha \otimes \boldsymbol{n}_\alpha$ and Hooke's law $\dot{\boldsymbol{S}}^e = \boldsymbol{K} : \dot{\boldsymbol{E}}^e$, where $\boldsymbol{S}^e$ is the second Piola-Kirchhoff stress tensor based on the configuration obtained after plastic deformation and $\boldsymbol{E}^e$ the elastic Green-Lagrange strain tensor, the following equation can be derived (see Harren [3]):

$$\dot{\boldsymbol{n}} = \hat{\boldsymbol{N}} : \dot{\boldsymbol{F}} - \sum_\alpha \dot{\gamma}^\alpha \boldsymbol{Y}_\alpha \tag{8}$$

with

$$\begin{aligned}
&\hat{N}_{ijkl} = F^{p-1}_{ia} F^e_{jb} K_{abcd} F^{p-1}_{kc} F^e_{ld} + \delta_{ik}\delta_{jl}\,, \\
&\boldsymbol{A}_\alpha = (\boldsymbol{F}^{eT}.\boldsymbol{F}^e.\boldsymbol{m}_\alpha \otimes \boldsymbol{n}_\alpha)^S\,, \qquad \boldsymbol{B}_\alpha = (\boldsymbol{m}_\alpha \otimes \boldsymbol{n}_\alpha.\boldsymbol{S}^e)^S\,, \\
&\boldsymbol{Y}_\alpha = \boldsymbol{F}^{-p}.(\boldsymbol{K} : \boldsymbol{A}_\alpha + 2\boldsymbol{B}_\alpha).\boldsymbol{F}^{eT}\,, \qquad \boldsymbol{S}^e = \boldsymbol{F}^{e-1}.\boldsymbol{F}.\boldsymbol{n}.\boldsymbol{F}^{e-T}.
\end{aligned} \tag{9}$$

Note that $\hat{\boldsymbol{N}}$ in (8) exhibits diagonal symmetry (i.e., $\hat{N}_{ijlk} = \hat{N}_{klji}$).

The Schmid law reads at finite strain:

$$f_\alpha = \boldsymbol{m}^e_\alpha . \boldsymbol{\tau} . \boldsymbol{n}^e_\alpha - \tau^c_\alpha = 0\,, \quad \boldsymbol{m}^e_\alpha = \boldsymbol{F}^e . \boldsymbol{m}_\alpha\,, \quad \boldsymbol{n}^e_\alpha = \boldsymbol{n}_\alpha . (\boldsymbol{F}^e)^{-1}\,, \tag{10}$$

where $\boldsymbol{\tau} = det(\boldsymbol{F})\boldsymbol{\sigma}$ is the Kirchhoff stress. The resolved shear stress used in the Schmid law is the projection of the surface traction $\boldsymbol{\tau}.\boldsymbol{n}^e_\alpha$ not on the initial ($\boldsymbol{m}_\alpha$) but on the current slip direction $\boldsymbol{m}^e_\alpha$. This choice is physically questionable and has significant consequences on the symmetry of the tangent modulus for the single crystal. We still use the hardening law (4) for the critical resolved shear stress. The slip rate $\dot{\gamma}^\alpha$, which is a linear function of $\dot{\boldsymbol{F}}$, can be derived from the consistency condition and reintroduced into (8), so as to derive the expected constitutive equation for the single crystal:

$$\begin{aligned} &\dot{\boldsymbol{n}} = \boldsymbol{N} : \dot{\boldsymbol{F}}\,, \\ &\boldsymbol{N} = \hat{\boldsymbol{N}} + \sum_{\alpha,\beta} (H_{\alpha\beta} + 2\boldsymbol{B}_\alpha : \boldsymbol{A}_\beta + \boldsymbol{A}_\beta : \boldsymbol{K} : \boldsymbol{A}_\alpha)^{-1} \boldsymbol{Y}_\alpha \otimes \boldsymbol{Y}_\beta\,. \end{aligned} \tag{11}$$

Note that, even if $H_{\alpha\beta}$ is symmetric, the moduli $\boldsymbol{N}$ do not exhibit diagonal symmetry anymore. Since this fact can induce some difficulties for finite strain homogenization schemes which rely on the use of Green techniques, we decide arbitrarily to replace $H_{\alpha\beta}$ by $\hat{H}_{\alpha\beta}$ such that

$$\hat{H}_{\alpha\beta} = (H_{\alpha\beta})^S + \boldsymbol{B}_\beta : \boldsymbol{A}_\alpha - \boldsymbol{B}_\alpha : \boldsymbol{A}_\beta \tag{12}$$

in order to save the diagonal symmetry of $\boldsymbol{N}$.

3.3. Specific features of the affine approach at finite strain

Linearization of the constitutive stress-strain relations. The tangent moduli $\boldsymbol{N}$ are no more constant between two consecutive events because they depend on $\boldsymbol{F}(t,\boldsymbol{x})$, $\boldsymbol{F}^e(t,\boldsymbol{x})$, $\boldsymbol{n}(t,\boldsymbol{x})$. In view of an explicit prediction, we choose to linearize the constitutive equations at the beginning of the current step at time $t = t_n$:

$$\begin{aligned} \boldsymbol{n}(t,\boldsymbol{x}) &= \boldsymbol{N}(t_n,\boldsymbol{x}) : \boldsymbol{F}(t,\boldsymbol{x}) + \boldsymbol{p}(t_n,\boldsymbol{x})\,, \\ \boldsymbol{N}(t_n,\boldsymbol{x}) &= \boldsymbol{N}(\boldsymbol{F}(t_n,\boldsymbol{x}), \boldsymbol{F}^e(t_n,\boldsymbol{x}), \boldsymbol{n}(t_n,\boldsymbol{x}))\,, \\ \boldsymbol{p}(t_n,\boldsymbol{x}) &= \boldsymbol{n}(t_n,\boldsymbol{x}) - \boldsymbol{N}(t_n,\boldsymbol{x}) : \boldsymbol{F}(t_n,\boldsymbol{x})\,. \end{aligned} \tag{13}$$

These are the constitutive equations of a continuously heterogeneous linear thermoelastic composite under finite strain.

Approximation of piecewise uniformity. Once again, the problem is simplified through the approximation of piecewise uniformity of the moduli $\boldsymbol{N}(t_n,\boldsymbol{x})$ and the prestresses $\boldsymbol{p}(t_n,\boldsymbol{x})$, which depend on $\boldsymbol{x}$ through $\boldsymbol{F}(t_n,\boldsymbol{x})$, $\boldsymbol{F}^e(t_n,\boldsymbol{x})$ and $\boldsymbol{n}(t_n,\boldsymbol{x})$: in phase s, they are ascribed to take their value at $\boldsymbol{F}_s(t_n) =$

$\langle \boldsymbol{F}(t_n, \boldsymbol{x}) \rangle_{\Omega_s}$, $F_s^e(t_n) = \langle F^e(t_n, x) \rangle_{\Omega_s}$ and $n_s(t_n) = \langle n(t_n, x) \rangle_{\Omega_s}$ (with $\langle A \rangle_{\Omega_s}$ the average of A in phase s). This approximation results at step n in the following constitutive equations of the linear thermoelastic multiphase comparison material:

$$\begin{aligned} \boldsymbol{n}(t, \boldsymbol{x}) &= \boldsymbol{N}_s(t_n) : \boldsymbol{F}(t, \boldsymbol{x}) + \boldsymbol{p}(t_n)\,, \\ \boldsymbol{N}_s(t_n) &= \boldsymbol{N}(\boldsymbol{F}_s(t_n), \boldsymbol{F}_s^e(t_n), \boldsymbol{n}_s(t_n))\,, \\ \boldsymbol{p}_s(t_n) &= \boldsymbol{n}_s(t_n) - \boldsymbol{N}_s(t_n) : \boldsymbol{F}_s(t_n)\,. \end{aligned} \tag{14}$$

Homogenization. At finite strain, no minimum potential energy principle is available. Ball [1] showed the existence of a solution to the problem of energy potential minimization for a polyconvex potential. However, uniqueness is not ensured. Thus, only variational principles (see Ogden [11]) exist and estimates only can be derived. Here, we restrict ourselves to Hashin-Shtrikman-type estimates, including the self-consistent scheme for specific application to polycrystals. They are based on the use of Green techniques which can be discussed in the context of finite strains through the 9-dimensional Eshelby inclusion problem. In terms of variables $\boldsymbol{n}$ and $\boldsymbol{F}$, this problem is defined as follows:

$$\begin{aligned} &\boldsymbol{n}(\boldsymbol{X})_{ij,i} = 0 \quad \forall \boldsymbol{X}\,, \\ &\boldsymbol{n} = \boldsymbol{N}^0 : \boldsymbol{F} + \chi_I(\boldsymbol{X})\boldsymbol{n}^I\,, \qquad \boldsymbol{F} = \frac{\partial \boldsymbol{x}(\boldsymbol{X},t)}{\partial \boldsymbol{X}}\,, \\ &\lim_{\boldsymbol{X}\to\infty} \boldsymbol{F}(\boldsymbol{X}, t) = \bar{\boldsymbol{F}}\,, \end{aligned} \tag{15}$$

where $\chi_I(\boldsymbol{X})$ is the characteristic function of the ellipsoidal inclusion I and $\boldsymbol{n}^I$ is constant. The solution for $\boldsymbol{F}(\boldsymbol{X})$ is known to be uniform in I (Iwakuma and Nemat-Nasser [6]) and given by

$$\begin{aligned} \boldsymbol{F}^T(\boldsymbol{X}_0) &= \boldsymbol{P}^0 : \boldsymbol{n}^I + \bar{F} \quad \forall \boldsymbol{X}_0 \in I\,, \\ P^0_{ijkl}(\boldsymbol{X}_0) &= \int_I u_{lj,ik}(\boldsymbol{X} - \boldsymbol{X}_0)\, dV_X = \int_I \Gamma^0_{ijkl}(\boldsymbol{X} - \boldsymbol{X}_0)\, dV_X\,, \end{aligned} \tag{16}$$

where $\boldsymbol{u}(\boldsymbol{X} - \boldsymbol{X}_0)$ is the Green function for an infinite elastic medium with the moduli $\boldsymbol{N}^0$ and $\boldsymbol{\Gamma}^0$ is the associated strain-Green operator.

Opposite to the case of small strains, the skew-symmetric parts of $u_{lj,ik}$ with respect to (j, k) and (l, i) are taken into account in $\boldsymbol{\Gamma}^0$ which does not exhibit anymore minor symmetries. As mentioned above, some difficulties, related to the use of Green techniques at finite strain, arise when $\boldsymbol{N}^0$ does not exhibit anymore the diagonal symmetry. Although the uniqueness of the Green function is ensured, its existence is no more established, so that the derivation of $\boldsymbol{P}^0$ becomes problematic. That is why such a situation has been excluded in (3.2). Therefore, the solution (16) of the inclusion problem may be used to establish estimates for elastic or thermoelastic linear composites in terms of $\boldsymbol{n}$ and $\boldsymbol{F}$ in a way strictly similar to the one used for small strains. For the elasto-plastic affine treatment, associated to the linear self-consistent scheme, the concentration equations at the current step n can be, according to (14), approximated by

$$\begin{aligned} \boldsymbol{F}_s(t) &= \boldsymbol{A}_s(t_n) : \bar{\boldsymbol{F}} + \boldsymbol{a}_s(t_n) \qquad \forall t \in]t_n, t_{n+1}] , \\ \boldsymbol{A}_s(t_n) &= (\boldsymbol{I} + \boldsymbol{P}_{SC} : (\boldsymbol{N}_s(t_n) - \tilde{\boldsymbol{N}}_{SC}))^{-1} , \\ \boldsymbol{a}_s(t_n) &= (\boldsymbol{P}_{SC}{}^{-1} + \boldsymbol{N}_s(t_n) - \tilde{\boldsymbol{N}}_{SC})^{-1} : (\tilde{\mathbf{p}} - \mathbf{p}_s(t_n)) , \\ \boldsymbol{P}_{SC} &= \textstyle\int_I \boldsymbol{\Gamma}_{SC}(\boldsymbol{X} - \boldsymbol{X}_0)\, dV_X , \quad \tilde{\mathbf{p}} = \langle \mathbf{p}(t_n) : \boldsymbol{A}(t_n) \rangle , \end{aligned} \tag{17}$$

where $\boldsymbol{\Gamma}_{SC}$ is attached to the infinite medium with moduli $\boldsymbol{N}_{SC}$.

Variables updating. Specializing equation (17) to the linear composite of the former step at time $t = t_n$ and proceeding similarly for the linear comparison composite of the current step at time $t = t_{n+1}$ lead to

$$\Delta \boldsymbol{F}_s = \boldsymbol{A}_s(t_n)\Delta\bar{\boldsymbol{F}} + (\boldsymbol{A}_s(t_n) - \boldsymbol{A}_s(t_{n-1}))\bar{\boldsymbol{F}}(t_n) + \boldsymbol{a}_s(t_n) - \boldsymbol{a}_s(t_{n-1}), \tag{18}$$

where $\Delta \boldsymbol{F}_s = \boldsymbol{F}_s(t_{n+1}) - \boldsymbol{F}_s(t_n)$ and $\Delta\bar{\boldsymbol{F}} = \bar{\boldsymbol{F}}(t_{n+1}) - \bar{\boldsymbol{F}}(t_n)$. The expression (18) of the deformation gradient increment clearly distinguishes the affine treatment from Hill's incremental one: according to Hill's method, this expression would reduce to $\Delta \boldsymbol{F}_s = \boldsymbol{A}_s(t_n)\Delta\bar{\boldsymbol{F}}$. The macroscopic stress increment $\Delta\bar{\boldsymbol{n}}$ and then the local stress increment $\Delta \boldsymbol{n}_s$ can be derived from (18), (11) as a function of the macroscopic deformation. The lattice rotation is obtained as usual by subtracting the plastic deformation gradient increment from the total one.

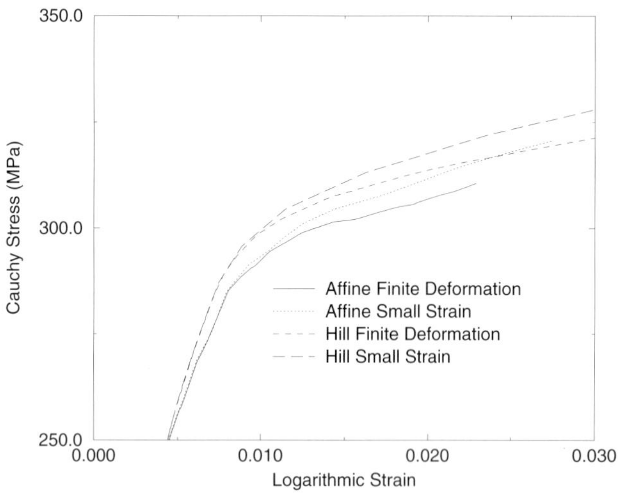

Figure 1. Tensile stress-strain curves of an untextured fcc polycrystal according to the incremental and affine models at small and finite strains.

4. Preliminary results and conclusion

Both the affine and the incremental approaches have been applied to derive the tensile stress-strain response of an untextured elastoplastic face-centred cubic polycrystal. Hill's method has been implemented by using a true mixed formulation as explained in section (3.1). The elastic behavior is assumed to be isotropic ($K = 44400$ MPA and $\mu = 30750$ MPA). The twelve octahedral slip systems are considered with the same critical resolved shear stress$\tau_\alpha^c = 100$ MPA $\forall\ \alpha$. Self and latent hardening are taken into account ($H_{\alpha\beta} = (1 - q)$ $h\delta_{\alpha\beta} + qh$) with $h = 200$ and $q = 0.3$. An initial quasi-isotropic texture (with ten orientations in the standard triangle) is considered.

Figure 1 provides the stress-strain tensile curves for both these approaches, for small as well as for large strains. We can see that the affine approach still yields softer predictions than Hill's one at finite strain. Moreover, both finite strain formulations yield softer predictions than their small strain counterparts. This result stems from the softening effect of the lattice rotations which are taken into account at finite strain.

According to these preliminary results, the affine approach can be thought to be able to improve significantly on Hill's incremental treatment for both small and large deformations. Nevertheless, future work is necessary to validate the affine finite strain formulation and especially to extend it to more general anisotropic initial textures.

References

[1] Ball, J.M. (1997). Convexity conditions and existence theorems in non linear elasticity. *Arch. Rat. Mech. Anal*, **63**, 337–403.

[2] Gilormini, P. (1996). A critical evaluation of various nonlinear extensions of the self-consistent model. In: *Micromechanics of plasticity and damage of multiphase materials*, Pineau and Zaoui (eds.), Kluwer Academic Publishers, 67–74.

[3] Harren, S.V. (1991). The finite deformation of rate-dependent polycrystals - I. A self-consistent framework. *J. Mech. Phys. Solids*, **39 3**, 345–360.

[4] Hill, R. (1972). On constitutive macro-variables for heterogeneous solids at finite strain. *Proc R. Soc. Lond. A*, **326**, 131–147.

[5] Hutchinson, J.W. (1976). Bounds and self-consistent estimates for creep of polycrystalline materials. *Proc. R. Soc. Lond. A*, **348**, 101–127.

[6] Iwakuma, T. and Nemat-Nasser, S. (1984). Finite elastic-plastic deformation of polycrystalline metals. *Proc R. Soc. Lond. A*, **394**, 87–119.

[7] Lebensohn, R. and Tomé, C.N. (1993). A self-consistent anisotropic approach for the simulation of plastic deformation and texture development of polycrystals: application to zirconium alloys. *Acta Metall. Mater.*, **41**, 2611–2624.

[8] Lipinski, P., Krier, J. and Berveiller, M. (1990). Elastoplasticite des metaux en grandes deformations. *Rev. Phys. Appl.*, **25**, 361–388.

[9] Masson, R., Bornert, M., Suquet, P. and Zaoui, A. (2000). An affine formulation for the prediction of the effective properties of nonlinear composites and polycrystals. *J. Mech. Phys. Solids*, **48**, 1203–1227.

[10] Molinari, A., Canova, G.R. and Ahzi S. (1987). A self-consistent approach of the large deformation polycrystal viscoplasticity. *Acta Metall.*, **35 12**, 2983–2994.

[11] Ogden, R.W. (1994). *Non linear elastic deformations*, Dover Publications, New York.

[12] Ponte Castañeda, P. (1991). The effective mechanical properties of nonlinear isotropic composites *J. Mech. Phys. Solids*, **39 1**, 45–71.

[13] Ponte Castañeda, P. (1996). Exact second-order estimates for the effective mechanical properties of nonlinear composite materials. *J. Mech. Phys. Solids*, **44 6**, 827–862.

A "Numerical Mesoscope" for the Investigation of Local Fields in Rate-Dependent Elastoplastic Materials at Finite Strain

H. Haddadi, C. Teodosiu
LPMTM-CNRS, Université Paris-Nord, Institut Galilée
Avenue J.-B. Clément, 93430 Villetaneuse Cedex, France
halim.haddadi@lpmtm.univ-paris13.fr, teodosiu@galilee.univ-paris13.fr

S. Héraud
Harting E.U.R.L.
181 avenue des Nations, Z.A.C. Paris-Nord II, 95972 Roissy CDG Cedex, France

L. Allais
Commissariat à l'Energie Atomique, CEREM/SRMA
Saclay, 91191 Gif-sur-Yvette Cedex, France

A. Zaoui
Laboratoire de Mécanique des Solides, CNRS-Ecole Polytechnique
91128 Palaiseau Cedex, France
zaoui@lms.polytechnique.fr

Abstract: We propose a "numerical mesoscope" which could be used for the analysis of the local mechanical fields over small critical areas of microheterogeneous materials, in order to predict the local initiation of specific deformation or damage mechanisms. The subdomain under investigation is embedded in a very large homogeneous matrix obeying the overall behavior of the studied material, as determined experimentally. This matrix is subjected to homogeneous stress or strain boundary conditions and the homogeneous elements of the subdomain and their interfaces are given their known or assumed constitutive behavior. A finite element analysis is then performed on the whole body by making use of different constitutive equations within the subdomain and in the surrounding matrix. The general methodology of this approach is reported and applied to a metallic rate-dependent elastoplastic polycrystal and to microheterogeneous subdomains consisting of given multicrystalline patterns whose grains obey crystalline elastoplastic constitutive equations of Schmid type at finite strain. Application to the intergranular creep damage of a stainless steel shows a good agreement between the largest computed normal stresses on the grain boundaries and the observed debonded boundaries of the actual material.

Keywords: Numerical simulation, polycrystals, rate-dependent elastoplasticity, finite strain, creep-damage.

C. Miehe (ed.), IUTAM Symposium on Computational Mechanics of Solid Materials at Large Strains, 311-320.

1. Principle of the "numerical mesoscope"

The knowledge of the local mechanical fields over small critical areas of microheterogeneous materials is generally needed for a better understanding and prediction of the initiation of specific deformation or damage mechanisms, such as twinning, phase transformation, grain boundary debonding, etc. Such regions can be in fact much smaller than the representative volume element (RVE), so that, unlike the RVE, they cannot be considered as subjected to homogeneous strain or stress boundary conditions. For example, the intergranular failure of metallic polycrystals depends on the mechanical state over a few adjacent grains, whereas the RVE can extend over millions of grains and the average strain or stress state in the RVE is of no help for predicting grain boundary debonding. On the other hand, direct numerical simulation of aggregates of several thousands of grains, i.e., large enough to be a good approximation of a RVE, in addition to being hardly tractable, cannot yield reliable predictions of debonding in actual polycrystals: although they are statistically representative of these polycrystals with respect to their crystallographic texture or their average grain shape, they cannot coincide with them at the scale of a few grains.

In what follows, by applying the concept of "numerical mesoscope" (Héraud et al. [6]), we consider the microheterogeneous subdomain under investigation as embedded with a perfect interface in a large homogeneous matrix. This matrix, which obeys the experimentally defined overall behavior of the studied material, is subjected to homogeneous stress or strain boundary conditions, so that, due to its own mechanical response, different mechanical conditions, which are expected to be more realistic, are transferred to the boundary of the considered subdomain. Adequate (known or assumed) local constitutive equations are defined at each point of the heterogeneous subdomain: their nature can be different from that of the matrix (e.g., for a polycrystal, they can be of a phenomenological type for the matrix and of crystalline Schmid type for the grains of the subdomain). Nevertheless, a single finite element analysis has to be performed on the whole body through some "numerical mesoscope", so as to yield the local mechanical state within the heterogeneous subdomain. Since this subdomain is not a RVE, the resulting analysis does not reduce to the classical *concentration* or *localization* procedure of usual homogenization schemes; in some sense, it could be seen as a *back-localization* analysis, which focuses on some specific critical areas within the RVE.

Since the same material is considered at the micro- and the macroscale, the local and the global constitutive equations used in the numerical analysis in the *inclusion* and in the *matrix*, respectively, have to be consistent with each other. In the following, this consistency condition is ensured through an auxiliary numerical computation: a large enough microheterogeneous RVE is generated and its overall response to a macroscopic loading is derived from the constitutive equations of the constituents. The consistency condition requires this response to coincide with the macroscopic response of the actual material. Alternative solutions, which would not need such auxiliary computations, could have recourse either to a micromechanical model predicting the overall behavior from the local one or to direct measurements of the displacement field at the boundary of the considered subdomain.

This approach is now focused on the case of rate-dependent elastoplastic polycrystals (Sect. 2) and then illustrated by an application (Sect. 3) to creep-damaged grain boundaries of a stainless steel, which has been experimentally studied in parallel by use of scanning electron microscopy (SEM) for the grain shapes and of electron back-scattering diffraction (EBSD) for the lattice orientations. Additional computations are reported in Sect. 4, in order to appreciate the pertinence of the proposed approach, and possible extensions are briefly discussed in conclusion.

2. "Numerical mesoscopy" and polycrystals

2.1. CONSTITUTIVE EQUATIONS

In view of the increasing diversity of potential applications to polycrystalline materials, rate-dependent elastoplasticity has been considered. In order to simplify the search for a satisfying consistency between the local and the overall constitutive equations, we have adopted similar forms for these equations at both scales. The resulting constitutive equations are derived from those proposed by Anand and associates (Anand [1], Brown et al. [3]) for the flow rule and by Kocks [7] for the description of the intragranular work-hardening. A finite strain formalism is needed, since, even for small macroscopic deformations, strains can be large at the local level. At the macroscopic scale, an isotropic formulation is adopted, for sake of simplicity. The strain-hardening of the material is described by a scalar internal state variable, called S. The plastic strain rate and the evolution equation of S are given by

$$\dot{\bar{E}}^{\mathrm{p}} = A\left[\sinh\left(\frac{\bar{\Sigma}}{S}\right)\right]^{\frac{1}{m}} \exp\left(-\frac{Q}{RT}\right), \tag{1}$$

$$\dot{S} = H_0 \left|\frac{S_{\mathrm{sat}} - S}{S_{\mathrm{sat}} - S_0}\right|^{a} \operatorname{sgn}(S_{\mathrm{sat}} - S_0)\, \dot{\bar{E}}^{\mathrm{p}}, \tag{2}$$

where $\dot{\bar{E}}^{\mathrm{p}}$ and $\bar{\Sigma}$ are the macroscopic von Mises equivalent strain rate and stress respectively, Q an apparent activation energy, T the absolute temperature, R the perfect gases constant; S_0 is the initial value of S and $S_{\mathrm{sat}} = \bar{S}(Z/A)^n$ its limit value, where $Z = \dot{\bar{E}}^{\mathrm{p}}\exp(Q/RT)$ denotes the Zener-Holomon parameter. Thus, apart from the elastic moduli, the material parameters are $A, m, Q/R, H_0, a, S_0, \bar{S}$ and n.

Similar equations are adopted at the crystal level for the relation between the plastic shear rate $\dot{\gamma}^{(s)}$ and the resolved shear stress $\tau^{(s)}$ on the slip system (s), namely

$$\dot{\gamma}^{(s)} = \alpha \left[\sinh \left(\frac{\tau^{(s)}}{\tau_c^{(s)}} \right) \right]^{\frac{1}{m}} \exp\left(-\frac{Q}{RT}\right) \operatorname{sgn}(\tau^{(s)}), \tag{3}$$

$$\dot{\tau}_c^{(s)} = \sum_u a^{(su)}\, h(z^{(u)}, \tau_c^{(u)})\, |\dot{\gamma}^{(u)}|, \tag{4}$$

with

$$h(z^{(u)}, \tau_c^{(u)}) = h_0(z^{(u)}) \left| \frac{\tau_{sat}^u(z^{(u)}) - \tau_c^{(u)}}{\tau_{sat}^u(z^{(u)}) - \tau_0^{(u)}} \right|^a \operatorname{sgn}(\tau_{sat}^{(u)}(z^{(u)}) - \tau_0^{(u)}), \tag{5}$$

$$\tau_{sat}^{(u)}(z^{(u)}) = \bar{\tau} \left(\frac{z^{(u)}}{\alpha} \right)^n, \qquad z^{(u)} = \dot{\gamma}^{(u)} \exp\left(\frac{Q}{RT}\right). \tag{6}$$

For sake of consistency with the macroscopic equations, material parameters $m, Q/R, a$ and n are the same at both levels. The hardening matrix $a^{(su)}$ may be used in the usual simple form $q + (1-q)\delta^{(su)}$, with $q \geq 1$, in order to take latent hardening into account. Further simplifications, which are considered in the sequel, consist of setting $h_0(z^{(u)}) = h_0$ and $\tau_0^{(u)} = \tau_0$, where h_0 and τ_0 are constants, i.e. independent of the slip system (u). The additional material parameters at the microscale then reduce to α, q, h_0, τ_0 and $\bar{\tau}$.

2.2. Numerical Integration

The finite element numerical treatment has been carried out by using an extension for the crystalline pattern elaborated by Héraud [5] of the 3D-code EPVIM3D developed by Haddadi [4]. At the local scale, a forward gradient scheme is used for the shear increments $\Delta\gamma^{(s)}$, while the shear stress increments $\Delta\tau^{(s)}$ and $\Delta\tau_c^{(s)}$ are linearized. The basic equation reads

$$\Delta\gamma^{(s)} = [(1-\beta)\,\dot{\gamma}^{(s)}(t) + \beta\,\dot{\gamma}^{(s)}(t+\Delta t)]\,\Delta t,$$

where β has a fixed value (around $2/3$). A first-order Taylor development is used for the derivation of $\dot{\gamma}^{(s)}(t+\Delta t)$. A similar treatment is used at the macroscopic scale. Each finite element has 8 nodes and 8 integration points, with use of reduced integration for the volumetric part of the deformation and of isoparametric shape functions. The time increments are optimized so as to be shorter whenever the intracrystalline shear rates are larger. Specific routines have been developed in order to derive the stress vectors on the grain boundaries from the stress tensors evaluated within the adjacent grains.

3. Application to a creep-damaged stainless steel

The foregoing concept of "numerical mesoscopy" has first been applied to the interpretation of grain-boundary debonding in a specimen of AISI316 stainless steel. This specimen was subjected to a creep test up to fracture, under 370 MPa at 550 °C during 395 hours and then cut along a longitudinal cross-section. A small area of about 50 grains and 100 μm diameter, exhibiting several debonded grain boundaries, has been selected in this section. The geometry, as well as the lattice orientations of 35 grains of the central multicrystal, have been determined through SEM observation and EBSD analysis. The problem was then to simulate this crystalline area as central pattern of our mesoscope, embedded in a large homogeneous matrix, subjected to the macroscopic loading of the actual creep test, and to compute the normal stresses on the grain boundaries of the pattern in order to look for any correlation between largest computed normal stresses and the most damaged boundaries.

Overall creep behavior. The constitutive equations of the studied material was first to be determined. This has been done by Héraud [5], using the experimental creep data obtained for this material at three different temperatures (550, 600 and 650 °C) and, for each temperature, under three different load levels (from 150 MPa to 420 MPa according to the temperature). A preliminary rough analysis has first been performed in order to enter initial parameters in the optimization code SIDOLO [8]. After a small number of iterations, we have got the following parameters (in addition to the Young modulus, $E = 143$ GPa, and the Poisson ratio, $\nu = 0.33$): $A = 1.96 \times 10^7$ s^{-1}, $m = 0.053$, $Q/R = 34200$ K, $H_0 = 3470$ MPa, $a = 0.8$, $S_0 = 36$ MPa, $\bar{S} = 102$ MPa and $n = 0.16$. The global constitutive equations (1) and (2) are thus determined.

Local constitutive equations. They have been identified from the fitting of the creep response of a simulated polycrystal whose grains obey equations (3) to (6) to the experimental data. This model polycrystal has first been defined for an isotropic copper polycrystal already studied by Anand et al. [2]. The result was a FEM model consisting of a cubic aggregate of 1000 cubic grains of the same size, where each grain is represented by one finite element. Application to our studied material and experimental data, through the same identification procedure as before, has led to the following local parameters: $\alpha = 0.5 \times 10^7$ s^{-1}, $h_0 = 920$ MPa, $\tau_0 = 37$ MPa, $\bar{\tau} = 120$ MPa and $q = 1.4$, to be used for the considered creep test at $T = 550$ °C under 370 MPa.

FEM computation. As a first approximation, a single three-dimensional layer of elements has been considered. The matrix area is about five times larger than the crystalline pattern. The grain boundaries are normal to the layer. The whole body has been meshed by use of the SIMAIL software from Simulog

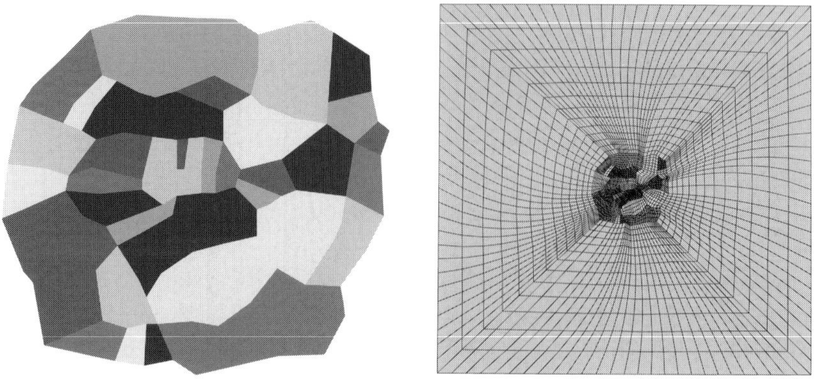

Figure 1. The 35-grain multicrystalline aggregate and the mesh of the whole mesoscopic domain.

(1145 elements for the multicrystal and 1430 for the matrix; see Figure 1). The four lateral faces, which are parallel to the creep direction, are supposed free, while the remaining two faces, which are perpendicular to the creep direction, are subjected to an imposed traction. In addition, any rigid-body displacement has been suppressed by adequate displacement boundary-conditions. For such boundary conditions, 15,786 degrees of freedom are needed (Héraud [5]).

Results. The simulation of the creep test has been performed up to277 hours, whereas fracture occurred after 395 hours in the actual test. The considered multicrystal proved to be stiffer than the matrix on average: he was less deformed and more stressed, which agrees with the fact that the selected pattern looked especially damaged. Among several local analyses of the stress and strain fields in the grains (Héraud [5]), special attention has been paid to the distribution of the stress vectors on the grain boundaries. Whereas no correlation was detected between the shear stress level and the damage amplitude on these boundaries, a definite one existed between the largest computed normal stresses and the experimentally debonded boundaries (compare e.g. the grain boundaries denoted by GB1, GB2 and GB3 in Figures 2 and 3). Although this conclusion can be appreciated as an encouraging evidence for the pertinence of the concept of "numerical mesoscope", further investigations have been performed in order to corroborate this statement.

4. Pertinence of the numerical mesoscope

As argued in the introduction, the main reason for developing such a complex simulation tool lies in the fact that any local mechanical analysis on subdomains which are smaller than the RVE cannot be sensibly performed through direct application of the macroscopic boundary conditions to the boundary of these

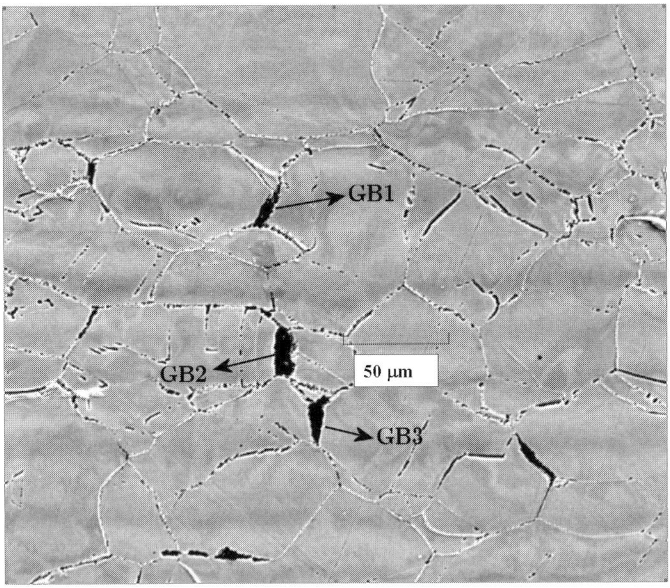

Figure 2. Location of debonded grain boundaries in the actual material.

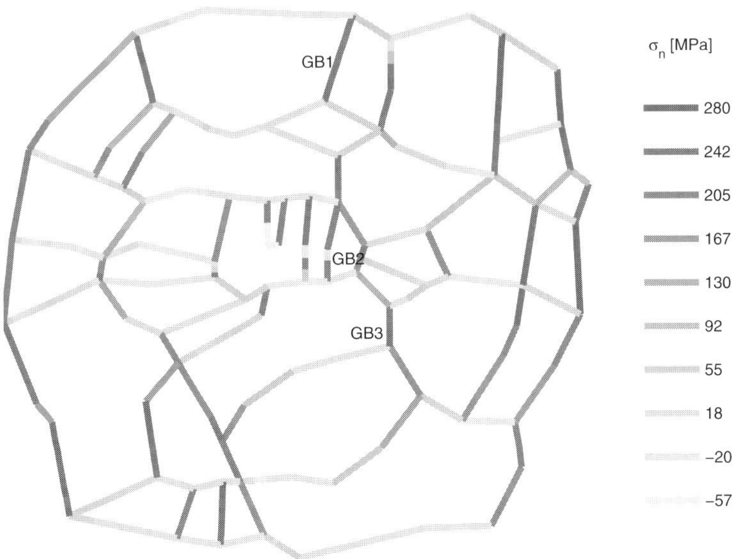

Figure 3. Distribution of the normal component of the stress vector on the grain boundaries of the simulated multicrystal.

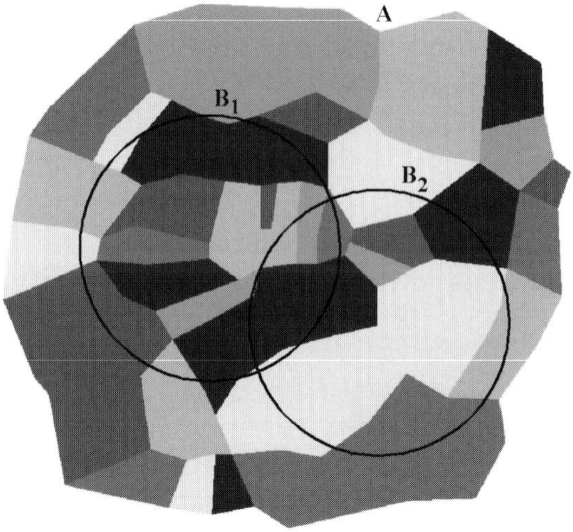

Figure 4. The initial 35-grain multicrystal (sample A) and the extracted samples B_1 and B_2.

subdomains. When experimental data of the actual displacement or traction fields on this boundary are not available, it is proposed to embed the subdomain of interest in a quasi-infinite matrix obeying the experimentally defined overall behavior and subjected to the macroscopic boundary conditions, so as to load the investigated subdomain in a more realistic way. Conversely, the pertinence of the numerical mesoscope can be appreciated *a posteriori* by comparing the actual displacement or traction fields on the boundary of the simulated subdomain, as they result from the whole FEM computation, with those which would be associated to the average macroscopic strain or stress tensors.

Such an analysis has been performed on the above reported application and it resulted in strong differences between these fields (Héraud [5]), which could have been considered as a sufficient justification of the rather lengthy numerical procedure required by the numerical mesoscope. However, this reasoning has been considered not enough conclusive, because the angular geometry of the multicrystal boundary, which results in strong local stress concentrations, may indeed affect the comparison with the rather smooth macroscopic boundary conditions.

That is the reason why further numerical analyses have been performed on subdomains cut out from the initial 35-grain multicrystal, which will be called sample A from now on. Two circular disks, say samples B_1 and B_2, have been extracted from sample A, both of them of the same size, but located at two different places in the multicrystal and consisting of a smaller number of grains, namely 15 and 12 grains, respectively (see Figure 4). In order to shorten the computation times for the same simulated creep test, some material parameters

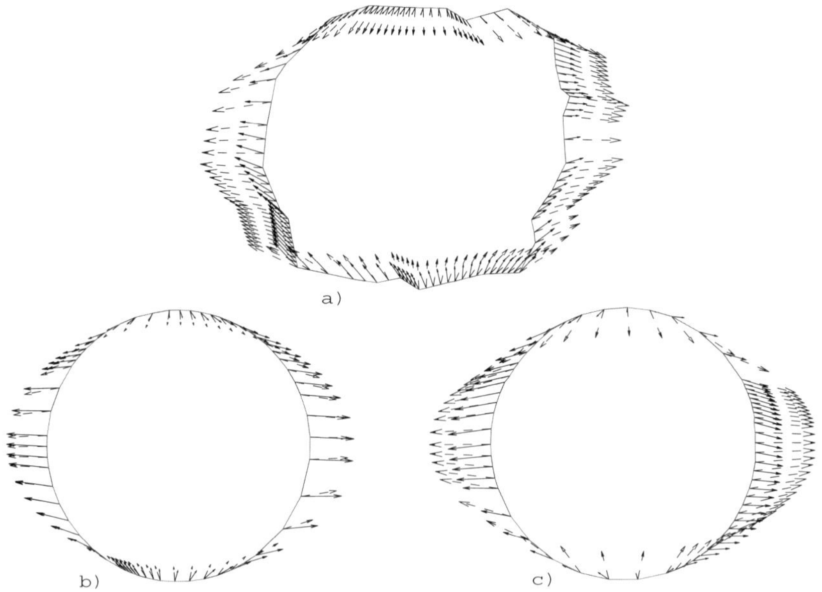

Figure 5. In-plane components of the relative displacements on the boundary of each of the three disk-shaped samples with respect to the centre of its top surface (solid vectors), compared with those derived from the respective average displacement gradients (dashed vectors). a) Sample A, b) sample B_1, and c) sample B_2.

where modified in a similar way for both the matrix and the constituent crystals, so as to lower the global and local creep compliances. After a given creep time, the volume average of the displacement gradient tensor, say $\boldsymbol{H}$, was calculated over each multicrystalline sample, as well as the relative displacements $\boldsymbol{u}$ of the boundary of each disk-shaped sample with respect to the centre of its top surface. Figure 5 shows the in-plane components of these relative displacements on the boundary of each of the three samples (solid vectors) compared with those derived from the average displacement gradient $\boldsymbol{H}$ (dashed vectores).

It is clearly seen from these results that the actual boundary conditions on the samples differ significantly from the macroscopic ones. Indeed, since the latter result from a homogeneous displacement gradient, the tips of the corresponding displacement vectors on the circular boundaries of samples B_1 and B_2 are smoothly distributed on ellipses, whereas the displacement fields resulted from the simulations on their boundaries vary more sharply from point to point. It is worth noting that these differences are more visible on sample B_2, which has been subjected to a larger plastic deformation, than on sample B_1, which was closer to an elastic state. Notwithstanding, even sample B_1 displays a significant change in the *direction* of the calculated displacement vectors with respect to the homogeneous boundary conditions, especially on the bottom of its circular boundary. It may be concluded that the ensemble of these results do substantiate the pertinence of the numerical mesoscope.

5. Conclusion

Further developments are obviously needed from this kind of "prototype". In addition to computational improvements (parallel computations, improved automatic meshing, etc.), two kinds of mechanical developments are currently in progress. On the one hand, a larger number of plane layers are to be considered, consisting either of crystal or of matrix material, with oblique grain boundaries. On the other hand, imperfect interfaces are developed within the crystalline pattern, in view of modeling debonding or sliding grain boundaries. It is expected that this will result in extended interpretation schemes, especially in the field of the identification of crystal constitutive equations from micromechanical measurements over grains, as well as in the complex field of the micromechanics of damage.

References

[1] Anand, L. (1985). Constitutive equations for hot-working of metals. *Int. J. Plasticity*, **1**, 213–231.

[2] Anand, L., Balasubramanian, S. and Kothari, K. (1996). Constitutive modeling of polycrystalline metals at large strains. In: *Large Plastic Deformation of Crystalline Aggregates*, C. Teodosiu (ed.), CISM Courses and Lectures No. 376, Springer Verlag, Wien, New York, 109–172.

[3] Brown, S.B., Kim, K.H. and Anand, L. (1989). An internal variable constitutive model for the hot working of metals. *Int. J. Plasticity*, **5**, 95–130.

[4] Haddadi, H. (1996). *Matériaux composites à matrice métallique: modélisation du comportement thermo-élastoviscoplastique et résolution de problèmes aux limites 3D.* PhD Thesis, Université Paris 13, France.

[5] Héraud, S. (1998). *Du polycristal au multicristal: élaboration d'un mésoscope numérique pour une analyse locale en élastoviscoplasticité.* PhD Thesis, Ecole polytechnique, Palaiseau, France.

[6] Héraud S., Allais L., Haddadi H., Marini B., Teodosiu C. and Zaoui A. (1998). Du polycristal au multicristal: vers un mésoscope numérique. *J. Phys. IV France*, 8:4-27-32.

[7] Kocks, U.F. (1987). Constitutive behaviour based on crystal plasticity. In: *Constitutive Equations for Creep and Plasticity*, A.K. Miller (ed.), Elsevier Appl. Sci., London, 1–88.

[8] SIDOLO, Notice d'utilisation (1993). Ecole des Mines de Paris, France.

Computational Mechanics of Heterogeneous Materials: Influence of Residual Stresses[1]

S. Schmauder, U. Weber, E. Soppa
Staatliche Materialprüfungsanstalt, University of Stuttgart
Pfaffenwaldring 32, 70569 Stuttgart, Germany
Siegfried.Schmauder@mpa.uni-stuttgart.de

Abstract: Heterogeneous materials such as Metal Matrix Composites (MMCs) of the type Al/SiC contain significant residual stresses due to different thermal expansion coefficients from the metal and ceramic constituents. They are believed to influence the mechanical properties of these materials to some extent – including some changes in their failure behavior. In this contribution, a physically based micromechanical approach is applied in order to clarify the influence of residual stresses on local as well as global properties of MMCs. A representative microstructural cut-out of an Al / 10% SiC-composite is meshed with finite elements in order to take phase boundaries into account. This mesovolume possesses all characteristic features of the material, such as volume fraction, distribution characteristics as well as shape of the particles. The deformation behavior of this microstructure is analyzed under large compressive external loading up to strains of about 10%. In addition, the failure behavior is modeled using Rice&Tracey's failure criterion which was recently shown to model microstructural failure to a good approximation. It is found that although residual stresses do have some impact on failure initiation in the microstructure, strains due to external loading are much more of importance in this respect. In order to illuminate the influence of particle shape and arrangement, artificial two-dimensional microstructures are analyzed as well. It is found that irregular particle shapes are much more prone to fracture in the matrix as compared to regular shapes and that particle alignments are not beneficial with respect to failure aspects. The distribution and maximum values of the damage parameter are shown. It is found that in all cases analyzed, damage follows the pattern of plastic deformation and is much less influenced by hydrostatic stresses than expected. Nevertheless, damage nucleates between clusters of particles where shear deformation as well as hydrostatic tensile stresses are concentrated in the matrix.

Keywords: Finite element method, MMC, damage parameter, residual stresses, computational mechanics.

1. Introduction

The role of residual stresses on the mechanical behavior of heterogeneous materials such as metal matrix composites (MMCs) has been known for a long time (Levy and Papazian [8]). The source of residual stresses is the difference in thermal expansion coefficients between the matrix phase and the inclusions when cooling down from processing to room temperature and during subsequent external loading. Thus, micromechanical models were applied in the recent past in order to quantify their impact on macroscopic (Zahl and McMeeking [14])

C. Miehe (ed.), IUTAM Symposium on Computational Mechanics of Solid Materials at Large Strains, 321-330.

as well as on microscopic (Dong et al. [4]) mechanical properties of MMCs. A main result of previous analyses showed that the overall mechanical behavior of MMCs is changed by several percents in the regime of small external strains while at large strains this influence vanishes (Leßle et al. [7]).

However, it is yet unclear, how far microstructural aspects such as phase volume fractions, as well as size and shape of particles influence local fracture initiation. These aspects are of importance when the microstructure shall be optimized in order to obtain for instance better forming capabilities.

The intention of this paper is thus to provide insights into the role of residual stresses on the global mechanical properties as well as on local properties of heterogeneous materials such as Al/SiC metal matrix composites.

2. Methods

The J2-flow theory of plasticity in conjunction with the von-Mises equivalent stress is used to characterize the rate-independent matrix material. The underlying geometry is a microstructure of size 100μm x 100 μm. Residual stresses will be considered as from cooling down from 500 °C (where no residual stresses are assumed to be present) to room temperature. In fact, due to the temperature dependence of the yield stress, residual stresses will be smaller than calculated. Symmetry boundary conditions on two sides and plane borders which can move freely or as forced by up to 10 % global compressive strain are applied.

Residual stresses in materials of the phases α and β can be given as average stresses in either phase, $\langle\sigma\rangle^\alpha$ and $\langle\sigma\rangle^\beta$, being composed of the average stresses σ^{II} over all grains of the corresponding phases. They are differentiated from peak stress values σ^{III} and overall stresses σ^I and add locally to the total stress value $\sigma = \sigma^I + \sigma^{II} + \sigma^{III}$.

In order to characterize damage in the microstructure, a damage parameter is used here which goes back to the theory of Rice and Tracey [10] considering the growth of a spherical void and which was modified in order to take more general void formation into account (Arndt [1]):

$$D = \frac{1}{A}\int e^{B\eta} d\varepsilon_{pl} \tag{1}$$

Here, $\eta = \sigma_H/\sigma_v$ denominates the stress triaxiality composed of σ_H (hydrostatic stress) and σ_v (von-Mises stress) while ε_{pl} represents the plastic strain and A and B are material constants.

A is usually chosen in such a way that $D = D_c = 1$ in case of local failure. This parameter was found to work very good for Al/SiC composites where it was successfully applied to simulate microstructural failure in agreement with experiments (Schmauder et al. [11]).

The constants A and B can be obtained for steel StE 690 as $A = 17.29$ and $B = 3.14$ from damage curves in Arndt [2] as well as for Co in WC/Co ($A = 1.41$, $B = 1.46$, see Hönle [6]) and for pure Al ($A = 51.42$, $B = 3.48$, see Zhu [15]). A and B can be derived by a combination of experiment and the calculation of the respective field quantities (η, ε_{pl}) for the case of the first appearance of voids (Sprock [12]).

Fig. 1 shows artificial microstructures in a 2D-model with square and round inclusions with the center of gravity at the same locations for comparison reasons. Other microstructures with differently orientated square and elliptical inclusions are meshed in the same manner. Inside the particles regular meshes with typically 30–40 finite elements are used, in the matrix the mesh is refined in the vicinity of the interface in order to take stress gradients into account.

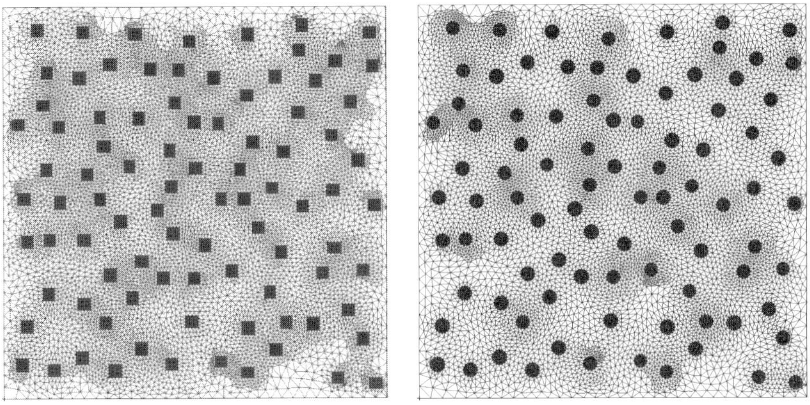

Figure 1. Square and round artificial microstructures (2D).

In order to generate and to mesh such microstructures, a separate program was set up and connected with MSC/PATRAN [9]. A typical mesh of these mesostructures contains ~20,000 triangular 6-noded elements with approximately 78,000 degrees of freedom.

In the following, results will be presented for different inclusion shapes and arrangements. The material analyzed is Al/SiC with elastic and elastic- plastic properties taken from Wulf [13] for Al/SiC. Residual stresses result from the different thermal expansion coefficients for Al (α_{Al} = 23.0E-06 1/K) and SiC (α_{SiC} =4.7E-06 1/K). As a measure for the propensities of failure, the maximum damage parameters which appear after cooling down and external straining, respectively, are calculated. The location of D_{max} has not necessarily to be identical in both cases (thermal- and thermo-mechanical loading).

Usually, only positive values of η in the matrix contribute to void formation. In this paper also negative values of η have been taken into account in eqn. (1), because strong plastic shearing in the matrix is supposed to contribute to damage in this case as well. Additionally, a comparison with the contribution of positive η-values will be made. Particle damage is not taken into account in this study as it appears rarely under compressive external loading in the MMCs under investigation.

3. Results

In the following, results are presented with elastic-plastic material parameters for the system Al / 10vol.% SiC and with values $A = 1.41$ and $B = 1.46$ as derived for Co in Hönle [6].

Fig. 2 (a–c) shows the distribution of the damage parameter value in compressive loaded microstructures with randomly arranged as well as with aligned square inclusions, the latter being loaded parallel or perpendicular to the inclusion alignments. Three observations can be made (it should be kept in mind that damage is due to the cooling process and external compressive loading up to $10\,\%$):

- The pattern of D-values reminds very much to the pattern of local shearing in random microstructures demonstrating the importance of shear for damage although on a low level as $D_{max} = 0.254$, only. This maximum value appears at one of the crossing points of the "shear bands".
- D_{max} is found to be more than twice as high in the case of aligned particles loaded perpendicular to the alignments with the maximum values found between the particles in the alignments.
- When aligned particles are loaded in the alignment direction, the maximum value D_{max} is nearly 10 times as high as in the case of loading perpendicular to the alignments and is again located between particles in the alignments.

The change to round and elliptical particles of the same areal fraction (Fig. 2, d–f) shows a similar D_{max}-value when elliptical particles are orientated perpendicular to the loading direction and reduced values for round particles and for elliptical particles orientated in loading direction while the location of the maximum value D_{max} is found to remain unchanged.

Similarly, the orientation of square inclusions can influence theD_{max}-value significantly leading to the lowest values for the 45°-orientation when shear is less hindered, and to intermediate values in case of the 30°-orientations (Fig. 3a, b).

As the deformation through plastic shearing is easier in case of a few large particles as compared to many small ones of the same area fraction (Fig. 2d and Fig. 3c), the value D_{max} is also found to be decreased in this case. This demonstrates again the importance of the plastic strain component in eqn. (1)

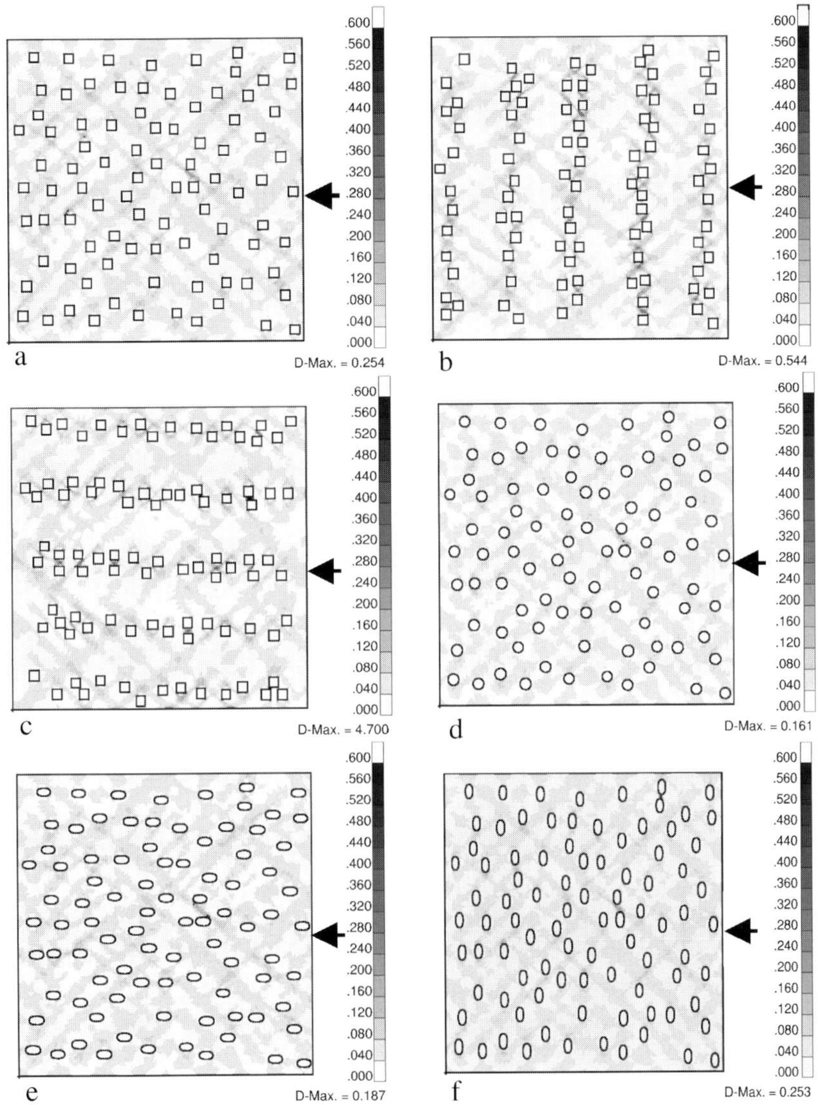

Figure 2. Damage parameter distributions in artificial microstructures: Influence of several particle distributions (a–c) and particle shapes (d–f).

due to local shearing. However, the opposite is true if the few large particles become clustered (Fig. 3d). In that case, local shearing is found in a narrow area between particles, which additionally experiences hydrostatic tensile stresses. Therefore, the D_{max}-value increases.

Next a real microstructural cut-out as given in Fig. 4a is automatically meshed with ~21.000 finite elements as shown in Fig. 4b.

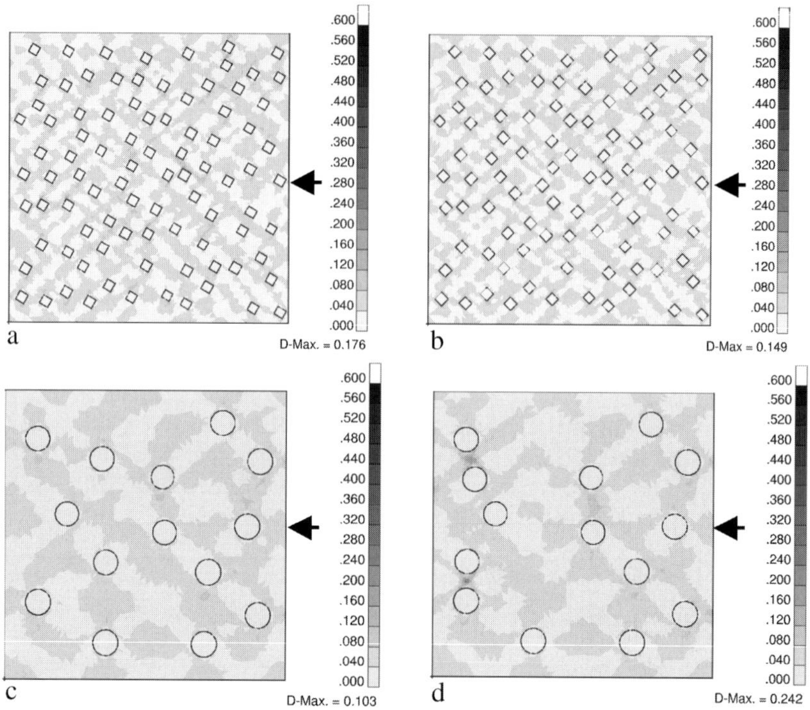

Figure 3. Damage parameter distributions in artificial microstructures: Influence of several particle orientations ($a = 30°$, $b = 45°$) and particle clustering (c, d).

The comparison (Fig. 4c) with square, circular and elliptical particles (Fig. 2) shows the following remarkable and important differences:

- Areas of clustered real particles give rise to increased failure probability, similar as for the case of clusters of ideal round particles.
- This situation starts nearly simultaneously at several locations within the microstructure.

Finally, residual stresses by themselves can be found to influence the matrix only in the close vicinity of the inclusions (Fig. 5). However, compared to the case of round inclusions the D_{max}-value is more than 10 times as high for the real microstructure due to the particle clusterings.

The results of the investigations are summarized in Fig. 6 where the maximum possible damage parameter values D_{max} are given for pure thermal and for thermo-mechanical loading as a function of the inclusion shape, arrangement and orientation and in comparison with the real microstructure. In the artificial microstructures only very low values of D_{max} are reached under thermal loading

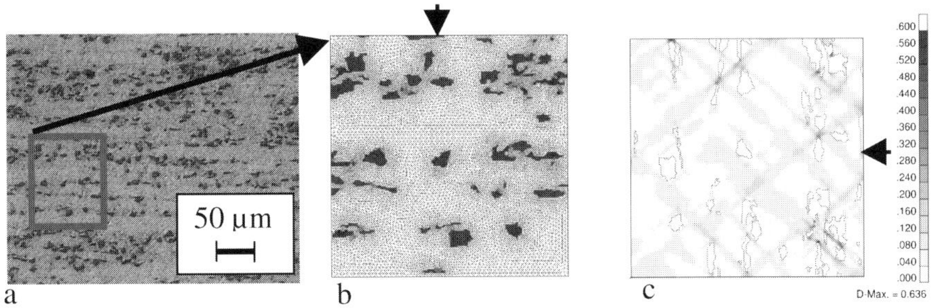

Figure 4. a) Real microstructural cut-out (Al/10vol.%SiC). b) Corresponding finite element mesh. c) Damage parameter distribution.

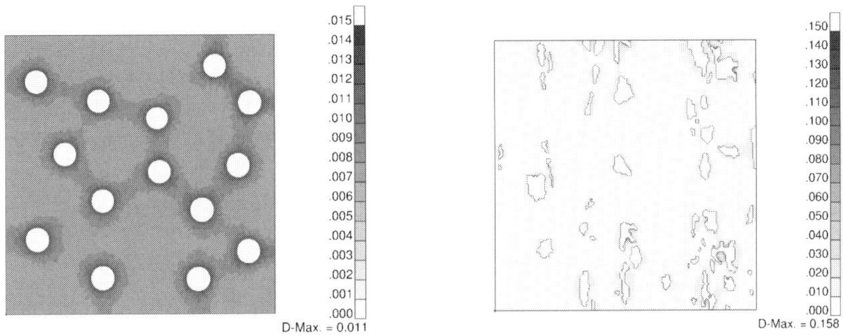

Figure 5. Damage parameter distribution due to residual stresses.

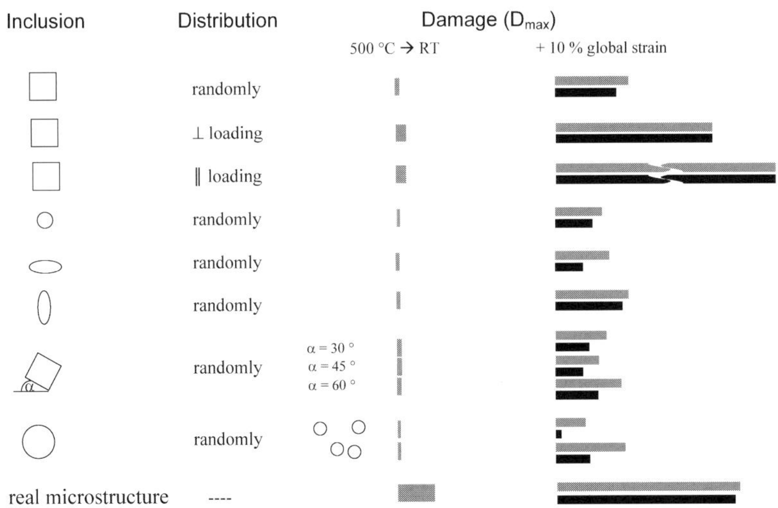

Figure 6. D_{max}-values as a function of inclusion shape, orientation, distribution and alignment (contribution of positive η-values are shown in black).

in contrast to the result for the real microstructure. Aligned quadratic inclusions are seen to be critical when loaded externally parallel to the alignments. The lowest damage potential in the matrix is found for randomly distributed inclusions, especially in case of round or favorably orientated (45°) square inclusions. The contribution of negative η-values is found to be neglectable in the case of residual stresses. In the case of external compressive loading, limited contributions of negative η-values are found for randomly distributed inclusions (see right column of Fig. 6).

If damage is allowed to occur in the real microstructure of Al / 10vol.% SiC, first damage is found to appear at 8 % global strain and spreads through part of the clustered area when the microstructure is further strained up to 10 % (Fig. 7). The influence of residual stresses during failure has been taken into account. It has to be noted that at 10 % global strain, damage also arises at several other locations inside the microstructure. A recent comparison for an Al / 20vol.% SiC composite has shown good agreement of calculated and observed crack path (Schmauder et al. [11]).

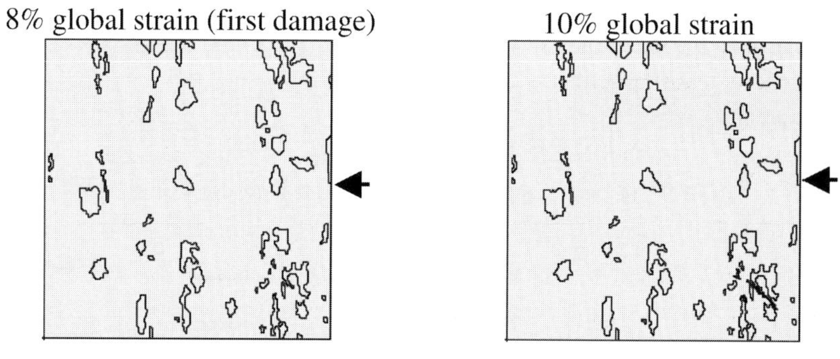

Figure 7. Damage evolution in an Al / 10vol.% SiC microstructure (see Fig. 4)

One of the fundamental problems in the analysis of metal matrix composites is to simulate the macroscopic behavior correctly on a microscopic basis. This process is frequently called homogenization and can be done using real microstructural cut-outs or simple representative volume elements (RVEs) containing just one inclusion. Real microstructures are appropriate to obtain additionally local field quantities present in the composite (Doumalin [5]). The advantage of single inclusion type RVEs are reduced computation times and modeling efforts. However, while real microstructural models automatically contain a large part of the complexity of the phase characteristics, simple RVEs can include the three-dimensionality in a much simpler way. A recently developed self-consistent numerical scheme was shown to be able in representing even the randomness of the particle distributions by some realistic averaging algorithm (Dong and Schmauder [3]). Moreover, a comparison with surface deformation characteristics showed even good agreement with local strain dis- tributions, as measured in experiments (Leßle et al. [7]).

4. Conclusions

In this contribution it was shown from computer simulations of artificial and real 2D-microstructures based on the modified Rice & Tracey damage parameter with respect to the influence of residual stresses that when high external compressive stresses are applied:

- random particle arrangements are less prone to damage in the matrix phase as compared to particle clustering such as in real microstructures
- local damage initiation is strongly influenced by shear accumulation in the matrix. Therefore, crossing points of shear bands in the vicinity of particles with additional high stress triaxialities show peak values of the damage parameter D
- facetted particles produce higher damage parameter values as compared to elliptical particles and to round particles (in this sequence for random particle arrangements)
- aligned facetted particulate MMCs are more prone to damage in the matrix and show a better performance in this respect if loading is applied in the direction of particle alignment
- facetted particles with facets aligned parallel to the shear direction behave better as compared to non-aligned facets
- the influence of residual stresses on damage is strongly dependent on the load capability of the composite, i.e. the ratio of residual to applied stresses
- negative η-values have a limited influence on the damage parameter D in the case of randomly distributed artificial inclusions.

Acknowledgments

We gratefully acknowledge funding of our work by the Deutsche Forschungsgemeinschaft (DFG) under contract SCHM-746/25-1/2/3.

Notes

1. An extended version of this manuscript was submitted for publication to *Computational Materials Science.*

References

[1] Arndt, J., Majedi, H. and Dahl, W. (1996). Influence of strain history on ductile failure of steel. *J. de Physique*, **IV C6**, 23–32.

[2] Arndt, J. (1997). *Experimentelle und rechnerische Untersuchungen zur Schädigung von Baustählen bei duktilem Versagen*, PhD Thesis, IEHK, RWTH Aachen, Band 3/97, Shaker Verlag.

[3] Dong, M. and Schmauder, S. (1996). Modeling of Metal Matrix Composites by a Self-Consistent Embedded Cell Model. *Acta metall. mater.*, **44**, 2465–2478.

[4] Dong, M., Leßle, P., Weber, U. and Schmauder, S. (1999). Mesomechanical Modeling of Composites Containing FGM Related Interpenetrating Microstructures Based on Micromechanical Matricity Models. In: *Materials Science Forum*, W.A. Kaysser (ed.), 308-311, 1000–1005.

[5] Doumalin, P. (2000). *Microextensométrie Locale par Corrélation d'Images Numériques*, PhD Thesis, École Polytechnique.

[6] Hönle, S. (1998). *Micromechanical Modelling of Deformation and Fracture of Graded WC/Co Hard Metals*, PhD Thesis, Max-Planck-Institut (MPI) für Metallforschung and University of Stuttgart, Bericht Nr. 67.

[7] Leßle, P., Dong, M., Soppa, E. and Schmauder, S. (1998). Simulation of Interpenetrating Microstructures by Self Consistent Matricity Models. *Scripta mater.* **38**, 1327–1332.

[8] Levy, A. and Papazian, M. (1991). Elastoplastic Finite Element Analyses of Short-Fiber-Reinforced SiC/Al Composites: Effects of Thermal Treatment. *Acta Metall. Mater.*, **39**, 2255–2266.

[9] MSC/PATRAN, Version 8.5, User's Manual (on-line), 1999.

[10] Rice, J. and Tracey, D.M. (1969). On the ductile enlargement of voids in triaxial stress fields. *J. Mech. Solids*, **17**, 201–217.

[11] Schmauder, S., Wulf, J., Steinkopff, Th. and Fischmeister, H.F. (1996). Micromechanics of Plasticity and Damage in an Al/SiC Metal Matrix Composite. In: *IUTAM Symposium on Micromechanics of Plasticity and Damage of Multiphase Materials, Sèvres-France, 29.8.-1.9.1995*, A. Pineau and A. Zaoui (eds.), Solid Mechanics and its Applications, Kluwer Academic Publishers, London, 255–262.

[12] Sprock A. (1995). *Numerische Untersuchungen lokaler und globaler Kriterien für die Initiierung von duktilen Bruchvorgängen*, PhD Thesis, IEHK, RWTH Aachen, Band 2/95, Shaker Verlag.

[13] Wulf, J. (1995). *Neue Finite-Elemente-Methoden zur Simulation des Duktilbruchs in Al/SiC*, Dissertation am Max-Planck-Institut (MPI) für Metallforschung und an der Universität Stuttgart, Reihe 18: Mechanik/Bruchmechanik, Nr. 173, VDI-Verlag GmbH, Düsseldorf.

[14] Zahl, D.B. and McMeeking, R.M. (1991). The Influence of Residual Stress on the Yielding of Metal Matrix Composites. *Acta Metall. Mater.*, **39**, 1117–1122.

[15] Zhu, K. unpublished research, Staatliche Materialprüfungsanstalt (MPA) Universität Stuttgart.

Nonlinear Waves in Solids and the Inverse Problems

J. Engelbrecht, A. Ravasoo, A. Salupere
Department of Mechanics and Applied Mathematics
Institute of Cybernetics, Tallinn Technical University
Akadeemia tee 21, 12618 Tallinn, Estonia
je@ioc.ee

Abstract: Reported in this paper are the results from the studies of nonlinear deformation waves in solids (i) with significant dispersion and (ii) with inhomogeneous properties and initial stress states. The analysis of wave fields at given properties and excitations (direct problems) gives insight for deriving the algorithms for solving the inverse problems. In this case the properties of the materials must be determined at given excitation and measured wave fields. The effects of nonlinearities are strong but give more information on the properties of materials or on the prestressed states of specimens.

Keywords: Nonlinear waves, dispersion, solitons, inverse problems.

1. Introduction

Classical continuum mechanics is based on macroscopic concepts and quantities, such as energy density, stress, strain, etc., usually presented within the framework of linear theories. Contemporary technology needs a more sophisticated framework in order to describe complicated effects due to intensive and short time loading, coupled fields, complex properties of materials, etc. Here, in this paper, wave motion is analyzed with main attention to effects caused by nonlinearity/ies and dispersion. Nonlinear effects in solids are mainly caused by geometry (large strain) or physics (stress-strain relations). For more detailed analysis the reader is referred to Maugin [10] and Engelbrecht [3]. Dispersion in its turn in wave-like processes is resulting from microstructural properties of materials, and is related to characteristic microlengths. Clearly this phenomenon is very complicated. The classical concept of homogeneity of a continuum can not be used any more. The first step towards generalization can be to account for continuous inhomogeneities of materials (Ravasso [15]) due to technology of manufacturing or internal stresses. In smaller scales, inhomogeneities can be accounted for as discrete elements (Maugin [9]). This idea, however, is closely related to lattice theories (Maugin [9]), and bridging these to continuum theories is especially useful for crystals. So in order to describe properly dispersive effects, the macroscopic and microscopic properties should both be taken into account simultaneously.

C. Miehe (ed.), IUTAM Symposium on Computational Mechanics of Solid Materials at Large Strains, 331-339.

Beside the phenomenological description, the character of problems should also be specified. It is possible to distinguish between two main classes of problems: direct and inverse. Direct problems involve the determination of wave fields at given material properties and known initial excitations. Inverse problems involve the determination of material properties at given initial excitations and measured wave fields. In technological applications the last problem is usually called Nondestructive Testing (NDT) (Blitz and Simpson [1]).

Obviously both direct and inverse problems for waves in solids with complicated properties are not trivial. First - correct mathematical models (all effects of the same order should be taken into account simultaneously), second - physical effects are to be analyzed and explained in the light of experiments, third - special algorithms are needed for solving the direct and inverse problems with proper accuracy.

In this paper, the main ideas for deriving the mathematical models of nonlinear waves in solids are presented in Section 2. Then two case studies are described. In Section 3, the direct problem for nonlinear waves in solids with complicated dispersive properties is solved. In Section 4, the inverse problem for determination of inhomogeneous properties and/or stress states is analyzed. The concluding Section 5 summarizes the results and envisages future studies and open interesting questions.

2. Mathematical modeling

2.1. Background

The classical equations of motion represent the conservation of momentum together with constitutive equations, and symmetry conditions. In order to take inhomogeneities and/or microstructure into account one should start from the analysis of scales. Several cases are to be distinguished:

- on the level of macrostructure spatial changes in density, elastic parameters, etc. are described by smooth functions and initial stress state by corresponding initial conditions (Ravasso [15]);
- mesoscopic theories where beside the macroscopic description the statistical distribution function for inhomogeneities is used; that can be useful for modeling dynamics in materials with microcracks (Muschik et al. [12]);
- lattice theories where a true microstructure including defects, grains, dislocations, vacancies, impurities, e.a. is easily introduced (Maugin [10]).

There are various continuum models introducing an internal length scale. The corresponding constitutive equations are analyzed, for example, in Mühlhaus [11], the mathematical models of waves in Engelbrecht and Braun [4].

2.2. MATHEMATICAL MODELS

Referring to the brief summary in Section 2.1 and the corresponding references, we focus our attention here to possible mathematical models. Two basic approaches are the following (Engelbrecht [3]):

- deriving the full set of equations that describe the whole wave field with all the waves taken into account;
- deriving the governing equations for single waves – evolution equations – that describe a part of a wave field with a certain asymptotical accuracy.

The case (i) is supported by growing computational capacities but even so sometimes difficult to be realized due to algorithmic complexity. As an example, the systems with observable and internal variables (Maugin [8]) are composed by hyperbolic and parabolic equations. The case (ii) is gaining much attention because of the general character of evolution equations, which usually (and for cases of our interest, always) are nonlinear. Leaving aside the details (for those, see Maugin [10], Engelbrecht [3]), we present here some important models.

If all the effects are described on the *macroscopic* level (case (i) above), then the mathematical model can be derived from the Cauchy's equations of motion (Maugin [10])

$$\frac{\partial}{\partial t}\mathbf{p}_R\,|_X - div_R\mathbf{T} = 0, \tag{1}$$

where $\mathbf{p}_R$ is the linear momentum per unit volume of the reference configuration and $\mathbf{T}$ the first Piola-Kirchhoff stress tensor.

The final model in terms of the displacement $u_i, i = 1, 2, 3$ is a system of the hyperbolic equations with possible small higher order terms (h.o.t.) describing dissipation, dispersion, inhomogeneity et al. If, for example, inhomogeneity of physical and geometrical properties and/or inhomogeneous predeformation is taken into account, then the 1D model is (Ravasso [15])

$$[f_1(X) + f_2(X)u_X]u_{XX} + f_3(X)u_X + f_4(X)u_X^2 - f_5(X)u_{tt} = 0, \tag{2}$$

where $X \equiv X_1, u \equiv u_1$ and $f_j(X), i = 1, ...5$ are smooth functions. Here and further, the index denotes the differentiation.

If dispersive effects are taken into account, then a 1D model with the appropriate scaling may be written as follows (Maugin [10])

$$c_0^2 u_{XX} + [F(u)]_{XX} + \beta u_{XXXX} - u_{XXXXXX} - u_{tt} = 0, \tag{3}$$

where c_0 is the velocity, $\beta > 0$ and $F(u)$ represents nonlinear potential.

If discontinuities in the form of microcracks are accounted for, then the 1D model reads (Rakotomana [13])

$$c_0^2(1 + ku_X)u_{XX} + c_0^2\chi_0 u_X - u_{tt} = 0, \qquad (4)$$

where k, χ_0 are the coefficients.

The models (2.2)–(2.4) describe two-wave fields, i.e. waves propagating to right and left in 1D case. The corresponding evolution equation can be easily derived by asymptotic methods (Engelbrecht [3]), these equations describe the single waves. From (2.3), for example, an evolution equation can be derived (Maugin [10], Salupere [19])

$$v_t + [P(v)]_x + d\, v_{xxx} + b\, v_{xxxxx} = 0. \qquad (5)$$

For shape-memory alloys, the nonlinearity is expressed by

$$P(v) = -\frac{1}{2}v^2 + \frac{1}{4}v^4. \qquad (6)$$

Here v is related either to u_x or u_t and independent coordinates x, t are given in a moving frame.

Another possibility, as noted above, is to separate*macroscopic* and *microscopic* effects. Then instead of (2.1) one should address the system (Gurtin and Podio-Guidugli [7], Cermelli and Pastrone [2])

$$\frac{\partial}{\partial t}\mathbf{p}_R\,|_X - div_R\mathbf{T} = 0 \qquad (7)$$

$$\frac{\partial}{\partial t}\mathbf{q}_R\,|_X - div_R\mathbf{S} - \mathbf{k} = 0 \qquad (8)$$

where $\mathbf{q}$ is the microstructural momentum, $\mathbf{S}$ is the microscopic stress, and $\mathbf{k}$ is the interaction force. The body forces are here neglected.

Assuming that both the macro- and microstructure are dissipative, a 1D model for waves in such a material exhibits the hierarchical properties (Engelbrecht et al. [5]). The corresponding evolution equation is:

$$[v_t + Qvv_x - Kv_{xx}]_x + \lambda^2[v_t + Rvv_x + Mv_x - Nv_{xx}] = 0 \qquad (9)$$

where Q, K, R, M, N are constants and λ^2 - a scale/input parameter. With λ^2 small, only the first part in brackets corresponding to macrostructure is of importance; with λ^2 large, the first part in brackets is negligible and the second part, i.e. effects of the microstructure are of importance. In Giovine and Oliveri [6], the similar 1D model is derived for granular media with dispersion. If in eq. (2.8), the operators are of the Burgers type inducingv_{xx}, then in the dispersive case the operators are of the Korteweg-de Vries type includingv_{xxx}:

$$[v_t + vv_x - N_1 v_{xxx}]_{xx} + \mu^2[v_t + vv_x + N_2 v_{xxx}] = 0. \qquad (10)$$

3. Direct problems for nonlinear waves

In solids with significant dispersion the 1D waves can be described by the KdV type equations, like (2.5) and (2.10). As widely known, in this case the effects of nonlinearity (generation of shocks) and dispersion can be balanced resulting in emerging solitary waves. In the standard KdV case this process is analyzed in detail, in more complicated cases (cf. eq. (2.5) with (2.6)), much is still open (Salupere et al. [20]). Here we continue the analysis of eq. (2.5) with (2.6) which govern the wave processes in microstructured alloys. Briefly – these are the processes, where the question of the possible balance of complicated nonlinearity (the fourth order elastic potential) and complicated dispersion (the third and the fifth order terms) is of interest.

Equations (2.5), (2.6) yield the following dispersion relation

$$\omega = k^3(bk^2 - d). \tag{11}$$

If $d > 0,\ b > 0$ then the normal and anomalous dispersion occurs depending on k. If $d > 0,\ b < 0$ then dispersion is normal over all the ranges of the wavenumber. These effects, combined with possible steepening of wave profiles due to nonlinearity, give rise to several wave types.

Equation (2.5) with (2.6) are solved at the following boundary and initial conditions

$$v(x,t) = v(x + 2\pi n, t), \quad n = \pm 1, \pm 2, ..., \tag{12}$$

$$v(x,0) = \sin x, \qquad 0 \leq x \leq 2\pi. \tag{13}$$

Equation (2.5) with (2.6) are nonintegrable and for the solutions under (3.2), (3.3) the pseudospectral method and the corresponding code are used. The numerical analysis is carried out for the following range of parameters:

$$0 \leq d_l \leq 2.4, \quad 0.4 \leq b_l \leq 4.4, \tag{14}$$

$$d_l = -\log d, b_l = -\log b. \tag{15}$$

The accuracy of the code and the time-step constants have been analyzed earlier (Salupere [18]).

The standard KdV equation solved under conditions (3.2), (3.4) shows the emergence of a soliton train. Here the situation is much more complicated. In what follows, the typical cases are briefly explained.

If $d > 0$ and $b < 0$, then dispersion is normal over all the wavenumbers and the train of negative solitons forms from conditions (3.4), (3.5) for any value of the dispersion parameters.

If $d > 0$ and $b > 0$, then depending on the wave number, the dispersion can be either normal or anomalous. The character of the solution depends essentially on values of the dispersion parameters:

- for relative high values of b, the fifth order effects dominate over those caused by the third order terms. Dispersion is normal only for rather small values of wavenumbers k. This results in a train of positive solitons;
- for relative high values of d, the opposite situation occurs and the dispersion is anomalous only for rather small values of wavenumber k. This results in a train of negative solitons;
- if neither the third nor the fifth order dispersive terms dominate, then the rivalry between normal and anomalous dispersion affects strongly the outcome. For stronger dispersion, a multiple soliton solution is formed. For weaker dispersion, however, the solution is chaotic.

Based on the analysis of wave fields at given conditions some conclusion can be drawn for solving the inverse problems. Clearly, the wave field excited by the most common (and technically easily realizable) harmonic initial field, is strongly dependent on values of dispersion parameters. The robust presentation leads then to a two-parameter system in the $d_l - b_l$ plane. Qualitatively, the character of wave profile shows the regions of the dispersion parameters. Quantitatively, one should determine the value of both parameters d_l and b_l from the measured wave field. For that not only the direct characteristics of wave profiles but also their spectral characteristics are at our disposal. The derivation of algorithms for solving inverse problems with suitable accuracy in physical terms is an important and challenging task. One could propose the following methods:

- harmonic excitation, spectral characteristics measured;
- solitary excitation, wave amplitudes and velocities measured.

The first method can easily be realized by using existing experimental techniques. The second could cause technical difficulties, but as a conjecture, its accuracy could be better because of clear physical evidence of interaction and radiation effects. This way or another, the realization of the ideas for solving the inverse problems characterized by strong effects of nonlinearity and dispersion, is a challenge.

4. Inverse problems

In many cases, the solution of an inverse problem for nonlinear equation of motion of material with complicated properties (2.2) is too complicated. More convenient is to pose the nondestructive material characterization problem on the basis of the solution to the equation of motion (2.2). This solution which is dependent on the initial and boundary conditions permits to develop the theoretical basis for NDT methods.

The basic idea of the NDT method presented here is to use the analytical or graphical relations wave characteristics (amplitude, velocity, etc.) versus material properties (density, elastic constants, etc.) composed on the basis of the solution to the equation of motion of material (2.2). After that, making use of the wave propagation measurement data the material property under consideration is determined resorting to the composed relations. Contrary to the linear or linearized description of the wave process utilized in NDT algorithms for conventional techniques, the additional information is obtained from the nonlinear effects of wave propagation. The considered method takes the small but finite deformations, the geometrical and the physical nonlinearity into account. It may be applied to nondestructive characterization of materials with complicated properties, to evaluation of various states of the material (predeformed state, temperature impact, etc.). The sources of information for NDT include nonlinear wave propagation effects, multi-wave simultaneous propagation, reflection and nonlinear interaction data, etc. The nondestructive measurement data may be obtained on the basis of various NDT techniques (time of flight technique, through transmission technique, two wave interaction technique, etc.). It should be emphasized that consideration of nonlinear effects of wave propagation enables essentially to enhance the possibilities of NDT.

The approach is illustrated by two cases:

- algorithm for nonlinear through transmission technique for nondestructive characterization of inhomogeneous materials (Ravasoo [16, 14]);
- ditto for two waves nonlinear interaction technique (Ravasoo and Lundberg [17]).

In both cases deformation of the medium is described on the basis of the*nonlinear theory of elasticity*. An attention is confined to small but finite deformations. To describe large or finite deformations, it is necessary to distinguish Lagrangian and Eulerian coordinate systems. Here in the governing equation (2.2) the Lagrangian coordinates are used.

The *geometrical nonlinearity* is involved by the Lagrangian deformation tensor and the *physical nonlinearity* corresponds to the nonlinear theory of elasticity with five elastic constants.

From the point of view of algorithms for ultrasonic nondestructive testing of materials with inhomogeneities caused by external effects, such as external loading for example, the geometrical nonlinearity of the problem must be taken into account. The geometrical nonlinearity enables to bound the stress fields caused by the external effects and the wave propagation. From the other side, it is proved that in many cases the geometrical nonlinearity and the physical nonlinearity has the same order of influence on the recorded data in NDT (on acoustoelastic effect see for example Ravasoo [14]).

5. Conclusions

Nonlinearity should always be treated together with other physical effects of the same order. Pure nonlinear effects obviously result in the generation of higher harmonics and the changes in energy spectra that can be used for gaining the additional information on the properties of solids or their prestressed stress states. However, the contemporary technology aims to producing complicated microstructured materials. In this case, dispersive effects are also important. The co-operative effects of nonlinearity and dispersion affect considerably the wave field. Beside solitary waves more complicated solitonic structures and even chaotic solutions can be observed. Inverse problems, i.e. the evaluation of material properties on the basis of the excited and measured wave field characteristics are extremely important tasks. Here the inverse problem is solved for materials with inhomogeneous properties and/or stress states. The inverse problem of highly dispersive materials is still a challenge, although the knowledge on direct problems already exists.

Acknowledgments

The authors appreciate the support from the Estonian Science Foundation.

References

[1] Blitz, J. and Simpson, G. (1996). Ultrasonic Methods of Non-Destructive Testing. *Chapman and Hall, London.*

[2] Cermelli, P. and Pastrone F. (1995). Influence of a dissipative microstructure on wave propagation. *In: J.L.Wegner and F.R.Norwood (eds), Nonlinear Waves in Solids*, ASME, 279–284.

[3] Engelbrecht, J. (1997). Nonlinear Wave Dynamics. *Complexity and Simplicity*, Kluwer, Dordrecht.

[4] Engelbrecht, J. and Braun, M. (1998). Nonlinear waves in nonlocal media. *Appl. Mech. Rev.*, **51**, 475–488.

[5] Engelbrecht, J., Cermelli, P. and Pastrone, F. (1999). Wave hierarchy in microstructured solids. In G.A.Maugin (ed.) *Geometry, Continua & Microstructure*, Hermann Publ., Paris, 99–111.

[6] Giovine, P. and Oliveri, F. (1995). Dynamics and wave propagation in dilatant granular materials. *Meccanica*, **30**, 341–357.

[7] Gurtin, M.E. and Podio-Guidugli, P. (1992). On the formulation of mechanical balance laws for structured continua. *Z. angew. Math. Phys.*, **43**, 181–192.

[8] Maugin, G.A. (1990). Internal variables and dissipative structures. *J. Non-Equilib. Thermodyn*, **15**, 173–192.

[9] Maugin, G.A. (1993). Material Inhomogeneities in Elasticity. *Chapman and Hall, London.*

[10] Maugin, G.A. (1999). Nonlinear Waves in Elastic Crystals. *Oxford Univ. Press.*

[11] Mühlhaus, H.-B. (1995). *Continuum Models for Materials with Microstructure*, John Wiley & Sons, Chichester et al.

[12] Muschik, W., Ehrentraut, H., and Papenfuss, C. (1999). Mesoscopic continuum mechanics. *G.A.Maugin (ed.), Geometry, Continua & Microstructure*, Hermann Publ., Paris, 19–60.

[13] Rakotomanana, L.R. (2000). Theoretical models of microcracked continua: discontinuity of scalar and vector fields. *Techn. Mechanik*, **20**, 159–168.

[14] Ravasoo, A. (1995). Ultrasonic nondestructive evaluation of inhomogeneous plane strain in elastic medium, *Res. Nondestr. Eval.*, **7**, 55–68.

[15] Ravasoo, A. (1999). Nonlinear longitudinal waves in inhomogeneously predeformed elastic media. *J. Acoust. Soc. America*, **106**, 3143–3149.

[16] Ravasoo, A. (1999). Nonlinear waves in characterization of inhomogeneous elastic material. *Mechanics of Materials*, **31**, 3, 205–213.

[17] Ravasoo, A., Lundberg, B. (2001). Nonlinear interaction of longitudinal waves in an inhomogeneously predeformed elastic medium. *Wave Motion*, **34**, 2, 225–237.

[18] Salupere, A. (1997). On the application of pseudospectral methods for solving nonlinear evolution equations and discrete spectral analysis. *Proc. 10th Nordic Seminar on Comp. Mech.*, Tallinn TU, 76–83.

[19] Salupere, A., Maugin, G.A. and Engelbrecht, J. (1997). Solitons in systems with a quartic potential and higher-order dispersion. *Proc. Estonian Acad. Sci. Phys. Math.*, **46**, 1/2, 118–127.

[20] Salupere, A., Engelbrecht, J. and Maugin, G.A. (2001). *Solitonic structures in KdV-based higher-order systems.* Wave Motion, **34**, 1, 51–61.

Advanced Numerical Methods

The Extended Finite Element and Level Set Methods for Non-Planar 3D Crack Growth

N. Moës

Laboratoire de mécanique et matériaux, Ecole centrale de Nantes

1 Rue de la Noe, 44321 Nantes, France

nicolas.moes@ec-nantes.fr

A. Gravouil

Laboratoire de Mécanique des Solides, Institut National des Sciences Appliquées de Lyon

34 Av. des Arts, 69621 Villeurbanne, France

anthony.gravouil@insa-lyon.fr

T. Belytschko

Department of Mechanical Engineering, Northwestern University

2145 Sheridan Road, Evanston, IL 60208, U.S.A

t-belytschko@northwestern.edu

Abstract: A methodology for treating non-planar three-dimensional cracks with geometries that are independent of the mesh is described. The method is based on the extended finite element method, in which the crack discontinuity is introduced as a Heaviside step function via a partition of unity. In addition, branch functions are introduced for all elements containing the crack front. The crack geometry is described by two signed distance functions (level sets), which in turn can be defined by nodal values. Consequently, no explicit representation of the crack is needed. A Hamilton-Jacobi equation is used to update the level sets as the crack grows. Numerical experiments show the robustness of the method in treating cracks with significant changes in topology. The method is readily extendable to inelastic fracture problems.

Keywords: Fracture, cracks, discontinuous approximation, finite elements, level sets.

1. Introduction

Three dimensional fracture analysis of engineering problem by standard finite element methods is still quite difficult because of the need to construct a mesh which conforms to both the crack surfaces and the surfaces of the component. If the crack surface is not aligned with the element boundaries, the displacement discontinuity and the traction conditions on the crack surface cannot be treated by standard finite element methods. Furthermore, for standard elements, the mesh must be designed so that it is substantially more refined around the crack than in the remainder of the model. The difficulties are further amplified when considering the growth of cracks, because then the model must be remeshed in

C. Miehe (ed.), IUTAM Symposium on Computational Mechanics of Solid Materials at Large Strains, 343-354.

the vicinity of the crack. In addition to this, it must be borne in mind that initial cracks in many locations of the component must be considered for a complete engineering analysis.

This paper summarizes a method presented in Moës et al. [17] and Gravouil et al. [11], in which a three dimensional version of the extended finite element method is developed and applied. The extended finite element method alleviates much of the burden associated with mesh generation for objects with cracks by not requiring the finite elements to conform to the crack surface. Moreover, it provides a convenient way for incorporating near-tip asymptotic fields, so that good accuracy can be obtained for elastic fracture with relatively coarse meshes around the crack.

The essential idea in X-FEM is to use a displacement field approximation that can model an arbitrary discontinuity and the near-tip asymptotic crack fields. As a consequence it is often not necessary to modify the mesh to consider a specific crack; at most, moderate refinement must be introduced around the crack to achieve engineering accuracy in elastic fracture mechanics.

The methodology was first presented in Belytschko and Black [2], Moës et al. [16] and Dolbow et al. [8]. It was shown that discontinuous functions can be used to enrich finite element approximations via the partition of unity concept introduced by Melenk and Babuška [15]. The resulting approximation can treat cracks that are arbitrarily aligned in the finite element mesh with great accuracy. The concept was generalized in Daux et al. [6] and in Belytschko et al. [3], which described the application of the concept to arbitrary discontinuities. Sukumar et al. [22] illustrated the potential of combining the extended finite element method with level sets by solving several problems involving inclusions and holes. In Stolarska et al. [21], the extended finite element methodology was combined with a level set method to provide a general method for growing cracks. All of the preceding papers dealt with two dimensional problems.

The first application of the extended finite element method to three-dimensional cracks was Sukumar et al. [24], who solved several planar crack mode I problems and showed that the method compared well with analytical and benchmark solutions. Subsequently Sukumar et al. [23] coupled the method with the fast marching method to solve several planar crack growth problems in three dimensions.

In this paper, the methodology is extended and modified so that it can handle arbitrary cracks in three dimensions. A key development that facilitates treatment of cracks in three dimensions is the description of crack geometry in terms of two signed distance functions. The displacement field is also described in terms of these signed distance functions. This enables us to construct a near-tip asymptotic field with a discontinuity that conforms to the crack even when it is curved or kinked near a tip. Furthermore, it eliminates the need for a surface model of the crack. As a consequence, no explicit representation of the crack is needed and the crack is entirely described by nodal data. Although the method will here be described for elastic fracture, it is not limited to linear problems and can easily be extended to nonlinear problems.

We cannot list references to all of the competing methods, but we list some recent papers in the following. The remeshing approach appears to be the most advanced for problems of an industrial type; recent accounts are given by Carter et al. [5] and Neto et al. [18]. Methods which rely on boundary element formulations combined with finite elements are given in Dhondt [7], Gerstle et al. [10] and Nishioka and Atluri [19]. Duarte et al. [9] used the partition of unity concept with the visibility criterion to develop methods for dynamic three dimensional crack growth. Three dimensional dynamic crack growth by the element free Galerkin method has been reported by Krysl and Belytschko [13]. The crack surface was represented by a set of triangular elements in 3D, which would be very awkward in a finite element method. The use of finite elements with embedded discontinuities also makes it possible to grow cracks in 2D without remeshing, see Bolzon and Corigliano [4] and Jirasek and Zimmermann [12] for recent works on the topic.

The methodology for treating arbitrary three dimensional cracks and their evolution by X-FEM are described in this and a companion paper. This paper focuses on the description of cracks in three dimensions in terms of level sets, the computation of the elastic solution and the stress intensity factors (SIFs), whereas the companion paper deals with the update of the level sets needed to model crack growth.

The outline of the paper is as follows. In section 2, the methods for defining the crack geometry and the displacement fields are described. The level sets update scheme is given in Section 3. Section 4 reports the numerical experiment of a cracked beam under bending.

2. Crack and displacement field description

We consider a body Ω with an outer surface Γ and interior crack surfaces Γ_{cr}. The crack can be treated as a single surface or as two surfaces: Γ_{cr}^{+} and Γ_{cr}^{-}. In the latter case, the initial crack surfaces are considered coincident and the outward normals to the surface of the crack are denoted by $\boldsymbol{n}^{+}$ and $\boldsymbol{n}^{-}$, respectively.

Stolarska et al. [21] described a crack geometry in 2D by two signed distance functions. We also use two signed distance functions to describe a crack in 3D as shown in Fig. 1. Note that the definition of the two level set functions is only needed in a neighborhood of the crack. The signed distance function $\phi(\boldsymbol{x})$ defines the surface of the crack. It is given by

$$\phi(\boldsymbol{x}) = \min_{\overline{\boldsymbol{x}} \in \Gamma_{\mathrm{cr}}^{\mathrm{ext}}} \|\boldsymbol{x} - \overline{\boldsymbol{x}}\| \operatorname{sign}\left(\boldsymbol{n}^{+} \cdot (\overline{\boldsymbol{x}} - \boldsymbol{x})\right) \tag{1}$$

where $\boldsymbol{x} = [x, y, z]$ and $\text{sign}(\cdot)$ is the sign function $\text{sign}(x) = +1$ if $x > 0$ and -1 if $x < 0$. We also need a smooth extension of the crack surface Γ_{cr}, denoted by $\Gamma_{\text{cr}}^{\text{ext}}$, which includes the entire crack surface, i.e. $\Gamma_{\text{cr}} \subset \Gamma_{\text{cr}}^{\text{ext}}$. The distance function (1) gives the shortest distance of any point $\boldsymbol{x}$ to the extension of the crack surface $\Gamma_{\text{cr}}^{\text{ext}}$. This corresponds to the orthogonal projection of $\boldsymbol{x}$ on $\Gamma_{\text{cr}}^{\text{ext}}$.

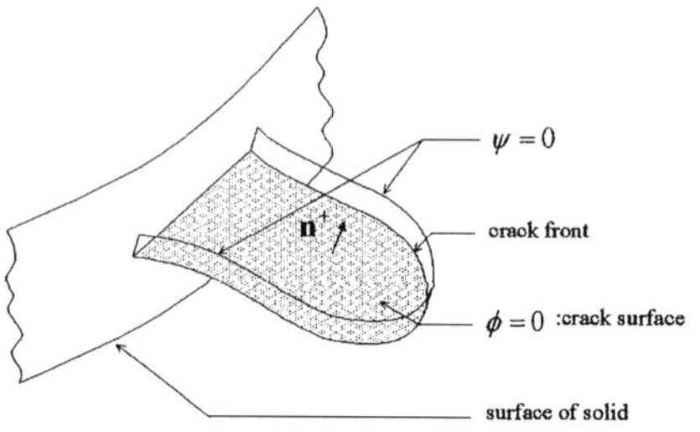

Figure 1. The two iso-zero level sets defining the crack location.

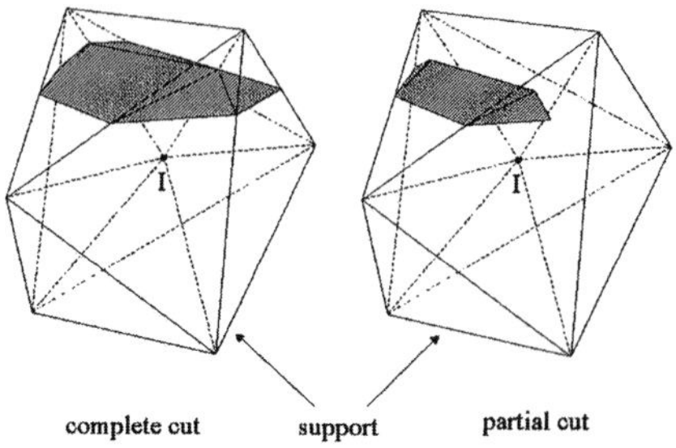

Figure 2. A node I whose support is completely and partially cut by the crack; the support is the volume of the elements connected to node I as shown.

Since the crack surface Γ_{cr} is a bounded surface with a crack front, it is also necessary to define the crack front. This is accomplished by using a second signed distance function, $\psi(\boldsymbol{x})$, which is approximately orthogonal to $\phi(\boldsymbol{x})$ so that $\nabla\phi \cdot \nabla\psi \sim 0$. The intersection of the surfaces $\psi(\boldsymbol{x}) = 0$ and $\phi(\boldsymbol{x}) = 0$ gives the crack front; we define the sign of ψ so that $\phi(\boldsymbol{x}) = 0$, $\psi(\boldsymbol{x}) < 0$ gives the crack surface Γ_{cr}.

The signed distance functions in this paper are approximated by the same shape functions as the displacement field. Therefore, in the computations the signed distance functions are given by

$$\phi = \sum_I N_I(\boldsymbol{x})\phi_I \tag{2}$$

$$\psi = \sum_I N_I(\boldsymbol{x})\psi_I \tag{3}$$

where N_I are the finite element shape functions and ϕ_I and ψ_I are the nodal values of the distance function. This enables the crack shape to be described entirely in terms of nodal values.

The displacement field $\boldsymbol{u}(\boldsymbol{x})$ for the body is decomposed into the continuous and discontinuous parts by

$$\boldsymbol{u} = \boldsymbol{u}_{\text{cont}} + \boldsymbol{u}_{\text{dis}} \tag{4}$$

where $\boldsymbol{u}_{\text{cont}}$ is continuous in Ω, whereas $\boldsymbol{u}_{\text{dis}}$ may have several surfaces of discontinuity in Ω. The locations of the discontinuities in $\boldsymbol{u}_{\text{dis}}$ are assumed to coincide with Γ_{cr}.

A standard finite element approximation is used for $\boldsymbol{u}_{\text{cont}}$, i.e.

$$\boldsymbol{u}_{\text{cont}} = \sum_{I \in \mathcal{N}} N_I(\boldsymbol{x})\boldsymbol{u}_I \tag{5}$$

where $\mathcal{N}$ is the set of all nodes in the mesh, N_I are the classical C^0 shape functions and $\boldsymbol{u}_I$ are displacement nodal degrees of freedom.

For the purpose of constructing the discontinuous field, the nodes are subdivided into three sets :

- $I \in \mathcal{N}_{\text{cut}}$: the set of nodes whose support (union of the elements connected to the node) are completely cut into two, i.e. bisected by the crack surface Γ_{cr}. An example of such a node is shown Figure 2.
- $I \in \mathcal{N}_{\text{branch}}$: the set of nodes whose support are partially cut by the crack surface Γ_{cr}, see Figure 2 for an example of such a node.
- $I \in \mathcal{N} - \mathcal{N}_{\text{cut}} - \mathcal{N}_{\text{cut}}$: the remaining nodes.

The discontinuous displacement fields are given as follows

$$\boldsymbol{u}_{\text{dis}} = \sum_{I \in \mathcal{N}_{\text{cut}}} N_I(\boldsymbol{x})H(\phi(\boldsymbol{x}))\boldsymbol{a}_I + \sum_{I \in \mathcal{N}_{\text{branch}}} \sum_{\alpha} N_I(\boldsymbol{x})B_\alpha(\phi(\boldsymbol{x}), \psi(\boldsymbol{x}))\boldsymbol{a}_{I\alpha}$$

In the above, $H(\cdot)$ is the Heaviside step function, $B_\alpha(.,.)$ are branch functions, and $\boldsymbol{a}_I$ and $\boldsymbol{a}_{I\alpha}$ are additional degrees of freedom for the displacement field. The branch functions are constructed in terms of the level sets functions

$$[B_\alpha] = [\sqrt{r}\sin\frac{\theta}{2}, \sqrt{r}\cos\frac{\theta}{2}, \sqrt{r}\sin\frac{\theta}{2}\sin\theta, \sqrt{r}\cos\frac{\theta}{2}\sin\theta] \tag{6}$$

where

$$r = \sqrt{\phi^2 + \psi^2}, \quad \theta = \tan^{-1}(\frac{\phi}{\psi}) \tag{7}$$

Note that the branch functions have been expressed in terms of the level set functions. By expressing the branch functions B_α in terms of the level set functions, it is guaranteed that the discontinuity always corresponds to $\phi = 0$ and $\psi < 0$, i.e. to the surface of the crack. The resulting field will not contain the exact basis of the asymptotic near field, but it is more important to construct the discontinuity in the correct place than to match the exact near-tip asymptotic field.

Only the first of the functions in Eq. (6) is discontinuous across $\phi = 0$. The others were added to improve the accuracy in elastic fracture problems. The above functions span the near-tip asymptotic solution for an elastic crack in two dimensions. In this study and previous studies, see Sukumar et al. [24], we have also found this basis to be quite accurate for three-dimensional cracks, although we have only considered smooth crack fronts.

This technique of adding asymptotic solutions through the partition of unity in finite elements can be considered an asymptotic matching technique. The displacement fields in the other elements provide the far-field, whereas the elements with the branch functions (6) provide the near field. The finite element procedure then matches these fields so that equilibrium is approximately satisfied.

Table 1. Scheme for level set update.

1 – extend V_ψ to the domain

$$\frac{\partial V_\psi}{\partial \tau} + \text{sign}(\psi)\frac{\nabla\psi}{\|\nabla\psi\|}\cdot\nabla V_\psi = 0 \quad \frac{\partial V_\psi}{\partial \tau} + \text{sign}(\phi)\frac{\nabla\phi}{\|\nabla\phi\|}\cdot\nabla V_\psi = 0 \tag{8}$$

2 – extend V_ϕ to the domain

$$\frac{\partial V_\phi}{\partial \tau} + \text{sign}(\phi)\frac{\nabla\phi}{\|\nabla\phi\|}\cdot\nabla V_\phi = 0 \quad \frac{\partial V_\phi}{\partial \tau} + \text{sign}(\psi)\frac{\nabla\psi}{\|\nabla\psi\|}\cdot\nabla V_\phi = 0 \tag{9}$$

3 – adjustment to prevent modification of previous crack surface

$$\overline{V}_\phi = H(\psi)\frac{V_\phi \psi}{V_\psi \Delta t} \tag{10}$$

4 – update and reinitialize the ϕ level set

$$\frac{\partial \phi}{\partial t} + \overline{V}_\phi\|\nabla\phi\| = 0 \quad \frac{\partial \phi}{\partial \tau} + \text{sign}(\phi)(\|\nabla\phi\| - 1) = 0 \tag{11}$$

5 – update the ψ level set

$$\frac{\partial \psi}{\partial t} + \overline{V}_\psi\|\nabla\psi\| = 0 \tag{12}$$

6 – orthogonalize and reinitialize the ψ level set

$$\frac{\partial \psi}{\partial t} + \text{sign}(\phi)\frac{\nabla\phi}{\|\nabla\phi\|}\cdot\nabla V_\psi = 0 \quad \frac{\partial \psi}{\partial \tau} + \text{sign}(\psi)(\|\nabla\psi\| - 1) = 0 \tag{13}$$

3. Level sets update

The initial level sets representing the crack location are given as data. For many cracks, level set functions which are not signed distance functions can easily be constructed. Then, they may be transformed into true signed distance function through a process called initialization, see Gravouil et al. [11]. The crack velocity on the crack front is given by

$$\boldsymbol{V} = V_\psi \boldsymbol{n}_\psi + V_\phi \boldsymbol{n}_\phi, \quad \boldsymbol{n}_\psi = \frac{\nabla\psi}{\|\nabla\psi\|}, \quad \boldsymbol{n}_\phi = \frac{\nabla\phi}{\|\nabla\phi\|} \tag{14}$$

where $\boldsymbol{n}_\psi$ and $\boldsymbol{n}_\phi$ are the unit normal vectors to the crack front and the crack surface, respectively. The computation of the velocity on the crack front is detailed in the next section. Table 1 summarizes the scheme for the level sets update. In steps 1 and 2, the crack front velocity is extended to all the nodes around the crack front. This extension procedure is classical in the level set method (Sethian [20]). The τ indicates a dummy time-like variable whereas the t symbol in the other steps of Table 1 indicates the true time. Equations (8) and (9) are solved in the "τ" time until V_ψ and V_ϕ are stationary, i.e. their derivatives with respect to τ is zero. The steps in Table 1 are explained in more details in Gravouil et al. [11].

The equations in Table 1 are Hamilton-Jacobi equations of the form

$$\frac{\partial f}{\partial t} + H(\nabla f, \boldsymbol{x}, t) = 0, \quad f(\boldsymbol{x}, 0) = f_0(\boldsymbol{x}) \tag{15}$$

where H is the Hamiltonian. In order to solve these equations they must be discretized in space and time. The level sets are approximated by finite elements:

$$\phi(\boldsymbol{x}, t) = \sum_I N_I(\boldsymbol{x})\phi_I(t), \quad \psi(\boldsymbol{x}, t) = \sum_I N_I(\boldsymbol{x})\psi_I(t) \tag{16}$$

The velocities are approximated by the same shape function:

$$V_\phi(\boldsymbol{x}, t) = \sum_I N_I(\boldsymbol{x})V_{\phi I}(t), \quad V_\psi(\boldsymbol{x}, t) = \sum_I N_I(\boldsymbol{x})V_{\psi I}(t) \tag{17}$$

The level set method has previously been applied to unstructured meshes by Barth and Sethian [1], and we used the same procedures.

4. Numerical experiment

A numerical experiment is performed for the growth of a crack in a beam under bending. The material properties are elastic and isotropic with Young's modulus $E = 280\,\mathrm{GPa}$ and Poisson's ratio $\nu = 0.3$. We consider fatigue crack growth

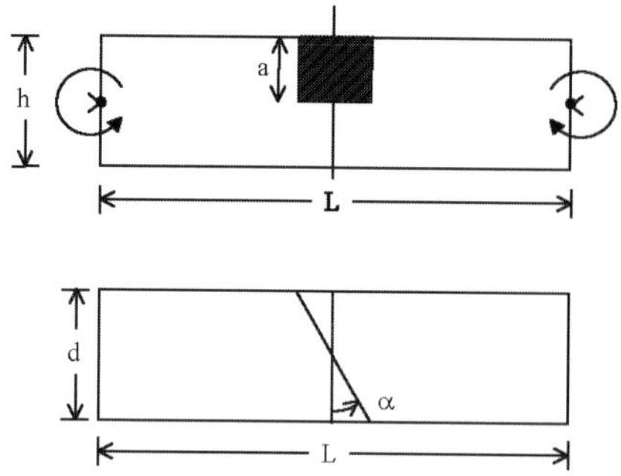

Figure 3. Side and top views of the beam with an initial crack.

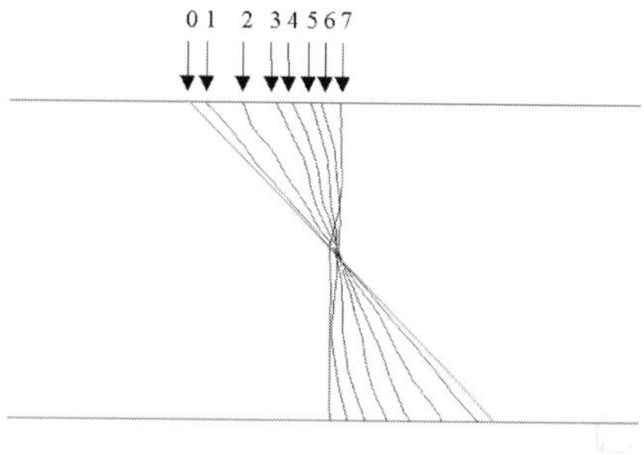

Figure 4. Evolution of the crack front in the beam seen from the top.

governed by the Paris law, which gives the rate of crack growth in mode I in terms of load cycles N by

$$\frac{da}{dN} = CG^m \tag{18}$$

where C is a constant to fit experimental results and$G$ is the maximum energy release rate. We consider the cycles N a time-like variable so the expression for the crack front velocity is

$$\boldsymbol{V} = CG^m(\cos\theta_c \boldsymbol{n}_\psi + \sin\theta_c \boldsymbol{n}_\phi) \tag{19}$$

where θ_c, the angle of the velocity to the plane tangent to the crack (or$\boldsymbol{n}_\phi$), is obtained by

$$\theta_c = 2\mathrm{arctan}\frac{1}{4}\left(\frac{K_I}{K_{II}} - \mathrm{sign}(K_{II})\sqrt{\left(\frac{K_I}{K_{II}}\right)^2 + 8}\right) \tag{20}$$

Thus, the crack growth direction depends on mode I and II stress intensity factors, whereas the crack speed on all three through the energy release rate$G$. In the numerical study, we chose $m = 1$ and $C = 1$.

We consider the beam shown in Figure 3 subjected to a bending load (see Lazarus et al. [14] for a theoretical study of this problem). The beam dimensions are $h = 0.02\,\mathrm{m}$, $l = 0.1\,\mathrm{m}$ and $d = 0.01\,\mathrm{m}$. An initial edge crack along a plane at an angle α to the plane of the cross-section is used to initiate crack growth with $a = 0.01\,\mathrm{m}$ and $\alpha = 45^o$ (see Figure 3). The initial geometry is discretized with an unstructured mesh of 2553 nodes and 13393 tetrahedrons. It is emphasized that the mesh does not conform to the crack geometry, and that the same mesh is used throughout the simulation. The crack is driven by Paris fatigue law with the maximum circumferential stress hypothesis. for the direction of crack propagation. From 15 to 25 points were used on the crack front to compute the velocity and the stress intensity factors. (The stress intensity factor computation is detailed in Moës et al. [17]). In this study, the beam is completely cut into two after 17 time steps. Figure 4 shows the evolution of the crack front from a top view. We can observe that the crack grows asymptotically to a plane orthogonal to the axis of the beam.

Figure 5 shows the crack after 12 time steps, and the vector velocity field on the crack front. The colors give the value of the signed distance function ψ which defines the signed distance from the crack front. An important remark concerns the topological properties of the method: with the level set approach, the crack does not need any specific treatment on the boundary, and the crack can completely cut the beam.

Figure 6 shows the evolution of a semi-circular crack. Again, we observe that the crack grows asymptotically to plane orthogonal to the axis of the beam. Here, the crack front first grows to the lateral free surfaces, then normal to the surface. No changes are needed in the algorithm to account for these topological changes. Other crack propagation examples may be found in Gravouil et al. [11].

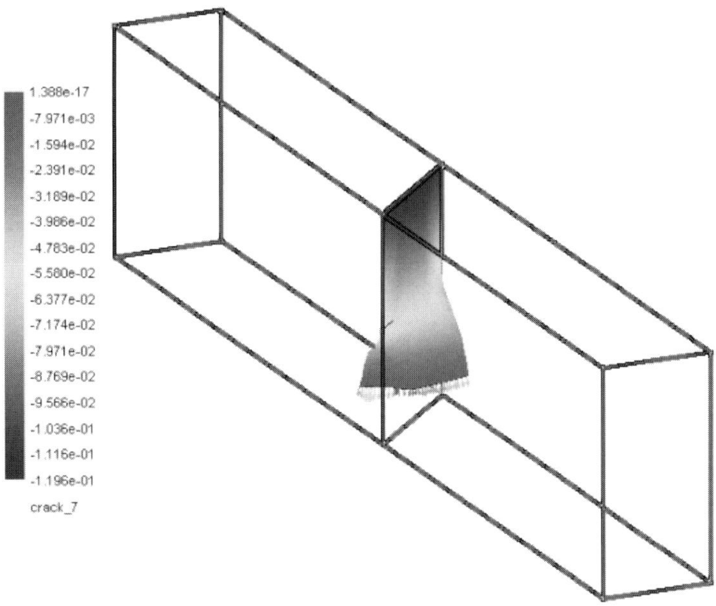

Figure 5. Crack location after 12 times steps.

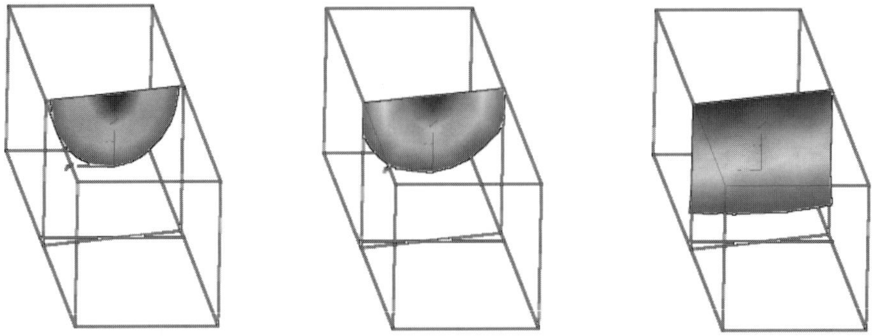

(a) (b) (c)

Figure 6. Evolution of a semi-circular crack in a beam.

5. Conclusions

A level set method for arbitrary non-planar cracks in three-dimensional bodies has been presented. The level set technique couples naturally to the extended finite element method, wherein the discontinuous and near-tip asymptotic fields are constructed through a partition of unity. The resulting combined method requires no explicit representation of the crack except in its visualization. Instead, the crack and its growth are described entirely in forms of nodal data. This simplifies the structure of the software and leads to great versatility in treating complex problems in crack growth.

References

[1] Barth, T.J. and Sethian, J.A. (1998). Numerical schemes for the Hamilton-Jacobi and level set equations on triangulated domains. *Journal of Computational Physics*, **145(1)**,1–40.

[2] Belytschko, T. and Black, T. (1999). Elastic crack growth in finite elements with minimal remeshing. *International Journal for Numerical Methods in Engineering*, **45(5)**, 601–620.

[3] Belytschko, T., Moës, N., Usui, S. and Parimi, C. (2001). Arbitrary discontinuities in finite elements. *International Journal for Numerical Methods in Engineering*, **50**, 993–1013.

[4] Bolzon, G. and Corigliano, A. (2001). Finite elements with embedded displacement discontinuity: a generalized formulation. *International Journal for Numerical Methods in Engineering*, **549(10)**, 1227–1266.

[5] Carter, B., Wawrzynek, P. and Ingraffea, A. (2000). Automated 3d crack growth simulation. *International Journal for Numerical Methods in Engineering*, **47**, 229–253.

[6] Daux, C., Moës, N., Dolbow, J., Sukumar, N. and Belytschko, T. (2000). Arbitrary branched and intersecting cracks with the eXtended Finite Element Method. *International Journal for Numerical Methods in Engineering*, **48**, 1741–1760.

[7] Dhondt, G. (1998). automatic 3-d mode i crack propagation calculations with finite elements. *International Journal for Numerical Methods in Engineering*, **41(4)**, 739–757.

[8] Dolbow, J., Moës, N. and Belytschko, T. (2000). Discontinuous enrichment in finite elements with a partition of unity method. *Finite elements in analysis and design*, **36**, 235–260.

[9] Duarte, C.A., Hamzeh, O.N., Liszka, T.J. and Tworzydlo, W.W. (2000). The element partition method for the simulation of three-dimensional dynamic crack propagation. *Comp. Meth. in Applied Mech. and Engrg.* In press.

[10] Gerstle, W.H., Martha, L. and Ingraffea, A.R. (1987). Finite and boundary element modeling of crack propagation in two- and three-dimensions. *Engineering with Computers*, **2**, 167–183.

[11] Gravouil, A., Moës, N. and Belytschko, T. (2001). Non-planar 3d crack growth by the extended finite element and level sets. part II: level set update. *International Journal for Numerical Methods in Engineering.* Submitted.

[12] Jirasek, T. and Zimmermann, T. (2001). Embedded crack model: Part I basic formulation. *International Journal for Numerical Methods in Engineering*, **50(6)**, 1269–1290.

[13] Krysl, P. and Belytschko, T. (1999). Element free Galerkin method for dynamic propagation of arbitrary 3-d cracks. *International Journal for Numerical Methods in Engineering*, **44(6)**, 767–800.

[14] Lazarus, V., Leblond, J.-B. and Mouchrif, S.-E. (2001). Crack front rotation and segmentation in mixed mode i+iii - part i: Calculation of stress intensity factors. *J. Mech. Phys. Solids*, **49(7)**, 1399–1420.

[15] Melenk, J. and Babuška, I. (1996). The partition of unity finite element method: Basic theory and applications. *Comp. Meth. in Applied Mech. and Engrg.*, **39**, 289–314.

[16] Moës, N., Dolbow, J. and Belytschko, T. (1999). A finite element method for crack growth without remeshing. *International Journal for Numerical Methods in Engineering*, **46**, 131–150.

[17] Moës, N., Gravouil, A. and Belytschko, T. (2001). Non-planar 3d crack growth by the extended finite element and level sets. part i: Mechanical model. *International Journal for Numerical Methods in Engineering.* Submitted.

[18] Neto, J.C., Wawrzynek, P., Carvalho, M., Martha, L. and Ingraffea, A. (2001). An algorithm for three-dimensional mesh generation for arbitrary regions with cracks. *Engineering with Computers*, **17**, 75–91.

[19] Nishioka, T. and Atluri, S. (1983). Analytical solution for embedded cracks, and finite element alternating method for elliptical surface cracks, subjected to arbitrary loading. *Engineering Fracture Mechanics*, **17**, 247–268.

[20] Sethian, J.A. (1999). *Level Set Methods & Fast Marching Methods: Evolving Interfaces in Computational Geometry, Fluid Mechanics, Computer Vision, and Materials Science.* Cambridge University Press, Cambridge, UK.

[21] Stolarska, M., Chopp, D.L., Moës, N. and Belytschko, T. (2001). Modelling crack growth by level sets and the extended finite element method. *International Journal for Numerical Methods in Engineering*, **51(8)**, 943–960.

[22] Sukumar, N., Chopp, D.L., Moës, N. and Belytschko, T. (2001). Modeling holes and inclusions by level sets in the extended finite element method. *Comp. Meth. in Applied Mech. and Engrg.* Accepted for publication.

[23] Sukumar, N., Chopp, D.L. and Moran, B. (2001). Extended finite element method and fast marching method for three-dimensional fatigue crack propagation. *Engineering Fracture Mechanics*. Submitted.

[24] Sukumar, N., Moës, N., Belytschko, T. and Moran, B. (2000). Extended Finite Element Method for three-dimensional crack modelling. *International Journal for Numerical Methods in Engineering*, **48(11)**, 1549–1570.

A Large Strain Discontinuous Finite Element Approach to Laminated Composites

G.N. Wells, J.J.C. Remmers, R. de Borst, L.J. Sluys
Koiter Institute Delft, Delft University of Technology
P.O. Box 5048, 2600 GA Delft, The Netherlands
R.deborst@lr.tudelft.nl

Abstract: A numerical model is presented for the geometrically nonlinear analysis of laminated composite materials. The formulation is derived in a consistent fashion from nonlinear continuum mechanics. The finite element formulation is briefly described, and a range of examples elucidate the performance and potential of the model. It is shown that geometric instabilities due to delamination can be captured with unstructured finite element meshes.

Keywords: Delamination, discontinuities, delamination buckling, finite elements, partition-of-unity.

1. Introduction

Laminated composite materials are being increasingly used for advanced engineering constructions. They can be customized for special applications and generally possess characteristics superior to more traditional materials. For aerospace applications, laminated composites are generally lighter, less sensitive to fatigue and corrosion resistant when compared to traditional aluminium alloys.

A key failure mechanism of laminated composites is delamination – the debonding of individual laminae. The proper modeling of this failure mechanism requires the explicit simulation of the debonding process. Previously, this has been performed using interface finite elements to capture the displacement jump between finite elements. A comprehensive review of interface element formulations and the related interface constitutive models can be found in Alfano and Crisfield [1].

Recently, new finite element-based numerical techniques have evolved that allow the simulation of arbitrary displacement discontinuities (Belytschko and Black [4], Moës et al. [6]). These methods have been extended to simulate cohesive cracks (Wells and Sluys [10]) and shear bands (Wells and Sluys [9]) under both quasi-static and impact loadings.

In this work, the derivation of a geometrically non-linear model for the simulation of delamination is reviewed. Following the formulation, a range of

C. Miehe (ed.), IUTAM Symposium on Computational Mechanics of Solid Materials at Large Strains, 355-364.

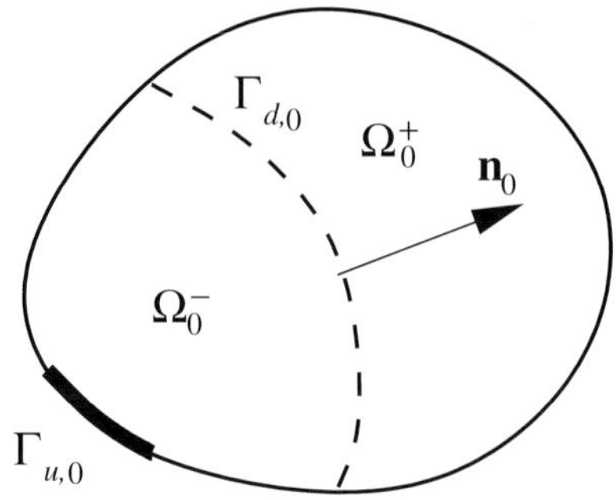

Figure 1. Material body crossed by a discontinuity surface (dashed line).

numerical examples are presented. The examples involve standard peel tests as well as a numerical experiment in delamination buckling.

2. Non-linear kinematic description

The key to the proposed formulation is the construction of appropriate kinematic relationships to describe delamination in a nonlinear setting. To do this, a body, crossed by a discontinuity is considered (figure 1). The position vector of a material point in the current (spatial) configuration $\mathbf{x}$ is given by:

$$\mathbf{x} = \mathbf{X} + \hat{\mathbf{u}} + \mathscr{H}_{\Gamma_d,0}\tilde{\mathbf{u}} \tag{1}$$

where $\mathbf{X}$ is the position vector of the point in the reference (material) configuration and the displacement vector of the material point is given by $\hat{\mathbf{u}} + \mathscr{H}_{\Gamma_d,0}\tilde{\mathbf{u}}$. The displacement is described in terms of a continuous component $\hat{\mathbf{u}}$, and a component which is discontinuous across the discontinuity surface, $\mathscr{H}_{\Gamma_d}\tilde{\mathbf{u}}$. The field $\tilde{\mathbf{u}}$ is continuous, with the jump provided by the Heaviside function $\mathscr{H}_{\Gamma_d,0}$, which is defined as being equal to unity on Ω^+ and zero elsewhere. For a laminated composite, $\mathscr{H}_{\Gamma_d,0}$ would be equal to unity on one side of an interface and zero on the other.

The deformation gradient $\mathbf{F}$ is obtained by taking the gradient of equation (1) with respect to the reference configuration. This leads to:

$$\mathbf{F} := \mathbf{I} + \nabla_{\mathbf{X}}\hat{\mathbf{u}} + \mathscr{H}_{\Gamma_d,0}\nabla_{\mathbf{X}}\tilde{\mathbf{u}} + \delta_{\Gamma_d}\left(\tilde{\mathbf{u}} \otimes \mathbf{n}_0\right) \tag{2}$$

where δ_{Γ_d} is the Delta function, located at the discontinuity surface, and $\mathbf{n}_0$ is the normal to the discontinuity surface in the reference configuration. It is important to note the presence of the Heaviside function in the expression for the deformation gradient. This implies that the deformation gradient from the left and right-hand sides of the discontinuity surface are *not* necessarily equal.

From Nanson's relationship, the vector to a surface in the current configuration is a function of the normal vector to the surface in the reference configuration and the deformation gradient. The lack of uniqueness of the deformation gradient at a discontinuity surface implies that the normal vector can have two values in the updated configuration. This asserts that geometric instabilities at an interface can develop, with individual laminae buckling. In works based on "embedded discontinuities" (Armero and Garikipati [3], Steinmann et al. [7], Larsson et al. [5]), it is required that $\tilde{\mathbf{u}}$ be a constant function, which leads to the elimination of the Heaviside function from the deformation gradient. This assumption precludes the development of geometric instabilities at an interface as the material faces of the discontinuity in the current configuration must remain parallel.

3. Variational formulation

In terms of the nominal stress $\mathbf{P}$, the weak equilibrium equation can be written as:

$$\int_{\Omega_0} \nabla_{\mathbf{X}} \mathbf{w} : \mathbf{P} \, \mathrm{d}\Omega_0 = \int_{\Gamma_{u,0}} \mathbf{w} \cdot \bar{\mathbf{T}} \, \mathrm{d}\Gamma_0 \qquad \forall \mathbf{w} \in W \tag{3}$$

where $\mathbf{w}$ is a displacement variation, W is the space of admissible displacement variations and $\bar{\mathbf{T}}$ is the nominal traction acting on part of the exterior boundary of Ω, Γ_u (see figure 1). Decomposing the function $\mathbf{w}$ in a similar fashion to the displacement field in equation (1) leads to:

$$\nabla_{\mathbf{X}} \mathbf{w} = \nabla_{\mathbf{X}} \hat{\mathbf{w}} + \mathscr{H}_{\Gamma_d,0} \nabla_{\mathbf{X}} \tilde{\mathbf{w}} + \delta_{\Gamma_d} \left(\tilde{\mathbf{w}} \otimes \mathbf{n}_0 \right) \tag{4}$$

where $\hat{\mathbf{w}}, \tilde{\mathbf{w}} \in W$. Inserting the above expression into equation (3) yields:

$$\begin{aligned} \int_{\Omega_0} \nabla_{\mathbf{X}} \hat{\mathbf{w}} : \mathbf{P} \, \mathrm{d}\Omega_0 + \int_{\Omega_0^+} \nabla_{\mathbf{X}} \tilde{\mathbf{w}} : \mathbf{P} \, \mathrm{d}\Omega_0 + \int_{\Omega_0} \delta_{\Gamma_d} \left(\tilde{\mathbf{w}} \otimes \mathbf{n}_0 \right) : \mathbf{P} \, \mathrm{d}\Omega_0 & \\ = \int_{\Gamma_{u,0}} \hat{\mathbf{w}} \cdot \bar{\mathbf{T}} \, \mathrm{d}\Gamma_0 + \int_{\Gamma_{u,0}} \mathscr{H}_{\Gamma_d,0} \tilde{\mathbf{w}} \cdot \bar{\mathbf{T}} \, \mathrm{d}\Gamma_0 \qquad \forall \hat{\mathbf{w}}, \tilde{\mathbf{w}} \in W & \end{aligned} \tag{5}$$

The Heaviside function has been eliminated from the above equation by changing the appropriate integration domains from Ω_0 to Ω_0^+. Using the integral property of the Delta function, the above variational statement can be expressed as:

$$\int_{\Omega_0} \nabla_{\mathbf{X}} \hat{\mathbf{w}} : \mathbf{P} \, \mathrm{d}\Omega_0 = \int_{\Gamma_{u,0}} \hat{\mathbf{w}} \cdot \bar{\mathbf{T}} \, \mathrm{d}\Gamma_0 \qquad \forall \hat{\mathbf{w}} \in W \tag{6a}$$

$$\int_{\Omega_0^+} \nabla_{\mathbf{X}} \tilde{\mathbf{w}} : \mathbf{P} \, \mathrm{d}\Omega_0 + \int_{\Gamma_{d,0}} \left(\tilde{\mathbf{w}} \otimes \mathbf{n}_0 \right) : \mathbf{P} \, \mathrm{d}\Gamma_0 = \int_{\Gamma_{u,0}} \mathscr{H}_{\Gamma_d,0} \tilde{\mathbf{w}} \cdot \bar{\mathbf{T}} \, \mathrm{d}\Gamma_0 \quad \forall \tilde{\mathbf{w}} \in W \tag{6b}$$

Using standard relationships from continuum mechanics, the variational statements in (6) can be pushed forward to the current configuration. This yields:

$$\int_{\Omega} \nabla_{\mathbf{x}} \hat{\mathbf{w}} : \sigma \, \mathrm{d}\Omega = \int_{\Gamma_u} \hat{\mathbf{w}} \cdot \bar{\mathbf{t}} \, \mathrm{d}\Gamma \qquad \forall\, \hat{\mathbf{w}} \in W \tag{7a}$$

$$\int_{\Omega^+} \nabla_{\mathbf{x}} \tilde{\mathbf{w}} : \sigma \, \mathrm{d}\Omega + \int_{\Gamma_d} \tilde{\mathbf{w}} \cdot \mathbf{t} \, \mathrm{d}\Gamma = \int_{\Gamma_u} \mathscr{H}_{\Gamma_d} \tilde{\mathbf{w}} \cdot \bar{\mathbf{t}} \, \mathrm{d}\Gamma \quad \forall\, \tilde{\mathbf{w}} \in W \tag{7b}$$

where $\mathbf{t}$ is the Cauchy traction acting at the discontinuity surface. The definition of the traction $\mathbf{t}$ requires the definition of the normal vector to the discontinuity surface in the current configuration. As discussed in the previous section, the push-forward of the normal vector involves the deformation gradient at the discontinuity surface which is not unique. It is chosen to relate the traction vector to the mid-surface between the two free surfaces. This corresponds to using $\hat{\mathbf{F}} + \frac{1}{2}\tilde{\mathbf{F}}$ to calculate the updated normal vector $\mathbf{n}$.

Linearization of the preceding variational statements is a lengthy process. The linearization of terms involving volume integrals follows standard arguments. The linearization of terms relating to the discontinuity surface requires some special considerations. For details of the linearization, the reader is referred to Wells et al. [8].

4. Finite element formulation

The finite element formulation involves interpolating the functions $\hat{\mathbf{u}}$ and $\tilde{\mathbf{u}}$ in equation (1) independently. The finite element formulation is similar to that developed by Belytschko and Black [4] and Moës et al. [6] for cracks in linear elastic media. The position vector of a material point in current configuration is given by:

$$\mathbf{x} = \mathbf{X} + \underbrace{\mathbf{Na}}_{\hat{\mathbf{u}}} + \mathscr{H}_{\Gamma_d,0} \underbrace{\mathbf{Nb}}_{\tilde{\mathbf{u}}} \tag{8}$$

In this manner, a displacement discontinuity can be simulated independently of the mesh structure. Discontinuities can pass through solid finite elements or along the boundary between elements. The discretization in equation (8) leads to a compatible formulation, with the order of the displacement jump interpolation along a discontinuity surface equal to the polynomial order of the underlying element. There are no restrictions on the type of underlying finite element. The practical implementation of the procedure involves the activation of extra degrees of freedom at nodes in the region of a discontinuity. If the support of an element is crossed, two extra degrees of freedom (in two dimensions) are active at a node, giving four degrees of freedom in total. It is emphasised that no new nodes are created. This is illustrated in figure 2 for an unstructured mesh composed of six-noded elements. At the highlighted nodal points, four degrees of freedom are active.

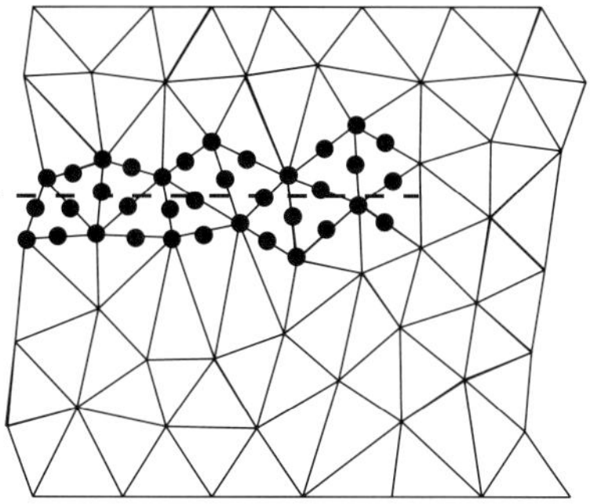

Figure 2. Section of unstructured mesh crossed by a discontinuity surface. At the highlighted nodal points, four degrees of freedom are active.

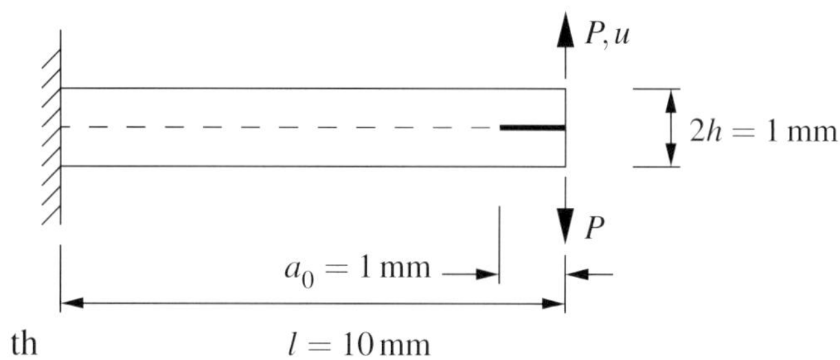

Figure 3. Geometry of a double cantilever beam peel test.

5. Numerical examples

The numerical examples in this section are chosen to illustrate the performance of the model for problems involving large deformations . For all examples, an isotropic hyperelastic response (compressible neo-Hookean) has been assumed for the laminae. There is however no restriction on the constitutive response and the formulation extends to any finite strain constitutive model.

5.1. PEEL TEST

Consider the double cantilever beam shown in figure 3. Both layers of the beam are made of the same material with Young's modulus $E = 100.0\ \mathrm{N/mm^2}$ and Poisson's ratio $\nu = 0.3$. The two beams are connected by an adhesive material with an ultimate peel traction $t_{\mathrm{ult}} = 1.0\ \mathrm{N/mm^2}$. The fracture toughness is equal to $G_c = 0.05$ N/mm. An initial delamination at the tip with length $a_0 = 1.0$ mm is assumed.

Debonding of the adhesive is simulated with a linear damage model. The traction in the shear direction after crack initiation is assumed to be zero. The ini-

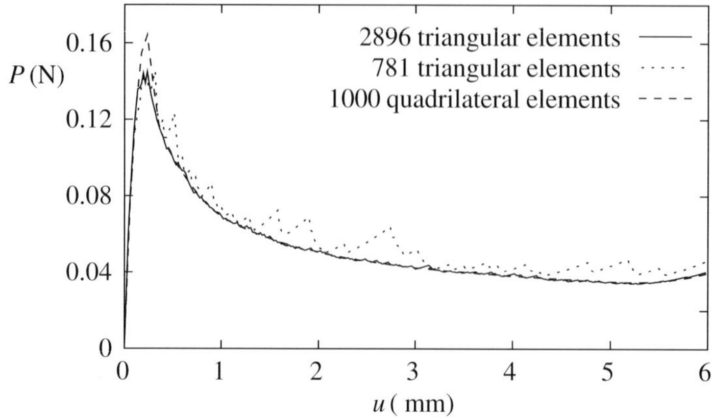

Figure 4. Load-displacement curve for the double cantilever beam peel test.

tial delamination is modeled as a traction free discontinuity. The simulations are performed for various meshes, with quadratic triangular and quadrilateral elements.

Figure 4 shows the tip displacement of the double cantilever beam as a function of the applied load. Since the stress gradient at the delamination front is high, a dense mesh is required to obtain accurate results. The load-displacement curve for the coarse mesh is globally correct. The oscillations in the response can be attributed to the large element size.

5.2. PEEL TEST FROM A STIFF SUBSTRATE

In the previous example, the normal vector to the discontinuity mid-surface remains constant. In the following example, the normal vector to the discontinuity mid-surface changes as delamination develops. Consider a laminate as shown in figure 5. The top layer is peeled off by a single force P. Both layers are made of the same material, with Young's modulus $E = 100.0\,\mathrm{N/mm^2}$ and Poisson's ratio $\nu = 0.3$. The ultimate peel traction of the adhesive is equal to $t_{\mathrm{ult}} = 1.0\,\mathrm{N/mm^2}$. After debonding, a perfectly plastic type response is assumed. The two layers are initially delaminated over a length $a_0 = 1.0\,\mathrm{mm}$.

The test is performed with two meshes, the first with unstructured triangular elements and the second with structured quadrilateral elements. The deformed meshes are shown in figure 6. Note that the deformations are not magnified. It is clear in figure 6 that the deformed configurations are the same for the two meshes, in spite of the obvious differences in structure. It should be mentioned that despite the elongated appearance of some elements, the volume change in the elements is close to zero. The elongated appearance is due to the presence of a discontinuity through the elements.

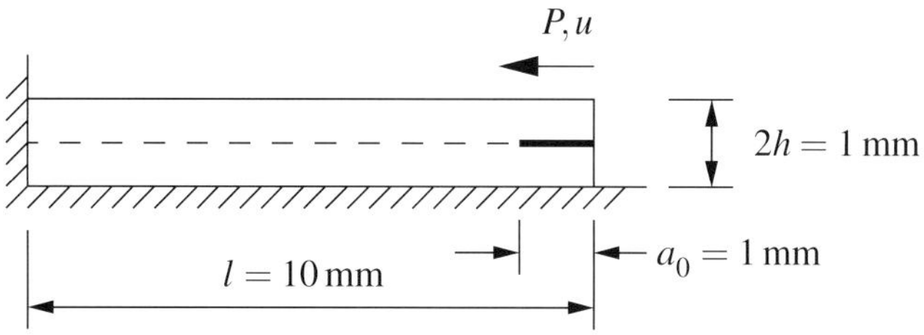

Figure 5. Geometry of specimen for peel test from a stiff substrate.

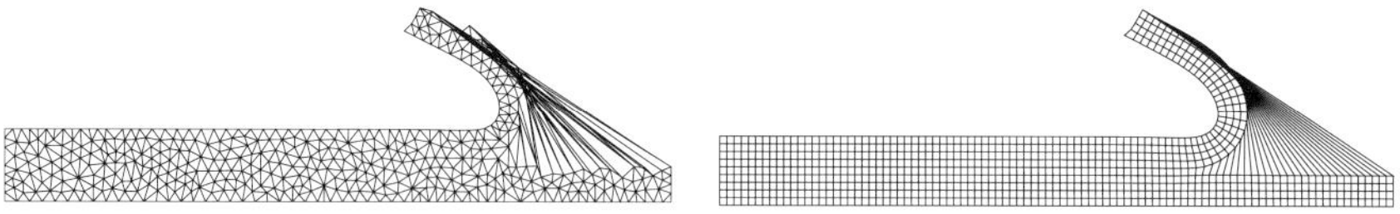

Figure 6. Deformed meshes for the peel test from a stiff substrate at $u = 4$ mm.

5.3. Delamination Buckling Test

Failure as a combination of delamination growth and structural instability is considered in the following example. This example was analyzed using conventional interface elements by Allix and Corigliano [2]. Figure 7 shows a double cantilever beam with an initial delamination length of $a_0 = 10.0$ mm. The beam is subjected to an axial compressive force $2P$. Two small perturbation forces $P_0 = 0.001$ N are applied to trigger the desired buckling mode. Both layers are made of the same material with Young's modulus $E = 135{,}000$ N/mm^2 and Poisson's ratio 0.18. The ultimate strength of the interlaminar bonding in mode-I is $t_{\mathrm{I,ult}} = 50.0$ N/mm^2. The fracture toughness is set to $G_c = 0.8$ N/mm. The critical load for local buckling of the beam, prior to delamination growth, can be calculated analytically using the equation for a single cantilever beam with length a_0 and height h. For the configuration in figure 7 and the adopted elastic material properties, the analytical buckling load is equal to:

$$P_{cr} = \frac{\pi^2 E h^3}{48 a_0^2} = 2.22\,\mathrm{N}. \tag{9}$$

The finite element mesh used for the analysis consists of 750 eight-noded quadrilateral elements, with three elements through the thickness and 250 elements in the length. In order to emphasize the performance of the model with respect to spatial discretization, the calculations have also been performed with a much coarser mesh that consists of just one element through the thickness and 100 elements in the length direction. Note that in the latter case, delamination is modeled (although crudely) with just one layer of elements!

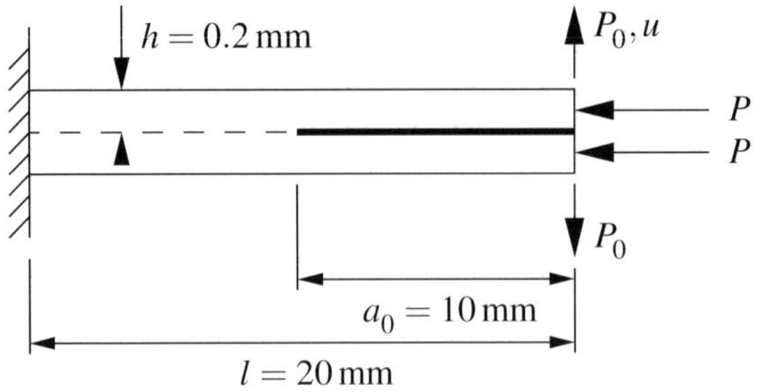

Figure 7. Double cantilever beam with initial delamination under compression.

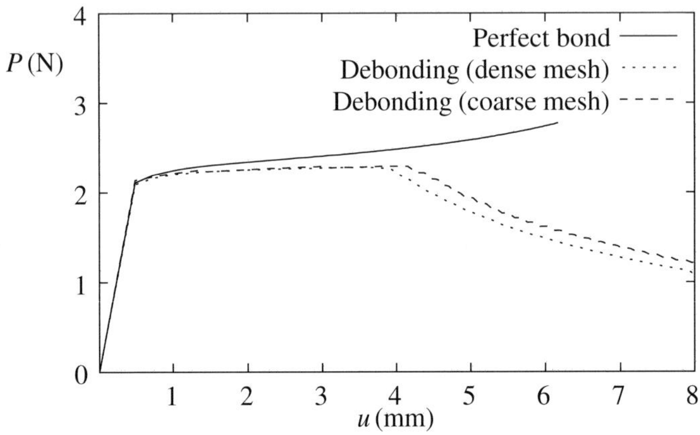

Figure 8. Load displacement curve for delamination buckling test.

Figure 8 shows the lateral displacement u of the beam as a function of the external force P. The load-displacement curve for a specimen with a perfect bond (no delamination) is given as a reference. The numerically calculated buckling load is in agreement with the analytical calculation. When delamination is allowed to develop, steady delamination propagation begins at a lateral displacement $u = 4\,\text{mm}$, which is evident in figure 8. From this point, material instabilities interact with geometric instabilities. Figure 9 shows the final deformation for the coarse mesh at a tip displacement $u = 6\,\text{mm}$.

6. Conclusions

A formulation for the geometrically non-linear analysis of laminated composite materials has been presented. The formulation uses concepts from non-linear continuum mechanics, and avoids premature discretization of the governing equations. This results in a transparent and consistent formulation, which highlights the necessary assumptions which can be clouded when con-

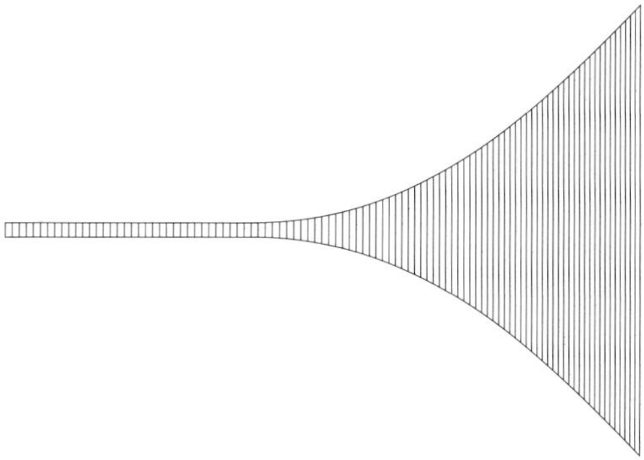

Figure 9. Deformation of the coarse mesh after buckling and delamination growth (tip displacement 6 mm).

sidering the governing equations in a discretized format. The numerical implementation allows displacement discontinuities to pass arbitrarily through solid finite elements.

References

[1] Alfano, G. and Crisfield, M.A. (2001). Finite element interface models for the delamination analysis of laminated composites: mechanical and computational issues. *International Journal for Numerical Methods in Engineering*, **50(7)**, 1701–1736.

[2] Allix, O. and Corigliano, A. (1999). Geometrical and interfacial non-linearities in the analysis of delamination in composites. *International Journal of Solids and Structures*, **36(15)**, 2189–2216.

[3] Armero, F. and Garikipati, K. (1996). An analysis of strong discontinuities in multiplicative finite strain plasticity and their relation with the numerical simulation of strain localization. *International Journal of Solids and Structures*, **33(20–22)**, 2863–2885.

[4] Belytschko, T. and Black, T. (1999). Elastic crack growth in finite elements with minimal remeshing. *International Journal for Numerical Methods in Engineering*, **45(5)**, 601–620.

[5] Larsson, R., Steinmann, P., and Runesson, K. (1998). Finite element embedded localization band for finite strain plasticity based on a regularized strong discontinuity. *Mechanics of Cohesive-Frictional Materials*, **4(2)**, 171–194.

[6] Moës, N., Dolbow, J., and Belytschko, T. (1999). A finite element method for crack growth without remeshing. *International Journal for Numerical Methods in Engineering*, **46(1)**, 131–150.

[7] Steinmann, P., Larsson, R., and Runesson, K. (1997). On the localization properties of multiplicative hyperelasto-plastic continua with strong discontinuities. *International Journal of Solids and Structures*, **34(8)**, 969–990.

[8] Wells, G.N., De Borst, R., and Sluys, L.J. (2001). A consistent geometrically non-linear approach for delamination. *International Journal for Numerical Methods in Engineering.* (accepted).

[9] Wells, G.N. and Sluys, L.J. (2001). Discontinuous analysis of softening solids under impact loading. *International Journal for Numerical and Analytical Methods in Geomechanics*, **25(7)**, 691–709.

[10] Wells, G.N. and Sluys, L.J. (2001). A new method for modelling cohesive cracks using finite elements. *International Journal for Numerical Methods in Engineering*, **50(12)**, 2667–2682.

Cohesive Zone Modeling of Crack Growth along Different Functionally Graded Joints between Two Materials

V. Tvergaard
Department of Mechanical Engineering, Solid Mechanics, Technical University of Denmark
Nils Koppels Allé, Building 404, 2800 Kgs. Lyngby, Denmark
viggo@fam.dtu.dk

Abstract: For crack growth in a functionally graded layer joining dissimilar elastic-plastic solids the effect of different grading profiles through the layer is investigated. The behavior is related to interface crack growth along a sharp interface between the two solids. The analyses are carried out for conditions of small scale yielding, and the boundary conditions applied on the outer edge of the region analyzed are displacements for the elastic oscillating stress singularity fields corresponding to a sharp interface. The fracture process is represented by a cohesive zone model, with the peak stress and the work of separation per unit area as basic parameters. Different parabolic variations of material properties through the graded layer are compared, for cracks growing along planes with the same local material properties.

Keywords: Cohesive zone modeling, crack growth, plasticity.

1. Introduction

The effect of plastic yielding on the resistance against crack growth along an interface joining dissimilar solids has been analyzed by Tvergaard and Hutchinson [10] for a case where the substrate is rigid, and Tvergaard [11] has analyzed cases where both materials joined at the interface are taken to be elastic-plastic. In these numerical studies the fracture process has been modeled by a traction-separation law, and it has been shown that the macroscopic work of fracture is much larger than the work absorbed by the local fracture process, so that a significant part of the fracture toughness results from the plastic work during crack growth. Analyses for different mode mixity have shown that the predicted fracture toughness is much increased when mode II loading gives a significant contribution at the crack tip, relative to the levels found for pure mode I loading at the tip, as also found experimentally (e.g. see Liechti and Chai [5], Cao and Evans [1]).

In a number of cases where two materials are joined, the properties vary gradually over a certain interval, rather than changing abruptly at a sharp interface. Such graded composite materials have been studied by Reiter et al. [7], and a broad overview of functionally graded composites or metals has been given by Mortensen and Suresh [6, 9]. Cracks in functionally graded materials have been

C. Miehe (ed.), IUTAM Symposium on Computational Mechanics of Solid Materials at Large Strains, 365-374.

studied by Erdogan [2] and Kim et al. [4]. Recently, in a study of crack growth along a graded interface layer between dissimilar materials, Tvergaard [12] has assumed linear variations of the material properties across the graded layer. In the present study these analyses are extended to also consider the sensitivity to nonlinear grading variations.

2. Problem formulation and numerical procedure

The crack in a functionally graded layer between two different elastic-plastic solids is illustrated in Fig. 1. Material No. 1 is mostly in the range of positive x^2-coordinates, while material No. 2 is mostly in the range of negative x^2-coordinates. Both materials are taken to be elastic-plastic with the true stress-logarithmic strain curve in uniaxial tension specified by

$$\epsilon = \begin{cases} \sigma/E & \text{for} \quad \sigma \leq \sigma_Y \\ (\sigma_Y/E)(\sigma/\sigma_Y)^{1/N} & \text{for} \quad \sigma \geq \sigma_Y \end{cases} \tag{1}$$

Here, σ_Y is the initial yield stress and N is the power hardening exponent, while E and ν are Young's modulus and Poisson's ratio, respectively. For material No. 1 the parameters are denoted E_1, ν_1, σ_{Y1} and N_1, while for material No. 2 they are E_2, ν_2, σ_{Y2} and N_2. In the functionally graded material layer of initial width D the same material description is assumed, but the values of the four material parameters inside the layer are taken to vary nonlinearly between the properties of the two neighboring solids, as e.g. expressed for the initial yield stress by the equation

$$\sigma_Y(r) = \sigma_{Y2} + (\sigma_{Y1} - \sigma_{Y2})(r - \gamma r(1-r)), \tag{2}$$

where $r = (x^2 - D_1 + D)/D$ is a parameter ranging from 0 at the bottom of the layer to 1 at the top of the layer (see Figs. 1 and 2). For $\gamma = 0$ Eq. (2) represents a linear interpolation, while $\gamma \neq 0$ gives a parabolic variation of σ_Y through the layer (Fig. 2).

A small scale yielding formulation of the crack problem is used, assuming that the layer with D is small relative to the outer radius A_0 of the region analyzed. Then, the elastic crack tip singularity field corresponding to a sharp interface is sufficiently accurate far away from the crack tip. Following the formulation of Rice [8] (see also Tvergaard and Hutchinson [10]) the crack has tractions acting on the interface, which are given in terms of the two stress intensity factor components, K_I and K_{II}, by

$$\sigma_{22} + i\sigma_{12} = (K_I + iK_{II})(2\pi r)^{-1/2} r^{i\epsilon}, \tag{3}$$

where r is the distance from the tip, $i = \sqrt{-1}$, and ϵ is an oscillation index given in terms of the second Dundurs' parameter β. The magnitude of stress intensity factors is specified as $|K| = \sqrt{K_I^2 + K_{II}^2}$.

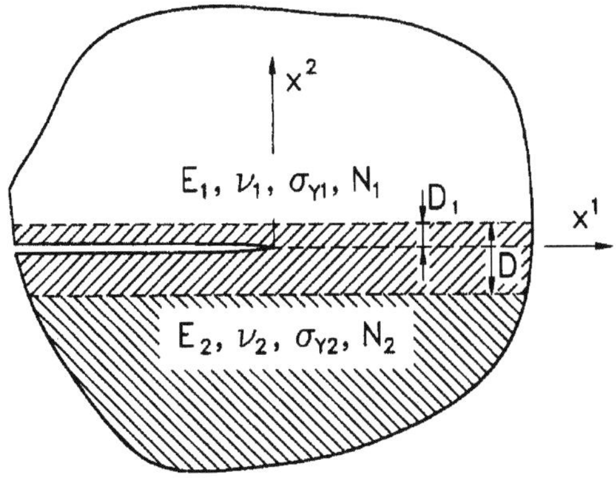

Figure 1. Crack in a functionally graded layer of thickness D between dissimilar elastic-plastic solids.

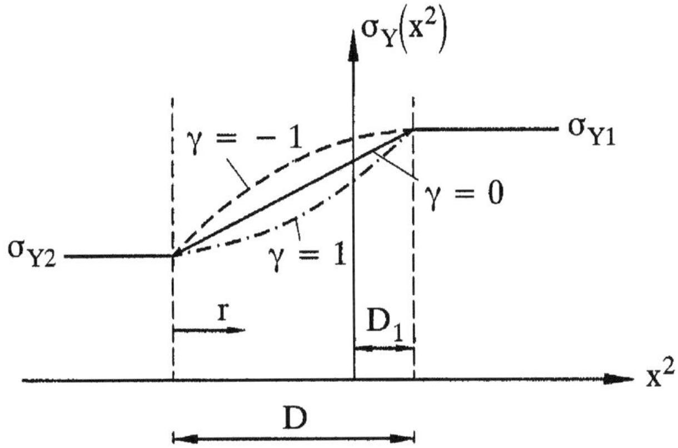

Figure 2. Assumed variations of initial yield stress across graded layer.

With a chosen reference length L, an L–dependent angular measure of mode mixity ψ is defined, such that $\tan\psi = \sigma_{12}/\sigma_{22}$ at $r = L$ on the interface. More details of the singularity field, with amplitude $|K|$, are specified in Tvergaard and Hutchinson [10].

In the traction-separation law used to represent the fracture process (see Tvergaard and Hutchinson [10]) δ_n and δ_t denote the normal and tangential components of the relative displacement of the crack faces across the interface (Fig. 3). When δ_n^c and δ_t^c are critical values of these displacement components and a single non-dimensional separation measure is defined as $\lambda = ((\delta_n/\delta_n^c)^2 + (\delta_t/\delta_t^c)^2)^{1/2}$ the tractions drop to zero when $\lambda = 1$. With $\sigma(\lambda)$

displayed in Fig. 3, a potential is defined as

$$\Phi(\delta_n, \delta_t) = \delta_n^c \int_0^\lambda \sigma(\lambda') d\lambda' \tag{4}$$

and the normal and tangential components of the traction acting on the interface in the fracture process zone are given by $T_n = \partial\Phi/\partial\delta_n$ and $T_t = \partial\Phi/\partial\delta_t$. The peak normal traction under pure normal separation is $\hat{\sigma}$, and the work of separation per unit area of interface is given by (4) with $\lambda = 1$, such that the function $\sigma(\lambda)$ in Fig. 3 gives the work $\Gamma_0 = \frac{1}{2}\hat{\sigma}\delta_n^c(1 - \lambda_1 + \lambda_2)$. It has been found that the two most important parameters characterizing the fracture process in this model are Γ_0 and $\hat{\sigma}$.

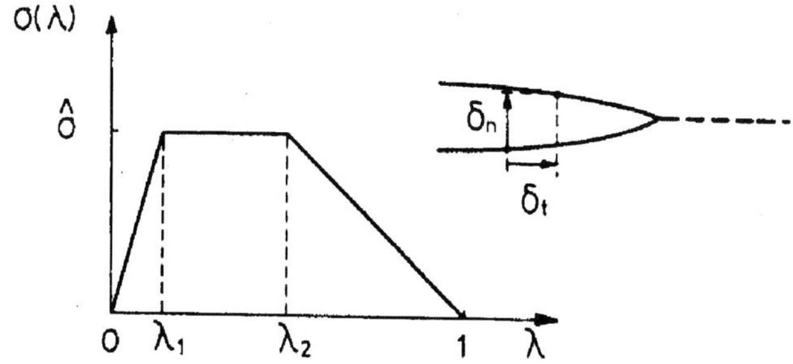

Figure 3. Specification of traction-separation relation.

A reference stress intensity factor and a corresponding length scale are defined as

$$K_0 = \left[\frac{1-\nu_1^2}{E_1} + \frac{1-\nu_2^2}{E_2}\right]^{-1/2} \left(\frac{2\Gamma_0}{1-\beta^2}\right)^{1/2}, \quad R_0 = \frac{1}{3\pi}\left(\frac{K_0}{\sigma_{Y1}}\right)^2. \tag{5}$$

Here, K_0 represents the value of $|K|$ needed to advance the interface crack in the absence of any plasticity, and the length R_0 scales with the size of the plastic zone in material No. 1, when $|K| \cong K_0$. The mode mixity angle ψ_0 that results from choosing the value R_0 for the reference length L will be used in the following for presenting results.

In the numerical analyses finite strains are accounted for, using a convected coordinate, Lagrangian formulation of the field equations, in which g_{ij} and G_{ij} are metric tensors in the reference configuration and the current configuration, respectively, with determinants g and G, and $\eta_{ij} = \frac{1}{2}(G_{ij} - g_{ij})$ is the Lagrangian strain tensor. The contravariant components τ^{ij} of the Kirchhoff stress tensor on the current base vectors are related to the components of the Cauchy stress tensor σ^{ij} and $\tau^{ij} = \sqrt{G/g}\,\sigma^{ij}$. Then, in the finite strain

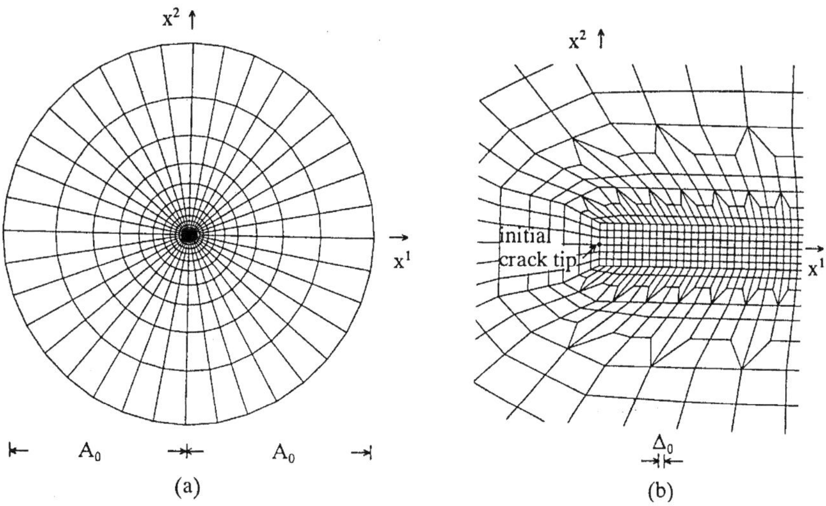

Figure 4. Mesh used for the crack growth analyses.

generalization of J_2-flow theory discussed by Hutchinson [3] an incremental stress-strain relationship is obtained of the form $\dot{\tau}^{ij} = L^{ijkl}\dot{\eta}_{kl}$.

The numerical solutions are obtained by a linear incremental method, using a finite element approximation of the displacement fields in the incremental principle of virtual work. Details of the crack growth procedure and the methods used to avoid numerical instabilities are given in Tvergaard [11, 12]. An example of the mesh used for the computations is shown in Fig. 4, where it is seen that a uniform mesh region is used in the range where crack growth is studied. The length of one square element inside the uniform mesh is denoted Δ_0, the initial crack tip is located at $x^1 = 0$, and 120 x 6 quadrilaterals are used in the uniform mesh. The quadrilateral elements are each build-up of four triangular, linear-displacement elements. The outer radius of the region analyzed is chosen to be $A_0/\Delta_0 = 80000$, in order that the plastic zone size should not exceed $A_0/10$.

The crack growth analyzed here is assumed to be parallel with the functionally graded layer, at distance D_1 below the upper edge of the layer (Fig. 1). In such a material configuration it is quite possible that the crack will grow out of the initial crack plane, but here the growth is restricted to remain on the initial crack plane, to get a parametric understanding of the effect of D_1/D and γ on the fracture toughness. As for crack growth along a sharp interface between dissimilar solids under mixed mode loading, the straight ahead crack growth for all combinations will only occur if the crack grows more easily in this plane than out of the plane.

3. Results

Both materials No. 1 and 2 are taken to be elastic-plastic in the cases to be analyzed here, and the same two materials are used for all the cases analyzed. The material parameters considered are $\sigma_{Y1}/E_1 = 0.003$, $\nu_1 = 1/3$ and $N_1 = 0.1$, while the substrate is specified by $E_2/E_1 = 2.0$, $\sigma_{Y2}/\sigma_{Y1} = 2.0$, $\nu_2 = 1/3$ and $N_2 = 0.1$. The peak stress in the cohesive zone model is specified by $\hat{\sigma}/\sigma_{Y1} = 4.0$. The differences between the cases studied lie in the width of the functionally graded layer, as specified by D/R_0, the location of the crack specified by D_1/D, and the nonlinearity of the grading specified by the value of γ in (2) and illustrated in Fig. 2.

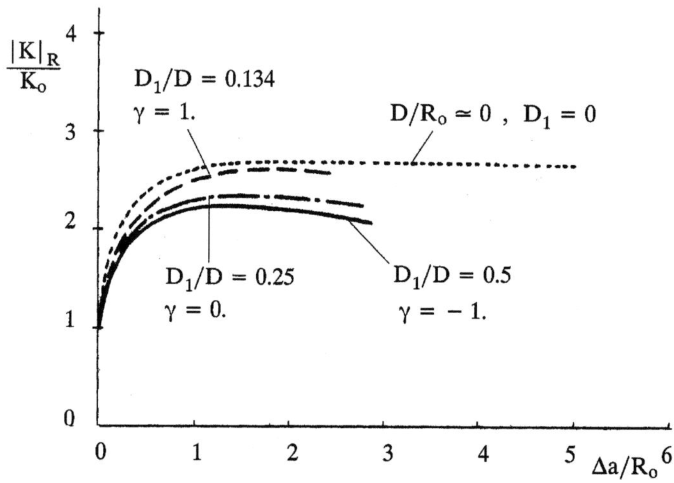

Figure 5. Crack growth resistance curves for different graded layers, where the values of σ_Y and E at the crack plane are 1.25 times those in material No. 1. Material parameters are $\hat{\sigma}/\sigma_{Y1} = 4.0$, $\sigma_{Y1}/E_1 = 0.003$, $E_2/E_1 = 2.0$, $\sigma_{Y2}/\sigma_{Y1} = 2.0$ and $D/R_0 = 1.109$, with the near-tip mode mixity specified by $\psi_0 = 3.95°$.

In the first cases considered the width of the graded layer is given by $D/R_0 = 1.109$. The mesh size in the uniform part of the mesh is chosen such that $D/\Delta_0 = 16$. Fig. 5 shows a comparison of crack growth resistance curves for three different graded layers, where the common feature is that the values of the parameters D_1/D and γ are chosen such that the values of σ_Y and E at the crack plane are 1.25 times those in material No. 1. Also, each of the resistance curves corresponds to the same near-tip mode mixity specified by $\psi_0 = 3.95°$. For comparison, the dotted curve in Fig. 5 is included corresponding to a sharp interface, where D is negligibly small. Now, for a linearly graded layer ($\gamma = 0$) the effect of different values of D_1/D, ranging from 0 to 1, has been analyzed previously (Tvergaard [12]) for this particular combination of materials on the two sides of the interface. There it was found that $D_1/D = 0$ gives much higher toughness than the sharp interface, that $D_1/D = 1$ gives much lower toughness, and that $D_1/D = 0.25$ gives a little lower toughness, as also shown by the dash-dotted curve in Fig. 5.

Relative to these previous results, the two resistance curves in Fig. 5 for non-zero values of γ show the effect of a nonlinear variation of the grading. For $\sigma_{Y1} < \sigma_{Y2}$, as in the cases analyzed here, $\gamma = -1.0$ gives a grading curve below the linear interpolation, and therefore the required value of D_1/D is as large as 0.5 in Fig. 5, which gives a slightly reduced toughness relative to the linear grading. This deviation is in the same direction as that found by increasing D_1/D for a linear grading, but the effect is much weaker in the nonlinearly graded material, where σ_Y and E are kept fixed at the crack plane while D_1/D is increased. The resistance curve for $\gamma = 1.0$ and $D_1/D = 0.134$ shows the opposite effect of somewhat increasing the toughness relative to the linearly graded case.

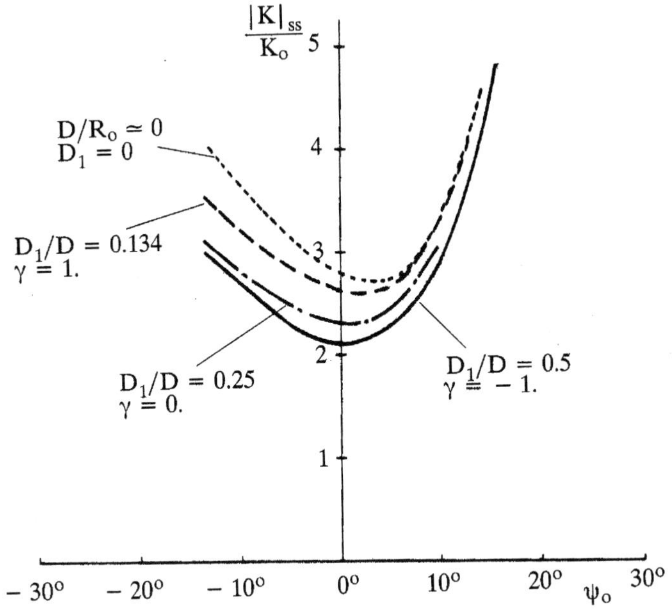

Figure 6. Steady-state toughness as a function of the local mixity measure ψ_0, for different graded layers, where the values of σ_Y and E at the crack plane are 1.25 times those in material No. 1. Material parameters are $\hat{\sigma}/\sigma_{Y1} = 4.0$, $\sigma_{Y1}/E_1 = 0.003$, $E_2/E_1 = 2.0$, $\sigma_{Y2}/\sigma_{Y1} = 2.0$ and $D/R_0 = 1.109$.

The influence of the mode mixity, ψ_0, is illustrated in Fig. 6 by showing the steady state toughness $|K|_{ss}$, defined here as the peak values of $|K|_R$ on resistance curves as those shown in Fig. 5 for one particular value of ψ_0. The result that increasing mode II contribution near the crack tip (ψ_0 deviating from zero) gives a strong increase of the fracture toughness, as seen in Fig. 6, has been found earlier for a sharp interface (Tvergaard and Hutchinson [10] and Tvergaard [11]) and for a linearly graded layer at the interface (Tvergaard [12]). Fig. 6 shows that this behavior is maintained when the grading varies nonlinearly through the graded layer.

Fig. 7 shows curves analogous to those in Fig. 6 for nonlinearly graded layers, but with the difference that the values of σ_Y and E at the crack plane are 1.125 times those in material No. 1 for the two upper curves, and 1.5 times those in material No. 1 for the two lower curves. In each case the difference between the two types of nonlinearities (γ equal to +1.0 or -1.0) is quite analogous to that shown in Figs. 5 and 6. The large difference between the levels of the steady state toughnesses for the two sets of curves is fully in agreement with the corresponding results found for a linearly graded layer at the interface (Tvergaard [12]).

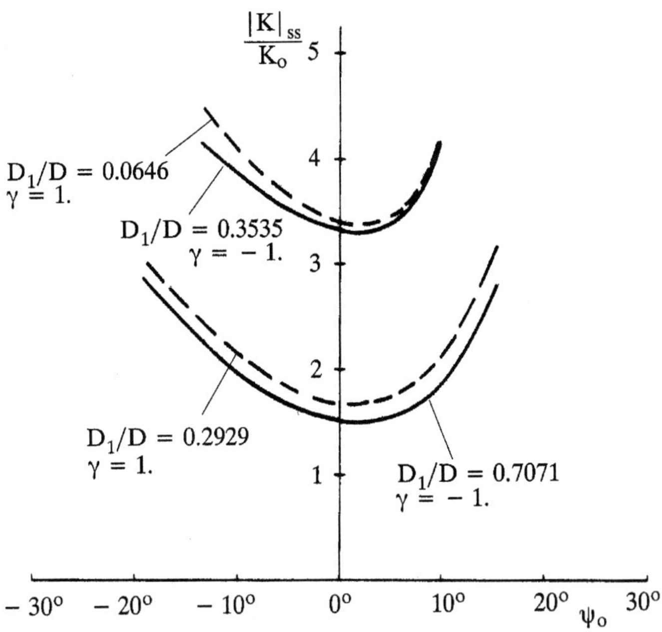

Figure 7. Steady-state toughness as a function of the local mixity measure ψ_0, for different graded layers, where the values of σ_Y and E at the crack plane are 1.125 or 1.5 times those in material No. 1. Material parameters are $\hat{\sigma}/\sigma_{Y1} = 4.0$, $\sigma_{Y1}/E_1 = 0.003$, $E_2/E_1 = 2.0$, $\sigma_{Y2}/\sigma_{Y1} = 2.0$ and $D/R_0 = 1.109$.

In Fig. 8 the main difference from the three previous figures is that the thickness of the graded layer is reduced by a factor four relative to the reference length, so that here $D/R_0 = 0.277$. The figure shows two sets of curves, corresponding to those in Fig. 6 and the lower curves in Fig. 7, i.e. for the values of σ_Y and E at the crack plane equal to 1.25 and 1.5 times those in material No. 1, respectively. There is still a clear difference between the two types of nonlinearities (γ equal to +1.0 or -1.0). But it is noted that the levels of the two sets of curves differ much less than found in the previous figures, as would be expected since all the curves have to coincide if the graded layer thickness D tends towards zero relative to the reference length R_0 .

It may be concluded based on the present results and those found earlier for linearly graded layers that the values of the material properties at the crack plane have a strong influence on the fracture toughness, such that a lower yield stress gives higher toughness. Also nonlinearities in the grading variation have an effect, but this effect is much smaller than that of the actual values of the properties at the crack plane.

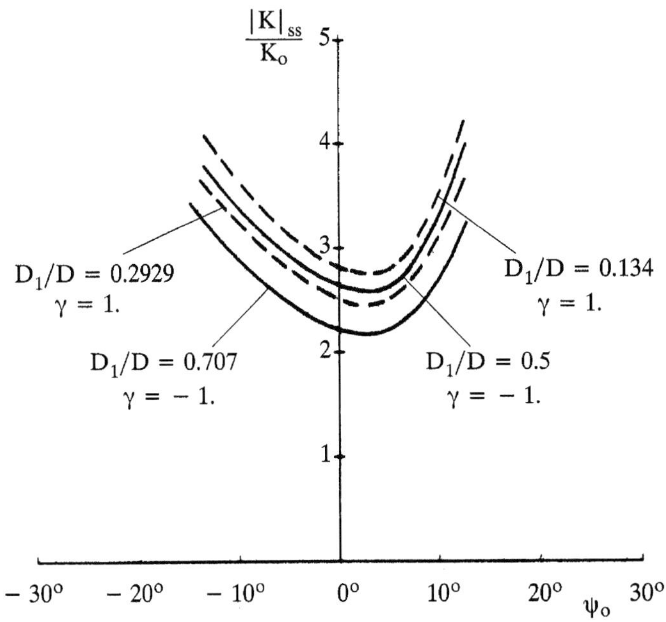

Figure 8. Steady-state toughness as a function of the local mixity measure ψ_0, for different graded layers, where the values of σ_Y and E at the crack plane are 1.25 or 1.5 times those in material No. 1. Material parameters are $\hat{\sigma}/\sigma_{Y1} = 4.0$, $\sigma_{Y1}/E_1 = 0.003$, $E_2/E_1 = 2.0$, $\sigma_{Y2}/\sigma_{Y1} = 2.0$ and $D/R_0 = 0.277$.

References

[1] Cao, H.C. and Evans, A.G. (1989). An experimental study of the fracture resistance of bimaterial interfaces. *Mech. Mater.*, **7**, 295–304.

[2] Erdogan, F. (1995). Fracture mechanics of functionally graded materials. *Composites Engineering*, **5**, 753–770.

[3] Hutchinson, J.W. (1973). Finite strain analysis of elastic-plastic solids and structures. *Numerical Solution of Nonlinear Structural Problems*, ASME, New York, **17**.

[4] Kim, A.S., Suresh, S. and Shih, C.F. (1997). Plasticity effects on fracture normal to interfaces with homogeneous and graded compositions. *Int. J. Solids Structures*, **34**, 3415–3432.

[5] Liechti, K.M. and Chai, Y.S. (1992). Asymmetric shielding in interfacial fracture under in-plane shear. *J. Appl. Mech.*, **59**, 295.

[6] Mortensen, A. and Suresh, S. (1995). Functionally graded metals and metal-ceramic composites: Part 1 Processing. *Int. Materials Reviews*, **40**, 239–265.

[7] Reiter, T., Dvorak, G.J. and Tvergaard, V. (1997). Micromechanical models for graded composite materials. *J. Mech. Phys. Solids*, **45**, 1281–1302.

[8] Rice, J.R. (1988). Elastic fracture mechanics concepts for interfacial cracks. *J. Appl. Mechanics*, **55**, 98–103.

[9] Suresh, S. and Mortensen, A. (1997). Functionally graded metals and metal-ceramic composites: Part 2 Thermo-mechanical behavior. *Int. Materials Reviews*, **42**, 85–116.

[10] Tvergaard, V. and Hutchinson, J.W. (1993). The influence of plasticity on mixed mode interface toughness. *J. Mech. Phys. Solids*, **41**, 1119–1135.

[11] Tvergaard, V. (2001). Resistance curves for mixed mode interface crack growth between dissimilar elastic-plastic solids. *J. Mech. Phys. Solids*, in press.

[12] Tvergaard, V. (2001). Theoretical investigation of the effect of plasticity on crack growth along a functionally graded region between dissimilar elastic-plastic solids. To appear in: *Engng. Fract. Mech.*, Special Issue on Fracture of Functionally Graded Materials.

Discrete Modeling of Cracking of Brittle Materials in Large Relative Motion and Localization Problem

A. Delaplace, A. Ibrahimbegovic
Laboratoire de Mécanique et Technologie, Ecole normale supérieure de Cachan
61 Avenue du Président Wilson, 94235 Cachan, France
ai@lmt.ens-cachan.fr

Abstract: We introduce a discrete model approach in order to study fracture of a reinforced-concrete structure. The "mesh" of the discrete model is constructed from a Voronoi tessellation, where particles are tied together by the cohesive forces represented by beams elements. The links between particles can break when their local strain reaches a random threshold, either in rotation or translation. In order to analyze large relative motion and associated localization mode, we use geometrically nonlinear Reissner's beam model to represent the links between particles, that take into account large relative motion. The proposed formulation allows to develop an efficient solution Newton-like procedure and to obtain easily the tangent stiffness operator. A numerical simulation of a bridge pear under flexion load is given in order to illustrate the presented model.

Keywords: Discrete model, Reissner's beam, large relative motion.

1. Introduction

Civil engineering materials typically exhibit a very complex behavior and strongly nonlinear response in the post-peak range. However, new building codes force an engineer to deal with these difficult problems and to provide a full description of structure failure, from the very beginning of the loading to the ultimate state. It is well known that classical continuum models have a lot of difficulty in dealing with this post-peak nonlinear response and related localization phenomena (e.g. see Benallal [1] and references therein). An alternative to use such models is to consider another approach based on discrete models (see, for instance, Herrmann and Roux [6] or Delaplace et al. [5]). These kinds of models are relevant for describing failure of quasi-brittle heterogenous materials like concrete, and very suitable for dealing with fragmentation process (van Mier [10]).

In this paper, we will focus on particle model, where the main idea is to map a set of rigid blocks on the structures (see, for instance, Cundall and Strack [4]). Initially, the cohesive forces hold the blocks together. When a certain load level is applied on the structure, the cohesive forces are overpowered with a local

C. Miehe (ed.), IUTAM Symposium on Computational Mechanics of Solid Materials at Large Strains, 375-383.

stress value passing a given threshold. At the later stage, the only interaction between the cells is by means of contact forces which are taken into account.

Usually, simple linear model like Euler-Bernoulli beams are chosen to represent the cohesive forces (Kun and Herrmann [9], Bolander and Saito [2]). Such an approach based on the classical linear theory ignores that the material can undergo large relative displacement in some areas, for instance near a crack tip. However, the classical beam behavior is based on small strains and cannot follow this large relative motion. Furthermore, reinforced concrete structures can undergo large deformation after peak load, due to steel reinforcement bar participation. For that reason, we propose in this paper to enrich the initial cohesive links in order to take into account eventual large relative motion. This goal is reached by joining blocks by enhanced beams capable of describing in an objective manner large relative motion. The failure criterion for the beam model concerns two kinds of mechanisms: the first one represents concrete behavior, with a quasi-brittle behavior and possible contact forces (crack closure effects), and the second one deals with reinforcement bars behavior.

The outline of the paper is as follows: in the next section, we will present the main ingredients bases of the discrete model. In Section 3, we introduce the kinematics of the beams in large relative motions. Finally, we will present an illustrative example in Section 4 and conclusions are in Section 5.

2. Discrete model

The main interest in dealing with discrete particle models is to introduce naturally a random micro-structure-like description of heterogenous materials. This aim is usually reached by using two main features:

- cohesive forces between two particles could have random characteristics, like a fracture threshold;
- the initial mesh is randomly chosen, with, for instance, a Voronoi tessellation.

Of course, these two could also be mixed. In order to increase computation efficiency, we use rigid body particles, i.e. particle shapes do not evolve during the loading, but an overlap of the cells is allowed and considered to generate contact forces. The mesh generation and the initial property of particles connections strongly influence the obtained result, and we start by describing these two points.

2.1. Mesh

Two kinds of meshes are usually generated to represent the microstructure: stack spheres and Voronoi tessellation (Figure 1). It is the last one which will allow more realistic representation of crack patterns, whereas the spherical particles will allow much faster computation. If we want to deal with a random meshing, we will only use Voronoi cells, which are able to represent classical elastic behavior, and in particular, Poisson's ratio effect.

We only study planar 2d discrete models. In that case, each particle has three degrees of freedom, two translations (u, v) and one rotation (θ). In the following, the initial position of particle i will be referred as (x_i, y_i).

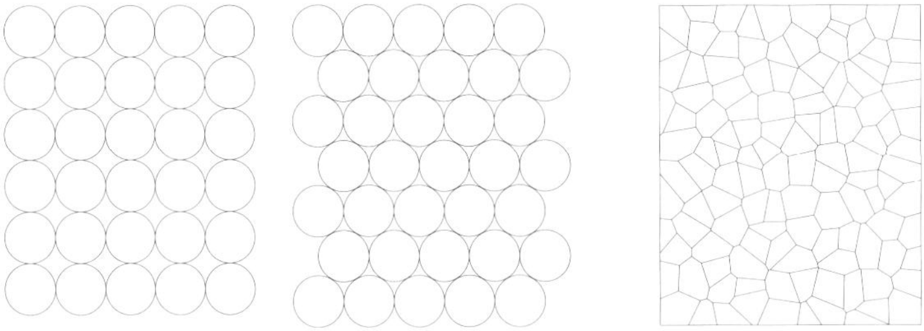

Figure 1. Samples of two meshes with spherical particles and with Voronoi tessellation.

2.2. COHESIVE FORCES

Cohesive forces are represented by beams which provide the corresponding stiffness due to axial and tangential relative displacements and relative rotations between particles. Each of these stiffness coefficients is initially activated, but would then disappear when the corresponding motion reaches a random threshold value. In particular, the following breaking criteria are chosen:

$$\max(\theta_i, \theta_j) \geq \theta_{ij}^{\max} \tag{1}$$

and

$$\frac{L_{ij}}{L_{ij}^0} \geq \varepsilon_{ij}^{\max} \tag{2}$$

where

$$L_{ij}^0 = \sqrt{(x_j - x_i)^2 + (y_j - y_i)^2}$$

is the initial distance between the two particles i and j and

$$L_{ij} = \sqrt{(x_j + u_j - x_i - u_i)^2 + (y_j + v_j - y_i - v_i)^2}$$

is the actual distance. In (1) and (2) above, $\varepsilon_{ij}^{\max}$ and $\theta_{ij}^{\max}$ are two threshold values chosen from two random probability distributions.

2.3. CONTACT FORCES

These forces are present and computed when two particles overlap. For this investigation, contact forces are assumed to be proportional to the distance between the particles. We note in passing that such a choice is quite different from usual discrete models, where contact forces are proportional to the overlapped area of the neighboring particles. This traditional approach is more expensive in computational time, particularly for Voronoi cells, and moreover it might no longer be justified when dealing with large relative motion.

3. Large relative motion of beams

Dealing with discrete models under large relative motions is facilitated by adding enhanced beams between cell centers. We use the beam model based on Reissner's theory which is developed within the framework of large displacement gradient theory (Ibrahimbegovic and Frey [7], Ibrahimbegovic and Mamouri [8]). We recall here the basic equations which governs the beam behavior. The position vector in beam deformed configuration is written as:

$$\varphi = \varphi_0 + \zeta \mathbf{t} = \left(\begin{array}{c} x+u \\ y+v \end{array} \right) + \zeta \left(\begin{array}{c} -sin(\alpha+\theta) \\ cos(\alpha+\theta) \end{array} \right) \tag{3}$$

The associated deformation gradient is:

$$\mathbf{F} = \left[\begin{array}{cc} \frac{\mathrm{d}x}{\mathrm{d}s} + \frac{\mathrm{d}u}{\mathrm{d}s} - \zeta \frac{\mathrm{d}\theta}{\mathrm{d}s} cos(\alpha+\theta) & -\sin(\alpha+\theta) \\ \frac{\mathrm{d}y}{\mathrm{d}s} + \frac{\mathrm{d}v}{\mathrm{d}s} - \zeta \frac{\mathrm{d}\theta}{\mathrm{d}s} sin(\alpha+\theta) & \cos(\alpha+\theta) \end{array} \right] \tag{4}$$

We can write the last equation in an equivalent form as $\mathbf{F} = \mathbf{R}\mathbf{U}$, where $\mathbf{R}$ is the rotation matrix:

$$\mathbf{R} = \left[\begin{array}{cc} cos(\alpha+\theta) & -\sin(\alpha+\theta) \\ sin(\alpha+\theta) & \cos(\alpha+\theta) \end{array} \right] \tag{5}$$

From the classical polar decomposition theorem it follows that the generalized strain measures $\mathbf{\Sigma} = (\Sigma, K, \Gamma)$ proposed by Reissner, can be computed from the rotated strain measure $\mathbf{H} = \mathbf{U} - \mathbf{I}$, with $\mathbf{U} = \mathbf{R}^{\mathbf{T}}\mathbf{F}$ and $\mathbf{I}$ as the identity matrix:

$$\mathbf{H} = \left[\begin{array}{cc} \Sigma - \zeta K & \Gamma \\ \Gamma & 0 \end{array} \right] \tag{6}$$

The last result can be explicitly written as:

$$\Sigma = \cos(\alpha+\theta) \left(\frac{\mathrm{d}x}{\mathrm{d}s} + \frac{\mathrm{d}u}{\mathrm{d}s} \right) + sin(\alpha+\theta) \left(\frac{\mathrm{d}y}{\mathrm{d}s} + \frac{\mathrm{d}v}{\mathrm{d}s} \right) - 1$$

$$\begin{aligned}\Gamma &= -sin(\alpha+\theta)\left(\frac{\mathrm{d}x}{\mathrm{d}s}+\frac{\mathrm{d}u}{\mathrm{d}s}\right)+\cos(\alpha+\theta)\left(\frac{\mathrm{d}y}{\mathrm{d}s}+\frac{\mathrm{d}v}{\mathrm{d}s}\right) \\ K &= \frac{\mathrm{d}\theta}{\mathrm{d}s}\end{aligned} \tag{7}$$

The stress resultant $\mathbf{r} = (N, V, M)^T$ energy-conjugate to strain measures in (7) are then obtained from:

$$\mathbf{r} = \mathbf{C}\boldsymbol{\Sigma} \tag{8}$$

with $\mathbf{C} = \mathrm{diag}(EA, GA, EI)$ which are proportional to the length r_{ij} of the common boundary segment between two particles i and j (Figure 2).

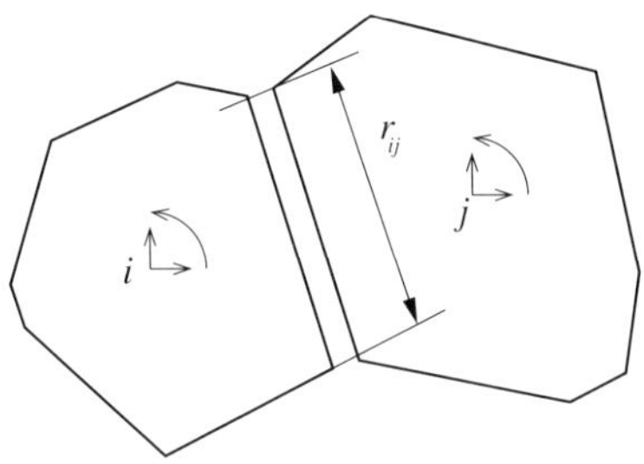

Figure 2. Two particles i and j which share the common segment of length r_{ij}.

The main advantage of this beam model pertains to the fact that the corresponding tangent stiffness operator is easily obtained which provides the quadratic convergence of the Newton-like solution procedure. This tangent operator can be split additively into two parts: a material one, $\mathbf{D_m}$, and a geometrical one, $\mathbf{D_g}$:

$$\mathbf{D_m} = \begin{bmatrix} \boldsymbol{\Lambda} \\ \mathbf{h}^T(\mathbf{a})\dfrac{\mathrm{d}\boldsymbol{\Lambda}}{\mathrm{d}\theta} \end{bmatrix} \mathbf{C} \left[\boldsymbol{\Lambda}^T; \frac{\mathrm{d}\boldsymbol{\Lambda}^T}{\mathrm{d}\theta}\mathbf{h}(\mathbf{a})\right] \tag{9}$$

$$\mathbf{D_g} = \begin{bmatrix} \mathbf{0} & \dfrac{\mathrm{d}\boldsymbol{\Lambda}}{\mathrm{d}\theta}\mathbf{r} \\ \mathbf{r}^T\dfrac{\mathrm{d}\boldsymbol{\Lambda}^T}{\mathrm{d}\theta} & \mathbf{h}^T(\mathbf{a})\dfrac{\mathrm{d}^2\boldsymbol{\Lambda}}{\mathrm{d}\theta^2} \end{bmatrix} \tag{10}$$

where we have introduced:

$$\boldsymbol{\Lambda} = \begin{bmatrix} \cos(\alpha+\theta) & -\sin(\alpha+\theta) & 0 \\ \sin(\alpha+\theta) & \cos(\alpha+\theta) & 0 \\ 0 & 0 & 1 \end{bmatrix} \text{ and } \mathbf{h}(\mathbf{a}) = \begin{pmatrix} \dfrac{\mathrm{d}x}{\mathrm{d}s}+\dfrac{\mathrm{d}u}{\mathrm{d}s} \\ \dfrac{\mathrm{d}y}{\mathrm{d}s}+\dfrac{\mathrm{d}v}{\mathrm{d}s} \\ \dfrac{\mathrm{d}\theta}{\mathrm{d}s} \end{pmatrix}$$

3.1. Discrete approximation

Given the previous equations, we can compute the corresponding discrete approximation of the strain from the displacements and rotations of the two particles i and j connected by a beam:

$$\Sigma = \frac{1}{L_{ij}^0}\left[(x_j - x_i) + (u_j - u_i)\right]\cos[\alpha_{ij}^0 + \frac{1}{2}(\theta_i + \theta_j)] + \frac{1}{L_{ij}^0}\left[(y_j - y_i) + (v_j - v_i)\right]\sin[\alpha_{ij}^0 + \frac{1}{2}(\theta_i + \theta_j)] - 1 \quad (11)$$

$$\Gamma = -\frac{1}{L_{ij}^0}\left[(x_j - x_i) + (u_j - u_i)\right]\sin[\alpha_{ij}^0 + \frac{1}{2}(\theta_i + \theta_j)] + \frac{1}{L_{ij}^0}\left[(y_j - y_i) + (v_j - v_i)\right]\cos[\alpha_{ij}^0 + \frac{1}{2}(\theta_i + \theta_j)] \quad (12)$$

$$K = \frac{1}{L_{ij}^0}(\theta_j - \theta_i) \quad (13)$$

where we have introduced the initial angle with respect to the reference frame which is computed according to $\alpha_{ij}^0 = \tan^{-1}(y_j - y_i)/(x_j - x_i)$.

4. Numerical example

We present here the computation of the failure of a bridge pear. The simple geometry and the simple loading allow to carry out a parametric analysis of the influence of different variables. We will focus here on the chosen probability distribution of the fracture thresholds. The material and geometric properties of the beam are chosen from the classical continuum representation of the material properties (Chang et al. [3]). Figure 3 represents the initial mesh, the boundary conditions and the loading.

Two probability distributions with the same average are tested. The first one is uniform whereas the second one is the classical Gaussian. As expected, the two failure modes are quite different. The main change concerns the appearance of macro-cracks. For the uniform distribution, a multi-crack pattern could be observed with one macro-crack which can appear and propagate until a strong link is reached to stop the propagation. Then, the next macro-crack will appear and start propagating until stopped and so on (Figure 4).

For a deterministic analysis, which can be formally considered as a probabilistic analysis with an equivalent Dirac delta distribution, we should take the same value for all thresholds. In this case, a single beam which breaks in the structure, leads to a stress localization with a single macro-crack which develops and

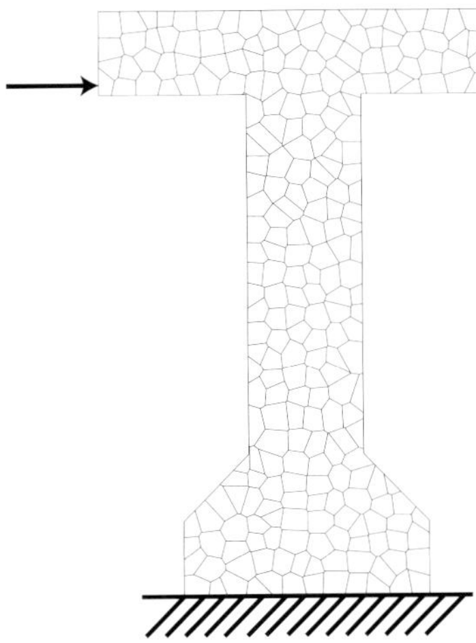

Figure 3. The initial mesh with the boundary conditions and the loading (imposed displacement in the direction represented by the arrow).

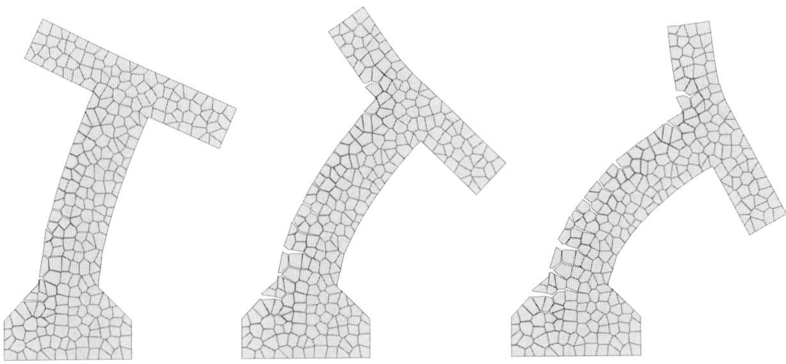

Figure 4. Three steps of deformation for a uniform distribution for the beam threshold values. A multi-crack pattern is found. The deformation is not amplified.

propagates until rupture. The Gaussian distribution will typically constitute an intermediate case approach more to one limit case or another depending upon its standard deviation. For the chosen small values of the standard deviation making the Gaussian distribution approaching the Dirac, figure 5 shows a single crack will lead to the rupture of the pear.

The force-displacement responses (Figure 6) of both computations reveal the different behaviors before the failure. As alluded to already, the Gaussian distribution with small standard deviation gives a more brittle behavior with smaller fracture strain and smaller damage zone than the uniform one with a well-developed fracture zone.

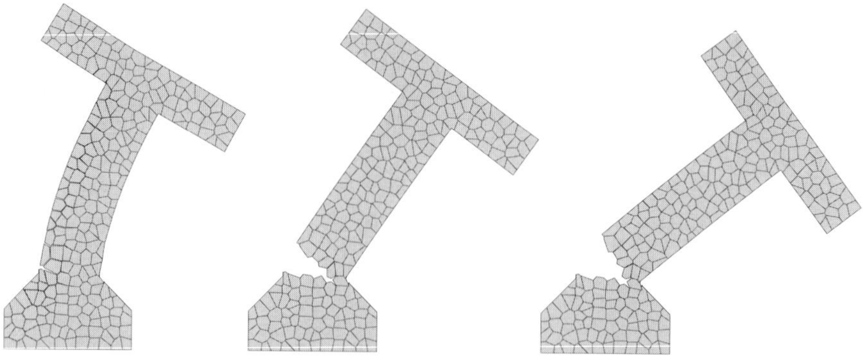

Figure 5. Three steps of deformation for a Gaussian distribution for beam threshold values. A single crack appears and leads to rupture.

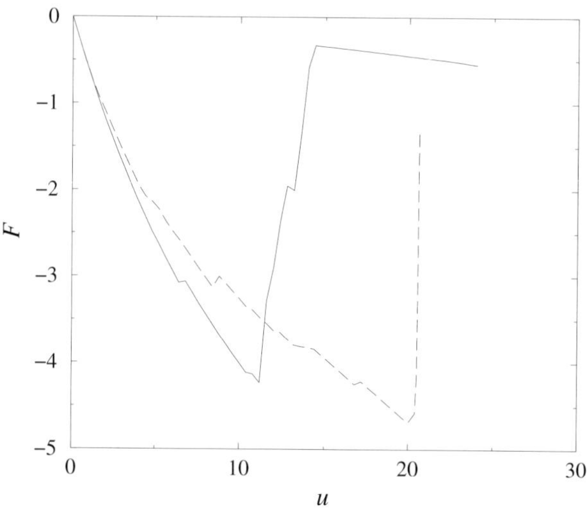

Figure 6. Evolution of the force versus the imposed displacement for both distributions. (——) corresponds to the Gaussian distribution, (– – –) corresponds to the uniform one.

5. Conclusion

In this work, we developed a discrete particle model which can take into account large motion of assembly typical of confined concrete behavior such as for reinforced concrete. The main motivation for this development is to provide a realistic representation of failure modes of such structures. Large relative displacement can occur locally near a strong discontinuity like a crack, or at a global level when the concrete resistance is broken and only reinforcement bars maintain resistance.

The discrete model is constructed by a Voronoi tessellation of the structures where each cell correspond to a single particle. The particles are joined by beams representing the cohesive forces among the particles that could break when local strain reaches a randomly selected threshold. The beam model allows large relative motion, and its behavior is based on Reissner's theory. The predictive capacities of the model shown on a simple example of a bridge pear, seem to be able to provide a realistic representation of the failure modes of a reinforced concrete structure.

References

[1] Benallal, A. ed. (2000). *Continuous damage and fracture*, (a book dedicated to Prof. J. Lemaitre), Cachan, Elsevier (Paris).

[2] Bolander, J.E. and Saito, S. (1998). Fracture analysis using spring networks with random geometry. *Engng. Fracture Mech.*, **61**, 569–591.

[3] Chang, C.S., Wang, T.K., Sluys, L.J. and van Mier, J.G.M. (2001). Fracture modeling with a microstructural mechanics approach. *Fracture mechanics of concrete structures*, de Borst et al (eds), A.A.Balkema Publishers, 27–34.

[4] Cundall, P.A. and Strack, O.D.L. (1979). A discrete numerical model for granular assemblies. *Geotechnique*, London, **29**, 47–65.

[5] Delaplace, A., Roux, S. and Pijaudier-Cabot, G. (1996). Progressive damage in discrete models and consequences on continuum modelling. *Journal of Physics and Mechanics of Solids*, **44**, 99–136.

[6] Herrmann, H.J. and Roux, S., (eds.) (1990). *"Statistical models for the fracture of disordered media"*, Elsevier Sci. Pub., Amsterdam.

[7] Ibrahimbegovic, A. and Frey, F. (1993). Finite element analysis of linear and nonlinear planar deformation of elastic initially curved beams. *International Journal for Numerical Methods in Engineering*, **36**, 3239–3258.

[8] Ibrahimbegovic, A. and Mamouri, S. (1999). Nonlinear dynamics of flexible beams in planar motion: formulation and time-stepping scheme for stiff problems. *Computer and Structures*, **70**, 1–22.

[9] Kun, F. and Herrmann, H.J. (1996). A study of fragmentation processes using a discrete element method. *Comp. Meth. Appl. Mech. Eng.*, **7**, 3–18.

[10] van Mier, J.G.M. (1997). Fracture mechanics of concrete. CRC Press, Boca Raton, Florida.

Goal-Oriented Error Control for Large Strain Viscoplasticity

F. Larsson, K. Runesson, P. Hansbo
Department of Applied Mechanics, Chalmers University of Technology
Hörsalsvägen 7, 41296 Göteborg, Sweden
fredrik.larsson@me.chalmers.se, kenneth.runesson@me.chalmers.se,
peter.hansbo@me.chalmers.se

Abstract: In this paper we investigate the possibility for, and characteristics of, reliable and efficient a posteriori error computation at the integration of the constitutive relations pertinent to large strain viscoplasticity. An important feature is the possibility to select "goal-oriented" error measures with great freedom. A key task is to identify and solve the pertinent dual problem.

Keywords: Adaptivity, error estimation, viscoplasticity.

1. Introduction

Within the computational mechanics community, considerable effort has been spent on the formulation and solution of initial-boundary-value problems involving viscoplastic response for large strains. However, it is only recently that significant attention has been paid to the error estimation and adaptive finite element strategies for problems with nonlinear dissipative material behavior, cf. Ladevèze et al. [7, 8], Rannacher and Suttmeier [18], Ramm et al. [3, 20] and Larsson et al. [9, 10]. A quite general strategy is to take advantage of the properties of a dual problem that represents a suitable linearization of the primal (actual physical) problem of interest along the solution path, cf. Johnson et al. [4, 6, 5], Rannacher et al. [1, 16, 17], Ramm et al. [2], Oden and Prudhomme [15, 13], Rüter et al. [14, 19], Larsson et al. [11, 12]. A major advantage is that it is possible to choose any goal-oriented error measure of engineering interest, and the data for the dual problem are chosen accordingly in such a fashion that an exact error representation can be obtained (at least formally).

In this paper, which is an extension of Larsson et al. [10] to large strains, we establish a finite element formulation in time and the framework for goal-oriented error estimation for the constitutive problem of viscoplasticity.

2. A canonical constitutive formulation of large strain (visco)plasticity

We introduce the multiplicative decomposition of the deformation gradient $\boldsymbol{F} = \bar{\boldsymbol{F}} \cdot \boldsymbol{F}^{\mathrm{p}}$, where $\bar{\boldsymbol{F}}$ represents the elastic deformation with respect to

C. Miehe (ed.), IUTAM Symposium on Computational Mechanics of Solid Materials at Large Strains, 385-394.

the "intermediate" configuration $\bar{\mathcal{B}}$, whereas $\boldsymbol{F}^{\mathrm{p}}$ represents the inelastic, or (visco)plastic, deformation with respect to the initial reference configuration $\mathcal{B}_0$. For simplicity, we shall consider only the case of perfect (visco)plasticity, which means that $\boldsymbol{F}^{\mathrm{p}}$ is the only internal variable in the free energy. We thus introduce the free energy per unit reference volume as $\Psi(\bar{\boldsymbol{C}})$, where $\bar{\boldsymbol{C}} \stackrel{\mathrm{def}}{=} \bar{\boldsymbol{F}}^{\mathrm{T}} \cdot \bar{\boldsymbol{F}} = [\boldsymbol{F}^{\mathrm{p}}]^{-\mathrm{T}} \cdot \boldsymbol{C} \cdot [\boldsymbol{F}^{\mathrm{p}}]^{-1}$ is the elastic right Cauchy-Green deformation tensor (with $\bar{\boldsymbol{C}} \stackrel{\mathrm{def}}{=} \boldsymbol{F}^{\mathrm{T}} \cdot \boldsymbol{F}$). From the Clausius-Duhem Inequality (CDI), we then arrive at the constitutive relation for the 2nd Piola-Kirchhoff stress $\boldsymbol{S}_2$ as

$$\boldsymbol{S}_2 = [\boldsymbol{F}^{\mathrm{p}}]^{-1} \cdot \bar{\boldsymbol{S}}_2 \cdot [\boldsymbol{F}^{\mathrm{p}}]^{-\mathrm{T}} \quad \text{with} \quad \bar{\boldsymbol{S}}_2 \stackrel{\mathrm{def}}{=} 2\frac{\partial \Psi}{\partial \bar{\boldsymbol{C}}}. \tag{1}$$

Moreover, we obtain the dissipation inequality

$$\mathcal{D}_0 \stackrel{\mathrm{def}}{=} \bar{\boldsymbol{T}} : \bar{\boldsymbol{L}}^{\mathrm{p}} \geq 0, \tag{2}$$

where $\bar{\boldsymbol{T}} \stackrel{\mathrm{def}}{=} \bar{\boldsymbol{C}} \cdot \bar{\boldsymbol{S}}_2$ is the Mandel stress with respect to $\bar{\mathcal{B}}$, and $\bar{\boldsymbol{L}}^{\mathrm{p}} \stackrel{\mathrm{def}}{=} \dot{\boldsymbol{F}}^{\mathrm{p}} \cdot [\boldsymbol{F}^{\mathrm{p}}]^{-1}$ is the "inelastic velocity gradient".

In order to establish the flow rule, we introduce the convex and smooth[1] quasistatic yield function $\Phi(\bar{\boldsymbol{T}})$ such that $\Phi(\boldsymbol{0}) < 0$. This function generates the convex set of plastically admissible states $\mathbb{E}$, defined as $\mathbb{E} \stackrel{\mathrm{def}}{=} \{\bar{\boldsymbol{T}} |\ \Phi(\bar{\boldsymbol{T}}) \leq 0\}$. We also introduce the overstress function $\eta(\Phi)$ in standard fashion:

$$\begin{aligned} \eta > 0,\ \eta' > 0 \quad & \text{when} \quad \Phi > 0, \\ \eta = 0 \quad & \text{when} \quad \Phi \leq 0, \end{aligned} \tag{3}$$

whereby we may postulate the viscoplastic flow rule as

$$\bar{\boldsymbol{L}}^{\mathrm{p}} = \mu \bar{\boldsymbol{M}}, \quad \mu = \frac{1}{t_*}\eta(\Phi), \quad \bar{\boldsymbol{M}} \stackrel{\mathrm{def}}{=} \frac{\partial \Phi}{\partial \bar{\boldsymbol{T}}}, \tag{4}$$

where t_* is a relaxation time.

3. Time-variational format

We consider the constitutive problem in (4) rewritten, with $\boldsymbol{q} \stackrel{\mathrm{def}}{=} \boldsymbol{F}^{\mathrm{p}} - \boldsymbol{I}$, as follows:

$$\dot{\boldsymbol{q}} - \mu \boldsymbol{g}(\boldsymbol{q}, t) = \boldsymbol{0}, \quad \boldsymbol{g}(\boldsymbol{q}, t) \stackrel{\mathrm{def}}{=} \bar{\boldsymbol{M}}(\boldsymbol{q}, t) \cdot (\boldsymbol{I} + \boldsymbol{q}), \tag{5}$$

$$t_* \mu - \eta(\boldsymbol{q}, t) = 0, \tag{6}$$

with the initial condition $\boldsymbol{q}(0) = \boldsymbol{0}$. The purpose is to compute $\boldsymbol{q}(t)$ and $\mu(t)$ for prescribed $\boldsymbol{F}(t)$ at any given time $t \in I = (0, T)$. With $\boldsymbol{F}^{\mathrm{p}} = \boldsymbol{I} + \boldsymbol{q}$ known, it is possible to compute $\bar{\boldsymbol{F}}$, $\bar{\boldsymbol{C}}$, $\boldsymbol{S}_2(\bar{\boldsymbol{C}})$, $\bar{\boldsymbol{T}}(\bar{\boldsymbol{C}})$ and $\boldsymbol{\tau} = \bar{\boldsymbol{F}} \cdot \boldsymbol{S}_2(\bar{\boldsymbol{C}}) \cdot \bar{\boldsymbol{F}}^{\mathrm{T}}$.

In order to give the appropriate variational format in time, corresponding to the strong format (5,6), we introduce a partition $0 = t_0 < t_1 < t_2 < \ldots < t_{N-1} < t_N = T$ of the time span I into time intervals $I_n = (t_{n-1}, t_n)$ of length $k_n = t_n - t_{n-1}$. We are then led to defining the spaces

$$\mathbb{Q}(I) = \{\boldsymbol{q}|\ \boldsymbol{q} \in \mathbb{H}^1(I_n) \cap \mathbb{L}_2(I)\}, \quad \mathbb{M}(I) = \mathbb{L}_2(I)$$

and we introduce, associated with each I_n, the form

$$\begin{aligned} a_n(\boldsymbol{q}, \mu; \boldsymbol{q}', \mu') &\stackrel{\text{def}}{=} \int_{I_n} \boldsymbol{q}' : [\dot{\boldsymbol{q}} - \mu \boldsymbol{g}(\boldsymbol{q}, t)]\, \mathrm{d}t + \boldsymbol{q}'(t_{n-1}^+) : [\![\boldsymbol{q}]\!](t_{n-1}) \\ &+ \int_{I_n} \mu' \left[t_* \mu - \eta(\boldsymbol{q}, t)\right]\, \mathrm{d}t, \end{aligned} \tag{7}$$

with $[\![\boldsymbol{q}]\!](t) \stackrel{\text{def}}{=} \boldsymbol{q}(t^+) - \boldsymbol{q}(t^-)$.
Upon introducing the notation $a = \sum_{n=1}^{N} a_n$, we formulate the time-variational equivalent of (5,6). However, due to the causality-principle, i.e. information *flows* forward in time, the variational problem can be replaced by the N problems: Find $\boldsymbol{q} \in \mathbb{Q}(I_n)$ and $\mu \in \mathbb{M}(I_n)$ such that, for each $n = 1, 2, \ldots, N$,

$$a_n(\boldsymbol{q}, \mu; \boldsymbol{q}', \mu') = 0 \quad \forall \boldsymbol{q}' \in \mathbb{Q}(I_n), \quad \forall \mu' \in \mathbb{M}(I_n). \tag{8}$$

4. Finite elements in time – discontinuous galerkin method

Introducing suitable finite element spaces $\mathbb{Q}_h \subset \mathbb{Q}$ and $\mathbb{M}_h \subset \mathbb{M}$, the approximate problem reads: Find $\boldsymbol{q}_h \in \mathbb{Q}_h$ and $\mu_h \in \mathbb{M}_h$ such that for $n = 1, 2, \ldots, N$,

$$a_n(\boldsymbol{q}_h, \mu_h; \boldsymbol{q}_h', \mu_h') = 0 \quad \forall \boldsymbol{q}_h' \in \mathbb{Q}_h, \quad \forall \mu_h' \in \mathbb{M}_h. \tag{9}$$

In general $a_n(\cdot, \cdot; \cdot, \cdot)$ is nonlinear in the first two arguments, and a number of iterative strategies for computing $\boldsymbol{q}_h$ and μ_h can be envisioned. In particular, (9) can be solved using Newton iterations:

$$\boldsymbol{q}_h^{(k+1)} = \boldsymbol{q}_h^{(k)} + \delta \boldsymbol{q}, \quad \mu_h^{(k+1)} = \mu_h^{(k)} + \delta \mu, \tag{10}$$

where the incremental problem reads: Find $\delta \boldsymbol{q}_h \in \mathbb{Q}_h$ and $\delta \mu_h \in \mathbb{M}_h$ such that for $n = 1, 2, \ldots, N$,

$$a_n'(\boldsymbol{q}_h^{(k)}, \mu_h^{(k)}; \delta \boldsymbol{q}_h, \delta \mu_h, \boldsymbol{q}_h', \mu_h') = 0 \quad \forall \boldsymbol{q}_h' \in \mathbb{Q}_h, \quad \forall \mu_h' \in \mathbb{M}_h. \tag{11}$$

The Gateaux derivative of a_n, denoted a_n', is defined by

$$\begin{aligned} a_n'(\boldsymbol{q}, \mu; \hat{\boldsymbol{q}}, \hat{\mu}, \boldsymbol{q}', \mu') &\stackrel{\text{def}}{=} \frac{\partial}{\partial \xi} a_n(\boldsymbol{q} + \xi \hat{\boldsymbol{q}}, \mu + \xi \hat{\mu}; \boldsymbol{q}', \mu')|_{\xi=0} \\ &= \int_{I_n} \boldsymbol{q}' : \left[\dot{\hat{\boldsymbol{q}}} + \mu \mathbf{D} : \hat{\boldsymbol{q}} - \boldsymbol{g} \hat{\mu}\right] \mathrm{d}t \\ &+ \boldsymbol{q}'(t_{n-1}^+) : [\![\hat{\boldsymbol{q}}]\!](t_{n-1}) + \int_{I_n} \mu' \left[t_* \hat{\mu} + \boldsymbol{c} : \hat{\boldsymbol{q}}\right] \mathrm{d}t, \end{aligned} \tag{12}$$

where we introduced the notation

$$\mathbf{D} \stackrel{\text{def}}{=} -\frac{\partial \boldsymbol{g}}{\partial \boldsymbol{q}}, \quad \boldsymbol{c} \stackrel{\text{def}}{=} -\frac{\partial \eta}{\partial \boldsymbol{q}}. \tag{13}$$

The tensors $\mathbf{D}$ and $\boldsymbol{c}$ will be employed later in the formulation of the dual problem.

Now, the dG(0)-method is defined by the spaces $\mathbb{Q}_h \subset \mathbb{Q}$ and $\mathbb{M}_h \subset \mathbb{M}$ of piecewise constant $\boldsymbol{q}_h(t)$ and $\mu_h(t)$, respectively, as follows:

$$\mathbb{Q}_h(I) \stackrel{\text{def}}{=} \{\boldsymbol{q}_h | \, \boldsymbol{q}_h|_{I_n} = {}^n\boldsymbol{q}_h\}, \quad \mathbb{M}_h(I) \stackrel{\text{def}}{=} \{\mu_h | \, \mu_h|_{I_n} = {}^n\mu_h\}.$$

Hence, each $\boldsymbol{q}_h(t) \in \mathbb{Q}_h$ and $\mu_h(t) \in \mathbb{M}_h$ are discontinuous at the nodal points t_n in the time-mesh, such that (9) can be rephrased explicitly as

$${}^n\boldsymbol{q}_h - {}^n\mu_h \int_{I_n} \boldsymbol{g}({}^n\boldsymbol{q}_h, t)\mathrm{d}t = {}^{n-1}\boldsymbol{q}_h, \tag{14}$$

$$k_n t_* {}^n\mu_h - \int_{I_n} \eta({}^n\boldsymbol{q}_h, t)\, dt = 0. \tag{15}$$

In practice, any quadrature rule can be chosen to compute the integrals in (14) and (15). The standard BE-rule is recovered if we set$t = t_n$.

5. A posteriori error estimates based on the dual solution

5.1. PRIMAL PROBLEM – SECANT REPRESENTATION

We wish to derive an *exact* representation of a suitable measure (to be defined later) of the time discretization error. In particular, we wish to control the error at $t = T$. To this end, we use a_n' in (12) and establish the secant form $A = \sum_{n=1}^{N} A_n$ where

$$A_n(\boldsymbol{q}, \mu, \boldsymbol{q}_h, \mu_h; \hat{\boldsymbol{q}}, \hat{\mu}, \boldsymbol{q}', \mu') \stackrel{\text{def}}{=} \int_0^1 a_n'(\bar{\boldsymbol{q}}(s), \bar{\mu}(s); \hat{\boldsymbol{q}}, \hat{\mu}, \boldsymbol{q}', \mu')\mathrm{d}s, \tag{16}$$

with

$$\bar{\boldsymbol{q}}(s) \stackrel{\text{def}}{=} \boldsymbol{q}_h + s\boldsymbol{e}_q, \quad \bar{\mu}(s) \stackrel{\text{def}}{=} \mu_h + s e_\mu. \tag{17}$$

Here, we also introduced the finite element errors $\boldsymbol{e}_q \stackrel{\text{def}}{=} \boldsymbol{q} - \boldsymbol{q}_h$ and $e_\mu \stackrel{\text{def}}{=} \mu - \mu_h$.

Remark 5.1 *Since $\boldsymbol{q}(0) = \boldsymbol{q}(t_0^-) = \mathbf{0}$ is fixed, then it follows that $\hat{\boldsymbol{q}}(t_0^-) = \mathbf{0}$ in* (16). □

Remark 5.2 *The quantities $\mathbf{D}$, $\boldsymbol{g}$ and $\mathbf{c}$ in (12) that are to be inserted in (16) depend on the continuous (unknown) solutions$\boldsymbol{q}$ and μ, which must be replaced (in practice) by suitable approximations; in the following we choose the finite element solutions $\boldsymbol{q}_h$ and μ_h.*□

5.2. Error measure – secant representation

We now turn to the error analysis and select a scalar error measure $E(\boldsymbol{q}, \boldsymbol{q}_h)$, which is a function that measures the difference between $\boldsymbol{q}$ and $\boldsymbol{q}_h$. Upon introducing the Gateaux-derivative of E, denoted E', defined as

$$E'(\boldsymbol{q}, \boldsymbol{q}_h; \hat{\boldsymbol{q}}) \stackrel{\text{def}}{=} \frac{\partial}{\partial \xi} E(\boldsymbol{q} + \xi \hat{\boldsymbol{q}}, \boldsymbol{q}_h)|_{\xi=0}, \tag{18}$$

we define the secant form (linear functional in $\hat{\boldsymbol{q}}$)

$$J(\boldsymbol{q}, \boldsymbol{q}_h; \hat{\boldsymbol{q}}) \stackrel{\text{def}}{=} \int_0^1 E'(\bar{\boldsymbol{q}}(s), \boldsymbol{q}_h; \hat{\boldsymbol{q}}) \mathrm{d}s, \quad \forall \hat{\boldsymbol{q}} \in \mathbb{Q}(I), \tag{19}$$

whereby it is concluded that $J(\bullet; \boldsymbol{e}_q) = E(\bullet)$.

5.3. Dual problem

We associate with A the dual bilinear form A^* such that $A^*(\bullet; \hat{\boldsymbol{q}}, \hat{\mu}, \boldsymbol{q}', \mu') = A(\bullet; \boldsymbol{q}', \mu', \hat{\boldsymbol{q}}, \hat{\mu})$. Letting J_n and A_n^* be the restrictions of J and A^*, respectively, to I_n, i.e. $J = \sum_{n=1}^N J_n$ and $A^* = \sum_{n=1}^N A_n^*$, we may formulate the dual problem: Find $\boldsymbol{\chi} \in \mathbb{Q}(I_n)$ and $\rho \in \mathbb{M}(I_n)$ such that for $n = N, N-1, ..., 1$,

$$A_n^*(\bullet; \boldsymbol{\chi}, \rho, \boldsymbol{\chi}', \rho') = J_n(\bullet; \boldsymbol{\chi}', \rho') \quad \forall \boldsymbol{\chi}' \in \mathbb{Q}(I_n), \quad \forall \rho' \in \mathbb{M}(I_n). \tag{20}$$

We thus conclude that $\boldsymbol{\chi}$ must be solved backwards in time from $t = T$.

5.4. Error representation

Having computed the dual solutions $\boldsymbol{\chi}$ and ρ from (20), we may obtain the error representation as follows:

$$\begin{aligned} E(\boldsymbol{q}, \boldsymbol{q}_h) &= \sum_{n=1}^N J_n(\boldsymbol{q}, \boldsymbol{q}_h; \boldsymbol{e}_q) \\ &= \sum_{n=1}^N A_n(\boldsymbol{q}, \mu, \boldsymbol{q}_h, \mu_h; \boldsymbol{e}_q, e_\mu, \boldsymbol{\chi}, \rho) = \sum_{n=1}^N E_n. \end{aligned} \tag{21}$$

It turns out that the error contributions from each time-step, E_n, become

$$\begin{aligned} E_n = &\int_{I_n} {}^n\mu_h \boldsymbol{g}({}^n\boldsymbol{q}_h, t) : \boldsymbol{\chi}^* \mathrm{d}t - [{}^n\boldsymbol{q}_h - {}^{n-1}\boldsymbol{q}_h] : \boldsymbol{\chi}^*(t_{n-1}^+) \\ &- \int_{I_n} [t_* \, {}^n\mu_h - \eta({}^n\boldsymbol{q}_h, t)] \, \rho^* \mathrm{d}t, \end{aligned} \tag{22}$$

where we used the pertinent Galerkin-orthogonality and introduced the notation

$$\begin{aligned} \boldsymbol{\chi}^*(t) &= \boldsymbol{\chi}(t) - \boldsymbol{\chi}_h'(t^+), & \boldsymbol{\chi}_h' &\in \mathbb{Q}_h(I), \\ \rho^*(t) &= \rho(t) - \rho_h'(t^+), & \rho_h' &\in \mathbb{M}_h(I). \end{aligned} \tag{23}$$

6. Computation of the dual solution

6.1. Discontinuous Galerkin method – basic method

As to the computation of the dual solutions χ and ρ, the basic method is to use the dG(0)-method and compute approximations χ_h and ρ_h on the *same* time-mesh as for the primal solutions $\boldsymbol{q}_h$ and μ_h, e.g. $\chi_h \in \mathbb{Q}_h(I)$. Hence, we compute ${}^n\chi_h$ and ${}^n\rho_h$ for $n = N, N-1, ..., 1$ such that (20) becomes

$$\underline{\boldsymbol{J}}^{\mathrm{T}}\, {}^n\underline{\boldsymbol{\Psi}}_h = {}^{n+1}\underline{\boldsymbol{Z}}_h \quad \text{with} \quad \underline{\boldsymbol{\Psi}} \stackrel{\text{def}}{=} \begin{bmatrix} \chi \\ \rho \end{bmatrix}, \quad \underline{\boldsymbol{Z}} \stackrel{\text{def}}{=} \begin{bmatrix} \chi \\ 0 \end{bmatrix}, \tag{24}$$

where $\underline{\boldsymbol{J}}$ is the Jacobian matrix associated with the primal problem.

Now, considering the (exact) error representation (22), we note that $E_n = 0$ if $\chi \in \mathbb{Q}_h(I)$ and $\rho \in \mathbb{M}_h(I)$. Hence, it is necessary to compute better approximations than χ_h and ρ_h in order for (22) to produce a useful result. We shall, therefore, compute "enhanced" dual solutions $\tilde{\chi}_h \in \tilde{\mathbb{Q}}_h(I)$ and $\tilde{\rho}_h \in \tilde{\mathbb{M}}_h(I)$ such that $\mathbb{Q}_h \subset \tilde{\mathbb{Q}}_h \subset \mathbb{Q}$ and $\mathbb{M}_h \subset \tilde{\mathbb{M}}_h \subset \mathbb{M}$. Subsequently, we device a few schemes for constructing $\tilde{\mathbb{Q}}_h$ and $\tilde{\mathbb{M}}_h$, which are all *piecewise linear* on each I_n and which may be continuous or discontinuous at the time nodes t_n.

6.1.1. *Smoothed dG(0)*

The simplest way of constructing $\tilde{\mathbb{Q}}_h$ (and $\tilde{\mathbb{M}}_h$) is to choose

$$\tilde{\chi}_h(t_{n-1}) \left[= {}^{n-1}\tilde{\chi}_h^+ = {}^{n-1}\chi_h^-\right] = {}^n\chi_h \tag{25}$$

as the nodal values of the continuous functions in $\tilde{\mathbb{Q}}_h(I)$.

6.1.2. *Improved smoothed dG(0)*

It can be shown that the exact solutions $\chi(t)$ (and $\rho(t)$) may have a jump at the time nodes such that $\chi(t_{n-1}^+) \neq \chi(t_{n-1}^-)$ in the special case of rate-independent plasticity (defined by $t_* = 0$). We then construct the discontinuous functions in $\tilde{\mathbb{Q}}_h$ such that

$$\tilde{\chi}_h(t_{n-1}^+) \stackrel{\text{def}}{=} {}^{n-1}\tilde{\chi}_h^+ = {}^n\chi_h, \quad \tilde{\chi}_h(t_n^-) \stackrel{\text{def}}{=} {}^n\tilde{\chi}_h^+ = \mathbf{P} : {}^{n+1}\chi_h, \tag{26}$$

where $\mathbf{P}$ is a projection such that $\mathbf{P} \neq \mathbf{I}$ in general. (The case $\mathbf{P} = \mathbf{I}$ obviously defines the Smoothed dG(0), as defined above.)

6.1.3. *Decoupled dG(1)*

The obvious way of increasing the accuracy of the solution is to use the dG(1)-method, which is one order higher than dG(0). However, in order to keep the computational effort as low as possible by computing only one extra value of χ (and ρ) for each I_n (like in the case of the Improved Smoothed dG(0)), we shall choose ${}^{n-1}\tilde{\boldsymbol{\chi}}_h^+ = {}^n\boldsymbol{\chi}_h$ and compute ${}^n\tilde{\boldsymbol{\chi}}_h^- = {}^n\boldsymbol{\chi}_h + {}^n\Delta\tilde{\boldsymbol{\chi}}_h$. Using the variational format (20), we may compute ${}^n\Delta\underline{\tilde{\boldsymbol{\Psi}}}_h$ from the set of equations

$$\frac{1}{2}\left[\underline{\boldsymbol{D}} + 2\underline{\boldsymbol{J}}^{\mathrm{T}}\right] {}^n\Delta\underline{\tilde{\boldsymbol{\Psi}}}_h = {}^{n+1}\underline{\boldsymbol{Z}}_h - \frac{1}{2}\left[\underline{\boldsymbol{D}} + \underline{\boldsymbol{J}}^{\mathrm{T}}\right] {}^n\underline{\boldsymbol{\Psi}}_h, \tag{27}$$

where we introduced the notation

$$\underline{\boldsymbol{D}} \stackrel{\mathrm{def}}{=} \begin{bmatrix} \mathbf{I} & \mathbf{0} \\ \mathbf{0} & 0 \end{bmatrix}, \quad \Delta\underline{\tilde{\boldsymbol{\Psi}}} \stackrel{\mathrm{def}}{=} \begin{bmatrix} \Delta\tilde{\boldsymbol{\chi}} \\ \Delta\tilde{\rho} \end{bmatrix}. \tag{28}$$

The principal characteristics of the resulting discontinuous functions are the same as those of the Improved Smoothed dG(0).

Remark 6.1 *The set of equations* (27) *has the same size as the one in* (24) *that defines the dG(0)-solution. Hence, the cost corresponds to that involved in the Improved Smoothed dG(0).* □

7. Prototype viscoplasticity model

The prototype model, chosen for the numerical investigations, has the following characteristics:

- Neo-Hooke hyperelasticity:

$$\bar{\boldsymbol{S}}_2 = G\bar{J}^{-\frac{2}{3}}\left[\boldsymbol{I} - \frac{\bar{I}_1}{3}\bar{\boldsymbol{C}}^{-1}\right] + K\bar{J}[\bar{J}-1]\bar{\boldsymbol{C}}^{-1}, \tag{29}$$

  where $\bar{I}_1 = \mathrm{tr}\,\bar{\boldsymbol{C}}$, $\bar{J} = \det\bar{\boldsymbol{C}}$ and where G and K are the (constant) elastic shear and bulk modulus, respectively.

- von Mises (quasistatic) yield function:

$$\Phi = \bar{T}_{\mathrm{e}} - \sigma_{\mathrm{y}}, \quad \bar{T}_{\mathrm{e}} \stackrel{\mathrm{def}}{=} \sqrt{\frac{3}{2}}|\bar{\boldsymbol{T}}_{\mathrm{dev}}| = \sqrt{\frac{3}{2}}|\boldsymbol{\tau}_{\mathrm{dev}}|, \tag{30}$$

  where σ_{y} is the quasistatic yield stress.

- Norton-type overstress function:

$$\eta(\Phi) = \left[\frac{< \Phi >}{\sigma_{\mathrm{c}}}\right]^{n_{\mathrm{c}}}, \tag{31}$$

  where σ_{c} is the creep modulus and n_{c} is the creep exponent.

8. Numerical example

In order to assess the proposed error estimation strategy, we consider the constitutive problem of "simple shear" on $I = [0, 1]$, defined by

$$F_{ij} = 1 \text{ for } i = j\,, \quad F_{12} = \gamma(t) = \frac{t}{T}\gamma_{\mathrm{A}}\,, \quad \text{other } F_{ij} = 0\,. \qquad (32)$$

The error measure is taken as $E(\boldsymbol{q}, \boldsymbol{q}_h) = (\boldsymbol{q}(1) - \boldsymbol{q}_h(1))_{12}$. The error estimates are subsequently evaluated in terms of the effectivity index $\eta \stackrel{\text{def}}{=} \tilde{E}/E$, where $\tilde{E}$ is the computed and E is the exact error. The material data are $K = G$, $\sigma_{\mathrm{y}} = 0.005\ G$, $\sigma_{\mathrm{c}} = 0.1\ G$, $n_{\mathrm{c}} = 1$ and $t_* = 1$ (viscoplastic response). Typical results are shown subsequently for different values of γ_{A} (corresponding to "small" strain characteristics for $\gamma_{\mathrm{A}} = 0.01$ and "large" strain characteristics for $\gamma_{\mathrm{A}} = 1$). The computed error contributions from each timestep, using the Improved Smoothed dG(0)-method, are shown in Figure 1, whereas the effectivity of the error estimates, using different strategies for computing the dual solution, is shown in Figure 2.

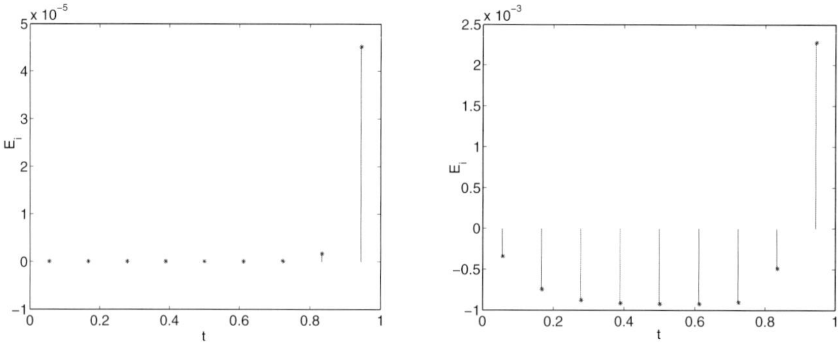

Figure 1. Error contributions from time-steps for $N = 10$ uniform time-steps. $\gamma_{\mathrm{A}} = 0.01$ (left) and $\gamma_{\mathrm{A}} = 1$ (right).

9. Conclusions and future work

A quite general framework for goal-oriented error control has been exploited for the constitutive problem of large strain viscoplasticity. In particular, we have investigated and compared different strategies for computing the dual solution accurately, which is essential in order to achieve good precision of the error estimate (defined as the effectivity index being close to unity). The present work is a (small but finite) step towards the goal of establishing a procedure for error control and adaptivity in coupled space-time for (visco)plasticity problems with large deformations.

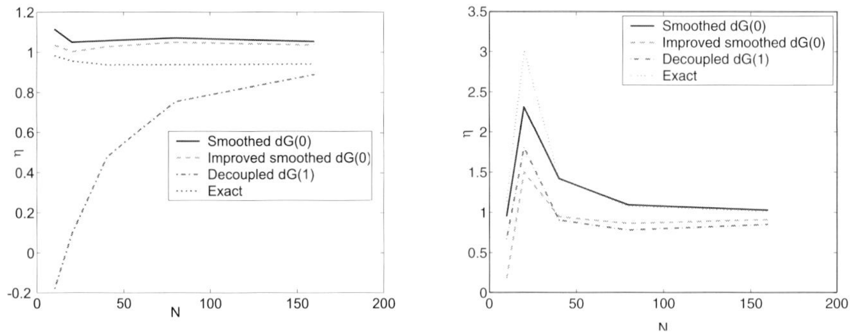

Figure 2. Effectivity index vs. number of uniform time-steps. $\gamma_A = 0.01$ (left) and $\gamma_A = 1$ (right).

Notes

1. Smoothness is assumed to avoid technical complexity of the resulting formulation.

References

[1] Becker, R. and Rannacher, R. (1996). A feed-back approach to error control in finite element methods: basic analysis and examples. *East-West J. Numer. Math.*, **4**, 237–264.

[2] Cirak, F. and Ramm, E. (1998). A posteriori error estimation and adaptivity for linear elasticity using the reciprocal theorem. *Computer Methods in Applied Mechanics and Engineering*, **156**, 351–362.

[3] Cirak, F. and Ramm, E. (2000). A posteriori error estimation and adaptivity for elasto-plasticity using the reciprocal theorem. *International Journal for Numerical Methods in Engineering*, **47**, 379–393.

[4] Eriksson, K. and Johnson, C. (1991). Adaptive finite element methods for parabolic problems I: A linear model problem. *SIAM Journal on Numerical Analysis*, **28**, 43–77.

[5] Eriksson, K., Estep, D., Hansbo, P. and Johnson C. (1995). Introduction to adaptive methods for differential equations. *Acta Numerica*, 105–158.

[6] Johnson, C. and Hansbo, P. (1992). Adaptive finite element methods in computational mechanics. *Computer Methods in Applied Mechanics and Engineering*, **101**, 143–181.

[7] Ladevèze, P., Moës, N. and Douchin, B. (1999). Constitutive relation errors (visco)plastic finite element analysis with softening. *Computer Methods in Applied Mechanics and Engineering*, **176**, 247–264.

[8] Ladevèze, P. (2001). Constitutive relation errors for F.E. analysis considering (visco)plasticity and damage. *International Journal for Numerical Methods in Engineering*, **52**, 527–542.

[9] Larsson, F., Runesson, K. and Hansbo, P. (2001). Computation of goal-oriented a posteriori error measures in space-time finite elements for viscoplasticity. In: *Trends in Computational Structural Mechanics*, Wall, W.A., Bletzinger, K.-U., Schweizerhof, K. (eds.), CIMNE, Barcelona, 499–510.

[10] Larsson, F., Runesson, K. and Hansbo, P. Time finite elements and error computation for (visco)plasticity with hardening or softening. *International Journal for Numerical Methods in Engineering*, in print.

[11] Larsson, F., Hansbo, P. and Runesson, K. Space-time finite elements and an adaptive strategy for the coupled thermoelasticity problem. *International Journal for Numerical Methods in Engineering*, in print.

[12] Larsson, F., Hansbo, P. and Runesson, K. Strategies for computing goal-oriented a posteriori error measures in nonlinear elasticity. *International Journal for Numerical Methods in Engineering*, submitted.

[13] Oden, J.T. and Prudhomme, S. (2001). Goal-oriented error estimation and adaptivity for the finite element method. *Computers and Mathematics with Applications*, **41**, 735–756.

[14] Ohnimus, S., Rüter, M. and Stein, E. A global and local a posteriori error estimator in finite elasticity. *Computational Mechanics*, submitted.

[15] Prudhomme, S. and Oden, J.T. (1999). On goal-oriented error estimation for elliptic problems: application to the control of pointwise errors. *Computer Methods in Applied Mechanics and Engineering*, **176**, 313–331.

[16] Rannacher, R. (1999). Error control in finite element computations. An introduction to error estimation and mesh-size adaptation. *NATO ASI series Series C. Mathematical and physical sciences*, **536**, 247–278.

[17] Rannacher, R. (2001). *An optimal control approach to a posteriori error estimation in finite element methods.* Preprint 2001–14 (SFB 359), University of Heidelberg.

[18] Rannacher, R. and Suttmeier F.T. (1999). A posteriori error estimation and mesh adaptation for finite element models in elasto-plasticity. *Computer Methods in Applied Mechanics and Engineering*, **176**, 333–361.

[19] Rüter, M., Stein, E., Larsson, F., Hansbo, P. and Runesson, K. (2001). Strategies for goal-oriented error measures in nonlinear elasticity. Proceedings: *2nd European Conference on Computational Mechanics.*

[20] Steeb, H. and Ramm, E. (2000). A general framework for local error estimation applied to material nonlinear problems. Proceedings: *European Congress on Computational Methods in Applied Sciences and Engineering.*

Affine-Approximate Finite Element Methods and Stabilization Techniques in Elasticity

B.D. Reddy
Faculty of Science, University of Cape Town
PD Hahn Building, 7701 Rondebosch, South Africa
bdr@science.uct.ac.za

Abstract: The purpose of this work is to report on a method for overcoming the effects of severe mesh distortion in finite element analyses. The method entails the replacement of arbitrary quadrilaterals, in computations, by parallelograms that are close to the original elements, in a rigorous sense. The resulting approach is referred to as an affine-approximate method, and is developed in the context of enhanced assumed strains with low-order elements. Theoretical results on convergence of the method are presented, and its extension to problems involving Mindlin-Reissner plates is discussed.

Keywords: Affine approximations, four-noded quadrilateral, enhanced assumed strains.

1. Introduction

The inherent economy of four-noded quadrilaterals in two dimensions, and eight-noded hexahedral elements in three, results in these being popular choices of elements in finite element analyses. Unfortunately, they are not without their drawbacks. In problems of solid mechanics in which bending deformations dominate, analyses based on these elements exhibit poor accuracy, at least when coarse meshes are used. In addition, in the incompressible limit, or when the compressibility is small, locking behavior is experienced.

There is a vast literature that is devoted to the construction of methods which are intended to overcome these problems, while retaining the advantages of using low-order elements. Remedies include the use of underintegration and stabilization (see Belytschko and Bachrach [5] and Flanagan and Belytschko [8]). Another popular approach is that of enhanced assumed strains (Simo and Rifai [18]), which contains as a special case the nonconforming method of incompatible modes (Wilson et al. [22], Taylor et al. [20]). The method has been successfully extended to nonlinear problems (see, for example, Simo et al. [19]). It has been shown (Reddy and Simo [15]) that, for linear problems and for affine elements, that the enhanced strain method is stable and convergent. This result has been extended to isoparametric maps in Arunakirinathar and Reddy [2]. In more recent work (Braess et al. [6]), it has been shown that the method converges uniformly in the incompressible limit.

C. Miehe (ed.), IUTAM Symposium on Computational Mechanics of Solid Materials at Large Strains, 395-404.

The method of enhanced assumed strains exhibits a decline in accuracy with increase in element distortions. This problem leads to the notion of replacing the arbitrary quadrilateral by the affine element (a parallelogram) that is closest to it, in a manner that can be made precise. Such an element is known as the equivalent parallelogram in two dimensions, and the equivalent parallelepiped in three. This set of ideas has been proposed, and then tested numerically, first in the context of problems of linear elasticity in Küssner and Reddy [12] and subsequently for problems involving nonlinearly elastic materials in Reese et al. [16, 17]. In all cases the numerical results are encouraging, and suggest a significant improvement in efficency and accuracy when this approach is used, particularly in circumstances in which element distortions are significant.

The purpose of this work is, first, to report on an analysis that is aimed at verifying the good approximation properties of affine-approximate finite element methods. The key result is that the method converges at the optimal rate provided that the element distortion is of the order of mesh size, in the asymptotic limit. These results are presented in greater detail in Arunakirinathar and Reddy [3]. The second aim of this work is to present results on the extension of these methods to the analysis of Mindlin-Reissner plates, for which shear locking is a major challenge in the thin-plate limit. Preliminary results in this regard have appeared in Reddy and Ehl [14] and are discussed further here.

2. Formulation of the problem

Let Ω be a bounded domain in $\mathbb{R}^d$ $(d = 2, 3)$. We will make use of the space $L^2(\Omega)$ of square-integrable functions defined onΩ. The inner product and norm on this space are denoted respectively by $(\cdot,\cdot)_0$ and $\|\cdot\|_0$. We recall also the definition of the Sobolev space $H^1(\Omega)$, which is the Hilbert space of functions which, together with their first generalized derivatives, are members of$L^2(\Omega)$. We also define the space $H_0^1(\Omega)$ of functions in $H^1(\Omega)$ which vanish on the boundary, in the sense of traces. The seminorm $|\cdot|_1$ is a norm on $H_0^1(\Omega)$, equivalent to the standard norm $\|\cdot\|_1$.

Denote by $V := [H_0^1(\Omega)]^d$ the space of admissible displacements, and define the bilinear form $a(\cdot,\cdot)$ and linear functional $\ell(\cdot)$ by

$$a : V \times V \to \mathbb{R}, \qquad a(\boldsymbol{u}, \boldsymbol{v}) = \int_\Omega \mathbb{C}\boldsymbol{\epsilon}(\boldsymbol{u}) : \boldsymbol{\epsilon}(\boldsymbol{v})\, d\boldsymbol{x}, \tag{1}$$

$$\ell : V \to \mathbb{R}, \qquad \ell(\boldsymbol{v}) = \int_\Omega \boldsymbol{b} \cdot \boldsymbol{v}\, d\boldsymbol{x}. \tag{2}$$

Here $\boldsymbol{\epsilon} = \frac{1}{2}(\nabla\boldsymbol{u} + (\nabla\boldsymbol{u})^T)$ is the infinitesimal strain tensor, $\boldsymbol{u}$ is the displacement vector, $\mathbb{C}$ is the fourth-order elasticity tensor, and $\boldsymbol{b}$ is a prescribed body force vector. For convenience we assume that the displacement satisfies a homogeneous Dirichlet boundary condition.

The properties of $\mathbb{C}$ guarantee that $a(\cdot,\cdot)$ is symmetric, continuous, and V-elliptic. The standard variational problem is as follows.

Problem S. Given $\boldsymbol{b} \in V'$, find $\boldsymbol{u} \in V$ which satisfies

$$a(\boldsymbol{u},\boldsymbol{v}) = \ell(\boldsymbol{v}) \tag{3}$$

for all $\boldsymbol{v} \in V$. It is well known (see, for example, Ciarlet [7]) that Problem S has a unique solution, which depends continuously on the data.

Finite element approximations. We confine attention to plane situations, so that $d = 2$. The domain Ω is assumed to be polygonal, and a finite element mesh $\mathcal{T}$ of quadrilateral elements is constructed on Ω in the usual manner. A typical element K in $\mathcal{T}$ is generated by an isoparametric map F from a reference element $\hat{K} \equiv (-1,1)\times(-1,1)$. The mesh parameter h is defined by $h = \max_{K\in\mathcal{T}} \sup\{|\boldsymbol{x}-\boldsymbol{y}| : \boldsymbol{x},\boldsymbol{y} \in K\}$. We define basis functions $\hat{N}_A$ $(A = 1,\ldots,4)$ on $\hat{K}$ by

$$\hat{N}_A(\boldsymbol{\xi}) = \tfrac{1}{4}(1+\xi\xi_A)(1+\eta\eta_A),$$

where $\boldsymbol{\xi}_A \equiv (\xi_A,\eta_A)$ are the nodal coordinates on $\hat{K}$, with $(\xi_1\cdots\xi_4) = (1\ -1\ -1\ 1)$ and $(\eta_1\cdots\eta_4) = (1\ 1\ -1\ -1)$. Denote by Q_1 the space of bilinear functions spanned by $\hat{N}_A$. Then it is convenient to express the map F in the form

$$F : \hat{K} \to K, \qquad \boldsymbol{x} = F(\boldsymbol{\xi}) = \sum_{A=1}^{4} \boldsymbol{x}_A \hat{N}_A(\boldsymbol{\xi}) \tag{4}$$

in which $\boldsymbol{x}_A$ are the nodal points of K. The Jacobian matrix and determinant are denoted by $\boldsymbol{J}$ and j.

We define

$$V^h = \{\boldsymbol{v}_h \in V : (v_h)_i|_K \circ F \in Q_1\}. \tag{5}$$

The equivalent parallelogram. The equivalent parallelogram associated with a quadrilateral is obtained by perturbing the quadrilateral to obtain the parallelogram that is closest to it, in a precise sense. It has been shown in Arunakirinathar and Reddy [1] that the equivalent parallelogram $\tilde{K}$ associated with K is defined by the *affine* map $\tilde{F}$ obtained simply by discarding the bilinear terms in (4). That is, if we define the vector $\boldsymbol{k}$ by $\boldsymbol{k} = \frac{1}{4}(\boldsymbol{x}_1 - \boldsymbol{x}_2 + \boldsymbol{x}_3 - \boldsymbol{x}_4)$, then the map $\tilde{F}$ may be expressed in the form

$$\tilde{F}(\boldsymbol{\xi}) = F(\boldsymbol{\xi}) - \boldsymbol{k}\xi\eta = \sum_{A=1}^{4} N_A(\boldsymbol{\xi})\tilde{\boldsymbol{x}}_A$$

in which the nodal points $\tilde{\boldsymbol{x}}_A$ of the equivalent parallelogram are defined by

$$\tilde{\boldsymbol{x}}_A = \tfrac{3}{4}\boldsymbol{x}_A + \tfrac{1}{4}(\boldsymbol{x}_{A+1} - \boldsymbol{x}_{A+2} + \boldsymbol{x}_{A+3}), \quad A = 1, \ldots, 4 \text{ (modulo 4)}.$$

These notions are illustrated in Figure 1.

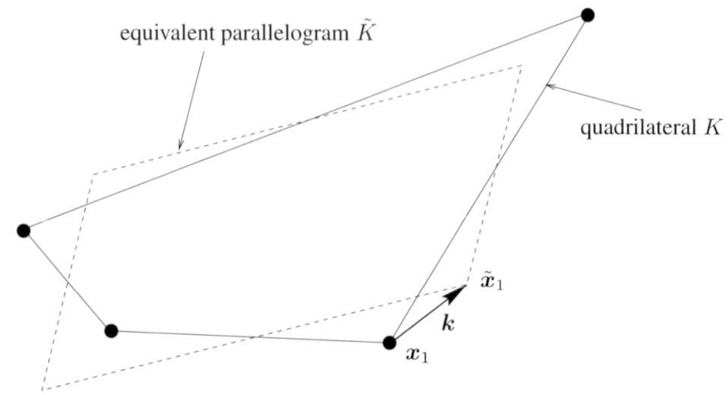

Figure 1. The equivalent parallelogram associated with a quadrilateral.

Suppose that the affine map from $\hat{K}$ to $\tilde{K}$ takes the form $\tilde{\boldsymbol{x}} = \boldsymbol{C}\boldsymbol{\xi} + \boldsymbol{c}$, in which $\boldsymbol{C}$ and $\boldsymbol{c}$ are respectively a constant matrix and vector; then the *distortion parameter* τ_K for element K is defined by

$$\tau_K = |\boldsymbol{C}^{-1}\boldsymbol{k}|. \tag{6}$$

The distortion parameter τ associated with a finite element mesh is defined by $\tau = \max_{K\in\mathcal{T}} |\tau_K|$. We define an *h-regular mesh* to be a finite element mesh for which $\tau = O(h)$.

The enhanced assumed strain problem. In this formulation, proposed in Simo and Rifai [18], the discrete strain $\boldsymbol{\epsilon}_h$ takes the form

$$\boldsymbol{\epsilon}_h = \boldsymbol{\epsilon}(\boldsymbol{u}_h) + \boldsymbol{\eta}_h, \tag{7}$$

the first term on the righthand side being the conventional infinitesimal strain, while the second term on the righthand side is the enhanced strain, which is required to have the property $\boldsymbol{\eta}_h \to \boldsymbol{0}$ as $h \to 0$.

In order to formulate the problem in weak form it is necessary to add to the spaces already defined the space Γ^h of enhanced strains, which is defined by

$$\Gamma^h := \{\gamma : \gamma_{ij} \in L^2(\Omega), \gamma_{ji} = \gamma_{ij}, \int_K \mathbb{C}\gamma|_K \, dxdy = \boldsymbol{0} \text{ for all } K \in \mathcal{T}\}. \tag{8}$$

In practice Γ^h will comprise functions of the form $\gamma = j^{-1}\hat{\gamma}$ on each element, in which the components of $\hat{\gamma}$ are simple polynomials defined on the reference element $\hat{K}$. A consequence of this definition is that $\int_{\hat{K}} \mathbb{C}\hat{\gamma}\, d\xi d\eta = \boldsymbol{0}$ (Arunakirinathar and Reddy [2]).

We set $\phi_h = (\boldsymbol{u}_h, \boldsymbol{\eta}_h)$ and $\boldsymbol{\psi}_h = (\boldsymbol{v}_h, \boldsymbol{\gamma}_h)$ for $\boldsymbol{u}_h,\ \boldsymbol{v}_h \in V^h$ and $\boldsymbol{\eta}_h,\ \boldsymbol{\gamma}_h \in \Gamma^h$, and define the product space $\Psi^h := V^h \times \Gamma^h$. The bilinear form $A : \Psi^h \times \Psi^h \longrightarrow \mathbb{R}$ is defined by

$$A(\boldsymbol{\phi}_h, \boldsymbol{\psi}_h) = \int_\Omega \mathbb{C}(\boldsymbol{\epsilon}(\boldsymbol{u}_h) + \boldsymbol{\eta}_h) : (\boldsymbol{\epsilon}(\boldsymbol{v}_h) + \boldsymbol{\gamma}_h)\, dxdy. \tag{9}$$

The weak formulation of the problem then takes the following form (Reddy and Simo [15], Simo and Rifai [18]).

Problem E^h. Find $(\boldsymbol{u}_h, \boldsymbol{\eta}_h) \in V^h \times \Gamma^h$ such that

$$A(\boldsymbol{\phi}_h, \boldsymbol{\psi}_h) = \ell(\boldsymbol{v}_h) \quad \text{for all } \boldsymbol{\psi}_h \in \Psi^h. \tag{10}$$

We have the following result (Reddy and Simo [15], Arunakirinathar and Reddy [2], Braess et al. [6]).

Theorem 2.1 *Let $\mathcal{T}$ be a regular mesh of quadrilaterals on a bounded polygonal domain $\Omega \in \mathbb{R}^2$. Let the space V^h be defined by (5), and the space Γ^h by (8). Assume, in addition, that*

- $\boldsymbol{\epsilon}(V^h) \cap \Gamma^h = \{\mathbf{0}\}$;
- *there exists a constant c_1 with $0 < c_1 < 1$ such that, for any $\boldsymbol{\gamma}_h \in \Gamma^h$, $\|P\boldsymbol{\gamma}_h\|_\Gamma \leq c_1\|\boldsymbol{\gamma}_h\|$, where P is the L^2-orthogonal projection onto $\boldsymbol{\epsilon}(V^h)$.*

Then there exists a unique solution to Problem E^h. Furthermore, if $\boldsymbol{u} \in [H^2(\Omega)]^2$, then there exists a constant $C > 0$, independent of h, such that

$$\|\boldsymbol{u} - \boldsymbol{u}_h\|_V + \|\boldsymbol{\eta}_h\|_\Gamma \leq Ch|\boldsymbol{u}|_{H^2}.$$

That is, the method is uniformly convergent in the incompressible limit.

Equivalence with underintegration plus stabilization. For the case in which the finite element mesh comprises elements which are parallelograms, it can be shown (Küssner and Reddy [12]) that, after the enhanced strain degrees of freedom have been condensed out at element level, the stiffness matrix takes the decoupled form $\boldsymbol{K} = \boldsymbol{K}_0 + \boldsymbol{K}_*$, in which $\boldsymbol{K}_0$ is the matrix corresponding to the constant part of the strain, and $\boldsymbol{K}_*$ is the rank two stabilization matrix for the plane strain problem, given by $\boldsymbol{K}_* = \lambda_7 \boldsymbol{e}_7 \boldsymbol{e}_7^{\mathrm{T}} + \lambda_8 \boldsymbol{e}_8 \boldsymbol{e}_8^{\mathrm{T}}$, in which $(\lambda_7,\ \boldsymbol{e}_7)$ and $(\lambda_8,\ \boldsymbol{e}_8)$ are the eigenpairs corresponding to the flexural modes.

3. Affine finite element approximations

An affine-approximate formulation is constructed by replacing integrals over arbitrary quadrilaterals by integrals over their equivalent parallelograms. Since the map from K to $\tilde{K}$ is affine, the associated Jacobian matrix and determinant are constant, and the integrals can be evaluated exactly.

Numerical results presented in Küssner and Reddy [12] in the case of linear elasticity, and in Reese et al. [16, 17] for problems involving nonlinear elasticity and finite deformations, show in addition that the results represent in many cases an improvement over those obtained by the conventional approach.

Define

$$A_{\tilde{K}}(\tilde{\boldsymbol{\chi}}_h, \tilde{\boldsymbol{\psi}}_h) = \int_{\tilde{K}} \tilde{\mathbb{C}}(\tilde{\boldsymbol{\epsilon}}(\tilde{\boldsymbol{w}}_h) + \tilde{\boldsymbol{\beta}}_h) : (\tilde{\boldsymbol{\epsilon}}(\tilde{\boldsymbol{v}}_h) + \tilde{\boldsymbol{\gamma}}_h)\, d\tilde{x} d\tilde{y} \tag{11}$$

and

$$\ell_{\tilde{K}}(\tilde{\boldsymbol{\psi}}_h) = \int_{\tilde{K}} \boldsymbol{b} \cdot \tilde{\boldsymbol{v}}_h \, d\tilde{x} d\tilde{y}, \tag{12}$$

where $\boldsymbol{\chi}_h = (\boldsymbol{w}_h, \boldsymbol{\beta}_h)$,

$$(\tilde{\nabla} \tilde{\boldsymbol{v}})_{ij} = \frac{\partial \tilde{v}_i}{\partial \tilde{x}_j}, \quad \tilde{\boldsymbol{\epsilon}}(\tilde{\boldsymbol{v}}) = \tfrac{1}{2}(\tilde{\nabla}\tilde{\boldsymbol{v}} + [\tilde{\nabla}\tilde{\boldsymbol{v}}]^T),$$

and superposed tildes on other quantities denote transformation to the domain $\tilde{K}$. Set

$$\tilde{A}(\boldsymbol{\chi}_h, \boldsymbol{\psi}_h) = \sum_{K \in \mathcal{T}} A_{\tilde{K}}(\tilde{\boldsymbol{\chi}}_h, \tilde{\boldsymbol{\psi}}_h) \text{ and } \tilde{\ell}(\boldsymbol{\psi}_h) = \sum_{K \in \mathcal{T}} \ell_{\tilde{K}}(\tilde{\boldsymbol{\psi}}_h). \tag{13}$$

Then the affine-approximate problem is the following.

Problem $\tilde{\mathrm{E}}^h$. Given $\boldsymbol{b} \in V'$, find $\boldsymbol{\chi}_h := (\boldsymbol{w}_h, \boldsymbol{\beta}_h) \in V^h \times \Psi^h$ which satisfy

$$\tilde{A}(\boldsymbol{\chi}_h, \boldsymbol{\psi}_h) = \tilde{\ell}(\boldsymbol{\psi}_h) \text{ for all } \boldsymbol{\psi}_h \in \Psi^h. \tag{14}$$

The following has been proved in Arunakirinathar and Reddy [3].

Theorem 3.1 *Let $\mathcal{T}$ be an h-regular finite element mesh of quadrilaterals, with the maximum distortion of quadrilaterals being bounded in the sense that $\tau \leq ch$ for some constant c, independent of h, as $h \to 0$. Let $\boldsymbol{\phi} = (\boldsymbol{u}, \boldsymbol{0}) \in \Psi$, where $\boldsymbol{u}$ is the solution to Problem S. Then Problem $\tilde{\mathrm{E}}^h$ has a unique solution $\boldsymbol{\chi}_h$ which satisfies*

$$\|\boldsymbol{\phi} - \boldsymbol{\chi}_h\|_{\Psi} \leq Ch,$$

the constant C depending on the geometry, on the material tensor $\mathbb{C}$, and on $\boldsymbol{u}$, but not on h.

4. Application to Mindlin-Reissner plates

In this section it is shown how the theory and methods described earlier may be extended to the case of Mindlin-Reissner plates. Consider a plate which occupies the domain $\Omega \times (-t/2, t/2)$. The plate midsurface displacement and rotation are denoted respectively by w and $\boldsymbol{\theta}$, and the curvature $\boldsymbol{\kappa}$ and shear strain $\boldsymbol{\gamma}$ are then given by $\boldsymbol{\kappa} = (\theta_{1,1}\ \ \theta_{2,2}\ \ \theta_{1,2}+\theta_{2,1})^T$ and $\boldsymbol{\gamma} = -\boldsymbol{\theta} + \nabla w$. The Kirchhoff thin plate limit corresponds to the situation in which$\theta_\alpha = w_{,\alpha}$ $(\alpha = 1, 2)$. With the bending stiffness defined by

$$\boldsymbol{D}_b = \begin{pmatrix} E_1 & E_2 & 0 \\ E_2 & E_1 & 0 \\ 0 & 0 & G \end{pmatrix} \tag{15}$$

where $E_1 = E/(1-\nu^2)$, $E_2 = \nu E_1$ and $G = E/(1+\nu)$, E and ν being Young's modulus and Poisson's ratio, respectively, the total potential energy of the plate is given by

$$\Pi = \tfrac{1}{2} \int_A \left[\tfrac{t^3}{12} \boldsymbol{\kappa}^T \boldsymbol{D}_b \boldsymbol{\kappa} + Gt\boldsymbol{\gamma}^T \boldsymbol{\gamma} \right] dA - \int_A fw \, dA.$$

The use of piecewise bilinear approximations for$w$ and θ leads to shear locking when the thickness is very small relative to the lateral dimensions of the plate (see, for example, Hughes [9]). Various remedies, for example, selective reduced integration (Hughes et al. [10]), assumed strains or mixed interpolation (MacNeal [13], Hughes and Tezduyar [11], Bathe and Dvorkin [4]), and linked interpolations (Tessler and Hughes [21]), have been proposed.

An attempt has also been made in Simo and Rifai [18] to overcome the problem of locking through an enhanced assumed strain approach, in which the discrete shear strain is given by $\boldsymbol{\gamma}_h = \boldsymbol{\gamma}(w_h, \theta_{1h}, \theta_{2h}) + \boldsymbol{G}\boldsymbol{\alpha}_h$; the displacements and rotations are approximated by bilinear functions on each element, and the matrix $\boldsymbol{G}$ is given by

$$\boldsymbol{G} = \begin{pmatrix} \xi & 0 & \xi\eta & 0 \\ 0 & \eta & 0 & \xi\eta \end{pmatrix} \tag{16}$$

on the reference element. The strain energy for the discrete problem then becomes

$$\tfrac{1}{2} \int_A \left[\tfrac{t^3}{12} \boldsymbol{\kappa}_h^T \boldsymbol{D}_b \boldsymbol{\kappa}_h + Gt(\boldsymbol{\gamma}_h + \boldsymbol{G}\boldsymbol{\alpha}_h)^T (\boldsymbol{\gamma}_h + \boldsymbol{G}\boldsymbol{\alpha}_h) \right] dA.$$

For arbitrary quadrilaterals it is necessary to use the transformation $\widetilde{\boldsymbol{G}} = (j/j_0)\boldsymbol{J}_0^{-T}\boldsymbol{G}$ in which $\boldsymbol{J}_0$ is the Jacobian at the centroid and $j_0 = \det \boldsymbol{J}_0$. This leads to a method that is identical, for rectangular elements, to those proposed in Hughes and Tezduyar [11] and Bathe and Dvorkin [4].

For arbitrary quadrilaterals the enhanced assumed strain approach exhibits sensitivity to distortions, and it is natural to enquire whether an affine-approximate approach is able to mitigate the effects of significant mesh distortion. The resulting approach is very similar to that described earlier in the context of problems of plane elasticity, and some examples of the performance of this approach may be found in Reddy and Ehl [14].

Insight into the behavior of this approach may be gained by carrying out a spectral analysis of a single element. Consider first a square element $(0,1) \times (0,1)$ with thickness $t = 0.01$ or $t = 0.001$, modulus of elasticity $E = 10^4$, and Poisson's ratio $\nu = 0.3$.

Table 1 shows the dependence of the eigenvalues λ_i on thickness t, for the standard element, selective reduced integration (SRI), and enhanced assumed strains (EAS). A locking free element should exhibit only two unbounded eigenvalues, which correspond to warping modes. For the standard element a total of eight eigenvalues become unbounded, in agreement with the known shear locking response of this element. Selective reduced integration minimizes the problem of shear locking and exhibits only two eigenvalues which become unbounded with decrease in thickness. But the formulation possesses two additional zero eigenvalues, over and above those corresponding to rigid body modes; these are associated with the zero-energy w-hourglass mode and in-plane twist mode, which indicate the well known rank deficiency of this element (Hughes [9]). The enhanced assumed strain method exhibits four eigenvalues going to infinity, thus leading to a formulation that exhibits mild locking.

Table 1. Dependence on eigenvalues on thickness for a square element.

Standard		*SRI*		*EAS*	
$h = 0.01$	$h = 0.001$	$h = 0.01$	$h = 0.001$	$h = 0.01$	$h = 0.001$
$48.08{*}10^2$	$48.08{*}10^4$	$48.08{*}10^2$	$48.08{*}10^4$	$48.08{*}10^2$	$48.08{*}10^4$
$48.08{*}10^2$	$48.08{*}10^4$	$48.08{*}10^2$	$48.08{*}10^4$	$48.08{*}10^2$	$48.08{*}10^4$
$28.85{*}10^2$	$28.85{*}10^4$	0.12	0.12	$28.85{*}10^2$	$28.85{*}10^4$
$3.21{*}10^2$	$3.21{*}10^4$	0.06	0.06	$3.21{*}10^2$	$3.21{*}10^4$
$3.21{*}10^2$	$3.21{*}10^4$	0.06	0.06	0.12	0.12
$3.21{*}10^2$	$3.21{*}10^4$	0.04	0.04	0.06	0.06
$1.07{*}10^2$	$1.07{*}10^4$	0.04	0.04	0.06	0.06
$1.07{*}10^2$	$1.07{*}10^4$	0.00	0.00	0.04	0.04
0.06	0.06	0.00	0.00	0.04	0.04
0.00	0.00	0.00	0.00	0.00	0.00
0.00	0.00	0.00	0.00	0.00	0.00
0.00	0.00	0.00	0.00	0.00	0.00

Table 2 reveals significant differences when the node at (1,1) is translated to (3,3). Results are shown for the case $t = 0.01$, and for the standard element, selective reduced integration, enhanced assumed strains, and enhanced assumed strains with the affine approximation (EAS-AFF). There is little difference between the eigenvalues corresponding to EAS and EAS-AFF, the latter also displaying a mild tendency towards locking.

Table 2. Eigenvalues for quadrilateral obtained by moving the node at (1,1) to (3,3).

Standard	*SRI*	*EAS*	*EAS-AFF*
$144.99*10^2$	$144.23*10^2$	$139.11*10^2$	$144.95*10^2$
$66.95*10^2$	$41.67*10^2$	$51.69*10^2$	$43.77*10^2$
$38.21*10^2$	0.28	$37.94*10^2$	$27.97*10^2$
$9.62*10^2$	0.13	$8.82*10^2$	$9.57*10^2$
$9.23*10^2$	0.10	0.26	0.27
$9.14*10^2$	0.06	0.13	0.12
$2.95*10^2$	0.02	0.06	0.06
$2.83*10^2$	0.00	0.06	0.03
0.19	0.00	0.02	0.01
0.00	0.00	0.00	0.00
0.00	0.00	0.00	0.00
0.00	0.00	0.00	0.00

Work in progress. From the results presented here and in Reddy and Ehl [14], it is clear that affine approximations for the plate problem require further investigation. Of particular interest is the question of the choice of a basis for the enhanced degrees of freedom. It can be shown that, for the basis (16), the resulting shear strains do not satisfy the constraint of continuous transverse shear along the edges for the case of an arbitrary quadrilateral (see, for example, Tessler and Hughes [21] for a discussion of this criterion). Preliminary studies indicate that the criterion is met if $\boldsymbol{G}$ is augmented by at least two modes, so that it takes the form

$$\boldsymbol{G} = \begin{pmatrix} \xi & 0 & \xi\eta & 0 & 1-3\eta^2 & 0 \\ 0 & \eta & 0 & \xi\eta & 0 & 1-3\xi^2 \end{pmatrix}.$$

The consequences of such a modification are currently being investigated.

References

[1] Arunakirinathar, K. and Reddy, B.D. (1995). Some geometrical results and estimates for quadrilateral finite elements. *Comput. Methods Appl. Mech. Engrg.*, **122**, 307–314.

[2] Arunakirinathar, K. and Reddy, B.D. (1995). Further results for enhanced strain methods with isoparametric elements. *Comput. Methods Appl. Mech. Engrg.*, **127**, 127–143.

[3] Arunakirinathar, K. and Reddy, B.D. (2002). A stable affine-approximate finite element method. To appear in: *SIAM Journal on Numerical Analysis*.

[4] Bathe, K.J. and Dvorkin, E.N. (1985). A four-node plate bending element based on Mindlin/Reissner plate theory and a mixed interpolation. *International Journal for Numerical Methods in Engineering*, **21**, 367–383.

[5] Belytschko, T. and Bachrach, W. (1986). Efficient implementation of quadrilaterals with high coarse-mesh accuracy. *Comput. Methods Appl. Mech. Engrg.*, **54**, 279–301.

[6] Braess, D., Carstensen, C. and Reddy, B.D. (2001). Uniform convergence and a posteriori estimators for the enhanced strain finite element method. In review.

[7] Ciarlet, P.G. (1978). *The Finite Element Method for Elliptic Problems*, North-Holland, Amsterdam.

[8] Flanagan, D.P. and Belytschko, T. (1981). A uniform strain hexahedron and quadrilateral with orthogonal hourglass control. *Int. J. Numer. Meth. Engng.*, **17**, 679–706.

[9] Hughes, T.J.R. (1978). *The Finite Element Method – Linear Static and Dynamic Finite Element Analysis*, Prentice-Hall, New Jersey.

[10] Hughes, T.J.R., Taylor, R.L. and Kanoknukulchai, W. (1977). A simple and efficient element for plate bending. *Int. J. Numer. Meth. Engng.*, **10**, 1529–1543.

[11] Hughes, T.J.R. and Tezduyar, T.E. (1981). Finite elements based upon Mindlin plate theory with particular reference to the four-node bilinear isoparametric element. *Journal of Applied Mechanics*, **48**, 587–596.

[12] Küssner, M. and Reddy, B.D. (2001). The equivalent parallelogram and parallelepiped, and their application to stabilized finite elements in two and three dimensions. *Computer Methods in Applied Mechanics and Engineering*, **190**, 1967–1983.

[13] MacNeal, R.H. (1982). Derivation of element stiffness matrices by assumed strain distributions. *Nuclear Engineering and Design*, **70**, 3–12.

[14] Reddy, B.D. and Ehl, E. (2001). Enhanced strain finite elements for Mindlin-Reissner plates. In: *Structural Engineering, Mechanics and Computation*, Zingoni A. [ed.], Elsevier, Amsterdam, 27–38.

[15] Reddy, B.D. and Simo, J.C. (1995). Stability and convergence of a class of enhanced strain methods. *SIAM Journal on Numerical Analysis*, **32**, 1705–1728.

[16] Reese, S., Küssner, M. and Reddy, B.D. (1999). A new stabilization technique for finite elements in non-linear elasticity. *Int. J. Numer. Meth. Engng.*, **44**, 1617–1652.

[17] Reese, S., Wriggers, P. and Reddy, B.D. (2000). A new locking-free brick element technique for large deformation problems in elasticity. *Computers and Structures*, **75**, 291–304.

[18] Simo, J.C. and Rifai, S. (1990). A class of mixed assumed strain methods and the method of incompatible modes. *Int. J. Numer. Meth. Engng.*, **29**, 1595–1638.

[19] Simo, J.C., Armero, F. and Taylor, R.L. (1993). Improved version of assumed enhanced strain tri-linear elements for 3D finite deformation problems. *Computer Methods in Applied Mechanics and Engineering*, **110**, 359–366.

[20] Taylor, R.L., Beresford, P.J. and Wilson, E.L. (1976). A non-conforming element for stress analysis. *International Journal for Numerical Methods in Engineering*, **10**, 1211–1220.

[21] Tessler, A. and Hughes, T.J.R. (1983). An improved treatment of transverse shear in the Mindlin-type four-node quadrilateral element. *Computer Methods in Applied Mechanics and Engineering*, **39**, 311–335.

[22] Wilson, E.L., Taylor, R.L. et al. (1973). Incompatible displacement models. In: *Numerical and Computer Models in Structural Mechanics*, Fenves S.J., Perrone N., Robinson A.R. and Schnobrich W.C. [eds.], Academic Press, 43–57.

Advanced Computational Applications

Numerical Investigation of High-Pressure Hydraulic Hoses with Steel Wire Braid

H. Rattensperger, J. Eberhardsteiner, H.A. Mang
Institute for Strength of Materials, Vienna University of Technology, Karlsplatz 13/202, 1040 Vienna, Austria
Herbert.Rattensperger@tuwien.ac.at

Abstract: In this work a model for numerical simulation of hydraulic hoses subjected to high pressure is presented. Within the framework of the finite element (FE) method for matrix material (rubber) and the reinforcement (wire or textile braid) of this composite structure are considered separately. A new biaxial membrane test provides experimental data for determination of material parameters for rubber treated as a hyperelastic material. The reinforcement is represented by rebar elements, their mechanical behavior is accounted for by means of a fabric lattice model.

Keywords: Hydraulic hose, hyperelasticity, woven material, fabric lattice model.

1. Introduction

High-pressure hydraulic hoses are used in industry for power transmission in steering, drive and brake systems and for fluid transport. Such hoses must be tight, flexible, and resistant against high inside pressure. Threaded fittings are used to adapt the connection of high-pressure flexible hoses to the required equipment. Hydraulic hoses are composite structures of rubber with embedded multi-layered wire or textile reinforcement. Depending on the technical requirements, this reinforcement can be arranged in multi-spiral form with an even number of layers with parallel fibers or in braided form with plainted fiber bundles. The influence of the interaction between individual fiber bundles on the global stiffness of the reinforcement is significant.

The purpose of this work is to develop a realistic numerical simulation model for different types of high-pressure hydraulic hoses. The goals of this research are realistic prediction of the deformation and stress response under service loads, determination of the bursting pressure, and optimization of the crimping of fittings. For that purpose, well-suited experimental tests for identification of the material parameters of rubber are essential. An important objective of this work is to provide reliable material data of the individual component materials under realistic loading conditions as well as of the composite structure.

Numerical modeling of the reinforcement represents the main part of this contribution. The concept for the mathematical description of the mechanical

C. Miehe (ed.), IUTAM Symposium on Computational Mechanics of Solid Materials at Large Strains, 407-416.

behavior of the woven fabric reinforcement is based on a trapezoidally-shaped fabric lattice model.

Finally, the developed analysis model is calibrated and verified by means of numerical simulations of a hydraulic hose under high internal pressure and of the crimping of a hose fitting.

2. Material behavior of rubber

The inside lining and the cover of hydraulic hoses consist of rubber materials. For the numerical representation of this hyperelastic material a host of constitutive models is available in the literature, e.g., constitutive models by Mooney-Rivlin, Ogden, van der Waals, or the tube model, and several polynomial laws. In general, the material parameters of these constitutive models are identified on the basis of uniaxial tension and compression tests. Material models with such parameters often fail when attempting to describe stress states as occurring in engineering applications. To avoid serious problems, more complex material tests are needed (Johannknecht and Jerrams [2]). For that reason, a biaxial membrane test was developed.

Figure 1 shows the middle section of the axi-symmetric specimen used for equibiaxial material tests. The shape of the specimen was optimized by numerical simulations ensuring a pure membrane state of stress in the vicinity of the apex.

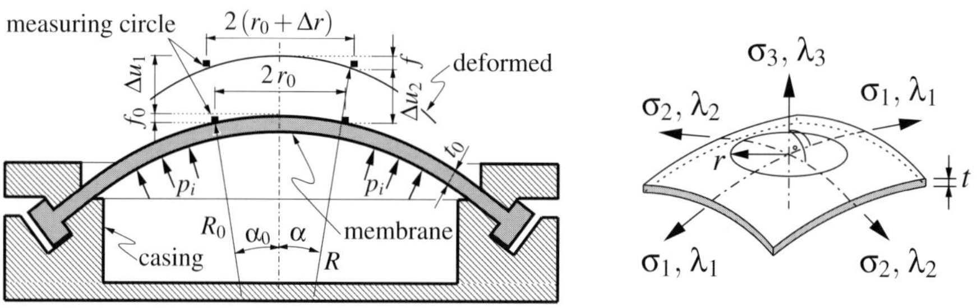

Figure 1. Axisymmetric biaxial membrane specimen.

The membrane specimen is clamped in a steel casing. It is subjected to increasing internal pressure p_i. The resulting displacements Δu_1 and Δr of a marked circle with a diameter of $2r_0$ in its undeformed configuration and the change of the rise, Δu_2, are measured. With these quantities, the states of membrane strain and stress at the apex in terms of the stretches λ_i, $i = 1, 2, 3$ and of the stress components σ_i, $i = 1, 2, 3$ (see Fig. 1) can be determined as follows:

$$\lambda_1 = \lambda_2 = \frac{R\,\alpha^{rad}}{R_0\,\alpha_0^{rad}}, \quad \Delta V = \lambda_1\,\lambda_2\,\lambda_3 - 1 = 0, \quad \lambda_3 = \frac{1}{\lambda_1\,\lambda_2}, \qquad (1)$$

$$\sigma_1 = \sigma_2 = \frac{R}{2\,t_0\lambda_3}p_i \qquad \sigma_3 = 0\,, \tag{2}$$

using the condition for incompressibility $\Delta V = 0$ and the following expression for the radius of curvature:

$$R = \frac{1}{2(f_0 + \Delta u_1 - \Delta u_2)}\left[(r_0 + \Delta r)^2 - (f_0 + \Delta u_1 - \Delta u_2)^2\right]. \tag{3}$$

The subscript "0" refers to quantities of the undeformed specimen.

The identified material parameters were validated by comparison of experimentally and numerically obtained states of deformation of rubber under realistic stress conditions. For that purpose, a wire braid reinforced hydraulic hose was partially peeled and subjected to different levels of internal pressure. This loading resulted in significant rubber deformations between the woven wire bundles, determined from surface profile measurements by means of a laser sensor.

3. Numerical modeling of wire braid reinforcement

Hydraulic hoses contain multi-spiral wires or wire braided reinforcement in order to sustain high inside pressure. The load-carrying mechanism of these types of reinforcement differ significantly. The fibers in a multi-spiral hose are only subjected to uniaxial stress. Between the fiber bundles in a wire braid hose, however, there is considerable interaction. Several concepts are available for numerical representation of the braid stiffness.

One mode of treatment of woven fabric reinforcement in FE analysis is the assumption of a homogenized composite material. This assumption is based on the choice of a suitable anisotropic constitutive model. The equivalent material parameters are identified by means of experiments with fabric composites or by the so-called unit-cell method. In the latter case a representative volume element is discretized using the properties of the different materials. Numerical analyses of characteristic deformation states yield equivalent material parameters, which are then introduced into a constitutive law for a homogeneous anisotropic material as part of a global FE model. A unit cell concept considering nonlinear material behavior was suggested by Reese [4].

Separate consideration of matrix material (rubber) and reinforcement (textile or steel wire braid) is the fundamental idea of the so-called fabric lattice model. In a paper by Kato et al. [3] the stiffness of the wire braid at the integration point level was integrated analytically. This resulted in restrictions regarding consideration of nonlinearities. In this work, a fabric lattice model based on a type of rebar layers is suggested. This allows to take many forms of nonlinear material behavior into account. It also permits numerical treatment of complex geometric situations, e.g., the simulation of the crimping of fittings.

3.1. MECHANICAL CONCEPT OF FABRIC LATTICE MODEL

In the present work a textile or wire braid is represented in the context of FE analysis by two rebar layers. Each layer has stiffness in only one of the two plaiting directions. Respective rebar layers are interacting. The rebar stresses and the braid stiffness are determined by means of the proposed local fabric lattice model.

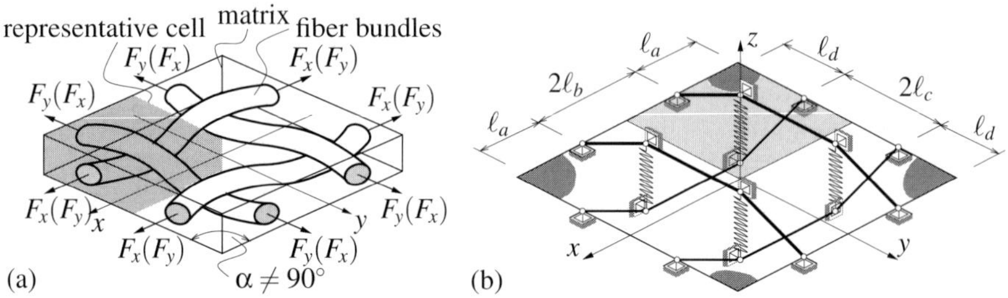

Figure 2. (a) Detailed structure of a braid. (b) System of the fabric lattice model.

Figure 2a shows the detailed structure of a wire or textile braid. Obviously, the forces acting in the fibers of one plaiting direction are influenced by the fiber forces in the other direction. This is reflected by notations such as $F_x(F_y)$ in Fig. 2a. The shaded quarter of the model is considered as a representative cell. One half of a bundle of wires of this representative cell is treated as a bar in the fabric lattice model (Fig. 2b). Nonlinear springs between the bars represent the mutual obstruction of the wire bundles. The fabric lattice has no shear stiffness. Hence, shear forces are carried by the rubber material. The proposed fabric lattice model is not restricted to orthogonal bundles of wires. This is important for realistic numerical simulations of high-pressure hydraulic hoses reinforced by wire braids.

3.2. IMPLEMENTATION OF THE FABRIC LATTICE MODEL

Because of symmetry only a quarter of the system illustrated in Fig. 2b needs to be taken into account (see Fig. 3). Consideration of plastic deformations in the wire or textile reinforcement as well as the nonlinear spring behavior requires an incremental-iterative solution algorithm for the local fabric lattice model in order to determine the stresses and the stiffness of the wire braid. The respective local nonlinear analysis must be performed in addition to the global nonlinear FE analysis.

3.2.1. Transfer from global to local quantities

The first step of the numerical simulation of the wire braid is the transfer of state variables from the global FE analysis (Green strain tensor $\mathbf{E}$, $\Delta\mathbf{E}$,

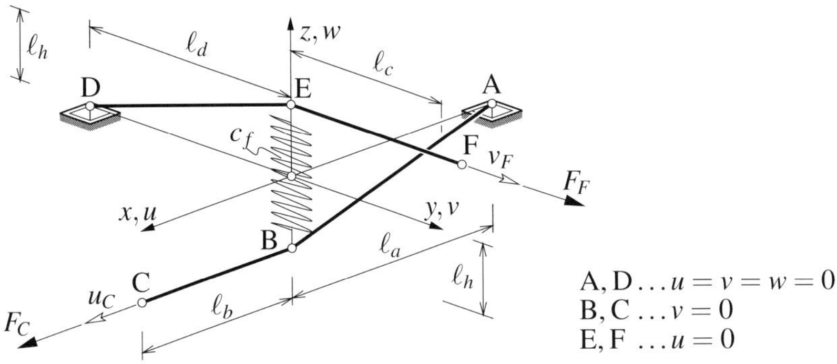

Figure 3. Considered quarter of fabric lattice model.

2nd Piola-Kirchhoff stress tensor **S**) to corresponding quantities of the local fabric lattice model (nodal displacement vector **u**, $\Delta\mathbf{u}$, vector of forces **F**). Figure 3 refers to the situation for the iteration step i of increment $m+1$ of the global nonlinear FE analysis of the hydraulic hose (see Table 1).

Table 1. Fabric lattice model – global and local variables.

global		local	nonlinear local analysis (increm./iter)	local		global
$\mathbf{E}^{m+1,i}$ $\mathbf{S}^{m+1,i}$ $\Delta\mathbf{E}^{m+1,i+1}$	$\Rightarrow$	$\mathbf{u}^{m+1,i}$ $\mathbf{F}^{m+1,i}$ $\Delta\mathbf{u}^{m+1,i+1}$		$\mathbf{F}^{m+1,i+1}$ $\Delta\mathbf{F}^{m+1,i+1}$ $\mathbf{K}^{m+1,i+1}$	$\Rightarrow$	$\mathbf{S}^{m+1,i+1}$ $\Delta\mathbf{S}^{m+1,i+1}$ $\mathbf{C}^{m+1,i+1}$

The uniaxial (stress) state in a rebar layer is characterized by $\mathbf{S} \equiv S_{11} = S$. The corresponding strain quantities are $\mathbf{E} \equiv E_{11} = E$, $\Delta\mathbf{E} \equiv \Delta E_{11} = \Delta E$, and $\lambda_1 = \lambda$, denoting the stretch. Hence, standard transfer relations between tensor components and the respective nodal displacements and forces of the fabric lattice model can be used. The displacements u_C and Δu_C of node C (see Fig. 3) are obtained as

$$\begin{aligned} u_C^{m+1,i} &= (\lambda^{m+1,i} - 1)(l_a + l_b), \\ \Delta u_C^{m+1,i+1} &= (\lambda^{m+1,i+1} - \lambda^{m+1,i})(l_a + l_b) \end{aligned} \tag{4}$$

with the stretches of the rebar layer given as

$$\begin{aligned} \lambda^{m+1,i} &= \sqrt{1 + 2E^{m+1,i}}, \\ \lambda^{m+1,i+1} &= \sqrt{1 + 2(E^{m+1,i} + \Delta E^{m+1,i+1})}. \end{aligned} \tag{5}$$

The nodal force F_C, acting at node C, follows from

$$F_C^{m+1,i} = \lambda^{m+1,i} A_0\, S^{m+1,i}\,, \tag{6}$$

where A_0 denotes the initial section area of the rebar layer and S is the 2nd Piola-Kirchhoff axial stress of the bar ending at node C of the fabric lattice model. The quantities associated with the y-direction of the lattice model are transformed analogous to Eqs. (4)–(6).

3.2.2. Elements of the local fabric lattice model

The fabric lattice model consists of bars and nonlinear springs. The bars represent the fibers of the woven braid, made of steel wires or textile materials.

In this work, elasto-plastic steel wire braids are considered. A characteristic stress-strain diagram for alloyed and cold-drawn steel wires, as used for high-pressure hydraulic hoses, is shown in Fig. 4a. The yield criterion used in this work,

$$f = |\tilde{S}| - \left[f_{sy} \left(0.79 + 50.3\sqrt{\alpha(0.0118 - \alpha)} \right) \right] , \quad (7)$$

was proposed by Dilger [1]. The respective hardening function is depicted in Fig. 4b. In Eq. (7), $\tilde{S}$ represents the stress component in the direction of the reinforcement, f_{sy} denotes the yield stress and α is a strain-like internal variable. Even if the bars of the fabric lattice are exposed to large displacements, the

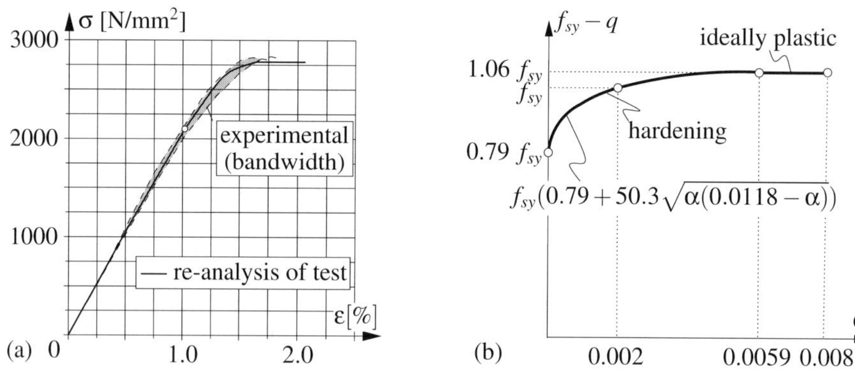

Figure 4. (a) Bandwidth of σ-ε diagrams of steel wire. (b) Hardening function generated from σ-ε diagram.

corresponding strains in the bars are small. This is a consequence of the material behavior of the steel wires (see Fig. 4). Therefore, an additive decomposition of the strains into an elastic and a plastic part is admissible. Thus, $\tilde{E} = \tilde{E}^e + \tilde{E}^p$. The evolution equations for the plastic strains $\tilde{E}^p$ and the internal variable α are given as

$$\dot{\tilde{E}}^p = \dot{\gamma}\partial_{\tilde{S}} f \quad \text{and} \quad \dot{\alpha} = \dot{\gamma}\partial_q f \quad (8)$$

where $\dot{\tilde{E}}^p$ denotes the plastic strain rate, q is a stress-like internal variable and $\dot{\gamma} \geq 0$ represents a Lagrangian multiplier. The time integration of the evolution equation (8) is performed in a strain-driven manner. Considering the strain increment $n+1$, the stress $\tilde{S}_{n+1}$ and the elasto-plastic tangent material stiffness $\tilde{C}_T = d\tilde{S}_{n+1}/d\tilde{E}_{n+1}$ for the given strain state $\tilde{E}_{n+1}$ are required. For a plastic load step an implicit integration scheme is used, where the update of the plastic strains and the internal variable are given by

$$\tilde{E}^p_{n+1} = \tilde{E}^p_n + \Delta\gamma_{n+1}\,\partial_{\tilde{S}} f\left(\tilde{S}_{n+1}, q_{n+1}\right), \tag{9}$$

$$\alpha_{n+1} = \alpha_n + \Delta\gamma_{n+1}\,\partial_q f\left(\tilde{S}_{n+1}, q_{n+1}\right). \tag{10}$$

The system of equations for the vector of unknowns of a plastic load step $(\mathbf{x} = \lfloor \tilde{E}^p, \alpha \mid \Delta\gamma \rfloor^T)$ is obtained as

$$\mathbf{R}(\mathbf{x}_{n+1}) = \left\{ \begin{array}{l} -\tilde{E}^p_{n+1} + \tilde{E}^p_n + \Delta\gamma_{n+1}\,\partial_S f(\tilde{S}_{n+1}, q_{n+1}) \\ -\alpha^p_{n+1} + \alpha^p_n + \Delta\gamma_{n+1}\,\partial_q f(\tilde{S}_{n+1}, q_{n+1}) \\ \hline f(\tilde{S}_{n+1}, q_{n+1}) \end{array} \right\} = \mathbf{0} \tag{11}$$

with $\tilde{S} = \tilde{C}(\tilde{E} - \tilde{E}^p)$ and $q = q(\alpha)$. Eq. (11) is solved by the Newton scheme. The time integration algorithm for the evaluation of the elasto-plastic tangent $\tilde{C}_T$ is linearized in closed form as proposed by Simo and Taylor [5].

For consideration of the mutual obstruction of the bundles of wires a nonlinear elastic spring is used to connect points B and E of the fabric lattice model (see Fig. 3). This spring is treated as a virtual bar with the same cross-section as the one of the wire bundles. The stress-strain law for this virtual bar is shown in Fig. 5. Its initial stiffness C^0 is set equal to the initial stiffness of the matrix material of the hydraulic hose. The material stiffness of the wire is taken as the stiffness C^1 in zone 3 of the stress-strain law. In zone 2 the stiffness of the virtual bar is represented by a parabolic function.

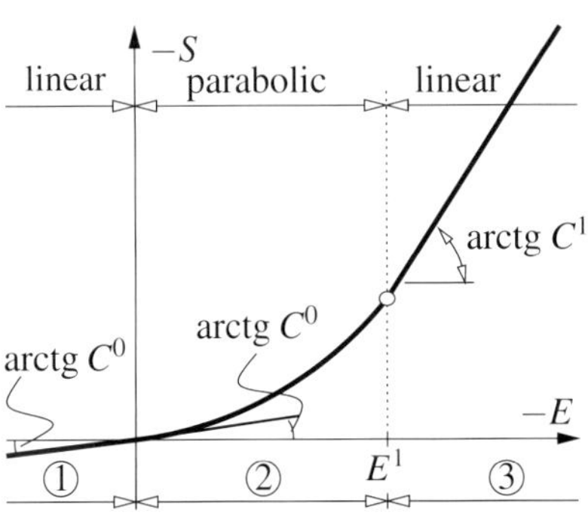

Figure 5. Stress-strain law for a virtual bar representing the nonlinear spring.

3.2.3. Transfer from local to global quantities

The result of the computation on the level of the local fabric lattice model are the tangent stiffness K of the fabric lattice, the force F_C and its increment ΔF_C, acting at the node at which displacement increments are applied. According to Table 1, these quantities must be transformed to corresponding quantities on the level of the rebar layers of the global FE model: the tangent stiffness C, the 2nd Piola-Kirchhoff stress S and its increment ΔS. The tangent stiffness C is obtained through

$$C^{m+1,i+1} = \frac{(l_a + l_b)}{(\lambda^{m+1,i+1})^2 A_0} K^{m+1,i+1}, \tag{12}$$

where $l_a + l_b$ is the length of the lattice model in the x-direction (see Fig. 3). The stress S and its increment ΔS are obtained as

$$S^{m+1,i+1} = \frac{F_C^{m+1,i+1}}{\lambda^{m+1,i+1} A_0}, \tag{13}$$

$$\Delta S^{m+1,i+1} = \frac{F_C^{m+1,i+1}}{\lambda^{m+1,i+1} A_0} - \frac{F_C^{m+1,i}}{\lambda^{m+1,i} A_0}. \tag{14}$$

The quantities associated with the y-direction of the lattice model are transformed in an analogous manner.

4. Verification of simulation model

The numerical model of a wire braid reinforcement was verified by means of simulation of a representative part of a high-pressure hydraulic hose under internal pressure. Figure 6 contains an axisymmetric model of the considered part of the hose. It also shows the boundary conditions. The reinforcement consists of two wire braids, represented by four rebar layers. The internal pressure, causing also axial tension in the hose, is increased until the bursting pressure is reached. During the simulation the change of the hose volume and the stress of the reinforcement are evaluated. Moreover, a braid and a spiral reinforcement are compared under the assumption of the same amount of reinforcement. Figure 7a shows that the increase in volume of a wire braid hose is larger than the one of a multi-spiral reinforced hose under the same internal pressure, and that the bursting pressure is lower as compared to the one of a multi-spiral reinforced hose.

In a relatively complicated numerical simulation the crimping of a hose fitting and the interactions between fitting, rubber and reinforcement are considered. Figure 8 shows an axisymmetric deformed model of a multi-spiral hose with a crimped fitting. For the crimping process, the hose is first partially peeled at its end. Then a cylindrical metallic ferrule is pushed over this end and a nipple

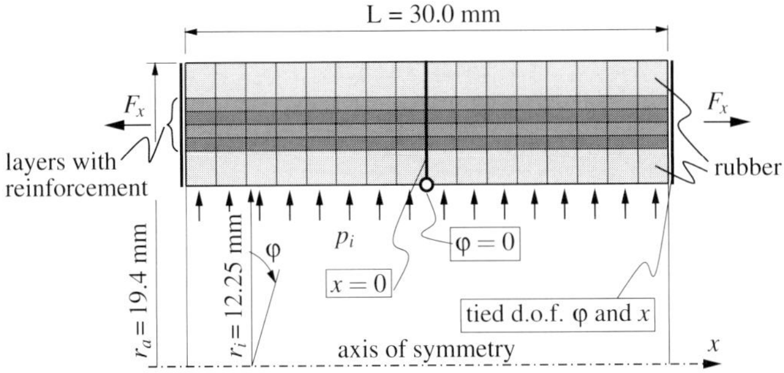

Figure 6. FE model of a representative part of a hydraulic hose.

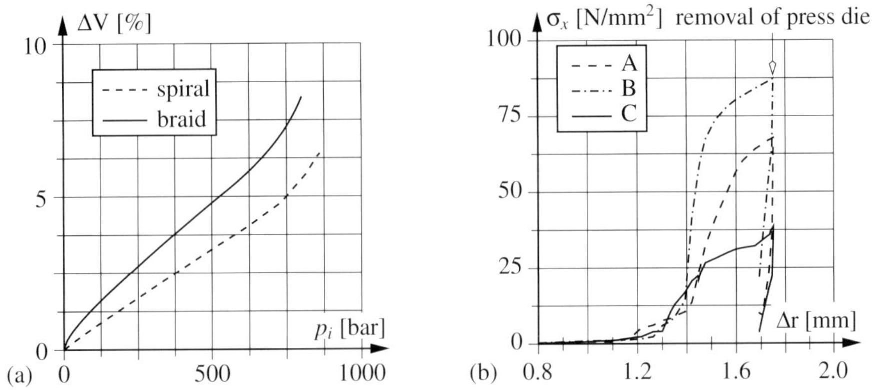

Figure 7. Results of numerical simulations. (a) Change of hose volume versus internal pressure. (b) Pressing path versus axial stress in consequence of crimping of a hose fitting

is inserted in the hose. Thereafter the ferrule is pressed onto the hydraulic hose by means of a press die. In Figure 7b the resulting axial stresses are shown for selected nodes A, B, C (see Fig. 8) as a function of the pressing path. The descending parts of the diagrams refer to unloading of the press die. A comparison of the simulations of the deformed structure with the results of an experiment shows very good agreement.

5. Conclusion

The objective of the described research project was to develop a suitable numerical analysis model for high-pressure hydraulic hoses in order to predict the deformation and stress response under service loads and the bursting pressure of such hoses. A local nonlinear fabric lattice model was proposed for consideration of steel wire braids representing the reinforcement of hydraulic hoses. This model was applied to a special type of rebar layers. The results from

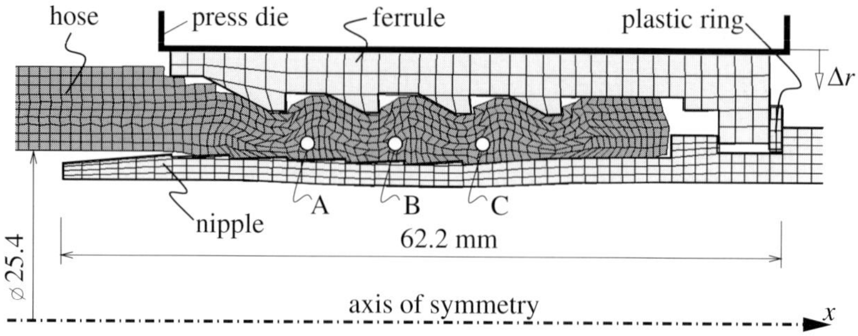

Figure 8. Deformed FE model of the crimping of a hose fitting.

numerical analysis of a representative part of a hose under internal pressure and the simulation of the crimping of a hose fitting have confirmed the suitability of the suggested model for numerical investigation of wire braids. The presented study will be extended to textile braids.

References

[1] Dilger, W. (1966). Veränderlichkeit der Biege- und Schubsteifigkeit bei Stahlbetontragwerken und ihr Einfluß auf die Schnittkraftverteilung unter Traglast bei statisch bestimmter Lagerung. *Deutscher Ausschuß für Stahlbeton*, **179**.

[2] Johannknecht, R. and Jerrams, S.J. (1999). The need for equi-biaxial testing to determine elastomeric material properties. *Proceedings of the First European Conference on Constitutive Models for Rubber*, Vienna, Austria.

[3] Kato, S., Yoshino, T. and Minami, H. (1999). Formulation of constitutive equations for fabric membranes based on the concept of fabric lattice model. *Engineering Structures*, **21**, 691–708.

[4] Reese, S. (2001). Meso-macro modelling of fibre-reinforced composites exhibiting elastoplastic material behaviour. *CD-ROM Proceedings of the European Conference on Computational Mechanics (ECCM 2001)*, Cracow, Poland.

[5] Simo, J. and Taylor, R. (1986). A return mapping algorithm for plane stress elastoplasticity. *International Journal of Numerical Methods in Engineering*, **22(3)**, 649–670.

Finite Deformation Fracture of Tires

S. Govindjee
Structural Engineering, Mechanics and Materials, University of California at Berkeley
721 Davis Hall, Berkeley, CA 94720-1710, U.S.A.
govindjee@ce.berkeley.edu

Abstract: The recent well publicized failures of the Firestone ATX tire has brought to light the poor state of research into the basic physical causes of such failures and the poor state of research into analytic methods suitable for analyzing such failures. The essence of problems of this nature lies in the finite deformation fatigue fracture of elastomers at large strains. The definition of failure criteria is further complicated by the issue of material aging. The computational problem itself is rather demanding due to the truly 3-D nature of the tire system under load. In particular, the computation of the failure forces (the singular energy momentum tractions) is rather challenging due to the spinning reference state normally utilized in tire analysis. In this paper we present a suitable numerical formulation for computing energy release rates in cracked tires and apply this to the question of the effect of varying inter-belt ply gauges. It is found that the energy release rates increase in the example tire for an increasing inter-belt gauge which seems to be in contradiction to the typical field experience of longer life for larger gauges. The contradiction however can be explained through a size effect argument related to the process zone around the crack-tip which causes the critical energy release rate to increase with gauge at a rate faster than the available energy release rate.

Keywords: Tire mechanics, energy release rates, finite deformation, steady state spinning.

1. Introduction

In assessing the durability of tires a common physical phenomena to be investigated for steel belted radial passenger tires is belt edge cracking. The physical quantity of interest in such assessments is the energy release rate. This quantity was introduced by Eshelby [2, 3] to characterize the driving force upon a singularity in an elastic field and is intimately related to the pioneering fracture studies by Ingles [8], Griffith [7], Irwin [9], and Rivlin and Thomas [20]. In the context of fracture, this driving force or energy release rate is often associated with the J-integral criteria (Rice [19], Budiansky and J.R. [1]). Under common conditions the two are synonymous.

For the case of a spinning tire in contact with a roadway the computation of the energy release rate is a difficult task given the present state of technology. Omitting experimental methods, we are forced into the usage of numerical approximations such as the finite element method (see e.g. Zienkiewicz and Taylor [23]). In fact, such a methodology has been effectively used for the computation of energy release rates in standard mechanical situations (see

C. Miehe (ed.), IUTAM Symposium on Computational Mechanics of Solid Materials at Large Strains, 417-424.

e.g. Li et al. [13] and Kishimoto et al. [11]). Even though these methodologies have existed for some time their direct application to the spinning tire in the open peer reviewed literature is amazingly scarce. In this paper we review an extension of Steinmann's method (Steinmann [21] and Steinmann et al. [22]) for finite deformation elasticity to the case of an elastic spinning tire in steady rotation as was done recently in Govindjee [5]. Standard stress analysis in a steady spinning frame of reference emanates from the early work of Lynch [14] and Oden and Lin [18] were apparently the first to give a complete detailed presentation of a suitable method in finite deformation; see also Le Tallec and Rahier [12] and Govindjee and Mihalic [6].

Below we give a summary of the theory for computing release rates in a spinning frame of reference (as was presented in Govindjee [4]) and then apply it to the question of the influence of inter-belt gauge on available energy for crack propagation.

2. Energy release rate

The issue of an energy release rate computation comes down to the computation of the following expression:

$$G_A = \lim_{\epsilon \to 0} \int_{\Gamma(\epsilon)} \Sigma^o_{AB} N^o_B \, d\Gamma(\epsilon) , \tag{1}$$

where $\Sigma^o_{AB} = W(\boldsymbol{F}^o)\delta_{AB} - F^o_{iA} P^o_{iB}$ is Eshelby's energy-momentum tensor, W is the strain energy density, F^o_{iA} is the deformation gradient, and P^o_{iA} is the first Piola-Kirchhoff stress tensor. The surface $\Gamma(\epsilon)$ with normal N^o_A surrounds the crack tip and is of radius ϵ. The result G_A is the crack tip driving force or energy release rate. Note that the result is zero unless the integrand is singular. Steinmann's method is based upon the strong form presentation of Maugin (see e.g. Maugin [15], Chap. 8) and upon the subtle observation that the quantity in Eq. (1) is in fact closely related to the "external work" terms in the weak form of the balance equation for the energy-momentum field. In the above and throughout we employ indicial notation with lower case Latin indices for the spatial frame and upper case Latin indices for the reference frame. The superscript o indicates quantities that are associated with the usual continuum mechanics definitions which are expressed in terms of a spatial and referential frame. Later on we will employ Greek indices for quantities expressed in a rotating frame and also introduce transformed counterparts to our usual continuum mechanics machinery; these will appear with the same symbols but without the superscript o.

3. Strong form of energy-momentum balance

The starting point for exploiting the observation of Steinmann [21] and Steinmann et al. [22] is to look at the governing balance equation for the energy-momentum tensor which can be expressed as:

$$\Sigma^o_{BA,A} - F^o_{iB}B^o_i = -F^o_{iB}\bar{\rho}\ddot{\phi}^o_i\,. \tag{2}$$

In the above, B^o_i represents any body forces (per unit reference volume), $\bar{\rho}$ the reference density of the body, and ϕ^o_i is the deformation (motion) map. The relation follows directly from linear momentum balance and an assumption on the smoothness of the motion. In what follows we will assume that $B^o_i = 0$ since the only direct body forces in the tire are gravitational and these can be safely neglected.

4. Weak form of energy-momentum balance

The weak form of this relation can be expressed with the aid of an arbitrary test function v_B and integration over the reference configuration of the body denoted as $\mathcal{B}$. After integrating by parts one has that

$$\int_{\mathcal{B}} v_{B,A}\Sigma^o_{BA} - \int_{\mathcal{B}} v_B F^o_{iB}\bar{\rho}\ddot{\phi}^o_i = \int_{\partial\mathcal{B}} v_B\Sigma^o_{BA}N^o_A\,. \tag{3}$$

The term on the right-hand side (RHS) is clearly related to the RHS of Eq. (1). In fact, Steinmann makes the observation that the RHS of Eq. (3) is the regular part of the "external virtual work" associated with the balance of energy momentum. For cases where this quantity is zero, the computation of the left-hand side of Eq. (3) will simply give the singular part. The reason the singular term does not appear explicitly on the RHS is that we have assumed (as is customary) that no singularities were present during the application of the divergence theorem when deriving Eq. (3). In the case of the tire, the belt edge cracks are open when the cracks are rather short and possibly for points outside of the footprint; in theses cases the RHS will be zero. For longer cracks and in the footprint the crack faces are closed and driven in a shear type mode. In this case, the RHS of Eq. (3) (representing the non-singular part) will be non-zero. In particular, if we assume frictionless crack faces and a crack face contact pressure field p, then we will have

$$RHS = \int_{\partial\mathcal{B}} v_B p N^0_A \delta_{AB} + v_B W N^0_B\,. \tag{4}$$

In order to compute the energy release rate (the singular part of the energy-momentum traction) this term needs to be subtracted from the LHS of Eq. (3). The feature that make this methodology attractive is that the machinery necessary for its computation is already built into almost all finite element programs. After solving for the deformed configuration of a body, one merely needs to evaluate the integrals in Eq. (3) to determine the energy release rates for any given crack tip.

5. Weak form of energy-momentum balance in a spinning frame

The main complication in utilizing Eq. (3) is the inertial terms. However, in the case of steady state spinning this obstacle can be overcome in a straightforward manner utilizing the formalism developed by Oden and Lin [18]; see also Le Tallec and Rahier [12] and Govindjee and Mihalic [6]. In this formalism, a third coordinate system is introduced that rotates with the tire. In this frame of reference points are located by their coordinates X_α where $X_\alpha = R_{\alpha B}(t)X^o_B$ and $R_{\alpha B}(t)$ represents the steady rotation of the tire about its axis of circular symmetry. The mapping from X_α to x_i is given by ϕ_i. The deformation gradient from this spinning frame of reference to the spatial frame is given by $F_{i\alpha} = F^o_{iA}R_{\alpha A}$. Since we assume the tire is in a steady state rotation we have for the acceleration:

$$\ddot{\phi}^o_i = x_{i,\alpha\beta}\Omega_{\beta\gamma}X_\gamma\Omega_{\alpha\delta}X_\delta + x_{i,\alpha}\Omega_{\alpha\gamma}\Omega_{\gamma\delta}X_\delta\,, \tag{5}$$

where $\Omega_{\alpha\beta} = \dot{R}_{\alpha B}R_{\beta B}$ is the spin rate of the tire. It is also noted that the surface normals are mapped as $N_\alpha = R_{\alpha A}N^o_A$.

For isotropic materials we also have that $W(\boldsymbol{F}^o) = W(\boldsymbol{F}\boldsymbol{R}) = W(\boldsymbol{F})$ and through an abuse of notation that $W(\boldsymbol{C}^o) = W(\boldsymbol{C})$, where $C_{\alpha\beta} = F_{i\alpha}F_{i\beta}$. With these results we can now define an energy-momentum tensor relative to the spinning frame as $\Sigma_{\alpha\beta} = R_{\alpha A}\Sigma^o_{AB}R_{\beta B} = W(\boldsymbol{C})\delta_{\alpha\beta} - C_{\alpha\gamma}S_{\gamma\beta}$, where $S_{\alpha\beta}$ represents the second Piola-Kirchhoff stress tensor measured by the deformation from the rotating frame to the current frame. If we now re-examine Eq. (3), then we find that the singular part of the energy-momentum traction is given by

$$\begin{aligned} G &= \int_{\mathcal{B}} R_{\beta B}v_{B,\alpha}\Sigma_{\beta\alpha} - \int_{\partial\mathcal{B}} R_{\beta B}v_B\Sigma_{\beta\alpha}N_\alpha \\ &- \int_{\mathcal{B}} \bar{\rho}R_{\beta B}v_B F_{i\beta}\delta_{i\alpha}(\Omega^2)_{\alpha\theta}X_\theta \\ &+ \int_{\mathcal{B}} \bar{\rho}R_{\beta B}v_{B,\delta}F_{i\beta}\Omega_{\alpha\gamma}X_\gamma u_{i,\alpha}\Omega_{\delta\theta}X_\theta \\ &+ \int_{\mathcal{B}} \bar{\rho}R_{\beta B}v_B F_{i\beta,\delta}\Omega_{\alpha\gamma}X_\gamma u_{i,\alpha}\Omega_{\delta\theta}X_\theta\,, \end{aligned} \tag{6}$$

where we have assumed that the boundary $\partial\mathcal{B}$ possesses circular symmetry.

To practically execute these integrals we now introduce a finite element approximation by expressing the motion ϕ_i as an element of a finite dimension functions space; *i.e.* as $\phi_i = \sum_a x_i^a M^a(\boldsymbol{X})$ where x_i^a are the nodal degrees of freedom and M^a is the shape function associated with node a. Likewise we approximate the space of test functions such that $v_B = \sum_a v_B^a M^a(\boldsymbol{X})$. From the arbitrariness of the test function we can now express the energy release rate for the steady rotation case upon any node a as:

$$
\begin{aligned}
G_\beta^a &= \int_{\mathcal{B}} M_{,\alpha}^a \Sigma_{\beta\alpha} - \int_{\partial\mathcal{B}} M^a \Sigma_{\beta\alpha} N_\alpha - \int_{\mathcal{B}} \bar{\rho} M^a F_{i\beta} \delta_{i\alpha} (\Omega^2)_{\alpha\theta} X_\theta \\
&+ \int_{\mathcal{B}} \bar{\rho} M_{,\delta}^a F_{i\beta} \Omega_{\alpha\gamma} X_\gamma u_{i,\alpha} \Omega_{\delta\theta} X_\theta \\
&+ \int_{\mathcal{B}} \bar{\rho} M^a F_{i\beta,\delta} \Omega_{\alpha\gamma} X_\gamma u_{i,\alpha} \Omega_{\delta\theta} X_\theta \,, \qquad (7)
\end{aligned}
$$

In using these relations one needs to also consider the fact that the tire is composed of materials that are not isotropic. In particular the bead is effectively transversely isotropic and the steel belts are orthotropic. For the general conditions of transverse isotropy and orthotropy the above results do not hold. However, the anisotropy of the tire is very special in that the tire possess a global geometric invariance with respect to the spin of the tire. In this special case, and in this special case only, the expressions above also hold for general anisotropy without modification.

6. Application

The sequence of computational steps in computing the energy release rate at the tip of a circumferential belt edge crack is as follows. First one models the tire in circular symmetry. This implies that the cross tread grooves can not be modeled; only the circumferential tread grooves can be modeled. For a given tire inflation, speed, and axle force the deformation of the tire is computed by solving the linear momentum balance equations is a steady rotating frame using the method of Oden and Lin [18] as presented by Govindjee and Mihalic [6]. Once the deformation state has been computed it is a straightforward post-processing step to compute G_β^a using the relations from the previous section. For simplicity, we only include the dominant terms from Eq. (7) –*viz.* the first three. We also assume that the crack face normal terms from the top and bottom faces are self canceling; *i.e.* Eq. (4) is approximately zero.

As an example we consider what happens to the release rate at the tip of an inter-belt crack in a typical steel-belted radial tire when the inter-belt gauge is increased. The tire model consists of over 15k elements and contains roughly 45k equations. The model itself represents no particular tire but is representative of a large passenger vehicle tire. The model represents the entire tire without recourse to approximate symmetry planes, the roadway as a frictionless rigid contact surface, and a inter-belt crack of 10 mm starting at the end of the number 1 belt in the tire.

In Fig. 1 we show the effect of a varying inter-belt gauge on the energy release rate (in the 3-direction which is the dominant direction) at the crack tip as a function of angular position. 0 radians corresponds to the center of the footprint. The computation is for an inflation pressure of 242 kPa, 120 kph road speed, and a vertical load of 3.9 kN. The slight asymmetry about 0 radians is due to the orthotropic steel belts. As can been seen in the figure the energy release rate at the tip of the crack increases with increasing gauge as does the increment in energy release rate computed as $G_3(0) - G_3(\pi)$ (though to a lesser extent). This result is seemingly at odds with the typical field experience that an increase in the inter-belt gauge increases tire durability. This field experience leads to the typical argument that increasing the gauge increases the compliance of the inter-belt layer and in a constant strain system one will experience a drop in the stored energy density thus leading to a drop in the energy release rate and an increase in durability. The present computation however indicates that this traditional line of reasoning is flawed. With respect to the line of reasoning concerning a constant strain system, a better line of reasoning would be:

1 The energy state of the inter-belt material is dominated by a deformation controlled process that involves the bending of an elastomer shell into a flat configuration.

2 This energy density (per unit area) will be approximately proportional to the cube of the thickness of the shell.

3 Thus an increase in the thickness of the shell (*i.e.* the inter-belt gauge) will result in an increase in the strain energy stored and hence also the energy release rate for crack propagation.

One is then still faced with reconciling the increased energy release rate with the field experience of increased durability. A plausible means of effecting this reconciliation is to bring into the picture a size effect argument. It has been noted in the literature that some of the classical fracture toughness experiments are noticeably dependent upon specimen size (in certain ranges). For example, Kadir and Thomas [10] show that for pure shear SBR test specimens there is a dramatic dependency upon specimen size in the neighborhood of 1 mm thick sheets. Similar findings for SBR are also noted in the works of Mazich et al. [16] and Morman et al. [17]. In all of these papers the noted finding is that the critical energy release rate increases with specimen thickness. The general argument of Kadir and Thomas [10], which also likely applies here even though one normally has a NR system, is that there are characteristic sizes associated with the fracture process zone. When the fracturing specimen is below this size

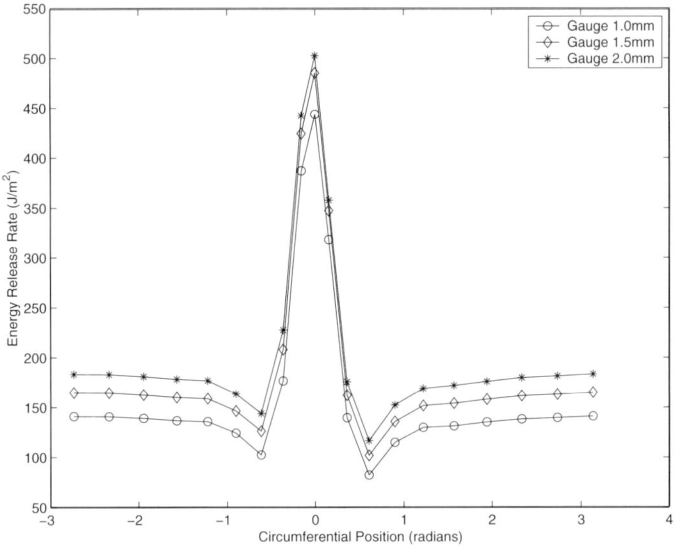

Figure 1. Energy release rate as a function of angular position for a 10 mm crack at 120 kph, inflation pressure of 242 kPa, and a vertical load of 3.9 kN. 0 radians corresponds to the center of the footprint.

then one is unable to fully develop the process zone thus leading to a lower critical energy release rate. The reported results for SBR show a large increase in the critical energy release rate with increasing thickness and one that is much larger than the computed increase for the energy release rate at the crack tip from our FEA computation. Thus it is argued that the likely explanation for the observed field experience is *not* that the crack tip release rate has decreased but *rather* that at typical gauge sizes the critical energy release rate is increasing faster than the crack tip energy release rate – all due to a size effect associated with the crack tip process zone.

References

[1] Budiansky, B. and J.R., R. (1973). Conservation laws and energy-release rates. *Journal of Applied Mechanics*, **40**, 201–203.

[2] Eshelby, J. (1956). The continuum theory of lattice defects. *Seitz, F. and Turnbull, D., editors, Solid State Physics, Advances in Research and Applications*, **III**, 79–119. Academic Press.

[3] Eshelby, J. (1975). The elastic energy-momentum tensor. *Journal of Elasticity*, **5**, 321–335.

[4] Govindjee, S. (2001). Finite deformation energy release rate computations in a steady rolling tire. *Proceedings of the Tenth International Congress of Fracture*, Elsevier Press, CD-ROM.

[5] Govindjee, S. (2001). Firestone tire failure analysis. Technical report.

[6] Govindjee, S. and Mihalic, P. (1998). Viscoelastic constitutive relations for the steady spinning of a cylinder. *Technical Report UCB/SEMM-98/02*, University of California Berkeley, Department of Civil Engineering.

[7] Griffith, A. (1920). The phenomena of rupture and flow in solids. *Philosophical Transactions of the Royal Society of London*, **221A**, 163–197.

[8] Ingles, C. (1913). Stresses in a plate due to the presence of cracks and sharp corners. *Transactions of the Institute of Naval Architects*, **55**, 219–241.

[9] Irwin, G. (1948). Fracture Dynamics. *American Society for Metals*. 147–166

[10] Kadir, A. and Thomas, A. (1981). Tear behavior of rubbers over a wide range of rates. *Rubber Chemistry and Technology*, **54**, 15–23.

[11] Kishimoto, K., Aoki, S., and Sakata, M. (1980). Dynamic stress intensity factors using $\hat{\text{j}}$-integral and finite element method. *Engineering Fracture Mechanics*, **13**, 387–394.

[12] Le Tallec, P. and Rahier, C. (1994). Numerical models of steady rolling for non-linear viscoelastic structures in finite deformations. *International Journal for Numerical Methods in Engineering*, **37**, 1159–1186.

[13] Li, F., Shih, C., and Needleman, A. (1985). A comparison of methods for calculating energy release rates. *Engineering Fracture Mechanics*, **21**, 405–421.

[14] Lynch, F. D.S. (1969). A finite element method of viscoelastic stress analysis with application to rolling contract problems. *International Journal for Numerical Methods in Engineering*, **1**, 379–394.

[15] Maugin, G. (1999). *The thermomechanics of nonlinear irreversible behaviors*. World Scientific Publishing Co.

[16] Mazich, K., Morman, K.J., Oblinger, F., Pan, T., and Killgoar, P.J. (1989). The effect of specimen thickness on the tearing energy of a gum vulcanizate. *Rubber Chemistry and Technology*, **62**, 850–862.

[17] Morman, K.J., Mazich, K., Oblinger, F., and Zhang, F. (1992). Critical tearing energy in a circumferentially-cracked circular cylinder of rubber under finite deformation. *International Journal of Fracture*, **53**, 129–157.

[18] Oden, J. and Lin, T. (1986). On the general rolling contact problem for finite deformations of a viscoelastic cylinder. *Computer Methods in Applied Mechanics and Engineering*, **57**, 297–367.

[19] Rice, J. (1968). A path independent integral and the approximate analysis of strain concentration by notches and cracks. *Journal of Applied Mechanics*, **35**, 379–386.

[20] Rivlin, R. and Thomas, A. (1953). Rupture of rubber 1. Characteristic energy for tearing. *Journal of Polymer Science*, **10**, 291–318.

[21] Steinmann, P. (2000). Application of material forces to hyperelastic fracture mechanics. I. Continuum mechanical setting. *International Journal of Solids and Structures*, **37**, 7371–7391.

[22] Steinmann, P., Ackermann, D., and Barth, F. (2001). Application of material forces to hyperelastic fracture mechanics. II. Computational setting. *International Journal of Solids and Structures*, **38**, 5509–5526.

[23] Zienkiewicz, O. and Taylor, R. (2000). *The Finite Element Method, Volume 1*. Butterworth and Heinemann, 5th edition.

Computational Strategies for Inelastic Solids at Large Strains: Some Recent Issues with an Industrial Application

D. Perić, W. Dettmer
Department of Civil Engineering, University of Wales Swansea
Singleton Park, Swansea SA2 8PP, United Kingdom
d.peric@swansea.ac.uk, cgdettme@swansea.ac.uk

Abstract: This work is concerned with issues related to computational modelling of large deformations of inelastic solids at large strains with particular reference to large scale industrial applications. A thermodynamically consistent framework has been presented for a unified description of inelastic solids incorporating combined elastic, viscoelastic and elasto(visco)plastic constitutive behaviours. It is argued that the successful finite element based solution strategy for complex industrial problems involving large deformations of inelastic solids at large strains requires rigorous treatment of both theoretical and algorithmic issues, and invariably involves some form of adaptive finite element method.

Keywords: Finite element method, nonlinear solid mechanics, viscoelasticity, elastoplasticity, adaptive strategy.

1. Introduction

This text briefly discusses some aspects of computational strategies for strongly nonlinear solid mechanics problems that are associated with finite element simulation of large deformations of viscoelastic/elastoplastic solids in the presence of finite strains.

For a large number of practically relevant solid mechanics problems involving large elasto-plastic strains, the *optimal mesh configuration* changes continually throughout the deformation process. Often, due to element distortion, rapid solution degradation is experienced that precludes achievement of any solution for the region of large deformations of practical interest. Moreover, additional complexities emerge when material failure takes place and the numerical model has both to indicate the correct site of fracture onset and to efficiently re-mesh the critical zones. Therefore, the introduction of adaptive mesh refinement processes is crucial for the solution of large scale industrial problems, which necessitates: (i) specification of an appropriate error estimation criterion, (ii) development of a strategy for adapting the mesh based on the error distribution (iii) automatic mesh generation tools, and (iv) appropriate transfer operators (Ortiz and Quigley IV [5], Lee and Bathe [3], Perić et al. [6, 9] and Perić and Owen [7]).

C. Miehe (ed.), IUTAM Symposium on Computational Mechanics of Solid Materials at Large Strains, 425-434.

To illustrate the computational strategy described in this work simulation of an industrially relevant problem is presented. For further details and a range of numerical examples we refer to Perić and Owen [7], Perić et al. [9] and references therein.

2. The constitutive model for large inelastic deformations at finite strains

Some important aspects of constitutive modelling of inelastic solids at finite strains are briefly reviewed in this section. From the outset the attention is restricted to isotropic solids.

2.1. Multiple split of the deformation gradient

The chief hypothesis underlying the present approach to modelling of large deformations of inelastic solids at finite strains is *the multiplicative split* of the deformation gradient into elastic and inelastic parts $\boldsymbol{F} = \boldsymbol{F}^e \boldsymbol{F}^i$. This assumption admits the existence of a local unstressed *intermediate configuration.* Due to its suitability for the computational treatment of large deformations of inelastic solids at finite strains, the hypothesis of multiplicative decomposition has been widely employed in the computational mechanics literature in modelling both elasto-plastic (Eterovic and Bathe [2], Simo [13], Cuitño and Ortiz [1], Perić et al. [8]) and viscoelastic solids (Simo [12], Reese and Govindjee [11])(see also Miehe and Keck [4] for an alternative point of view). A useful generalization of the above multiplicative model may be obtained by considering the multiplicative decomposition of the deformation gradient $\boldsymbol{F}$ in terms of $k = 1, 2, ..., n_{\text{rheol}}$ rheological components arranged in parallel. The relationships

$$\boldsymbol{F} = \boldsymbol{F}_1^e \boldsymbol{F}_1^i = \boldsymbol{F}_2^e \boldsymbol{F}_2^i = \boldsymbol{F}_3^e \boldsymbol{F}_3^i = ... \tag{1}$$

then represent generalization of the multiplicative split of a single rheological component. In this case the free energy of a material model, can be represented as the sum of the free energies of the rheological components, i.e.

$$\psi = \psi_1(\boldsymbol{b}_1^e, \boldsymbol{\xi}_1) + \psi_2(\boldsymbol{b}_2^e, \boldsymbol{\xi}_2) + \psi_3(\boldsymbol{b}_3^e, \boldsymbol{\xi}_3) + ... \, , \tag{2}$$

where $\boldsymbol{b}_k^e$ denotes the left Cauchy-Green tensor defined as $\boldsymbol{b}_k^e := \boldsymbol{F}_k^e \boldsymbol{F}_k^{eT} = \boldsymbol{F}(\boldsymbol{F}_k^p)^{-1}(\boldsymbol{F}_k^p)^{-T}\boldsymbol{F}^T$, while $\boldsymbol{\xi}_k$ denotes the set of variables associated with dissipative mechanisms corresponding to the rheological component k.

By considering the Clausius-Duhem inequality $\boldsymbol{\tau} : \boldsymbol{d} - \dot{\psi} \geq 0$ and by employing the elastic isotropy and the properties of the inner product, the following relationship can be obtained

$$\left(\boldsymbol{\tau} - \sum_k 2 \frac{\partial \psi_k}{\partial \boldsymbol{b}_k^e} \boldsymbol{b}_k^e\right) : \boldsymbol{d} + \\ \sum_k 2 \frac{\partial \psi_k}{\partial \boldsymbol{b}_k^e} \boldsymbol{b}_k^e : \left(-\tfrac{1}{2} (\mathscr{L}_v \boldsymbol{b}_k^e)\, \boldsymbol{b}_k^{e\,-1}\right) - \sum_k \frac{\partial \psi_k}{\partial \boldsymbol{\xi}_k} \cdot \dot{\boldsymbol{\xi}}_k \geq 0\,, \tag{3}$$

where $\mathscr{L}_v \boldsymbol{b}^e := \boldsymbol{F} \{\partial_t [\boldsymbol{F}^{-1} \boldsymbol{b}^e \boldsymbol{F}^{-T}]\} \boldsymbol{F}^T = \dot{\boldsymbol{b}}^e - \boldsymbol{l} \boldsymbol{b}^e - \boldsymbol{b}^e \boldsymbol{l}$ is the so-called *Lie derivative* of the left Cauchy-Green tensor, and $\boldsymbol{l} := \dot{\boldsymbol{F}} \boldsymbol{F}^{-1}$ is the velocity gradient.

The standard argument then yields the relation

$$\boldsymbol{\tau} = \sum_k \boldsymbol{\tau}_k \qquad \text{where} \qquad \boldsymbol{\tau}_k := 2 \frac{\partial \psi_k}{\partial \boldsymbol{b}_k^e} \boldsymbol{b}_k^e, \quad k = 1, 2, ..., n_{\text{rheol}}\,. \tag{4}$$

In order for the inequality (3) to be satisfied, each rheological component must satisfy the non-negative dissipation requirement, that is,

$$\boldsymbol{\tau}_k : \left(-\tfrac{1}{2} (\mathscr{L}_v \boldsymbol{b}_k^e)\, \boldsymbol{b}_k^{e\,-1}\right) - \frac{\partial \psi_k}{\partial \boldsymbol{\xi}_k} \cdot \dot{\boldsymbol{\xi}}_k \geq 0, \qquad k = 1, 2, .., n_{\text{rheol}}. \tag{5}$$

2.2. GENERAL MATERIAL MODELS

Inspection of constitutive relations given in Section 2.1 suggests a strategy for designing a generalized material model for large deformations of inelastic solids by combining in parallel arrangement constitutive components that follow different constitutive laws. Type and number of components to be used in description of a given material will of course depend on the constitutive properties of the material at hand, and will need to be evaluated by a certain material identification procedure. By considering a combination of a single elastic component, n_{ve} viscoelastic (Maxwell) and n_{ep} plastic (Prandtl) elements, the complete set of constitutive equations to be used in description of a generic material may be expressed as:

$$\boldsymbol{\tau} = \boldsymbol{\tau}_\infty + \overset{\text{(Maxwell)}}{\sum_k^{n_{\text{ve}}}} \boldsymbol{\tau}_k + \overset{\text{(Prandtl)}}{\sum_k^{n_{\text{ep}}}} \boldsymbol{\tau}_k$$

Elastic spring

$$\boldsymbol{\tau}_\infty = 2\frac{\partial \psi_\infty}{\partial \boldsymbol{b}}\boldsymbol{b}$$

Viscoelastic components $(k = 1, 2, .., n_{\text{ve}})$

$$\boldsymbol{\tau}_k = 2\frac{\partial \psi_k}{\partial \boldsymbol{b}_k^e}\boldsymbol{b}_k^e$$

$$-\tfrac{1}{2}(\mathscr{L}_v \boldsymbol{b}_k^e)\,\boldsymbol{b}_k^{e\,-1} = \frac{1}{2\eta_k}\text{dev}(\boldsymbol{\tau}_k) \tag{6}$$

Plastic components $(k = 1, 2, .., n_{\text{ep}})$

$$\boldsymbol{\tau}_k = 2\frac{\partial \psi_k}{\partial \boldsymbol{b}_k^e}\boldsymbol{b}_k^e \qquad q_k = -\frac{\partial \psi_k}{\partial \xi_k}$$

$$-\tfrac{1}{2}(\mathscr{L}_v \boldsymbol{b}_k^e)\,\boldsymbol{b}_k^{e\,-1} = \dot{\gamma}_k \frac{\partial \phi_k}{\partial \boldsymbol{\tau}_k} \qquad \dot{\xi}_k = \dot{\gamma}_k \frac{\partial \phi_k}{\partial q_k}$$

$$\phi_k := \phi(\boldsymbol{\tau}_k, q_k)$$

$$\phi_k \le 0 \qquad \dot{\gamma}_k \ge 0 \qquad \phi_k \dot{\gamma}_k = 0$$

3. Numerical integration of the constitutive equations

In the context of a finite element schemes for solution of path dependent problems, given the values of the variables $\{\boldsymbol{\tau}_n, \boldsymbol{b}_n^e, \boldsymbol{\xi}_n\}$ at the beginning of a generic increment $[t_n, t_{n+1}]$, an algorithm for integration of the above constitutive equations is required to obtain the updated values $\{\boldsymbol{\tau}_{n+1}, \boldsymbol{b}_{n+1}^e, \boldsymbol{\xi}_{n+1}\}$ at the end of the increment.

3.1. THE EXPONENTIAL MAP

The crucial point in the derivation of the algorithm adopted in this work is the discretization of the constitutive laws for the inelastic evolution laws given in (6), by means of a fully implicit exponential approximation (Weber and Anand [15], Eterovic and Bathe [2], Simo [13]). By restricting attention to a single inelastic constitutive component of a generalized material model (6), the case of the elasto-plastic evolution law, and a conventional backward Euler approximation, the exponential approximation leads to the following incremental constitutive update of the elastic left Cauchy-Green tensor $\boldsymbol{b}^e$:

$$\boldsymbol{b}^e_{n+1} = \exp\Big(-2\Delta\gamma\,\partial_{\boldsymbol{\tau}}\phi\Big)\;\boldsymbol{b}^{e\text{ trial}}_{n+1}\,. \tag{7}$$

where $\boldsymbol{b}^{e\text{ trial}}_{n+1} := \boldsymbol{f}\boldsymbol{b}^e_n\boldsymbol{f}^T = \boldsymbol{F}\,\boldsymbol{C}^{i\;-1}_n\boldsymbol{F}^T$ denotes the so-called trial elastic left Cauchy Green tensor and $\boldsymbol{C}^i_n = \boldsymbol{F}^{i\,T}_n\boldsymbol{F}^i_n$.
Similar expressions can be obtained for numerical integration of the evolutions laws for finite strain viscoplasticity and viscoelasticity (Reese and Govindjee [11], Simo and Miehe [14]).

3.2. Formulation in the principal directions

It may easily be shown that for the case of isotropic materials the incremental update given by (7) takes place at fixed principal axes fully defined by the trial elastic left Cauchy-Green tensor $\boldsymbol{b}^{e\text{ trial}}_{n+1}$. This implies that the optimal implementation strategy for solution of a generally non-linear local update (7) is provided by formulation in terms of principal values of the tensors in (7). By performing the spectral decomposition and taking the logarithms of the resulting expressions, the incremental local update of the finite strain elasto-plastic model given in (6) can be expressed, in terms of the principal values of the elastic Eulerian logarithmic strain tensor, as

$$\varepsilon^e_{A\,n+1} = \varepsilon^{e\text{ trial}}_{A\,n+1} - \Delta\gamma\;\partial_{\tau_A}\phi|_{n+1}\,, \tag{8}$$

which has the same format of the update formula for the elastic strains of the standard return mapping algorithms of the infinitesimal theory (Eterovic and Bathe [2], Cuitiño and Ortiz [1], Perić et al. [8], Simo and Miehe [14], Simo [13]). Similar arguments can be applied in formulation of the constitutive updates for other rheological components, including both viscoelastic and viscoplastic constitutive laws. It should be emphasized that, as a consequence of the exponential mapping in the implicit integration of the inelastic evolution laws, the incompressibility constraint is carried over exactly to the incremental rules (7).

4. Adaptive solution update

Some computational issues related to the adaptive solution strategies for large inelastic solids at finite strains are briefly discussed in this section.

4.1. Error indicators

An elementary procedure for the error estimation may be defined by the replacement of the exact values of variables and relevant derivatives of the problem by some post-processed values obtained from the available finite element solution and the problem data. Such procedures have proved to be effective for linear problems, and their extension to elasto-plastic solids has been found to be an appropriate solution to the adaptive solution of a variety of strongly nonlinear solid mechanics problems (Perić et al. [10, 9]). However, for problems where the requirements of the physical problem necessitate a sharp resolution of high gradients of the state variables over a small region of the mesh, which, as a final outcome of the process, typically incorporates material failure along a line or a surface in two- or three-dimensional problems, respectively, an extension of the above error indicators has proved to be advantageous. In order to effectively model material failure the essential idea is to correlate the adaptive procedure to the underlying failure mechanism. Therefore, an error measure based on *rate of fracture indicators* is described in detail by Perić et al. [9], based on the principle that the adaptive procedure should not only capture the progression of the plastic deformation but also provide refined meshes at regions of possible material failure.

4.2. Transfer operations for evolving meshes

The history dependent nature of the process necessitates transfer of all relevant problem variables from the old mesh to the new one, as successive remeshing is applied during the process simulation. Hence, some suitable means for transferring the state variables between meshes, or *transfer operators*, needs to be defined. Several important aspects of the transfer operation have to be addressed (Ortiz and Quigley IV [5], Lee and Bathe [3], Perić et al. [6]): (i) consistency with the constitutive equations, (ii) requirement of equilibrium (which is fundamental for implicit FE simulation), (iii) compatibility of the history-dependent internal variables transfer with the displacement field on the new mesh, (iv) compatibility with evolving boundary conditions, (v) minimization of the numerical diffusion of transferred state fields.

To briefly describe the transfer operation, let us define a state array ${}^h\boldsymbol{\Lambda}_n = ({}^h\boldsymbol{u}_n, {}^h\boldsymbol{F}_n, {}^h\boldsymbol{\tau}_n, {}^h\boldsymbol{b}^e_n, {}^h\boldsymbol{\xi}_n)$ where ${}^h\boldsymbol{u}_n, {}^h\boldsymbol{F}_n, {}^h\boldsymbol{\tau}_n, {}^h\boldsymbol{b}^e_n, {}^h\boldsymbol{\xi}_n$ denote values of the displacement, deformation gradient, stress tensor, the elastic left Cauchy-Green tensor, and a vector of internal variables at time t_n for the mesh h. Assume, furthermore, that the estimated error of the solution ${}^h\boldsymbol{\Lambda}_n$ respects the prescribed criteria, while these are violated by the solution ${}^h\boldsymbol{\Lambda}_{n+1}$. In this case a new mesh $h+1$ is generated and a new solution ${}^{h+1}\boldsymbol{\Lambda}_{n+1}$ needs to be computed. As the backward Euler scheme is adopted the internal variables ${}^{h+1}\boldsymbol{\xi}_n$ for a new mesh $h+1$ at time t_n need to be evaluated. In this way the state ${}^{h+1}\widetilde{\boldsymbol{\Lambda}}_n = (\bullet, \bullet, \bullet, {}^{h+1}\boldsymbol{b}^e_n, {}^{h+1}\boldsymbol{\xi}_n)$ is constructed, where a symbol $\widetilde{\ }$ is used to denote a reduced state array. It should be noted that this state characterizes the history of the material and, in the case of a fully implicit scheme, provides sufficient information for computation of a new solution ${}^{h+1}\boldsymbol{\Lambda}_{n+1}$.

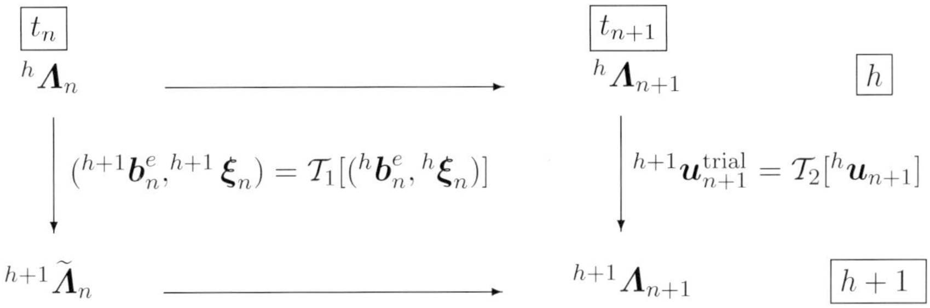

Figure 1. Transfer operator diagram.

Conceptually, Figure 1 summarizes a typical transfer operation that includes both, the mapping of the internal variables and mapping of the displacement field. The implementation of the given general transfer operation is, for the case of evolving finite element meshes composed of constant strain triangles, provided in Perić et al. [6, 9]. We point out that, for general unstructured and non-nested meshes, the transfer operation step involves certain amount of numerical diffusion.

5. Numerical verification

To illustrate the scope of the described computational strategy numerical solutions of an industrially relevant problem is provided in this section.

Example: Stamping of a 3D soft solid. The initial tool/workpiece set-up of the problem is shown in Figure 2 together with the corresponding meshes. A von Mises elasto-plastic material model is employed to describe the behaviour of the workpiece. The tools are assumed to be rigid, with a Coulomb friction describing the frictional behaviour at the interfaces.

Remeshing is, obviously, an essential tool for a successful simulation of this process due to the large elasto-plastic deformations involved and very high strains in the near corner regions. In this simulation, the mesh size used in description of the workpiece varies between about 30,000 and 150,000 pressure stabilized tetrahedral elements, in the initial and final stages of the simulation, respectively. It should be noted that due to very sharp curvatures in the problem geometry, very fine resolution is required in the corner regions.

Figure 3 presents the finite element mesh evolution and distribution of the equivalent plastic strain at several stages of punch displacement. As the punch advances, the excess material flows around the cavity edges to form the so-

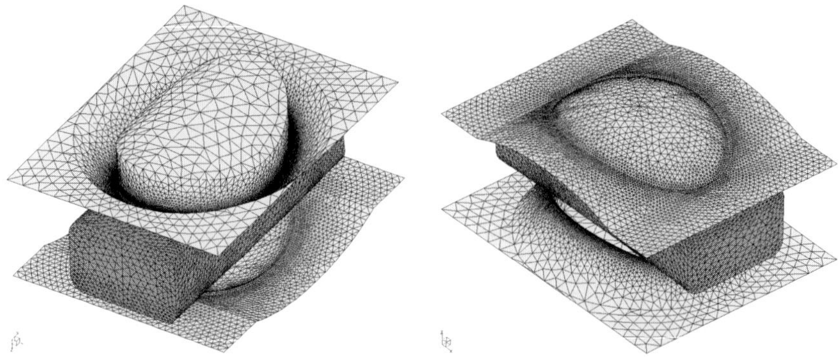

Figure 2. Stamping of a 3D soft solid: Initial mesh views.

called *flash*, which is is ultimately removed from the final product. Clearly the simulation of stamping operation can provide an important tool for optimizing this important industrial process.

6. Summary

A thermodynamically consistent framework has been presented for a unified description of inelastic solids incorporating combined elastic, viscoelastic and elasto(visco)plastic constitutive behaviours. It is argued that the presented approach offers optimal algorithmic framework for numerical analysis of a class of complex inelastic processes widely encountered in practice. Some computational aspects of adaptive strategies for strongly nonlinear solid mechanics problems are briefly discussed and their use illustrated by means of an industrially relevant numerical example involving extremely large deformations.

Acknowledgments

This work was supported by the EPSRC of UK through Grant GR/N21604, awarded to the University of Wales Swansea, and this support is gratefully acknowledged.

References

[1] Cuitiño, A. and Ortiz, M. (1992). A material-independent method for extending stress update algorithms from small-strain plasticity to finite plasticity with multiplicative kinematics. *Engng. Comp.*, **9**, 437–451.

[2] Eterovic, A. and Bathe, K.-J. (1990). A hyperelastic based large strain elasto-plastic constitutive formulation with combined isotropic-kinematic hardening using the logarithmic stress and strain measures. *Int. J. Num. Meth. Engng.*, **30**, 1099–1114.

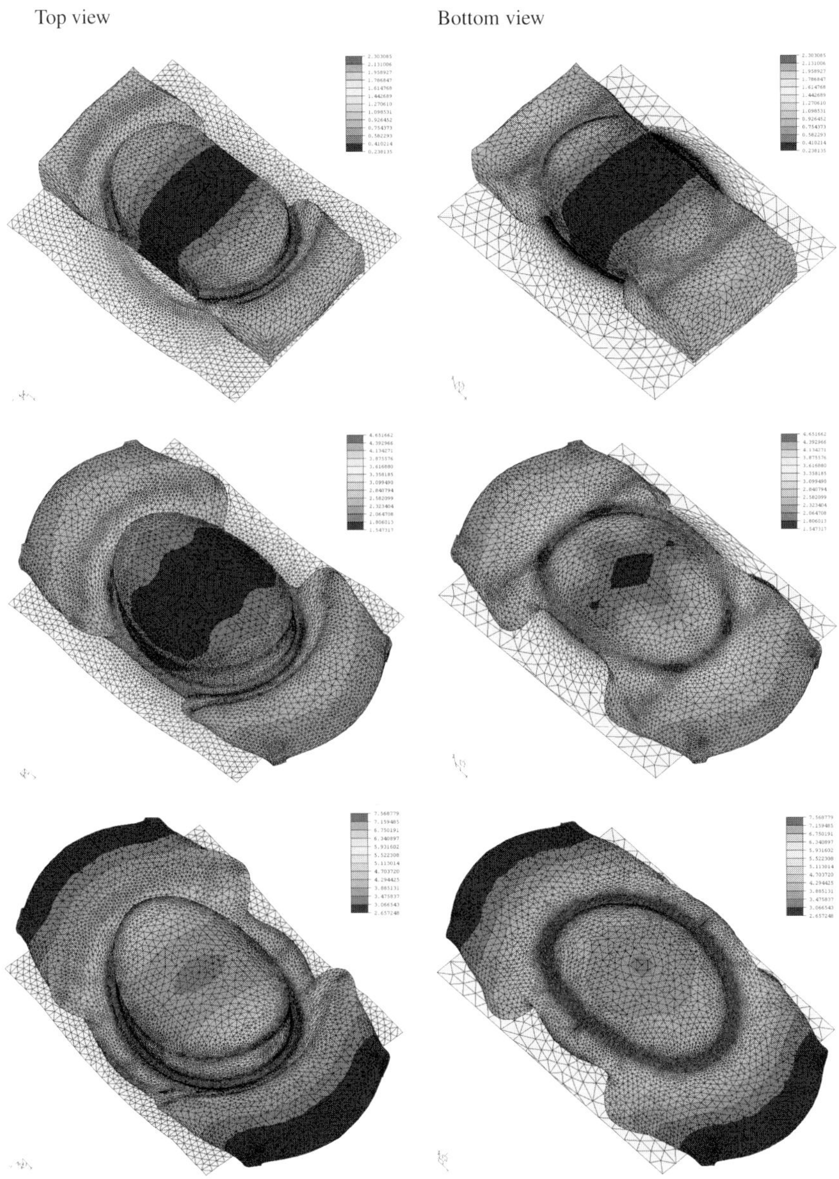

Figure 3. Stamping of a 3D soft solid: Finite element meshes and equivalent plastic strain distribution at different stages of the stamping process.

[3] Lee, N.-S. and Bathe, K.-J. (1994). Error indicators and adaptive remeshing in large deformation finite element analysis. *Fin. Elem. Anal. Design*, **16**, 99–139.

[4] Miehe, C. and Keck, J. (2000). Superimposed finite elastic-viscoelastic-plastoelastic stress response with damage in filled rubbery polymers. experiments, modelling and algorithmic implementation. *J. Mech. Phys. Solids*, **48**, 323–365.

[5] Ortiz, M. and Quigley IV, J. (1991). Adaptive mesh refinement in strain localization problems. *Comp. Meth. Appl. Mech. Engng.*, **90**, 781–804.

[6] Perić, D., Hochard, C., Dutko, M. and Owen, D. (1996). Transfer operators for evolving meshes in small strain elasto–plasticity. *Comp. Meth. Appl. Mech. Engng.*, **137**, 331–344.

[7] Perić, D. and Owen, D. (1998). Finite-element applications to the nonlinear mechanics of solids. *Rep. Prog. Phys.*, **61**, 1495–1574.

[8] Perić, D., Owen, D. and Honnor, M. (1992). A model for finite strain elasto-plasticity based on logarithmic strains: Computational issues. *Comp. Meth. Appl. Mech Engng.*, **94**, 35–61.

[9] Perić, D., Vaz Jr., M. and Owen, D. (1999). On adaptive strategies for large deformations of elasto-plastic solids at finite strains: Computational issues and industrial applications. *Comp. Meth. Appl. Mech. Engng.*, **176**, 279–312.

[10] Perić, D., Yu, J., and Owen, D. (1994). On error estimates and adaptivity in elastoplastic solids: Application to the numerical simulation of strain localization in classical and cosserat continua. *Int. J. Num. Meth. Engng.*, **37**, 1351–1379.

[11] Reese, S. and Govindjee, S. (1998). A theory of finite viscoelasticity and numerical aspects. *Int. J. Solids Struct.*, **35**, 3455–3482.

[12] Simo, J. (1987). On a fully three-dimensional finite-strain viscoelastic damage model: Formulation and computational aspects. *Comp. Meth. Appl. Mech. Engng.*, **60**, 153–173.

[13] Simo, J. (1992). Algorithms for static and dynamic multiplicative plasticity that preserve the classical return mapping schemes of the infinitesimal theory. *Comp. Meth. Appl. Mech. Engng.*, **99**, 61–112.

[14] Simo, J. and Miehe, C. (1992). Associative coupled thermoplasticity at finite strains: Formulation, numerical analysis and implementation. *Comp. Meth. Appl. Mech. Engng.*, **98**, 41–104.

[15] Weber, G. and Anand, L. (1990). Finite deformation constitutive equations and a time integration procedure for isotropic, hyperelastic-viscoplastic solids. *Comp. Meth. Appl. Mech. Engng.*, **79**, 173–202.

Reliability-Based Analysis of Large Deformations in Metal Forming Operations

M. Kleiber, J. Knabel, J. Rojek, R. Stocki
Institute of Fundamental Technological Research, Polish Academy of Sciences
Swietokrzyska 21, 00-049 Warsaw, Poland
mkleiber@ippt.gov.pl

Abstract: Methodology developed for reliability calculations of structures is applied to estimate reliability of sheet metal forming operations, typical mechanical problems characterized with large deformations. Forming Limit Diagrams (FLD) used in the industrial practice as a criterion of material breakage in the manufacturing process are treated as the limit state function for reliability analysis. We try to quantify intuitive terms of probability of failure/success of forming operations given some uncertainty of parameters characterizing a forming process like friction parameters or blankholding force. Since the employment of the gradient-based reliability techniques is very much limited due to numerical noise introduced by the explicit dynamic algorithm used to perform sheet stamping simulation, the method of Adaptive Monte Carlo simulations and Response Surface method were chosen for reliability assessment.

Keywords: Sheet stamping simulation, reliability analysis, Adaptive Monte Carlo technique, response surfaces.

1. Introduction

Sheet metal forming operations are typical mechanical problems characterized with large deformations. Numerical simulation can be successfully applied to evaluate product manufacturability and predict possible defects. Material breakage, wrinkling and shape defects due to springback are most frequent defects in sheet metal forming operations. Sheet forming operations are characterized with a significant scatter of the results. Small differences in the contact conditions or material properties, for instance, can lead to significant changes in the deformation state of the sheet.

An appropriate tool to analyze probability of point-wise material failure in sheet metal forming operations is offered by the theory of reliability. The essence of this approach consists in development of an effective methodology for assessment of probability of violating some pre-defined limit states given complete or partial information on the probability density functions for parameters whose randomness is likely to have an influence on the behavior in question.

The objective of this paper is to combine the two advanced computational methodologies: the sheet forming simulation and the system reliability assess-

C. Miehe (ed.), IUTAM Symposium on Computational Mechanics of Solid Materials at Large Strains, 435-444.

ment techniques. The forming limit curve (FLC) is used to construct the limit curve for the reliability analysis.

2. Finite element modeling of sheet forming operations

In a typical model of sheet forming operation the workpiece in the form of a blank and three basic tools, die, punch and blankholder, are considered. The tools can be treated as rigid bodies, which enforce the blank deformation by the contact action under prescribed kinematics. Both normal and frictional contact interaction between tools and sheet should be taken into account. Friction has great influence on the forming process. The sheet undergoes complex deformation processes characterized by large displacements and large strains.

Different formulations can be used in the simulation of sheet forming operations. Within the finite element method the analysis of these processes can be performed employing either the so called flow and solid models and dynamic or quasistatic approaches (cf. Oñate et al. [12], Chenot and Bay [2]). In the present work the so-called explicit dynamic formulation has been used in the finite element simulation. Because of its efficiency in the analysis of large-scale systems the explicitly integrated dynamic approach has become very popular in sheet stamping simulation. The method is based on the solution of the discretized equations of motion written in the current configuration in the following form:

$$\mathbf{M}\ddot{\mathbf{a}} + \mathbf{C}\dot{\mathbf{a}} = \mathbf{p} - \mathbf{f}\,, \tag{1}$$

where $\mathbf{M}$ and $\mathbf{C}$ are the mass and damping matrices, $\ddot{\mathbf{a}}$ and $\dot{\mathbf{a}}$ are the vectors of nodal accelerations and velocities, $\mathbf{p}$ and $\mathbf{f}$ are the vectors of external loads and internal forces, respectively. The element internal force vector is calculated from the relation

$$\mathbf{f} = \int_V \mathbf{B}^T \boldsymbol{\sigma}\,\mathrm{d}V\,, \tag{2}$$

where $\boldsymbol{\sigma}$ is the Cauchy stress tensor.

Contact forces due to interaction between the sheet and tools are included in the vector of external load $\mathbf{p}$. Contact constraints in the normal direction are enforced using the penalty method and the Coulomb law is employed to evaluate frictional forces.

3. Constitutive model used

The constitutive model used in any analysis of metal forming process must represent properly complex deformation and elasto-plastic properties of the material. The elasto-plastic constitutive model used in our analysis is that developed in García Garino [4], and presented in García Garino and Oliver [5]. In the description of large elasto-plastic deformations the multiplicative decomposition of the deformation gradient tensor $\mathbf{F}$ into its elastic $\mathbf{F}^{\mathrm{e}}$ and plastic part $\mathbf{F}^{\mathrm{p}}$ is assumed. The model is developed in the stress-free intermediate configuration, and then all the constitutive relationships are transformed to the deformed configuration.

In the deformed configuration the following additive decomposition of the Almansi strain tensor $\mathbf{e}$ into the elastic and plastic parts, $\mathbf{e}^{\mathrm{e}}$ and $\mathbf{e}^{\mathrm{p}}$, respectively, is obtained:

$$\mathbf{e} = \mathbf{e}^{\mathrm{e}} + \mathbf{e}^{\mathrm{p}} \tag{3}$$

The Almansi strain tensor, and its elastic and plastic parts, $\mathbf{e}$, $\mathbf{e}^{\mathrm{e}}$ and $\mathbf{e}^{\mathrm{p}}$, can be expressed by the deformation gradient tensor $\mathbf{F}$ and its elastic $\mathbf{F}^{\mathrm{e}}$ and plastic part $\mathbf{F}^{\mathrm{p}}$ in the following form:

$$\begin{aligned} \mathbf{e}^{\mathrm{e}} &= \tfrac{1}{2}(\mathbf{I} - \mathbf{F}^{\mathrm{e}-\mathrm{T}}\,\mathbf{F}^{\mathrm{e}-1}), \\ \mathbf{e}^{\mathrm{p}} &= \tfrac{1}{2}(\mathbf{F}^{\mathrm{e}-\mathrm{T}}\,\mathbf{F}^{\mathrm{e}-1} - \mathbf{F}^{-\mathrm{T}}\,\mathbf{F}^{-1}), \\ \mathbf{e} &= \tfrac{1}{2}(\mathbf{I} - \mathbf{F}^{-\mathrm{T}}\,\mathbf{F}^{-1}) \end{aligned} \tag{4}$$

The implementation of the general model has been simplified by the assumption that elastic strains are small, which for metals is fully justified. This allows us to assume a simple form of the elastic part of the free energy function ψ^{e}

$$\psi^{e}(\mathbf{e}^{\mathrm{e}}) = \lambda \mathrm{tr}(\mathbf{e}^{\mathrm{e}})^2 \;+\; \mu\,(\mathbf{e}^{\mathrm{e}} : \mathbf{e}^{\mathrm{e}}) \tag{5}$$

where λ and μ are the Lamé constants. With this form of the elastic potential the Kirchhoff stress tensor $\boldsymbol{\tau}$ is obtained as:

$$\boldsymbol{\tau} = \frac{\partial \psi^{e}(\mathbf{e}^{\mathrm{e}})}{\partial \mathbf{e}^{\mathrm{e}}} = \lambda\,\mathrm{tr}(\mathbf{e}^{\mathrm{e}}) \;+\; 2\mu\,\mathbf{e}^{\mathrm{e}} \tag{6}$$

In the present implementation of the model the associated flow rule is assumed with the Huber-Mises yield criterion. The stress-strain curve is taken in the following form

$$\sigma_Y = K(a + \bar{\varepsilon}^{\mathrm{p}})^n \tag{7}$$

where σ_Y is the yield stress and K, a and n are the material constants.

4. Reliability problem formulation

Parameters describing the sheet metal forming process posses random nature. These parameters called the basic variables and constitute a vector $\mathbf{X} = [X_1, X_2, \ldots, X_n]^{\mathrm{T}}$ whose samples $\mathbf{x} = [x_1, x_2, \ldots, x_n]^{\mathrm{T}}$ belong to the Euclidian space. The probability measure is defined by the joint probability density function $f_{\mathbf{X}}(\mathbf{x})$ (PDF) of the random vector $\mathbf{X}$. The failure criterion, due to material fracture etc., is usually expressed by the equation $g(\mathbf{x}) = 0$, called the limit state surface. It divides the space into two regions: the failure domain $\Omega_f = \{\mathbf{x} : g(\mathbf{x}) \leq 0\}$ and the safe domain $\Omega_s = \{\mathbf{x} : g(\mathbf{x}) > 0\}$. Hence, the failure probability of the structural system is determined by the following integral:

$$P_f = \mathbb{P}[\mathbf{X} \in \Omega_f] = \mathbb{P}[g(\mathbf{X}) \leq 0] = \int_{\Omega_f} f_{\mathbf{X}}(\mathbf{x}) \, \mathrm{d}\mathbf{x}, \tag{8}$$

where $\mathbb{P}[A]$ means the probability of the random event A.

In practice, when the number n of X_i can be large, involving possibly a finite element numerical procedure, the direct integration appears to be impractical. Therefore, some approximate methods have been developed to assess the P_f.

In the commonly used approach the problem of the reliability calculation is appropriately transformed, $\mathbf{U} = \mathbf{T}(\mathbf{X})$ (see e.g. Madsen et al. [7], Doliński [3] and Melchers [10]), into the standard normal space where PDF $f_{\mathbf{U}}(\mathbf{u}) = \prod_{i=1}^{n} \varphi(u_i)$ becomes the product of the n one-dimensional standard normal PDF's of random variables $U_i = T_i(\mathbf{X})$. The limit state condition is also transformed $g(\mathbf{x}) = 0 \rightarrow h(\mathbf{u}) = g[\mathbf{T}^{-1}(\mathbf{u})] = 0$. Then, the axial symmetry of the PDF $f_{\mathbf{U}}(\mathbf{u})$ assures for any linear function $l(\mathbf{u}) = \beta - \boldsymbol{\alpha}^{\mathrm{T}}\mathbf{u} = 0$, the following equality to be true

$$P_f = \mathbb{P}[l(\mathbf{U}) \leq 0] = \int_{\{\boldsymbol{u}:l(\boldsymbol{u})\leq 0\}} f_{\mathbf{U}}(\mathbf{u}) \, \mathrm{d}\mathbf{u} = \Phi(-\beta), \tag{9}$$

where the coefficients, $-\alpha_i, i = 1, 2, \ldots, n$, are the components of the normalized gradient of the hyperplane $l(\mathbf{u}) = 0$, i.e. $\boldsymbol{\alpha}^{\mathrm{T}}\boldsymbol{\alpha} = 1$, $\beta = \mathrm{sign}[l(\mathbf{0})]\delta$ is the signed distance δ between the hyperplane and the origin and $\Phi(\cdot)$ is the standard normal distribution. Thus, the linear approximation of the transformed limit state surface $h(\mathbf{u}) = 0$ in the point closest to the origin (so-called design point $\mathbf{u}^*$) provides a simple estimate of the failure probability of structural system

$$P_f = \mathbb{P}[h(\mathbf{U}) \leq 0] \approx \mathbb{P}[l(\mathbf{U}) \leq 0] = \Phi(-\beta), \tag{10}$$

where β is called the reliability index. This approach is called the first order reliability method (FORM). The reliability index $\beta = \mathrm{sign}[h(\mathbf{0})]\delta^*$ is determined as a solution of the following optimization problem:

$$\delta^* \equiv \|\mathbf{u}^*\| = \min\|\mathbf{u}\| \quad \text{subject to: } \; h(\mathbf{u}) \leq 0. \tag{11}$$

Two of the most efficient nonlinear optimization algorithm, namely the Rackwitz-Fiessler algorithm (Abdo and Rackwitz [1]) and NLPQL algorithm (Schittkowski [13]) are considered in practice (cf. Liu and Der Kiureghian [6] for comparison of various methods).

Unfortunately, to use the above gradient-based optimization techniques, the limit state function $g(\mathbf{x})$ (LSF) must be differentiable. The function chosen here as a failure criterion in the reliability analysis of sheet metal forming operations fails to meet this requirement.

Insensitive to this requirement are simulation methods (see Melchers [9]). Fundamental one is the crude Monte Carlo approach. In real life problems where the expected P_f is really small, to get the accurate result, with, say coefficient of variation $\nu_{\widehat{P}_f} = 5\%$, it is required to perform millions of simulations. This computational cost is certainly not acceptable.

To overcome this problem it was decided to use the so-called adaptive Monte Carlo (AMC) method. This alternative approach consists in seeking the design point during sampling by "moving" the sampling density based on the information from the previous samples (used in example version of AMC is implemented in COMREL-TI, the part of the STRUREL package, see [14]).

Another, more efficient method is the response surface method (RSM) based on Rackwitz–Fiessler recursive formula

$$\mathbf{u}^{(k+1)} = \frac{1}{\|\nabla h(\mathbf{u}^{(k)})\|^2} \left[\nabla h^T(\mathbf{u}^{(k)})\, \mathbf{u}^{(k)} - h(\mathbf{u}^{(k)})\right] \nabla h(\mathbf{u}^{(k)}) \tag{12}$$

with the following convergence criterion

$$\left| u_i^{(k+1)} - u_i^{(k)} \right| \leq \varepsilon \quad \text{and} \quad |\, g(\mathbf{x}^*)\,| \leq \varepsilon \quad \text{for all variables,} \tag{13}$$

where ε is the assumed tolerance and $\mathbf{x}^* = \mathbf{T}^{-1}(\mathbf{u}^*)$. On each step, instead of an unknown function $h(\mathbf{u})$ (and $\nabla h(\mathbf{u})$), the following least square linear approximation is used

$$\hat{h}(\mathbf{u}) = b_0 + \sum_{i=1}^{n} b_i u_i \quad \text{and then} \quad \nabla \hat{h}(\mathbf{u}) = [b_0, b_1, ..., b_n]\,. \tag{14}$$

It is based on the points placed with assumed distance $\triangle^k \mathbf{u}$ around the central point and additional points from previous steps taken from spherical vicinity of the k-th central point (Fig. 1). RSM estimates LSF in vicinity of the design point, and then standard, gradient reliability optimization method (FORM) is used to search *exact* design point on LSF approximation.

5. Reliability analysis of sheet metal forming operations

There are different uncertainties in the sheet metal forming process and material parameters that lead to uncertainties in the results of practical realization and numerical analysis for a given process. Real values of virtually every such parameter, like material constants, friction coefficients, geometric dimensions, blankholder force, etc. are known to have a certain scatter around their nominal

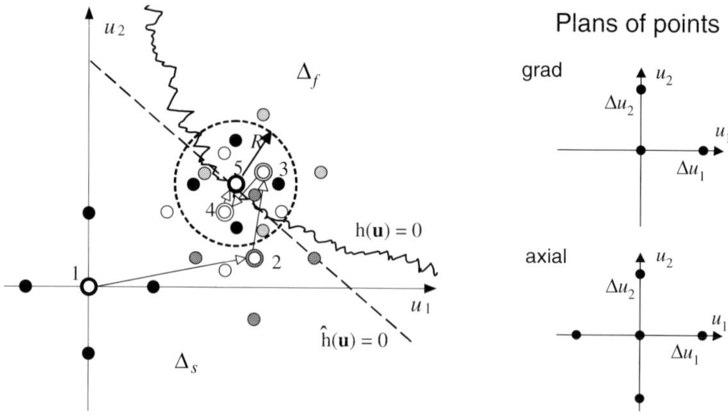

Figure 1. Idea of response surface method.

values – adopting the latter ones to describe the process may lead to essential deviations in terms of the simulation results to be obtained. The forming process is very sensitive to friction conditions. Small variation in lubrication can affect the path of deformation of the sheet. Variation of sheet thickness and material properties affects the forming process strongly as well. This indicates that we can treat a sheet metal forming operation as a process with some uncertainties.

Possibility of material fracture in sheet metal forming operations is estimated in practice using forming limit diagrams (FLD). In a typical forming limit diagram (Fig. 2), major principal strain values are plotted against minor principal strain values. Usually logarithmic or engineering strain measures are used. Points representing strain states all over the deformed sheet are confronted with the forming limit curve (FLC). FLC is supposed to represent the boundary between the strain combinations which produce instability (above the curve) and/or fracture and those that are permissible in forming operations (below the curve).

Results of sheet forming operations exhibit significant scatter, therefore any forming limit curve can be regarded as bounding the safe zone with some probability only. The safe zone is considered as the one where failure is highly improbable, while the failure zone is regarded as the one defining strain states with a high probability of failure. Usually between the two zones, safe and failure, a critical zone (marginal zone) is introduced (Fig. 2), with the probability of failure high enough so that the strain state cannot be considered safe. Our present work is aimed to quantify these qualitative notions.

We take advantage of the forming limit diagrams and define LSF as the signed minimal distance from the FLC of the point corresponding to principal strains in the given finite element (Fig. 3). In the adopted sign convention the minus sign is for the points above the curve.

Depending on the realization of the vector of random variables the "cloud" of the points on the forming limit diagram can take a different shape which implies that different points in the sheet metal may be located close to the FLC. Adding to it the shape of the FLC (piecewise linear with vertices) and some "numerical noise" introduced by using the explicit dynamic approach in the finite element analysis it appears unjustified to assume that the LSF is differentiable. Therefore, very efficient gradient-base optimization methods cannot be used in reliability analysis, and time consuming simulation method should be applied or alternative response surface methods have to be investigated and employed.

6. Numerical illustration

Deep drawing of a square cup, the benchmark problem at the Numisheet '93 conference (Makinouchi et al. [8]) has been analyzed. The aluminium square blank of dimensions 150×150 mm and 0.81 mm thick has the following properties: Young's modulus $E = 71$ GPa, Poisson's ratio $\nu = 0.33$, stress-strain curve $\sigma = 576.79(0.01658 + \varepsilon^p)^{0.3593}$ MPa. The Coulomb friction coefficient $\mu = 0.162$ and the blankholding force 19.6 kN are taken. Definition of the tool geometry can be found in Makinouchi et al. [8]. In the analysis the symmetry conditions have been used and only one quarter of the problem has been analyzed. This clearly imposes certain constraints on variability of random variables to be considered.

In Makinouchi et al. [8] the results for the punch travel of 15 mm have been reported. Figure 4 presents deformed shape with contours of logarithmic thickness strains. The results obtained are in a good agreement with the experimental and numerical results reported in Makinouchi et al. [8]. The major and minor

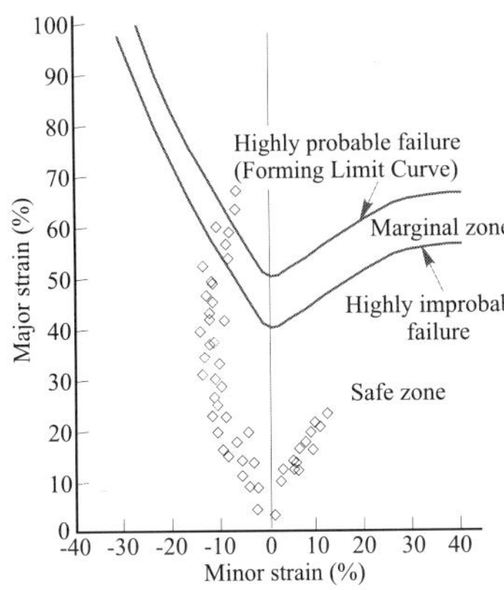

Figure 2. Typical forming limit diagram.

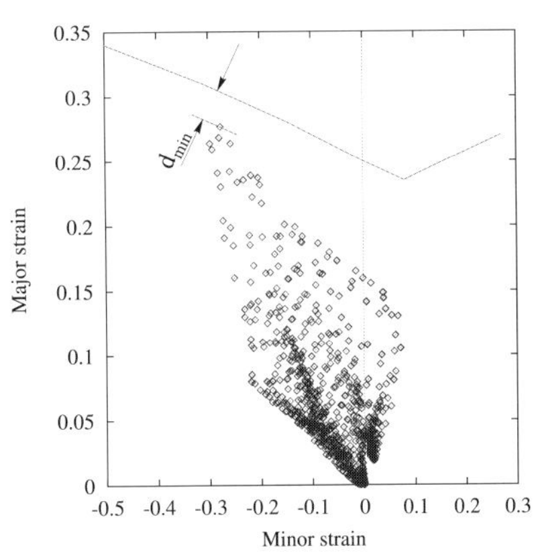

Figure 3. Definition of the LSF.

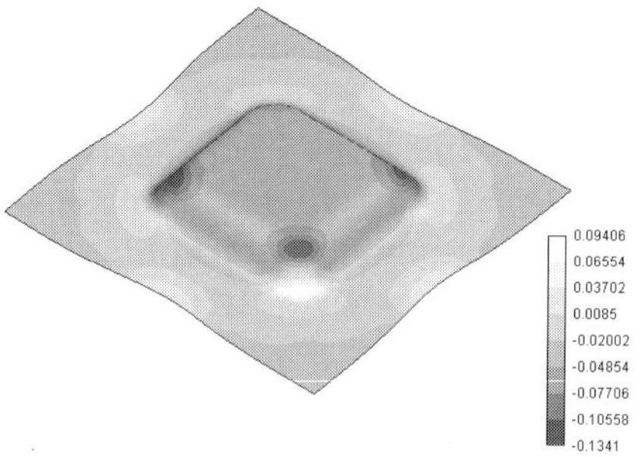

Figure 4. Distribution of logarithmic thickness strain at 15 mm depth of drawing.

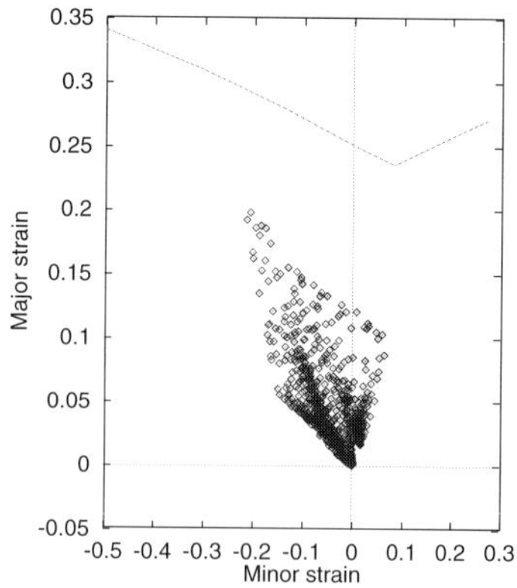

Figure 5. Forming limit diagram for 15 mm depth of drawing.

strains are plotted in the forming limit diagram in Fig. 5. Here the strains are still far from FLC and no danger of failure is expected. Continuing drawing the failure conditions are first met for the punch travel of 20 mm. This is in agreement with experimental results, in Nakamachi [11] the failure in laboratory tests at punch stroke of 19 mm is reported.

Reliability analysis has been carried out for the drawing depth of 15 mm. The stochastic description of the system involves 6 random variables. They are as follows: the initial thickness of the sheet metal, the material constant K in the constitutive relation (7), blankholder force and three friction coefficients – between sheet metal and punch, die and blankholder, respectively. The random variables are assumed to be lognormally distributed with mean values and standard deviations shown in Table 1.

The friction coefficients, represented by the last three random variables X_4–X_6, are assumed to be correlated with the same correlation coefficient ρ. To investigate the influence of this correlation on the probability of failure the reliability analysis employing AMC technique was performed for various values of ρ. The results are presented in Table 2. The probability of failure may vary even by two orders of magnitude depending on the value of correlation coefficient when the latter changes from $\rho = 0$ (independent variables X_4–X_6) to $\rho = 1$ (full dependence). Since the sheet metal surface properties are most frequently distributed in a manner close to uniform, more realistic seems to be the assumption of the strong correlation of the values of friction coefficients.

All gradient based algorithms would encounter difficulties while locating a design point $\mathbf{u}^*$ (cf. (11)). However, the globally linear LSF can be approximated quite accurately by the linear function. Thus it was interesting to employ the FORM approach computing the gradients of LSF by finite differences (pre-

Table 1. Random variables.

Var.	*Distribution type*	*Mean value*	*Std. dev.*	*Description*
X_1	lognormal	0.8 mm	0.04 mm	initial thickness
X_2	lognormal	$5.7679 \cdot 10^8$ Pa	$3.0 \cdot 10^7$ Pa	material constant
X_3	lognormal	4900 N	50 N	blankholder force
X_4	lognormal	0.162	0.015	frict. coeff. sheet–punch
X_5	lognormal	0.162	0.015	frict. coeff. sheet–die
X_6	lognormal	0.162	0.015	frict. coeff. sheet–blankholder

Table 2. Results of metal forming reliability analysis.

Method	ρ	P_f	$\beta = -\Phi^{-1}(P_f)$	*No. of simul.*	$\nu_{\hat{P}_f}$	ε
AMC	0	$4.07 \cdot 10^{-7}$	4.93	9264	5%	
AMC	0.3	$6.51 \cdot 10^{-6}$	4.36	4326	5%	
AMC	0.9	$4.62 \cdot 10^{-5}$	3.91	3325	5%	
AMC	1	$5.06 \cdot 10^{-5}$	3.89	2826	5%	
CD	0.9	$5.35 \cdot 10^{-5}$	3.874	82		
RS grad	0.9	$6.01 \cdot 10^{-5}$	3.846	67		0.1
RS axial	0.9	$4.75 \cdot 10^{-5}$	3.903	110		0.1
RS grad	0.3	$8.94 \cdot 10^{-6}$	4.29	127		0.1
RS axial	0.3	$6.51 \cdot 10^{-6}$	4.36	130		0.1

formed by central differences (CD)) using large perturbations to avoid problems with the "noisy" character of the function.

More general remedy in reliability analysis of sheet metal forming processes seems to be RSM. The number of points taking place in LSF approximation increases during optimization procedure, making process stable, and insensitive to LSF local discontinuities, cf. Table 2.

7. Conclusions

Adaptive Monte Carlo simulation proves useful for reliability assessment of sheet metal forming operations but for reasons of efficiency it appears suitable for small and medium size problems only. Very high sensitivity of the probability of local necking type failure with respect to the correlation of friction coefficients has been observed. In view of the frequent lack of sufficient data on random variables correlations this aspect requires further studies. The probability of local necking type failure turns out very sensitive with respect to correlation of friction coefficients. Response surface technique allows to obtain assessment of P_f in LSF function in the noise effect case, when it is impossible to use the standard gradient methods and computational cost is much lower than in simulation techniques for low number of variables. There is a need to develop methods that would improve the RSM for second order correction in probability of failure assessment.

References

[1] Abdo, T. and Rackwitz, R. (1990). Reliability of uncertain structural systems. *Proc. Finite Elements in Engineering Applications*, 161–176, Stuttgart, INTES GmbH.

[2] Chenot, J.-L and Bay, F. (1998). An overview of numerical modelling techniques. *J. Mater. Proc. Technology*, **80-81**, 8–15.

[3] Doliński, K. (1983). First-order second-moment approximation in reliability of structural systems: critical review and alternative approach. *Structural Safety*, **1**, 211–231.

[4] García Garino, C. (1993). *A Numerical Model for the Analysis of Large Elasto-plastic Deformations of Solids*, PhD thesis, Universitat Politecnica de Catalunya, Barcelona, (in Spanish).

[5] García Garino, C. and Oliver, J. (1992). A numerical model for elastoplastic large strain problems. *Computational Plasticity*, D.R.J. Owen et al. (eds.)

[6] Liu, P.-L. and Der Kiureghian, A. (1991). Optimization algorithms for structural reliability. *Structural Safety*, **9**, 161–177.

[7] Madsen, H.O., Krenk, S. and Lind, N.C. (1986). *Methods of Structural Safety*. Prentice-Hall, Englewood Cliffs, New Jersey.

[8] Makinouchi, A., Nakamachi, E., Oñate, E. and Wagoner, R.H. (1993). In: *Proceedings of the 2nd International Conference NUMISHEET 93: Numerical Simulation of 3-D Sheet Metal Forming Processes – Verification of Simulation with Experiment.* Isehara, Japan.

[9] Melchers, R.E. (1992). Simulation in time-invariant and time-variant reliability problems. In: *Reliability and Optimization of Structural Systems '91, Proc. 4th IFIP WG 7.5 Conf., Munich, 11–13 September 1991*, R. Rackwitz and P. Thoft-Christensen (eds.), Springer-Verlag, 39–82.

[10] Melchers, R.E. (1999). *Structural Reliability Analysis and Predictions, 2nd Ed.*, Wiley.

[11] Nakamachi, E. (1993). Sheet forming process characterization by static-explicit anisotropic elastic-plastic finite element simulation. In: *Proceedings of the 2nd International Conference NUMISHEET 93: Numerical Simulation of 3-D Sheet Metal Forming Processes – Verification of Simulation with Experiment*, Isehara, Japan.

[12] Oñate, E., Rojek, J. and García Garino, C. (1995). NUMISTAMP: A research project for assessment of finite element models for stamping processes. *Journal of Materials Processing Technology*, **50** (1–4), 17–38.

[13] Schittkowski, K. (1985). *User's guide for the nonlinear programming code nlpql. Handbook to optimization program package NLPQL*, University of Stuttgart, Institute for Informatics, Germany.

[14] *STRUREL: Structural Reliability Analysis Program Package, User's manual.* RCP GmbH, Barer Strasse 48, Munich, Germany, 1999.

Development of Crystal Plasticity Design System to generate a High-Strength and High-Formability Material

E. Nakamachi
Osaka Institute of Technology, Department of Mechanical Engineering
5-16-1 Omiya, Asahiku, Osaka, Japan
nakamati@med.oit.ac.jp

Abstract: A crystal plasticity finite element (FE) method, by employing a crystalline ODF result and a newly developed "Orientation Probability Assignment Method", has been developed to design a dual phase high-strength and high-formability sheet metal. At first, the texture effect on the earing and strain localization of two steel sheet metals, such like a "soft phase" mild and a "hard phase" high strength steel, was studied. Second, an optimum design was attempted to find an artificial crystal texture of dual phase steel sheet, which shows high strength and high formability, through VDI benchmark sheet forming process simulations.

Keywords: Textural target design, crystalline plasticity, finite element analysis, dual phase steel sheet, soft phase, hard phase.

1. Introduction

Recently, the finite element (FE) simulation technology, based on the crystalline plasticity theory, has revealed a great progress in the assessment of strength and formability, and further the material design of sheet metal [1, 2, 5, 7, 9, 10, 11, 12, 13, 14]. Through FE analyses and experimental observations, it was confirmed that the initial and plastic deformation induced crystal textures could correlate with the occurrences of defects, such like tearing, earing, and springback in the sheet metal forming operation. The practical interest, in controlling crystallographic textures as a means of improving formability, focuses on application to a high-strength steel sheet, which is driven by the need of the automotive and steel industries for weight reduction. In this study, (1) I develop the textural target design method to improve formability of a high-strength steel sheet, the definition of the design parameter and the methodology to represent a designed texture, (2) I apply this design system to find a best texture by combining "hard phase" texture and "soft phase texture", through more complicated sheet forming process simulations by employing VDI benchmark problem [14].

C. Miehe (ed.), IUTAM Symposium on Computational Mechanics of Solid Materials at Large Strains, 445-454.

2. Modeling of dual phase steel employing measured textures

2.1. ODF analyses of "soft phase" and "hard phase" steel sheets

The pole figures and orientation distribution functions (ODF) of a "hard phase" high-strength steel (CP800) and a "soft phase" mild steel (DDQ) are shown in Figures 1a, 1b and 2a in terms of constant density surfaces in Euler angle coordinate space (Gelin and Picart [5]). The typical fibers and selected orientations, which can be related to the properties of anisotropy and yield loci, are shown in Figure 2c. The main fiber of a "soft phase" steel is so called γ fiber (Lücke and Hölscher [9]), in which $\langle 111 \rangle$ directions are parallel to the sheet normal direction, $\{111\}$ planes lay randomly on the sheet plane and the skeleton line is nearly a straight line from $(111)[1\bar{1}0]$ to $(111)[\bar{1}\bar{1}2]$ as shown in Figure 2a. On the other hand, a "hard phase" steel has a α fiber texture and very strong $\{001\}\langle 110 \rangle$ orientation as well as $\{113\}\langle 110 \rangle$, $\{233\}\langle 311 \rangle$ and $\{111\}\langle 110 \rangle$ orientations, but almost no γ fiber texture, as shown in Figure 2b.

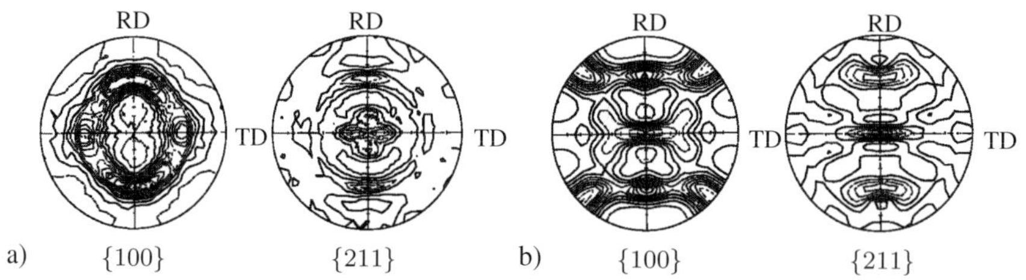

Figure 1. Pole figures of a) "soft phase" steel DDQ and b) "hard phase" steel CP800.

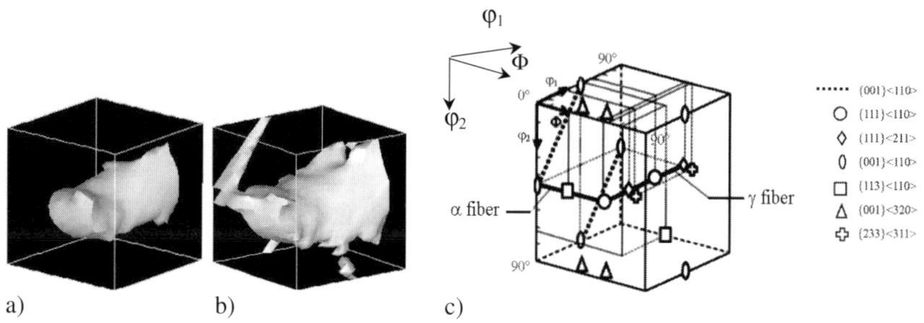

Figure 2. 25%-maximum density surfaces of ODFs of a) "soft phase" and b) "hard phase" steels and c) texture map with fibers and selected orientations in Euler angle coordinate space.

R-values and yield loci were calculated by using the crystallographic (ODF) Taylor-Bishop-Hill approach [3]. Those are shown in Fig. 3 and 4. The average $\bar{R}$ values, $\bar{R} = (R_0 + 2R_{45} + R_{90})/4$, are 0.88 for "hard phase" and 1.55 for "soft phase." It was found that $\{111\}$ texture – γ fiber – affects to increase $\bar{R}$ value and improve formability, while $\{001\}$ texture is harmful (Wilson [15]).

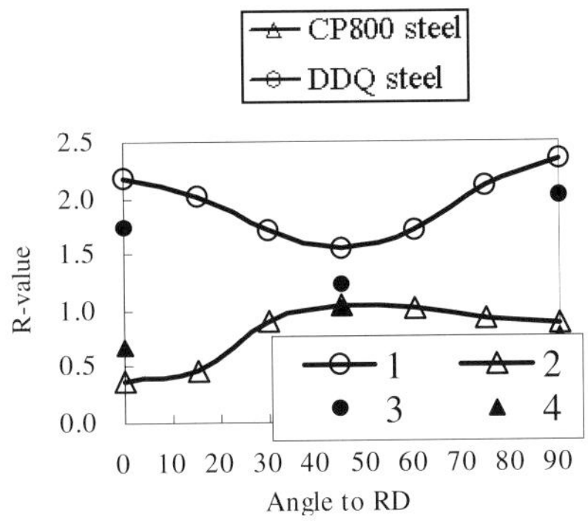

Figure 3. R-values predicted by popLA [6]. (1,2: FE analyses. 3,4: Experiment).

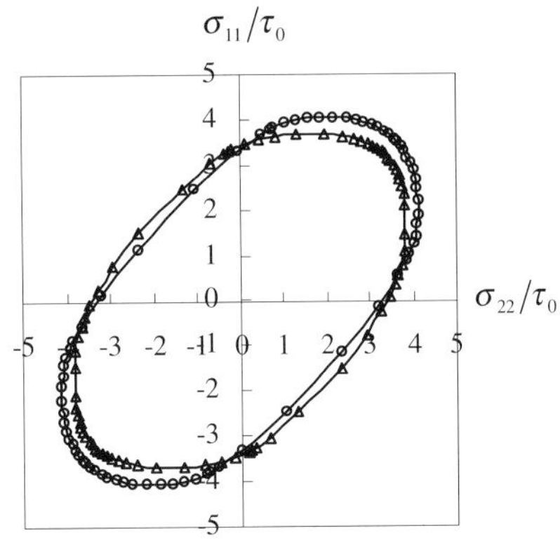

Figure 4. Bishop-Hill yield loci calculated by popLA.

2.2. ELASTIC / CRYSTALLINE VISCOPLASTIC CONSTITUTIVE EQUATION

The objective rate of Kirchhoff stress tensor $\hat{\boldsymbol{\Pi}}$ is given by Nakamachi and Dong [10]:

$$\hat{\boldsymbol{\Pi}} = \hat{\boldsymbol{\Pi}}^* - \sum_{a=1}^{N} \boldsymbol{\beta}^a \dot{\gamma}^a = \boldsymbol{C}^e : \boldsymbol{d} - \sum_{a=1}^{N} \boldsymbol{R}^a \dot{\gamma}^a, \tag{1}$$

$$\boldsymbol{d}^p = \sum_{a=1}^{N} \dot{\gamma}^a \boldsymbol{P}^a, \quad \boldsymbol{\omega}^p = \sum_{a=1}^{N} \dot{\gamma}^a \boldsymbol{W}^a, \quad \dot{\gamma}^a = \dot{a} \left[\frac{\tau^a}{g^a}\right] \left[\left|\frac{\tau^a}{g^a}\right|\right]^{(1/m)-1}. \tag{2}$$

$\dot{\gamma}^a$ means the shear strain rate on the slip system a, $\hat{\boldsymbol{\Pi}}^*$ the co-rotational rate of the Kirchhoff stress, $\boldsymbol{C}^e$ the elastic moduli, $\boldsymbol{d}$ the deformation rate tensor, $\boldsymbol{d}^p$ the plastic deformation rate tensor, $\boldsymbol{\omega}^p$ the plastic spin tensor, respectively. τ^a is the resolved shear stress on the slip system a, g^a the reference shear stress, $\dot{a}$ the reference shear strain rate, and m the coefficient of strain rate sensitivity. The conventional hardening-evolution equation is introduced (Peirce et al. [13], Asaro and Needleman [1]):

$$\dot{g}^a = \sum_{b=1}^{N} h_{ab} \left|\dot{\gamma}^b\right|, \quad h_{ab} = q_{ab} \frac{d\tau(\gamma)}{d\gamma} + (1 - q_{ab}) \frac{d\tau(\gamma)}{d\gamma} \delta_{ab},$$
$$\tau = \tau_0 + (\tau_S - \tau_0) \tanh\left(\frac{H_0 \gamma}{\tau_S - \tau_0}\right) \tag{3}$$

h_{ab} means the coefficient of hardening evolution, obtained by τ-γ relationship. q_{ab} means coefficients of interaction between slip system a and b. H_0, τ_0 and τ_S mean the initial hardening ratio, the critical (initial) and saturated resolved shear stresses.

2.3. PARAMETER IDENTIFICATION OF HARDENING EVOLUTION EQUATION

Table 1 shows material parameters of "soft phase" (DDQ) and "hard phase" (CP800) steel sheets. Those values are identified by employing the stress-strain relationships as shown in Figure 5. The "soft phase" steel shows higher strength, but smaller elongation.

Table 1. Material properties.

		Soft phase (DDQ)	*Hard phase (CP800)*
initial critical shear stress	τ_0	66.0 MPa	285.5 MPa
initial hardening ratio	H_0	206.0 MPa	246.0 MPa
saturated critical shear stress	τ_S	156.0 MPa	375.0 MPa
Young's modulus	E	210.0 GPa	
density	ρ	$7.8 \times 10^{-9}\,\mathrm{MPa/(mm/s)^2}$	
Poisson's ratio	ν	0.3	
reference strain rate	$\dot{a}$	$0.05\,\mathrm{s}^{-1}$	
strain rate sensitivity exponent	m	0.03	
elastic anisotropy parameters	c_{11}	237.0 GPa	
	c_{12}	141.0 GPa	
	c_{44}	116.0 GPa	

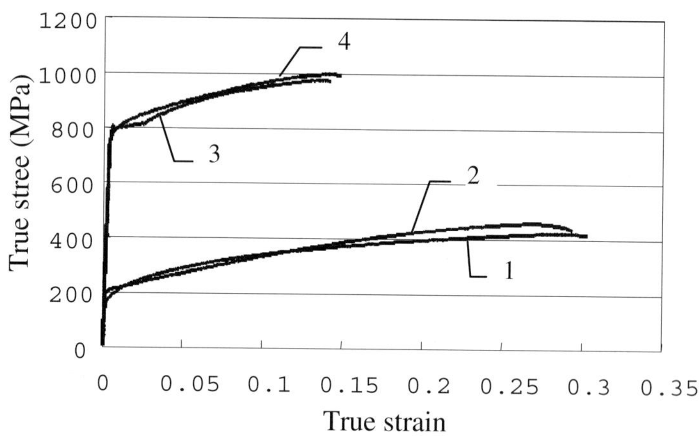

Figure 5. True stress-strain relationships, Nakamachi and Dong [10]. 1,3: Experiment DDQ and CP800. 2,4: FE results DDQ and CP800.

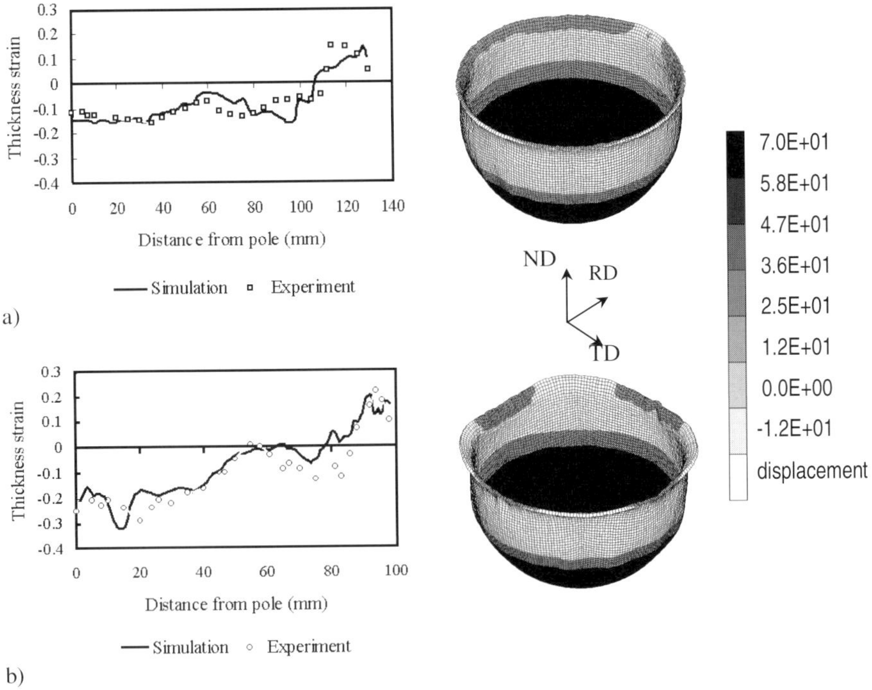

Figure 6. Comparisons of "earing" deformation and thickness strain distributions with a) "soft phase" and b) "hard phase" materials, at 70 mm punch travel in the deep drawing process.

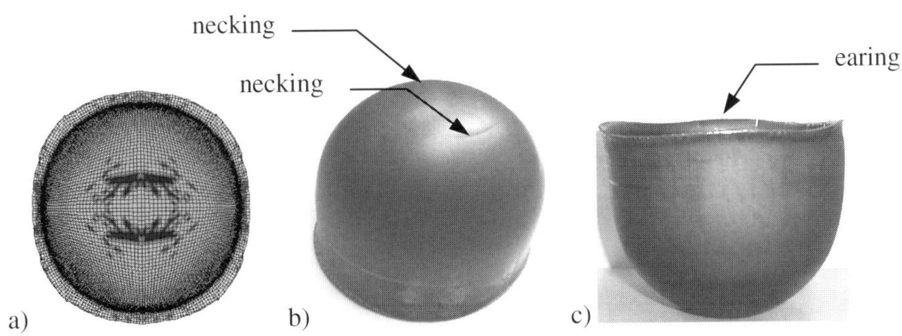

Figure 7. Comparisons of thickness strain distributions and localization with FE results and experimental ones, in case of the "hard phase" steel sheet (CP800). a) FE result at 70mm punch travel. b) Experimental results, given by Professor J. Danckert, Aalborg University, Denmark.

3. Textural target design

In this "textural target design" system, an artificial γ fiber texture, which shows very good formability, is blended to the "hard phase" steel sheet.

3.1. Analytic representation of the artificial γ fiber texture

3.1.1. Scattered orientation angle ω_0

To establish a textural design of a dual phase high strength steel sheet, the artificial γ fiber – "soft phase" texture – is defined by employing a rotationally symmetric Gaussian distribution of crystal orientation, which has a certain deviation angle and a certain scatter width ω_0. It distributes along the skeleton line from $(111)[1\bar{1}0]$ to $(111)[\bar{1}\bar{1}2]$ in Euler angle coordinate space. A rotationally symmetric Gaussian distribution of orientations is generated around one discrete ideal orientation g_0, such like Euler angles, $\Phi = 54.7°$ and $\varphi = 45°$. Here, three different orthogonal reference systems were used; the sample axes $C(X, Y, Z)$, $\langle 100 \rangle$ axes of the ideal g_0 orientation $C_0(x, y, z)$, and $\langle 100 \rangle$ axes of sample g_1 crystal $C_1(x', y', z')$. The transformation from C_0 to C_1 is executed easily by using a single rotation ω about a specific axis of rotation $\boldsymbol{d}$. The orientation of axis $\boldsymbol{d}$ can be characterized by the spherical coordinates δ, Ψ, as shown in Figure 8a. The rotational angle ω is specified as (Lequeu et al. [8])

$$p(\omega) = p(0)\exp\left[-\frac{1}{2}\left(\frac{\omega}{\omega_0}\right)^2\right], \tag{4}$$

where ω is an orientation rotation angle from the discrete ideal orientation, ω_0 the scatter width, $p(0)$ the distribution density of the ideal orientation and $p(\omega)$ the relative Gaussian probability distribution density. The transformation matrix between C and C_0 is also defined easily (Bunge [4]).

Figure 8b shows $\{100\}$ and $\{111\}$ pole figures of designed artificial γ fiber texture, where 300 orientations were defined with the scatter width $\omega_0 = 15°$. Inversely, the Euler angles of the orientations $g(\varphi_1, \Phi, \varphi_2)$ of the artificial texture can be obtained by employing the transfer matrix between C_1 and C.

3.1.2. Volume fraction $V_f^{(i)}$

In this study, the whole Euler angle space is divided into 5832 regular cubic orientation space elements with a grid of 5 degree. $N^{(i)}$ and N denote the numbers of orientations $g(\varphi_1, \Phi, \varphi_2)$ at each orientation space element and the total number of crystallites in the sample, respectively. If the sample consists of well-defined crystallites in whose interiors exists a uniform orientation, the fraction of crystallite numbers, $N_f^{(i)} = \frac{N^{(i)}}{N}$, approximates to the volume fraction $V_f^{(i)}$ in each orientation space element (Nakamachi et al. [11]).

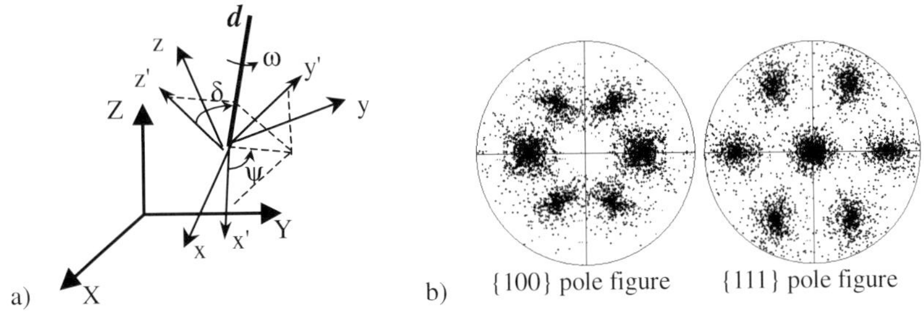

Figure 8. Coordinate definition and pole figures of designed texture by employing the scatter width $\omega_0 = 15°$. a) $C(X, Y, Z), C_0(x, y, z), C1(x', y', z')$. b) Pole figures of artificial γ fiber.

3.2. TEXTURAL DESIGN ALGORITHM

A dual phase high-strength steel sheet is designed by blending an artificial γ texture with the "hard phase" steel, in order to improve the formability. The volume fraction of dual phase steel, $V^{(i)}_{f\ NEW}$, was expressed as

$$V^{(i)}_{f\ NEW} = \frac{1}{1+k}\left(V^{(i)}_{f\ 1} + kV^{(i)}_{f\ 2}\right), \tag{5}$$

where $V^{(i)}_{f\ 1}$ and $V^{(i)}_{f\ 2}$ are volume fractions of "hard phase" texture and the artificial γ fiber texture in the ith orientation space element, respectively. k is the blending coefficient. An artificial steel sheet is designed by employing two design parameters, such like ω_0 and k. This blending coefficient k corresponds to the $\{111\}/\{001\}$ volume fraction ratio R_{vf},

$$R_{vf} = \sum_{i=1}^{M_{(111)}} V^{(i)}_{f\ (111)} \Bigg/ \sum_{j=1}^{M_{(001)}} V^{(j)}_{f\ (001)}, \tag{6}$$

$$V^{(i)}_{f} = \int_{\varphi_2^{(i)}}^{\varphi_2^{(i)}+d\varphi_2} \int_{\Phi^{(i)}}^{\Phi^{(i)}+d\Phi} \int_{\varphi_1^{(i)}}^{\varphi_1^{(i)}+d\varphi_1} \frac{f(g)}{8\pi^2} \sin\Phi\, d\varphi_1\, d\Phi\, d\varphi_2, \tag{7}$$

where $M_{(111)}$ and $M_{(001)}$ are the number of the orientation space elements containing $\{111\}$ and $\{001\}$ components, respectively; $V^{(i)}_{f\ (111)}$ and $V^{(i)}_{f\ (001)}$ are the volume fractions of $\{111\}$ and $\{001\}$ components in the ith orientation space element, respectively; $f(g)$ is the orientation distribution function expressed by Euler angle series expansion of harmonic function (Bunge [4]). Figures 9a–e show the constant density surfaces of the volume fraction of new dual phase steel sheets in cases of $k = 0.0, 0.02, 0.05, 0.10$ and 0.20 while $\omega_0 = 20°$.

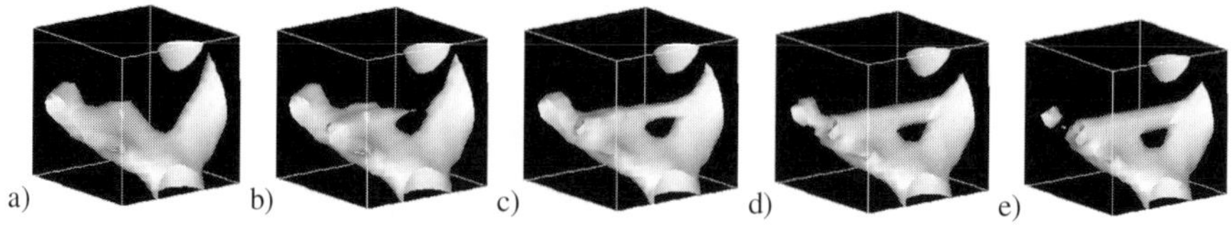

Figure 9. 20 %-maximum density surfaces of the volume fraction of new high-strength steel sheets. a) $k = 0.0$ (conventional), b) $k = 0.02$, c) $k = 0.05$, d) $k = 0.10$, e) $k = 0.20$.

3.3. Finite element modeling

An orientation probability assignment method was employed in polycrystalline finite element modeling (Nakamachi et al. [11]). A set of orientations obtained from the designed texture in terms of Euler angles are randomly selected from the set of orientation probability function and assigned to the FE integration points one by one. In this modeling, one FE integration point is corresponding to one orientation, which can rotate individually during deformation.

4. Numerical analyses

4.1. Texture optimization

Optimization of texture was executed to find a high-strength steel sheet with the best formability, which was found in the feasible region of the design parameter space. The scatter width ω_0 and the blending coefficient k were selected as the design parameters. For the objective function, the maximum absolute value of thickness strain $|\varepsilon_t|_{max}$, which means a maximum thinning ratio, was employed to evaluate the formability.

The LDH formability test (Numisheet'96 Benchmark problem, Lee et al. [7]) was employed to assess the formability of the designed textures and to search the optimum texture. I adopted the hardening evolution parameters of CP800, as shown in Table 1, for these artificial "hard phase" high-strength steel. To maintain the strength of "hard phase" steel, the ratio of blending coefficient$k$ of "γ fiber" texture was limited less than 0.25, and the scatter width ω_0 was fixed as 20° (Nakamachi et al. [11]). In the FE simulation, 1125 8-node SRI (selected reduced integration) solid elements are used. The grid method is employed to find a optimum condition. Five local minimum values of objective function, defined by $|\varepsilon_t|_{max}$ at punch travel 25 mm, were found in the interval $0.0 \leq k \leq 0.25$. Finally, it is confirmed that a global minimum point, the best condition, leads $k = 0.10$. Figure 10 shows deformed shape with thickness strain distribution in cases of best condition$k = 0.10$. The more favorable the texture, the higher the hardening and also the more the thickness uniformity.

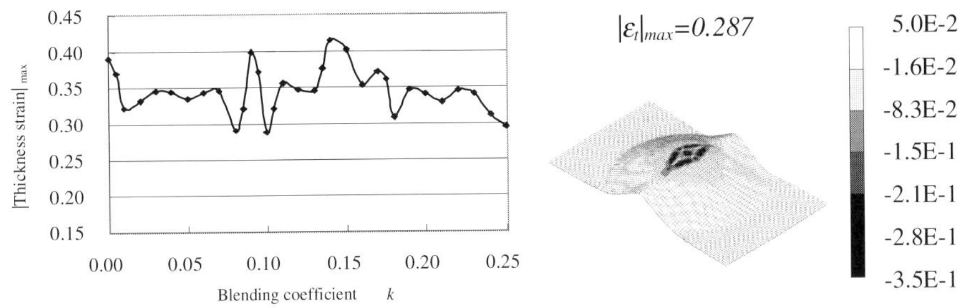

Figure 10. The variation of thickness strain with blending coefficients k, and deformed shapes with thickness strain distribution in case of $k = 0.10$.

4.2. VDI BENCHMARK TESTS

VDI benchmark test [14] was adopted to verify the availability of the textural target design algorithm to find an optimum texture. In this simulation, 4624 eight-node SRI solid elements were used. The material properties are the same as LDH test, which is listed in Table 1. It was confirmed that the optimum texture $k = 0.10$, which has been found in LDH test simulation as shown in Figures 10, shows best formability even in case of the complicated shape tool, as shown in Figure 11.

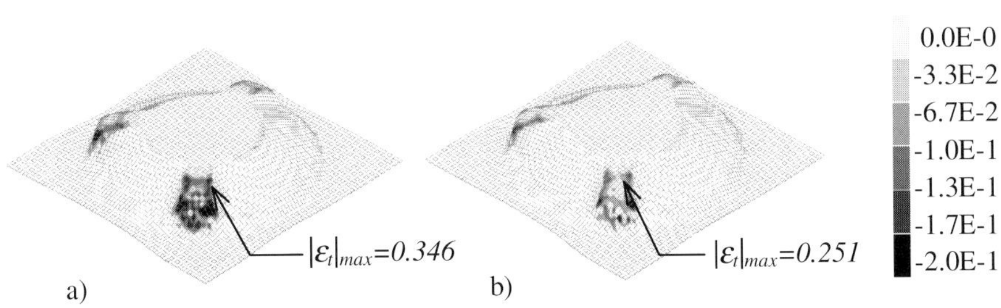

Figure 11. Deformed shapes with thickness strain distribution. a) $k = 0.0$ (CP800), b) $k = 0.10$.

Compared with the conventional high-strength steel (CP800), the improvement on strain localization is very pronounced as k is 0.10, the maximum absolute values of thickness strain $|\varepsilon_t|_{max}$ decrease by 27 %. The additional γ fiber texture moderate the strain localization and the optimum condition of artificial texture shows least thinning ratio of the sheet metal, as expected in LDH test.

5. Conclusions

The textural target design was proposed and applied to search the optimum texture of a dual phase high-strength steel sheet through the crystalline plasticity FE analyses. The artificial γ fiber texture was selected as a candidate to blend the "hard phase" steel sheet, which shows poor formability. For texture optimization, the scatter width and blending coefficient were selected as design parameters. The maximum absolute value of thickness strain, which was defined as objective function, was employed to evaluate the formability. The optimum texture was found in the feasible region by using FE analyses through LDH formability tests. The availability of the optimum textural design system was confirmed through the application to VDI sheet forming processes.

References

[1] Asaro, R.J. and Needleman, A. (1985). Texture development and strain hardening in rate dependent polycrystal. *Acta metallurgical*, **33**, 923–953.

[2] Beaudoin, A.J., Dawson, P.R., Mathur, K.K., Kocks, U.F. and Korzewa, D.A. (1994). Application of polycrystal plasticity to sheet forming. *Computer Methods in Applied Mechanics and Engineering*, **177**, 49–70.

[3] Bishop, J.F.W. and Hill, R. (1951). A theoretical derivation of the plastic properties of a polycrystalline face-centered metal. *Phil. Mag.*, **42**, 1298–1307.

[4] Bunge, H.J. (1982). *Texture analysis in material science*, London, Butterworths.

[5] Gelin, J.C. and Picart, P. (1999). *Proc. of the 4th Int. Conf. NUMISHEET'99*. Univ. of Franche-Comte and ENSMM Besancon, France.

[6] Kallend, J.S., Kocks, U.F., Rollett, A.D. and Wenk, H.R. (1991). Operational texture analysis. *Material Science and Engineering*, **A132**, 1–11.

[7] Lee, J.K., Kinzel, G.L. and Wagoner, R.H. (1996). *Proc. of the 3rd Int. Conf. NUMISHEET'96*. The Ohio State Univ.

[8] Lequeu, P., Gilormini, P., Montheillet, F., Bacroix, B. and Jonas, J.J. (1987). Yield surface for texture polycrystals-part I: crystallographic approach. *Acta Met.*, **35**, 439–451.

[9] Lücke, K. and Hölscher, M. (1991). Rolling and recrystallization textures of BCC steels. *Texture and Microstructures*, **14**, 585–596.

[10] Nakamachi, E. and Dong, X.H. (1997). Study of texture effect on sheet failure in a limit dome height test by using elastic/crystalline viscoplastic FE analysis. *ASME Journal of Applied Mechanics*, **64**, 519–524.

[11] Nakamachi, E., Xie, C.L. and Harimoto, M. (2000). Drawability assessment of BCC steel sheet by using elastic/crystalline viscoplastic finite element analyses. *International Journal of Mechanical Sciences*, **43**, 631–652.

[12] Nakamachi, E., Hiraiwa, K., Morimoto, H. and Harimoto, M. (2000). Elastic/crystalline viscoplastic finite element analyses of single- and poly-crystal sheet deformations and their experimental verification. *International Journal of Plasticity*, **16**, 1419–1441.

[13] Peirce, D., Asaro, R.J. and Needleman, A. (1983). Material rate dependence and localized deformation in crystalline solids. *Acta metallurgica*, **31**, 1951–1976.

[14] *Proc. of VDI Conference on FE-simulation of 3-D Sheet Metal Forming Processes in Automotive Industry*. (1991). VDI BERICHTE, Zürich, Switzerland, **894**.

[15] Wilson, D.V. (1969). Controlled directionality of properties in sheet metals. *Metallurgical reviews*, **14**, 175–188.

Forging Simulation incorporating Strain-Induced Phase Transformation using the Finite Volume Method

P.R. Ding
Asia-Pacific Services Division, MSC.Software Japan Ltd.
2-39, Akasaka 5-Chome, Minato-Ku, Tokyo, Japan
peiran.ding@mscsoftware.com

T. Inoue, S. Imatani
Department of Energy Conversion Science, Kyoto University
Yoshida-honmachi, Sakyo-ku, Kyoto 606-8501, Japan
inoue@energy.kyoto-u.ac.jp

D.Y. Ju
Department of Mechanical Engineering, Saitama Institute of Technology
Fusaiji 1690, Okabe, Saitama, Japan

E. de Vries
MSC.Software (E.D.C.) B.V.
Groningenweg 6, 2803 PV, Gouda, The Netherlands

Abstract: A finite volume computational method using Eulerian formulation to simulate forging is adopted in this paper. By means of this finite volume method, an approach based on the "metallo-thermo-mechanics" to simulate metallic structure, temperature and stress/strain coupled with strain-induced phase transformation has been developed. The material is considered as elastic-plastic and takes into account phase transformation effects on the yield stress. The temperature increase due to plastic deformation, heat conduction and thermal stress has been co-analyzed. Strain-induced phase transformation, latent heat, transformation stress and strain are included. The applicability of the developed method was confirmed through a backward extrusion. The strain rate dependence of the flow stress is formulated, and the effects of strain rate on the austenitic-martensite transformation and the aggregate flow stress are examined.

Keywords: Forging, finite volume method, strain-induced phase transformation, metallo-thermo-mechanics.

1. Introduction

In the numerical analysis of a forging process, only by taking account of ongoing structural changes, a reliable process simulation could be achieved. However, structural changes of material have often been neglected in most simulations

C. Miehe (ed.), IUTAM Symposium on Computational Mechanics of Solid Materials at Large Strains, 455-466.

of forging. In principle, the numerical analysis is usually performed without a sufficient metallurgical background.

A realistic simulation of the forging process involves a large quantity of factors influencing each other. Temperature changes induce thermal stresses and affect the material structure. Mechanical deformation may change the thermal boundary and generate heat by the dissipation of plastic work so that it influences the temperature and again changes the structure of the material. Besides an accurate knowledge of structural changes and thermal effect on the material and boundary, the formulation of the flow stress is of importance for a forging process. The flow stress is influenced largely by the initial structure and its ongoing evolution.

The knowledge available on structural changes of material during forming process has been gathered by earlier research [1, 2, 3, 4, 8, 9, 11, 13, 15]. Austenitic stainless steel SUS304 has been chosen as an example for the study. It is well known that for SUS304, an austenitic- martensite phase transformation takes place according to the strain and temperature during plastic deformation (Cortes et al. [2]). This phase transformation is called *strain-induced austenitic-martensite transformation*. It is developed at low temperatures and almost completely inhibited at higher temperatures within usual strain ranges. Kinetics of the strain-induced phase transformation has been developed during different studies [2, 4, 8, 9, 11, 13, 15].

Thus far, a number of noticeable and successful simulations [2, 3, 8, 13, 15], by employing the finite element method based on Lagrangian formulation, have been reported. However, those simulations are in most cases related to the forging processes of relatively simple shapes (two-dimensional models). One of the most notable barriers is the remeshing technology, which is the bottleneck for finite element methods in complicated three- dimensional forging.

More attention should be paid to the effects of strain rate on structural change and mechanical behavior. Hecker et al. [4] made a series of uniaxial and biaxial tests under different strain rates and shows that the amount of martensite decreases with increasing strain rate. Hecker concluded that the increased stability of the austenite resulted from adiabatic heating at high strain rates. Nevertheless, the numerical analysis of the effects of strain rate on structural change and mechanical behavior is still lacking.

In this paper, based on the framework of *metallo-thermo-mechanics* (Inoue et al. [6]), a finite volume method formulated in the Eulerian description has been introduced to simulate the three-dimensional forging process. The deforming workpiece flows through fixed finite volume meshes. The strain-induced austenitic-martensite phase transformation incorporated with forging is calculated. The incremental formulation of the phase transformation kinetics is derived by use of Tsuta and Cortes' model (Cortes et al. [2]).

A mixture rule is adopted to evaluate the aggregate flow stress of the austenite-martensite affected by the respective flow stresses and phase transformation. As a result of martensite phase transformation, the hardening increases rapidly with the deformation of the material during the forging process. We consider the structure evolution in order to understand ultimately the complete flow stress and other mechanical behavior.

This new approach has been developed and implemented into the commercial computer program MSC.SuperForge (Slagter et al. [14]). An analysis of backward extrusion is presented and compared with experimental results. The strain rate dependence of the flow stress is formulated, and the effects of strain rate on the austenitic-martensite transformation and the aggregate flow stress are examined.

2. Basic theory

2.1. Kinetics of strain-induced martensite transformation

For SUS304 stainless steel, we consider only two volume fractions: austenite ξ_A and martensite ξ_M, which are developed under the restriction:

$$\xi_A + \xi_M = 1. \tag{1}$$

The material parameters χ, such as thermal conductivity, specific heat, elastic modulus, yield stress, hardening modulus and so on, are assumed to be evaluated by the mixture law:

$$\chi = \chi_A \xi_A + \chi_M \xi_M. \tag{2}$$

In this paper, Tsuta and Cortes' model is adopted as the basis of the kinetics model of martensite transformation for SUS304. This model describes the volume of strain- induced transformation as a function of equivalent plastic strain $\bar{\varepsilon}^p$ and temperature T:

$$\xi_M = \left\{ 1 + \left[\frac{\bar{\varepsilon}^p}{Ce^{-D/T}} \right]^{-B} \right\}^{-1}, \tag{3}$$

where material parameters B, C and D are determined by experiments.

Since material deforms under high non-linear conditions, it becomes necessary to derive the kinetics equation to an incremental expression:

$$\dot{\xi}_M = \frac{\partial \xi_M}{\partial \bar{\varepsilon}^p} \dot{\bar{\varepsilon}}^p + \frac{\partial \xi_M}{\partial T} \dot{T}. \tag{4}$$

2.2. HEAT CONDUCTION EQUATION

Strain-induced phase transformation is strongly dependent on the temperature. The temperature field is affected by the heat generation converted from mechanical work and latent heat due to phase change. So the heat conduction equation is derived by considering the total energy balance while neglecting the small effect of heat generation due to elastic mechanical work (Inoue et al. [6]):

$$\rho c_p \dot{T} - \frac{\partial}{\partial x_i}\left(k\frac{\partial T}{\partial x_i}\right) - \sigma_{ij}\dot{\varepsilon}^p_{ij} + \sum \rho_I l_I \dot{\xi}_I = 0, \tag{5}$$

where c_p, k denote the specific heat, coefficient of heat conduction. ρ_I and l_I are the density and the latent heat produced by the progressive$I$-th constituent with volume fraction ξ_I. σ_{ij} is Cauchy stress tensor.

2.3. ELASTIC PLASTIC CONSTITUTIVE RELATIONS

Within the infinitesimal theory of kinematics, the total strain rate can be divided into elastic, plastic, and thermal strain rates and strain rates by structural dilatation due to phase transformation and transformation plasticity. The isotropic hardening rule of von Mises is used here as the yield function.

3. Numerical methodology

Under the Eulerian description, the governing equation, expressing the conservation of any physical quantity is formulated in the form:

$$\frac{\partial \boldsymbol{Q}}{\partial t} + \nabla \cdot \boldsymbol{F} = 0, \tag{6}$$

where $\boldsymbol{Q}$ is the variable and $\boldsymbol{F} = \boldsymbol{F}_{convection} + \boldsymbol{F}_{diffusion}$is the corresponding flux, consisting of convection and diffusion terms. Equation (6) gives the conservation laws of mass, momentum and energy, in which the variables and the corresponding fluxes are listed in Table 1. Hereν_i is the velocity vector; h_i is the heat flux; e_t is the total energy per unit of mass given by$e_t = e + 1/2\nu_j\nu_j$, with the internal energy per unit mass$e$. The body force is neglected throughout the present investigation.

By using Gauss' divergence theorem, Eq. (6) can be written in an integral form for a control volume V with a closed surface S, which is stationary within a total domain:

$$\frac{\partial}{\partial t}\int_V \boldsymbol{Q}\,\mathrm{dV} + \int_S \boldsymbol{F}\cdot\boldsymbol{n}\,\mathrm{dS} = 0. \tag{7}$$

When the governing equations are integrated in time by an explicit dynamic procedure, the complete system is solved by separating the diffusion flux and convection flux respectively into an acoustic step and an advection step.

Table 1. Variables and fluxes.

Conservation law	*Mass*	*Momentum*	*Energy*
$\boldsymbol{Q}$ variable	ρ	$\rho\nu_i$	ρe_t
$\boldsymbol{F}$ convection	$\rho\nu_i$	$\rho\nu_i\nu_j$	$\rho e_t\nu_i$
$\boldsymbol{F}$ diffusion	0	$-\sigma_{ij}$	$h_i - \sigma_{ij}\nu_j$

3.1. ACOUSTIC STEP

In the acoustic step only evolution of stress and impulse waves is provided. The contribution from the convection flux to the variation of the quantity in each volume will be taken care of in the advection step. Under the assumption that the mass density is constant in each finite volume element during one acoustic step, the conservation of mass is satisfied automatically. The volume integral form for the impulse is reduced from the conservation law of momentum to:

$$\int_V \frac{\partial \nu_i}{\partial t} dV = \frac{1}{\rho} \int_S \sigma_{ij} n_j dS, \tag{8}$$

where n_i is the unit normal vector on the boundary S of the control volume V. The values on faces of the finite volume elements can be obtained from solving the Riemann problem (Hirth [5]). Then the new velocity of the finite volume element is updated by using the velocity increment evaluated from Eq. (8).

3.2. ADVECTION STEP

In the advection step the material and corresponding properties are transported according to the velocity updated in the acoustic step without changing the value of the physical quantity associated with the material. The divergence of diffusion flux term in Eq. (6) disappears and mass, momentum and energy are transported as:

$$\frac{\partial(\rho\Phi)}{\partial t} + \frac{\partial(\rho\Phi\nu_i)}{\partial x_i} = 0. \tag{9}$$

Other properties (stress/strain, energy, material characteristics, e.g.) are transported in the same conservation form as Eq. (9). Here, Φ is any physical quantity per unit mass. The integral form to be used in the numerical advection scheme for the finite volume method reads:

$$\frac{D}{Dt} \int_V \rho\Phi dV = -\int_S \rho\Phi\nu_i n_i dS, \tag{10}$$

where material time derivative D/Dt is used for material particles transformation. Under the Eulerian description, D/Dt can be replaced by $\partial/\partial t$, so Eq. (7) also holds true in the advection step.

3.3. Update variables

The density is updated after the mass is transported. The strain rate tensor is calculated from the velocity increment evaluated in the acoustic step. Once computed, the strain and stress status is determined from the elastic-plastic constitutive relations. The temperature is determined by solving the heat conduction equation, and the phase transformation is calculated according to the new temperature and strain fields.

All the variables in the governing equations are evaluated in a Runge-Kutta time integration scheme (Hirth [5]).

3.4. Automatic surface tracking

The shape of the deforming material is explicitly defined and encapsulated by a geometric surface comprised of triangular facets. These facets are geometric entities rather than finite elements. The facet surface is constrained to move with the material in order to track the exact material surface and also precisely apply the boundary condition to the material. An algorithm called Resolution Enhancement Technology (RET) is employed to automatically refine the facet surface throughout the simulation to capture the continuously changing and increasing complexity of the deforming material. Another task of RET is to refine the material facet surface at locations of the die(s) which have significant curvature. The facet surface must be capable of *folding* around the features of the die surface. Principally speaking, RET is a kind of two-dimensional remeshing technology from geometry point of view (Slagter et al. [14]). Since the facets are geometric entities rather than finite elements, the common problems which are associated with finite elements based on remeshing algorithms, such as severe distortion and significant increase of computation time with a loss of accuracy, are avoided.

4. Simulations

4.1. Numerical model

4.1.1. Model description

Simulations on a backward extrusion are carried out. The experimental results given by Cortes and Tsuta [3] are employed to assess the reliability of the simulations. The material used in the experiment was SUS304 austenitic stainless steel. A specimen was machined into a cylindrical billet of 14.0 mm in diameter and 10.5 mm in height.

In order to verify the three-dimensional applicability of the program, the numerical model is generated in three dimensions shown in Figure 1. Because of the axi-symmetric condition a triangular prismatic mesh is used with a head

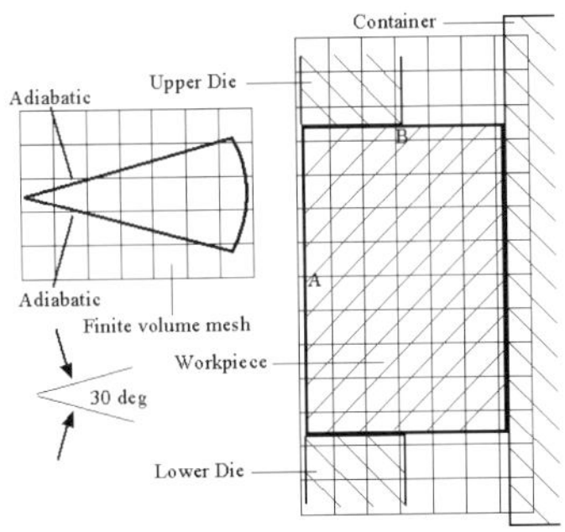

Figure 1. Schematic illustration of numerical mode with the points A and B.

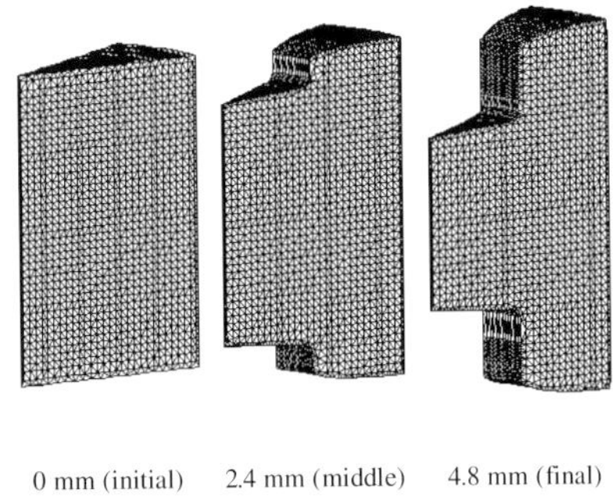

Figure 2. Deformation modes of billet.

angle of 30 deg. The dies are modeled as rigid bodies. The initial temperatures of the billet and dies are 15°C, and the steady temperature of the environment is taken as 15°C.

Investigations (Ludwigson and Berger [9]) have shown that the strain hardening for SUS304 stainless steel in austenite state exhibits flow curves that can be described as:

$$\bar{\sigma}_A = K_0\left(\dot{\bar{\varepsilon}}/\dot{\bar{\varepsilon}}_0\right)^m \left(\bar{\varepsilon}^p\right)^n, \tag{11}$$

where $\dot{\bar{\varepsilon}}$ is the equivalent strain rate, $\bar{\varepsilon}_0$ is the reference strain rate equal to $5.0 \times 10^{-4}\ \mathrm{s}^{-1}$ and m is the strain rate sensitivity as 0.013 (Tomita et al. [15]). The contribution of martensite on the flow stress is defined according to Ludwigson and Berger [9]:

$$\bar{\sigma}_M = C(\xi_M)^Q. \tag{12}$$

Here, the material parameters K_0, n, C and Q are experimentally determined to be 566.75 MPa, 0.18, 1268.6 MPa and 1.2798, respectively.

Table 2. Physical properties of workpiece and die.

Heat transfer coefficient, $\mathrm{W/(m^2K)}$	Workpiece/die	1.74×10^4
	Ambient	10.40
Thermal expansion coefficient, $1/\mathrm{K}$	Austenite	1.50×10^{-5}
	Martensite	1.10×10^{-5}
Volumetric dilatation due to structural change		7.30×10^{-4}
Latent heat, $\mathrm{J/kg}$		1.50×10^4
Intensity of transformation plasticity, $\mathrm{MPa^{-1}}$		5.08×10^{-5}

The physical properties of the workpiece and dies are listed in Table 2. Other properties such as coefficient of heat conduction, specific heat and elastic modulus, which are dependent upon the content of each phase and temperature, are taken from Inoue et al. [6], Ju et al. [7] and Ohnaka [10].

4.1.2. *Friction formulation*

If two bodies in contact with each other have rough surfaces and are forced to move tangentially with respect to one another, frictional shear stresses will develop on the interface. Since the finite volume method simply tracks the *material volume fraction* per finite volume element, the orientation and exact position of the material surface are not adequately defined. Therefore, the friction condition is difficult to be accurately applied. In this research the material surface is explicitly defined and encapsulated by a geometric surface comprised of triangular facets. The friction condition can be defined accurately on the facet surface as illustrated in Figure 3. Here, a finite volume element (the square with thin line) is intersected by triangular facets connected with grid points (the dark circles).

In the forging process involving relatively low contact pressure between dry contact surfaces, Coulomb's friction model is most commonly used. If the frictional shear stress reaches a critical value, the workpiece will slip along the die with the shear stress:

$$\tau_f = \mu\sigma_n, \tag{13}$$

where μ is the coefficient of friction and σ_n denotes the normal stress on the workpiece-die interface.

On the contrary, in the forging process involving relatively high contact pressure, it is generally more appropriate to use the law of plastic shear friction. Here, if the frictional shear stress exceeds a constant fraction η of the yield stress in shear, the workpiece starts to slip:

$$\tau_f = \eta\frac{\sigma_y}{\sqrt{3}}, \tag{14}$$

where σ_y is the yield stress.

Two critical values are calculated according to Eq. (13) and Eq. (14), and the smaller one is adopted as the frictional force.

In this paper, simulations are carried out with a constant coefficient of friction μ equal to a constant plastic friction shear factor η as 0.1 (Osakada [12]).

4.2. SIMULATED RESULTS AND DISCUSSION

In order to study the effect of strain rate on the phase transformation and the aggregate flow stress, the simulations are carried out with the constant extruding

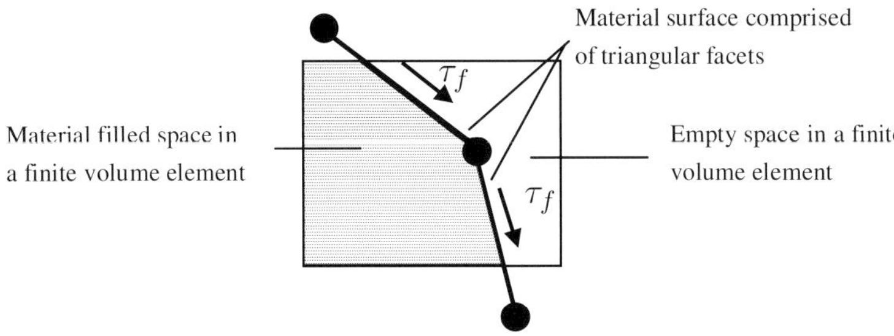

Figure 3. Illustration of friction condition applied on the facet surface.

velocities 0.0015 mm/s and 1.5 m/s on the upper die with respect to the fixed lower die. The velocity induces a low strain rate ($\sim 10^{-3}\mathrm{s}^{-1}$) and a high strain rate ($\sim 10^{3}\mathrm{s}^{-1}$). The stroke is 4.8mm.

4.2.1. Validity with experimental data

Figure 2 shows the deformation modes of the billets for some representative strokes. Comparison between measured and simulated martensite distributions (low strain rate only corresponding to the experiment) at the end of the process is shown in Figure 4. The volume fraction of martensite was measured by the magnetic balance method in the experiment. The simulation successfully duplicates the deformation of the workpiece and the martensite transformation.

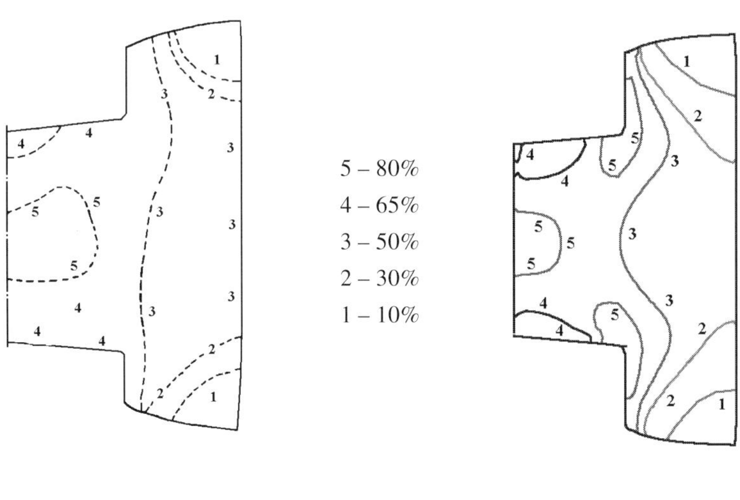

Figure 4. Experimental measured and calculated martensite distributions after backward extrusion.

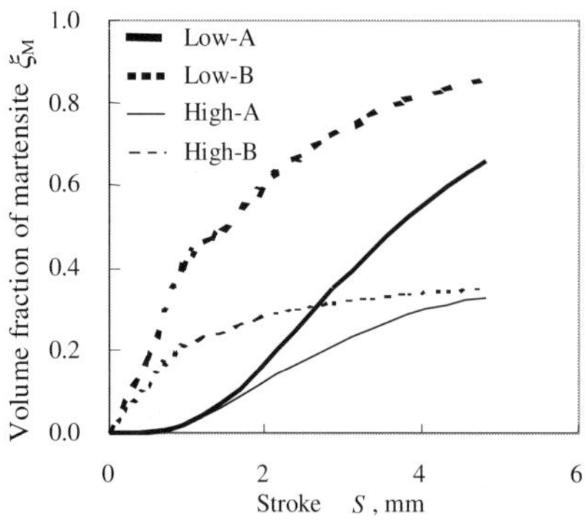

Figure 5. Variation of volume fraction of martensite with progressive stroke at Point A and B.

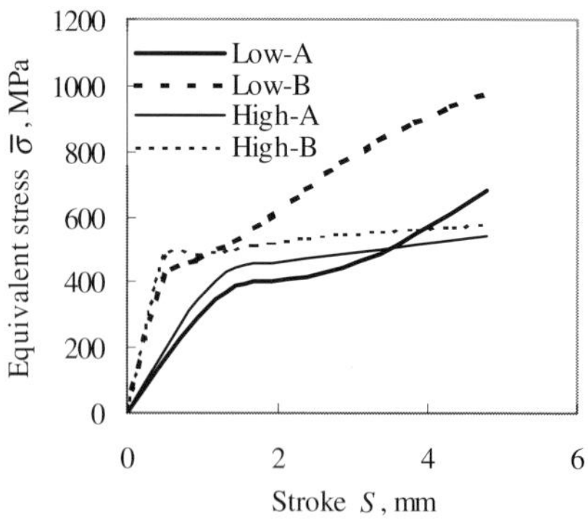

Figure 6. Variation of equivalent stress with progressive stroke at Point A and B.

4.2.2. *Effects of strain rate on volume fraction of martensite*

The simulated results of volume fraction of martensite are depicted in Figure 5 at points A and B during the forging stage. As shown in Figure 1, point A is located at the center of billet while point B is on the surface of the billet at the corner of the upper die.

Attention should be paid to the influence of strain rate on phase transformation of the SUS304 steel. The plastic strain with respect to the change of strain rate displays minor differences as long as the stroke is the same. But the temperature rises more readily in high strain rate than in low strain rate during the deformation. The martensite formation is found to be considerably less in high strain rate. The temperature increase resulting from the heat generated by the dissipation of plastic work is sufficient to suppress the austenitic-martensite phase transformation.

4.2.3. *Evolution of flow stress*

The simulated equivalent stresses at points A and B with progressive stroke are depicted in Figure 6. The figure shows that the aggregate flow stress of the austenite and martensite constituents is affected by the respective flow stresses and phase transformation. At low strain levels, the hardening behavior of the material is determined mainly by austenite. According to Eq. (11), higher strain rate causes stronger austenite hardening. The stresses at high strain rate are higher than the stresses at low strain rate during the beginning of the stroke. However, with the generation of martensite, the flow stress becomes highly dependent on the martensite content of the material. The low strain rate gives much higher stresses at the end of the simulation due to the larger amount of martensite. It shows that the structure evolution is very prerequisite for understanding the complete flow stress.

5. Conclusions

The finite volume method using Eulerian formulation is applied to simulating forging with strain-induced phase transformation. The incremental formulation of the kinetics equation is derived from Tsuta and Cortes' model. The coupled metallo-thermo- mechanics system to simulate metallic structure, temperature and stress/strain has been extended to large three-dimensional deformation. To verify this method a backward extrusion has been simulated. The simulation results show a good agreement in tendency with the experimental data.

The results show also that the effect of strain rate is very important on the structural change and mechanical behavior, and high strain rate results in sufficient adiabatic heating to inhibit the transformation and change the flow stress.

References

[1] Angel, T. (1954). Formation of Martensite in Austenitic Stainless Steels - Effects of Deformation, Temperature, and Composition, *J. Iron Steel Inst.*, **177**, 165–174.

[2] Cortes, R.J.A., Tsuta, T., Mitani, Y. and Osakada, K. (1992). Flow Stress and Phase Transformation Analysis in the Austenitic Stainless Steel under Cold Working – Part 1, Phase Transformation Characteristics and Constitutive Formulation by Energetic Criterion. *JSME Int. J.*, **I 35**, 201–209.

[3] Cortes, R.J.A. and Tsuta, T. (1993). Flow Stress and Phase Transformation Analysis in the Austenitic Stainless Steel under Cold Working – Part 3, Analysis for SUS 301, SUS 304, SUS 310 Steels and Experiments on Forward and Backward Extrusions. *Memoirs of the Faculty of Engineering*, Hiroshima University, **11-3**, 35–52.

[4] Hecker, S.S., Stout, M.G., Staudhammer, K.P. and Smith, J.L. 1982). Effects of Strain and Strain Rate on Deformation-Induced Transformation in 304 Stainless Steel - Part I. Magnetic Measurements and Mechanical Behavior, *Met. Trans.*, **A 13A**, 619–626.

[5] Hirth, C. (1990). Numerical Computation of Internal and External Flows. *Volume 2: Computational Methods for Inviscid and Viscous Flows*, John Wiley & Sons.

[6] Inoue, T., Nagaki, S., Kishino, T. and Monkawa, M. (1981). Description of Transformation Kinetics, Heat Conduction and Elastic-Plastic Stress in the Course of Quenching and Temperature of Some Steels. *Ing.-Arch.*, **50**, 315–327.

[7] Ju, D.Y., Inoue, T. and Matsui, H. (1993). Visco-plastic Behavior of SUS-304 Stainless Steel at Ultra-high Temperature - Description by Unified Constitutive Models. In W.B. Lee (eds.), *Advances in Engineering Plasticity and its Applications*, Elsevier Science Publishers, 521–528.

[8] Kawai, N., Saiki, H. and Hirate, H. (1976). Improvement of Stretchability of 18-8 Stainless-Steel Sheets due to Strain Induced Transformation. *J. of the JSTP*, **17**, 899–906 (in Japanese).

[9] Ludwigson, D.C. and Berger, J.A. (1969). Plastic Behavior of Metastable Austenitic Stainless Steels. *J. of the Iron and Steel Inst.*, **207**, 63–75.

[10] Ohnaka, I. (1985). Introduction of Heat Transfer and Solidification Analysis - Application on Casting Process, Maruzen Publisher, (in Japanese).

[11] Olson, G.B. and Cohen, M. (1975). Kinetics of Strain-Induced Martensitic Nucleation, *Met. Trans.*, **A 6A**, 791–795.

[12] Osakada, K. (1995). Stress, strain, temperature and defect generated in the material. In Japan Society of Technology Plasticity (eds.), *Forging Technology - Toward Net Shape Product*, Coruna Publishing Co., LTD., 66–147 (in Japanese).

[13] Shinagawa, K., Nishikawa, H., Ishikawa, T. and Hosoi, Y. (1990). Deformation-Induced Martensitic Transformation in Type 304 Stainless Steel during Cold Upsetting. *Iron and Steel*, **3**, 156–162 (in Japanese).

[14] Slagter, W.J., Florie, C.J.L and Venis, A.C.J. (1999). Advances in Three Dimensional Forging Process Modeling. *Proc. of the 15th National Conference on Manufacturing Research*, 73–78.

[15] Tomita, A., Harada, H. and Iwamoto, T. (1994). Numerical Simulation of Deformation Behavior of TRIP Steels. *Trans. JSME*, **A 60**, 166–173 (in Japanese).

Modeling and Characterization of Large Shear Strains at a Rail Surface

W. Daves
CDL-FMD and MCL Leoben, Montanuniversität Leoben
Franz-Josef-Straße 18, 8700 Leoben, Austria
daves@hal.unileoben.ac.at

F.D. Fischer
Institute of Mechanics, Montanuniversität Leoben
Franz-Josef-Straße 18, 8700 Leoben, Austria
fischer@unileoben.ac.at

W. Stadlbauer, E. Werner
CDL für moderne Mehrphasenstähle, Lehrstuhl für Mechanik A, TU München
85747 Garching, Germany
werner@lam.mw.tu-muenchen.de

Abstract: The surface and subsurface region of the head of railroad rails are subjected to extremely high and complex loads, leading to large shear strains within some 100 microns near the surface of a rail. The local state in this surface region is strongly influenced by the surface roughness of the rail and wheel, which must be included in the calculation of the local deformation and stress state. This influence is investigated by a two dimensional finite element study. A small part (micro-region) of a wheel is pressed onto a rail. The wheel and the rail are given a measured surface roughness. The wheel is loaded according to the average value of the contact pressure calculated by a three-dimensional finite element model and is made to slide along the rail. Furthermore, metallurgical scanning electron microscope investigations are presented, which characterize this local deformation state in the rail. These results show shear strains of 5 and higher. Also cracks along the deformation lines are detected. The large shear strains found by these investigations compare very well to those calculated by the finite element model.

Keywords: Wheel-rail contact, rough surface, numerical study, large shear strain, metallurgical investigation, characterization.

1. Introduction

To understand the microstructural changes in the surface region of railway rails one has to consider the extremely high loads which occur during the wheel-rail contact as well as the strain and stress state near the surface which causes a highly deformed, sometimes nanostructured, surface layer (Baumann [3],

C. Miehe (ed.), IUTAM Symposium on Computational Mechanics of Solid Materials at Large Strains, 467-475.

Baumann et al. [4, 5], Bürkle et al. [7], Lojkowski et al. [17, 18]). For the investigation of the processes occurring very close to the surface (within 100 microns depth) one needs to consider the surface roughness of the wheel and rail (Greenwood and Tripp [12], Greenwood and Williamson [11], Ogilvy [19], Peng and Bhushan [22, 20, 21], Robbe-Valloire et al. [25], Zhao and Chang [31], Tao et al. [28], Adams and Nosonovsky [2]).

The investigation of the local strain and stress state is not possible by means of experimental methods in a surface region during the wheel-rail contact. Therefore, the best approach for studying stress states in surface regions exposed to periodic contact appears to lie in realistic three-dimensional numerical simulations describing systems consisting of wheel, rail and bedding (Yan and Fischer [30]). As a second step, a direct connection between the periodically changing loading on the one hand and the resulting strain, wear and damage in the rail on the other hand must be achieved. Although reports on the wear and surface damage of railroad rails have been published, see e.g. Bhushan [6], Kapoor et al. [14], Kapoor [15], Franklin et al. [9], Pyzalla et al. [23], Ringsberg [24], no satisfactory solution for the above mentioned connection has yet been found.

Therefore within this work a model is proposed where the local state in this subsurface region can be calculated considering also the surface roughness of the rail and wheel. The investigation is done by a two dimensional finite element study.

For the validation of the model but also to characterize the large shear strains at a rail surface, metallurgical scanning electron microscope investigations are presented which characterize this local deformation state in the rail.

2. The local wheel-rail contact model

The main assumption of this model is that the strains at the surface can be calculated at each single micro-area where the wheel and rail are in contact.

It is known that the sliding of the wheel along the rail, during the wheel and the rail are in contact, is the main reason for the wear, damage and high plastic shear strains produced near the rail surface (Garnham and Beynon [10], Sato et al. [26], Tyfour and Beynon [29]). This sliding is called creepage. The longitudinal creepage C is defined by the following relation,

$$C = 2(VtVc)/(Vt + Vc), Vt > 0, Vc > 0. \tag{1}$$

Vt is the translational velocity of the wheel and Vc the circumferential velocity of the wheel, see also Carter [8] and Kalker [13].

This creepage is produced nearly always during acceleration of the driven wheels, braking of wheels and rolling in a curve.

For the model at hand one has to know the contact pressure as well as the creepage during the time of contact at one small area of the rail. Both values´can be taken from any calculation delivering the pressure and stick-slip time history of a wheel rail contact. This wheel micro-region is loaded by the average value of the contact pressure on this area, as calculated from a three–dimensional finite element model of a wheel-rail contact, see Figure 1 (Yan and Fischer [30]). It is assumed that within the modeled part of wheel and rail a constant nominal contact pressure is acting during the contact time, and also the creepage rate is taken as constant. It is possible to provide measured or calculated contact pressure values, however, for better comparability of the results, the nominal contact pressure is set to a typical value of 1000 MPa. Also the creepage is set to a common value of 1%, too.

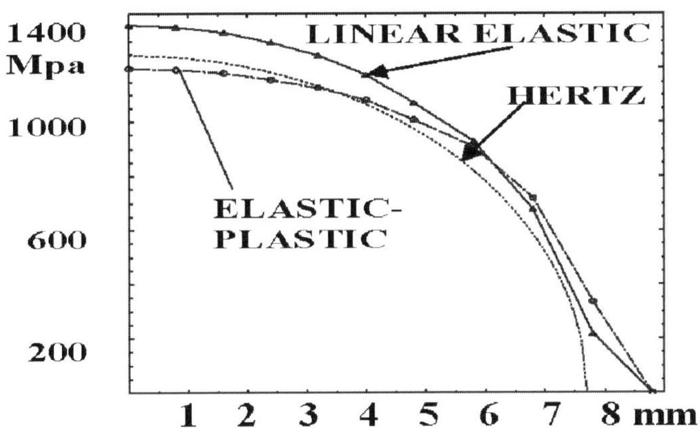

Figure 1. Different calculations of the contact pressure in longitudinal direction (half of the contact patch) by a three-dimensional finite element model (Yan and Fischer [30]).

The material data are taken for a new rail after roller straightening. For this rail steel cyclic tension-compression tests were made. The specimens were cut out of the rail head (Schleinzer [27]).

3. Finite element model

A two-dimensional finite element study is carried out where a small part of a wheel is pressed onto a rail.

In comparison to three-dimensional Finite Element Simulations of the wheel-rail contact only a micro region of the wheel and the rail is modeled, see Figure 2.

The wheel is assumed to have a certain surface roughness similar to that of the rail. The rail is given a measured surface roughness taken from a corrugated rail which was in use for several years, published by Knothe et al. [16].

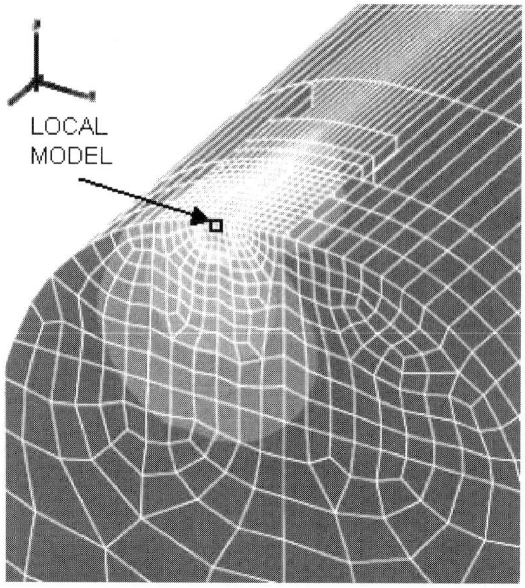

Figure 2. Local model with length of 1 mm in relation to a global three-dimensional modeling at a switch component.

The material model used is the one by Armstrong & Frederick implemented in the finite element code ABAQUS [1], which can describe non-linear kinematic hardening. This material model is used for the rail part only. The wheel is assumed to behave purely elastically. This is assumed since wheel-steels show a broad spectrum of plastic behavior and thus we are dealing with one extreme case. Thanks to this assumption, the calculation time is reduced, and the numerical contact problem can be solved more easily.

All results are calculated by two-dimensional generalized plane strain models where the boundaries at the sides (1 or longitudinal direction) and in the depth (2-direction) are fixed a sufficient distance from the investigated contact zone such that no variation of this distance produces any significant influence on the results. In Figure 3 a part of the finite element mesh is plotted. The roughness of the two contacting surfaces of the wheel and the rail part can be seen.

The number of elements in the model is 7900. Four-node generalized plane-strain elements are used for the investigations presented here. The elements at the surface have a size of about 5 microns.

4. Finite element results

Figure 4 shows the normal contact stress calculated by the micro-model under a normal load producing a nominal normal contact pressure of 1000 MPa. Using

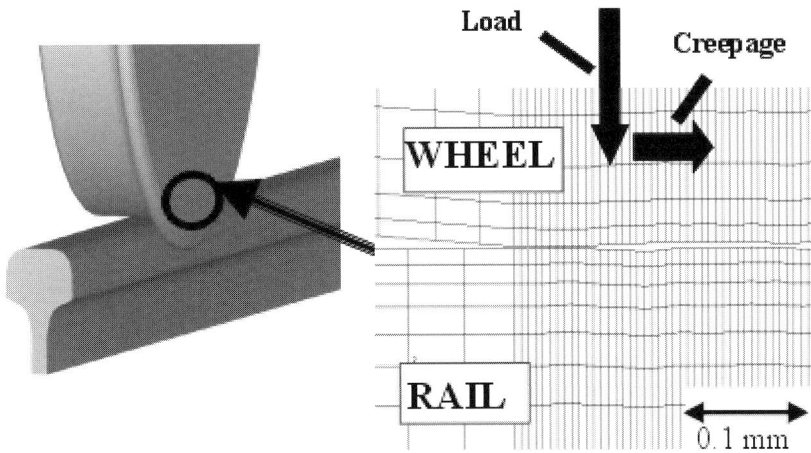

Figure 3. Part of the finite element mesh of wheel and rail near the contact line.

elasto-plastic material behavior the calculated normal stresses are a maximum of 2000 MPa at asperities in contact. This is twice the nominal stress applied and about 20% to 25% of that obtained from a calculation with purely elastic material behavior. The maximum plastic strains arising are no more than 0.06 for the normal contact case.

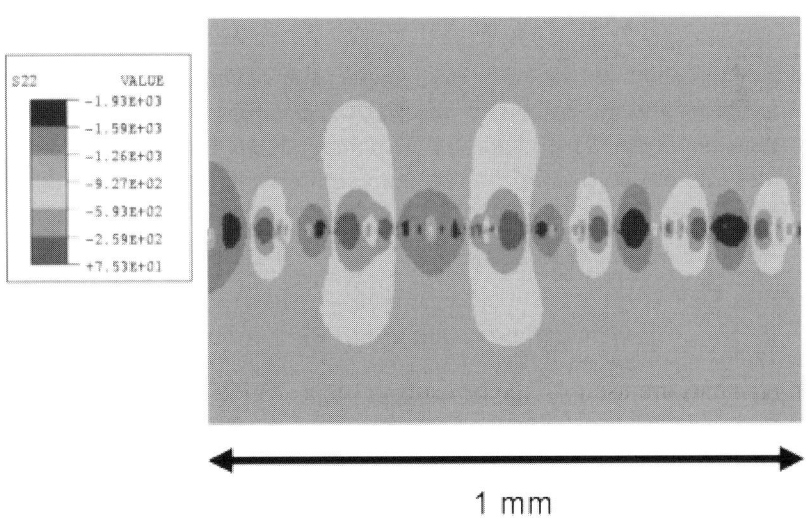

Figure 4. Normal contact stress calculated by the micro-model applying a force producing a nominal contact stress in the case of a flat surface of 1000 MPa.

A drastically changed picture is obtained if a creepage is introduced. Figure 5 shows that the shearing of a wheel relatively to the rail produces a very high shear strain at the asperities (higher than 2) and shear strains of about 0.25 within a depth of 50 microns. The plastic straining is restricted to a depth of about 100 microns. For this calculation the nominal normal load is set to 1000 Mpa, the creepage 1% and the friction coefficient 0.3.

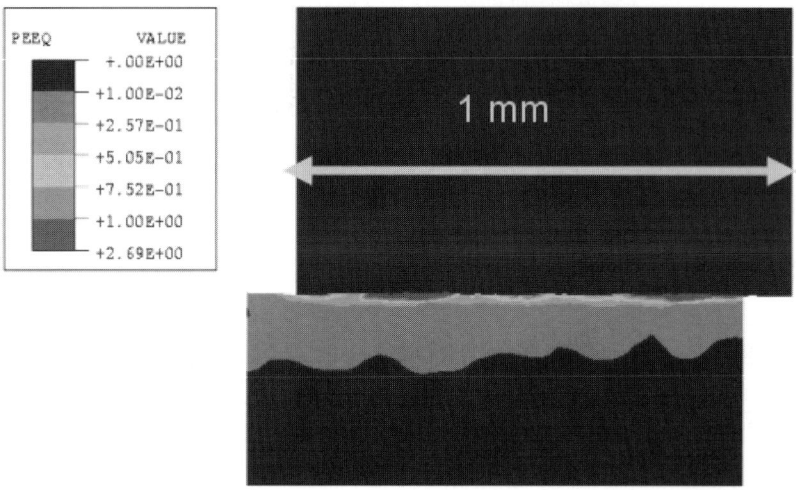

Figure 5. Plastic accumulated equivalent strain calculated assuming a creepage of 1%, with a nominal normal contact stress of 1000 MPa.

With the model at hand several loadings can be simulated. Figure 6 shows the deformation of the finite element mesh after 5 loadings in the same direction with the same parameters as above. Although several loadings are simulated the heavily deformed plastic zone remains limited to a depth of not more than 100 microns.

5. Characterization of large shear strains

In the following, two scanning electron microscope images are presented which characterizes this local deformation state in the rail. Figure 7 offers a very excellent picture of this extremely sheared surface especially in a range of 10 to 20 microns below the surface. A typical damage in this region is the initiation of surface cracks which for the most part coincide perfectly with the visible deformations lines, see Figure 8. Both Figures show shear strains of 3 up to 10 or higher in a range of 50 microns below the surface.

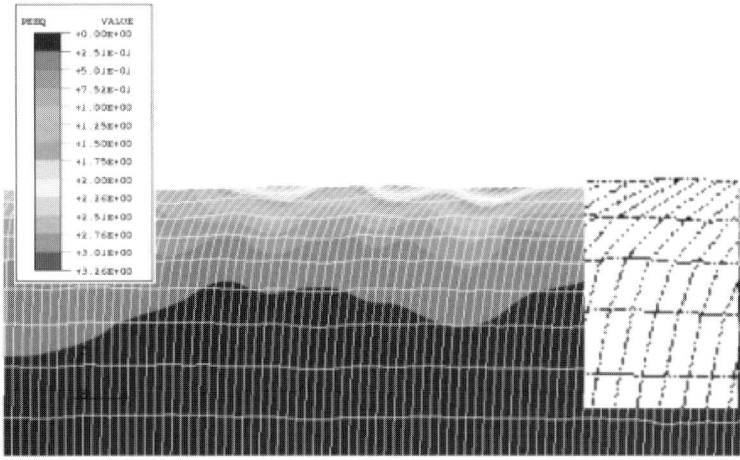

Figure 6. Finite element mesh showing the strongly sheared surface of a rail after several cycles, the height of the element layer at the surface is 5 micron.

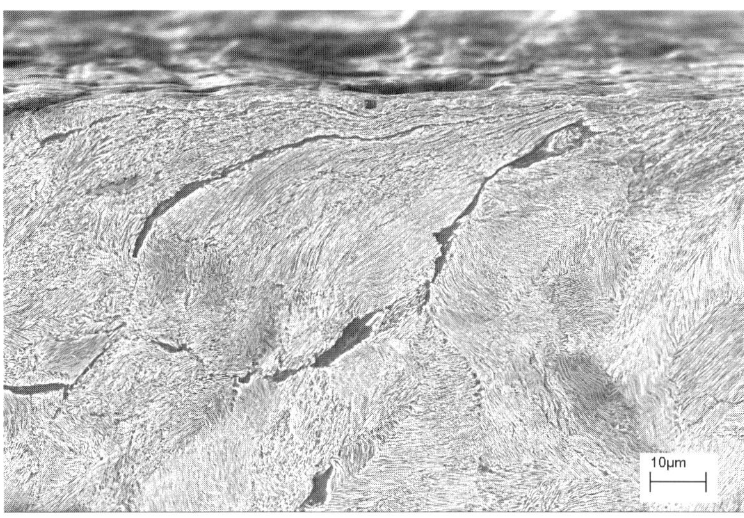

Figure 7. Typical deformation field near the surface of a rail, scanning electron microscope photograph.

6. Discussion

The very large shear strains found by these investigations compare very well to those calculated by the finite element model both in the amount and in the depth beneath the rail surface. A finite element model demonstrates only the early stage (about 100 loadings) of the cyclic shearing at the surface. Therefore,

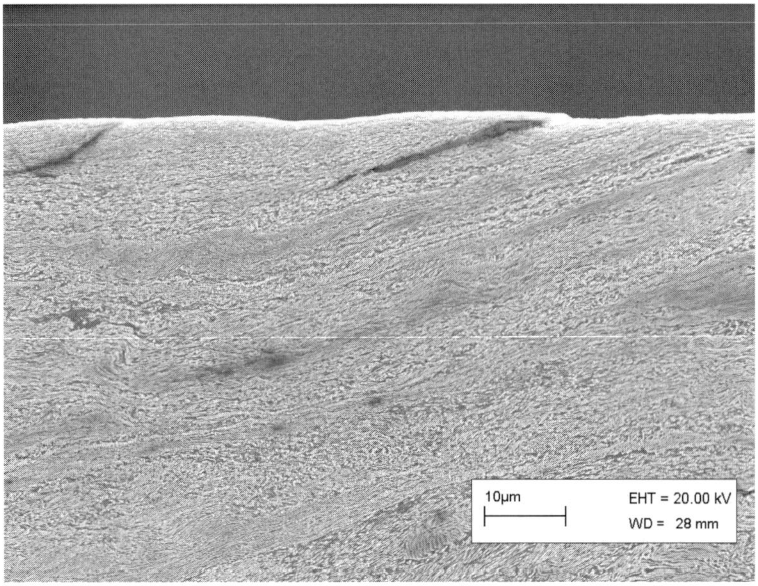

Figure 8. A surface crack initiated in the same direction as the visible deformation lines, scanning electron microscope photograph.

the amount of shear strain resulting from the calculations is obviously smaller than that shown by the metallurgical investigation on rails used for a long time (more than 10 years in the case represented here).

Both the metallurgical investigations and the numerical calculations show impressively that the extremely high shear strains near the surface of a rail have to be considered and included in all investigations and models of wear and damage at a rail surface.

The main progress applying the finite element model lies in the fact that the development of these extremely high shear strains can be modeled straightforward using very common assumptions about loading, creepage, material and friction coefficient. These results are obtained due to the introduction of an elasto-plastic rough surface contact into the finite element model which is a main improvement in calculating wheel-rail contacts.

Acknowledgments

The authors wish to express their thanks the Christian Doppler (CD) Society for founding the CD Lab for Functionally Oriented Materials Design and the CD Lab for "moderne Mehrphasenstähle" , the voestalpine Schienen GmbH, here especially Dr. P. Pointner, and the Material Center Leoben for their support of this research.

References

[1] ABAQUS Theory Manual Vers. 5.8, Hibbit, Karlsson & Sorensen, Inc., Providence, RI, USA, 1999.

[2] Adams, G.G. and Nosonovsky, M. (2000). *Tribol. Int.*, **33**, 431–442

[3] Baumann, G. (1997). Untersuchungen zu Gefügestrukturen und Eigenschaften der "Weißen Schichten" auf verriffelten Schienenlaufflächen. *PhD Thesis*. Verlag Dr. Köster, Berlin.

[4] Baumann, G., Fecht, H.J., Liebelt, S. (1996). *Wear*, **144**, 133–140.

[5] Baumann, G., Zhong, Y. and Fecht, H.J. (1996). *NanoStruct. Mater.*, **7**, 237–244.

[6] Bhushan, B. (1996). *Appl Mech. Rev.*, **49**, 275–298.

[7] Bürkle, G., Djahanbakhsh, M. and Fecht, H.J. (1999). *ZEV+DET Glas. Ann.*, **123**, 303–310.

[8] Carter, F.W. (1926). *Proc. R. Soc.*, **A112**, 151–157.

[9] Franklin, F.J., Widiyarta, I. and Kapoor, A. (2001). *Wear*, **251**, 949–955.

[10] Garnham, J.E. and Beynon, J.H. (1992). *Wear*, **157**, 81–109.

[11] Greenwood, J.A., Williamson, J.B.P. (1966). *Proc. R. Soc.*, **A295**, 300–319.

[12] Greenwood, J.A., Tripp, J.H. (1967). *J. Appl. Mech.*, **34**, 153.

[13] Kalker, J.J. (1991). *Wear*, **144**, 243–261.

[14] Kapoor, A., Williams, J.A. and Johnson, K.L. (1994). *Wear*, **175**, 81–92

[15] Kapoor, A. (1996). *Wear*, **212**, 119–130.

[16] Knothe, K. and Theiler, A. (1996). Proceedings of the 2nd Miniconf. on Contact Mechanics and Wear of Rail/Wheel Systems, Budapest, I. Zobory.

[17] Lojkowski, W., Djahanbakhsh, M., Bürkle, G., Gierlotka, S., Zielinski, W. and Fecht, H.J. (2001). *Mater. Sci. Eng.*, **A303**, 197–208.

[18] Lojkowski, W., Millman, W., Chugunova, S.I., Goncharova, I.V., Djahanbakhsh, M., Bürkle, G. and Fecht, H.J. (2001). *Mater. Sci. Eng.*, **A303**, 209–215.

[19] Ogilvy, J.A. (1991). *J. Phys. D. Appl. Phys.*, **24**, 2098–2109.

[20] Peng, W. and Bhushan, B. (2001). *Wear*, **249**, 741–760.

[21] Peng, W. and Bhushan, B. (2001). *J. Tribology*, **123**, 330–342.

[22] Peng, W. and Bhushan, B. (2002). *J. Tribology*, **124**, 46–61.

[23] Pyzalla, A., Wang, L., Wild, E. and Wroblewski, T. (2001). *Wear*, 901–907.

[24] Ringsberg, J.W. (2001). *Int. J. Fat.*, **23**, 575–586

[25] Robbe-Valloire, F., Paffoni, B. and Progri, R. (2001). *Mechanics of Materials*, **33**, 617–633.

[26] Sato, M., Anderson, P.M. and Rigney, D.A. (1993). *Wear*, 162–164, 159–172.

[27] Schleinzer, G. (2000). Residual Stress Formation During the Roller Straightening of Rails. *Fortschrittsberichte VDI*, Verlag Düsseldorf, Reihe 18, 251.

[28] Tao, Q., Lee, H.P. and Lim, S.P. (2001). *Wear*, **249**, 539–545.

[29] Tyfour, W.R. and Beynon, J.H. (1994). *Tribol. Int.*, 401–412.

[30] Yan, W. and Fischer, F.D. (2000). *Arch. Appl. Mech.*, **70**, 255–268.

[31] Zhao, Y. and Chang, L. (2001). *J. Tribology*, **123**, 857–864.

Author Index